HORMONES AND THEIR ACTIONS
PART II

Specific actions of protein hormones

New Comprehensive Biochemistry

Volume 18B

General Editors

A. NEUBERGER
London

L.L.M. van DEENEN
Utrecht

ELSEVIER
Amsterdam · New York · Oxford

Hormones and their Actions Part II

Specific actions of protein hormones

Editors

B.A. COOKE
Department of Biochemistry, Royal Free Hospital School of Medicine, University of London, Rowland Hill Street, London NW3 2PF, England

R.J.B. KING
Hormone Biochemistry Department, Imperial Cancer Research Fund Laboratories, P.O. Box No. 123, Lincoln's Inn Fields, London WC2A 3PX, England

H.J. van der MOLEN
Nederlandse Organisatie voor Zuiver-Wetenschappelijk Onderzoek (Z.W.O.), Postbus 93138, 2509 AC Den Haag, The Netherlands

1988
ELSEVIER
Amsterdam · New York · Oxford

© 1988, Elsevier Science Publishers B.V. (Biomedical Division)

All rights reserved. No part of this publication may be reproduced, stored in a retrieval system, or transmitted in any form or by any means, electronic, mechanical, photocopying, recording or otherwise, without the prior written permission of the Publisher, Elsevier Science Publishers B.V. (Biomedical Division), P.O. Box 1527, 1000 BM Amsterdam, The Netherlands.

No responsibility is assumed by the Publisher for any injury and/or damage to persons or property as a matter of products liability, negligence or otherwise, or from any use or operation of any methods, products, instructions or ideas contained in the material herein. Because of the rapid advances in the medical sciences, the Publisher recommends that independent verification of diagnoses and drug dosages should be made.

Special regulations for readers in the USA. This publication has been registered with the Copyright Clearance Center, Inc. (CCC), Salem, Massachusetts. Information can be obtained from the CCC about conditions under which the photocopying of parts of this publication may be made in the USA. All other copyright questions, including photocopying outside of the USA, should be referred to the Publisher.

ISBN 0-444-80997-X (volume)
ISBN 0-444-80303-3 (series)

Published by:

Elsevier Science Publishers B.V. (Biomedical Division)
P.O. Box 211
1000 AE Amsterdam
The Netherlands

Sole distributors for the USA and Canada:

Elsevier Science Publishing Company, Inc.
52 Vanderbilt Avenue
New York, NY 10017
USA

Library of Congress Cataloging in Publication Data
(Revised for vol. 2)

Hormones and their actions

  (New comprehensive biochemistry ; v. 18B-   )
  Includes bibliographies and index.
  1. Hormones--Physiological effect. 2. Hormones--
Physiology. I. Cooke, Brian A. II. King, R. J. B.
(Roger John Benjamin) III. Molen, H. J. van der.
IV. Series: New comprehensive biochemistry ; v. 18B, etc.
QD415.N48 vol. 18B, etc.   574.19'2 [612'.405]   88-16501
[QP571]
ISBN 0-444-80996-1 (pt. 1)
ISBN 0-444-80997-X (pt. 2)

Printed in The Netherlands

List of contributors

P.Q. Barrett, 93, 211
Yale University School of Medicine, New Haven, CT 06510, U.S.A.
L. Birnbaumer, 1
Baylor College of Medicine, Houston, TX 77030, U.S.A.
W.B. Bollag, 211
Yale University School of Medicine, New Haven, CT 06510, U.S.A.
A.M. Brown, 1
Baylor College of Medicine, Houston, TX 77030, U.S.A.
J. Codina, 1
Baylor College of Medicine, Houston, TX 77030, U.S.A.
P.M. Conn, 135
Department of Anatomy, Wright State University, School of Medicine, Dayton, OH 45435, U.S.A.
B.A. Cooke, 155, 163
Department of Biochemistry and Chemistry, Royal Free Hospital School of Medicine, University of London, Rowland Hill Street, London NW3 2PF, England
K.D. Dahl, 181
Department of Reproductive Medicine, School of Medicine, M-025, University of California, San Diego, La Jolla, CA 92093, U.S.A.
C. Denef, 113
Laboratory of Cell Pharmacology, Faculty of Medicine, University of Leuven, Campus Gasthuisberg, B-3000 Leuven, Belgium
J.H. Exton, 231
The Howard Hughes Medical Institute and the Department of Molecular Physiology and Biophysics, Vanderbilt University School of Medicine, Nashville, TN 37232, U.S.A.
W.J. Gullick, 349
Institute of Cancer Research, Chester Beatty Laboratories, Cell and Molecular Biology Section, Protein Chemistry Laboratory, Fulham Road, London, SW3 6JB, England
G.R. Guy, 47
Biochemistry Department, University of Birmingham, P.O. Box 363, Birmingham B15 2TT, England
P.J. Hornsby, 193
Department of Cell and Molecular Biology, Medical College of Georgia, Augusta, GA 30912, U.S.A.

M.D. Houslay, 321
Molecular Pharmacology Group, Department of Biochemistry, University of Glasgow, Glasgow G12 8QQ, Scotland
A.J.W. Hsueh, 181
Department of Reproductive Medicine, School of Medicine, M-025, University of California, San Diego, La Jolla, CA 92093, U.S.A.
L. Jennes, 135
Department of Pharmacology, University of Iowa, College of Medicine, Iowa City, IA 52242-1109, U.S.A.
N.C. Khanna, 63
Cell Regulation Group, Department of Medical Biochemistry, The University of Calgary, Calgary, Alberta, Canada T2N 4N1
C.J. Kirk, 47
Biochemistry Department, University of Birmingham, P.O. Box 363, Birmingham B15 2TT, England
R. Mattera, 1
Baylor College of Medicine, Houston, TX 77030, U.S.A.
H. Rasmussen, 93, 211
Yale University School of Medicine, New Haven, CT 06510, U.S.A.
F.F.G. Rommerts, 155, 163
Department of Biochemistry, The Medical Faculty, Erasmus University, Rotterdam, The Netherlands
M. Tokuda, 63
Cell Regulation Group, Department of Medical Biochemistry, The University of Calgary, Calgary, Alberta, Canada T2N 4N1
D.M. Waisman, 63
Cell Regulation Group, Department of Medical Biochemistry, The University of Calgary, Calgary, Alberta, Canada T2N 4N1
M.J.O. Wakelam, 321
Molecular Pharmacology Group, Department of Biochemistry, University of Glasgow, Glasgow G12 8QQ, Scotland
M. Wallis, 265, 295
Biochemistry Laboratory, School of Biological Sciences, University of Sussex, Falmer, Brighton BN1 9QG, England
A. Yatani, 1
Baylor College of Medicine, Houston, TX 77030, U.S.A.

Contents

List of contributors v

Chapter 1. G proteins and transmembrane signalling, by L. Birnbaumer, J. Codina, R. Mattera, A. Yatani and A.M. Brown. 1

1. Introduction 1
2. The G proteins identified by function and purification 5
   2.1. G_s, the stimulatory regulatory component of adenylyl cyclase 5
   2.2. Transducin (T), the light-activated GTPase 8
   2.3. G_i, the inhibitory regulatory component of adenylyl cyclase 9
   2.4. G_o, a PTX substrate with an α subunit of M_r 39 000 10
   2.5. G_ps, the regulatory components of phospholipase (PhL) activity 11
   2.6. G_k, the activator of 'ligand-gated' K^+ channels: mechanism of muscarinic regulation of atrial pacing 13
3. G proteins detected by ADP-ribosylation 17
   3.1. Labeling with CTX 18
   3.2. Labeling with PTX 19
4. G protein structure by cloning 20
   4.1. The α subunits 21
   4.2. The β subunits 31
   4.3. The γ subunit: its role as a membrane anchor 32
5. G protein mediation of receptor regulation of ion channels 32
   5.1. Effects of inhibitory receptors on K^+ channels in tissues other than heart atria . . . 32
   5.2. Inhibitory regulation of voltage-gated Ca^{2+} channels: direct or indirect involvement of a G protein? 35
   5.3. Stimulatory regulation of Ca^{2+} channels: direct G protein coupling in spite of regulation by cAMP-dependent protein kinase A 36
6. Concluding remarks 38
Acknowledgements 39
References 39

Chapter 2. Inositol phospholipids and cellular signalling, by G.R. Guy and C.J. Kirk 47

1. Introduction 47
2. Inositol phospholipids 48
3. Role of GTP-binding proteins in receptor-response coupling 50
4. Products of phosphatidylinositol 4,5-bisphosphate hydrolysis and their roles as second messengers in the cell 52
   4.1. Inositol trisphosphate and calcium mobilisation 52

4.2.	Diacylglycerol mobilisation and the activation of protein kinase C	52
5. Metabolism of the hydrolysis products of PtdIns 4,5-P_2		54
5.1.	Inositol trisphosphate	54
5.2.	Diacylglycerol	56
6. Fertilisation, proliferation and oncogenes		56
6.1.	Role of inositol lipid degradation	56
6.2.	Influence of ionophores and synthetic stimulators of protein kinase C	58
6.3.	Oncogenes	59
7. Release of arachidonic acid		59
7.1.	Mechanisms of arachidonate liberation	59
8. Summary		61
References		61

Chapter 3. The role of calcium binding proteins in signal transduction, by N.C. Khanna, M. Tokuda and D.M. Waisman — 63

1. Introduction		63
2. The calcium transient		65
3. Calcium binding proteins and signal transduction		67
4. Calcium binding proteins: structure and function		69
4.1.	Extracellular calcium binding proteins	70
4.2.	Membranous calcium binding proteins	70
4.3.	Intracellular calcium binding proteins	74
4.3.1.	The 'EF' domain family	74
4.3.2.	The annexin-fold family	77
4.3.3.	Miscellaneous calcium binding proteins	79
5. Calcium binding proteins and cellular function		80
5.1.	Muscle contraction	81
5.1.1.	Actin based regulation (skeletal and cardiac muscle)	81
5.1.2.	Myosin based regulation (smooth muscle)	82
5.2.	Metabolism	83
5.3.	Secretion and exocytosis	84
5.4.	Egg fertilization and maturation	86
5.5.	Cell growth and proliferation	87
References		89

Chapter 4. Mechanism of action of Ca^{2+}-dependent hormones, by H. Rasmussen and P.Q. Barrett — 93

1. Introduction		93
2. Cellular calcium metabolism		94
2.1.	Plasma membrane	95
2.2.	Endoplasmic reticulum	97
2.3.	Mitochondrial matrix	98
3. Mechanisms of Ca^{2+} messenger generation		99
4. Messenger calcium		99
4.1.	Coordinated changes in PI and Ca^{2+} metabolism	100
4.2.	Smooth muscle contraction	102
4.3.	Coordinate changes in cAMP and Ca^{2+} metabolism	103

| | | 4.3.1. | K⁺-mediated aldosterone secretion. | 103 |

        4.3.1. K^+-mediated aldosterone secretion. 103
        4.3.2. Control of hepatic metabolism by glucagon 105
5. Synarchic regulation 106
   5.1. Regulation of insulin secretion by CCK and glucose 106
6. Integration of extracellular messenger inputs 109
References 110

Chapter 5. Mechanism of action of pituitary hormone releasing and inhibiting factors, by C. Denef 113

1. The adenylate cyclase-cAMP system 114
   1.1. TRH 114
   1.2. VIP 114
   1.3. DA 115
   1.4. LHRH 116
   1.5. CRF 117
   1.6. Vasopressin 117
   1.7. GRF and SRIF 117
2. The Ca^{2+} messenger system 118
   2.1. TRH 118
   2.2. VIP 119
   2.3. DA 120
   2.4. LHRH 120
   2.5. CRF 121
   2.6. Vasopressin 122
   2.7. GRF and SRIF 122
3. The inositol polyphosphate-diacylglycerol-protein kinase C system 123
   3.1. TRH 123
   3.2. VIP 124
   3.3. DA 124
   3.4. LHRH 124
   3.5. CRF and vasopressin 125
   3.6. GRF and SRIF 126
4. Arachidonic acid derivatives 126
   4.1. TRH 126
   4.2. VIP 127
   4.3. DA 127
   4.4. LHRH 128
   4.5. CRF and vasopressin 128
   4.6. GRF and SRIF 129
5. Concluding remarks 130
References 130

Chapter 6. Mechanism of gonadotropin releasing hormone action, by L. Jennes and P.M. Conn 135

1. Introduction 135
2. Structure of GnRH 135
3. The biochemical properties of the GnRH receptor 137

4. Localization of the GnRH receptor 138
5. Role of receptor microaggregation. 139
6. Relationships between GnRH receptor number and cellular response 141
7. Second messenger systems 142
8. Calcium as a second messenger 143
9. Phospholipids 145
10. Diacylglycerols 146
11. GTP binding proteins 147
12. Protein kinase C 147
13. Conclusion 148
Acknowledgement 150
References 150

Chapter 7. The mechanisms of action of luteinizing hormone. I. Luteinizing hormone-receptor interactions, by B.A. Cooke and F.F.G. Rommerts. . . . 155

1. Introduction 155
2. The structure of LH 156
3. The LH receptor 157
    3.1. Purification and characterization 157
    3.2. Interaction of LH with its receptor 158
    3.3. LH receptor recycling and synthesis 159
    3.4. Regulation of LH receptors 160
References 161

Chapter 8. The mechanisms of action of luteinizing hormone. II. Transducing systems and biological effects, by F.F.G. Rommerts and B.A. Cooke 163

1. LH receptor transducing systems 163
    1.1. Formation of cyclic AMP 164
    1.2. The phosphoinositide cycle 164
    1.3. Arachidonic acid: release and metabolism to prostaglandins and leukotrienes . . 165
    1.4. Control and action of intracellular calcium 166
2. Steroidogenesis 166
    2.1. Second messengers 167
    2.2. Formation and possible roles of specific (phospho)proteins 168
    2.3. Control mechanisms in mitochondria 169
3. Desensitization and down regulation 171
    3.1. Uncoupling of the LH receptor from the adenylate cyclase system 171
    3.2. Reversal of desensitization 172
    3.3. Inhibition of steroidogenesis in LH desensitized cells 172
4. Other effects of LH 173
5. LH action on gonadal cells in perspective 175
References 178

*Chapter 9. Mechanism of action of FSH in the ovary, by K.D. Dahl and
A.J.W. Hsueh* 181

1. Introduction 181
2. Biochemistry of FSH 181
   2.1. α, β subunits 181
   2.2. Carbohydrate content 181
3. FSH receptors in target cells 182
   3.1. Radioligand receptor assay 183
   3.2. Agonistic and antagonistic effects of FSH analogs 183
4. Activation of the protein kinase A pathway 184
   4.1. Coupling between the FSH receptor and adenylate cyclase 184
   4.2. Stimulation of protein kinase A 185
5. FSH induction of granulosa cell differentiation 185
   5.1. LH and PRL receptors, and β-adrenergic responsiveness 185
   5.2. Lipoprotein receptors 186
   5.3. Gap junction and microvilli formation 186
6. FSH stimulation of steroidogenic enzymes 186
   6.1. Aromatase induction 187
      6.1.1. Enzyme induction 187
      6.1.2. Two-cell, two-gonadotropin theory 187
      6.1.3. Granulosa cell aromatase bioassay for FSH 188
   6.2. Induction of cholesterol side-chain cleavage enzymes 188
   6.3. Induction of the 3β-hydroxysteroid dehydrogenase enzyme 188
7. FSH stimulation of inhibin biosynthesis 188
8. FSH stimulation of tissue-type plasminogen activator 189
9. Conclusion 190
References 190

*Chapter 10. The mechanism of ACTH in the adrenal cortex, by
P.J. Hornsby* 193

1. ACTH and the cyclic AMP intracellular messenger system 193
   1.1. The intracellular messenger for ACTH 193
   1.2. Spare cyclic AMP generating capacity and its function 194
   1.3. The interaction of the ACTH receptor with adenylate cyclase 194
   1.4. Cyclic AMP-dependent protein kinase in the adrenal cortex 195
   1.5. The pathway of biosynthesis of steroids in the adrenal cortex 195
      1.5.1. The enzymes of steroidogenesis 195
      1.5.2. Zonation of steroidogenesis 196
   1.6. The regulation by ACTH of the rate-limiting step of steroidogenesis, the conversion of
        cholesterol to pregnenolone 197
      1.6.1. Nature of the rate-limiting step: limitation on cellular movement of
             cholesterol 197
      1.6.2. Supply of cholesterol to the precursor pool available for steroidogenesis . . . 198
      1.6.3. ACTH regulation of the cholesterol pool 198
      1.6.4. Regulation of the rate-limiting step by cyclic AMP-dependent protein kinase . . 199
   1.7. The regulation of the synthesis of the steroidogenic enzymes by ACTH 200

	1.7.1.	The integration of the short- and long-term actions of ACTH to provide increased steroidogenesis	200

        1.7.1. The integration of the short- and long-term actions of ACTH to provide increased steroidogenesis 200
        1.7.2. Mechanism of enzyme induction by cyclic AMP 201
  1.8. Indirect action of ACTH on growth and metabolism 202
2. Interaction of the ACTH/cyclic AMP system with other hormones and intracellular messengers 203
  2.1. Zonal differences 203
  2.2. Interactions at adenylate cyclase 204
  2.3. ACTH and cyclic GMP 205
  2.4. ACTH and the calcium intracellular messenger system 206
        2.4.1. Zonal differences 206
        2.4.2. The calcium second messenger system in the adrenal cortex 206
        2.4.3. Cyclic AMP phosphodiesterase 206
  2.5. ACTH and protein kinase C 207
        2.5.1. C-kinase in the adrenal cortex: presence and steroidogenic effects . . 207
        2.5.2. Coordinate regulation of adrenal enzyme synthesis by A- and C-kinases . 207
        2.5.3. Mechanisms for regulation of steroidogenic enzymes that differ in activity between the different zones 207
        2.5.4. The origin of zonation in the cortex 208
References 209

Chapter 11. Mechanism of action of angiotensin II, by P.Q. Barrett, W.B. Bollag and H. Rasmussen 211

1. Introduction 211
2. AII receptors 212
  2.1. Regulation of receptor affinity 212
  2.2. Regulation of receptor number 213
3. Receptor–guanine nucleotide interactions 214
4. Transducing enzyme activation 214
  4.1. Adenylate cyclase 215
  4.2. Phospholipase C 216
        4.2.1. Activation via G-proteins 216
        4.2.2. Substrate(s) 216
        4.2.3. Products (second messengers) 217
5. AII-induced changes in calcium metabolism 219
  5.1. Intracellular calcium concentration 219
  5.2. Calcium mobilization 219
  5.3. Total cell calcium 220
  5.4. Calcium entry 220
6. Integration of signals and cellular response 222
  6.1. Initiation of response 223
  6.2. Maintenance of response 224
  6.3. Temporal relationship of the two phases 226
References 228

Chapter 12. Mechanisms of action of glucagon, by J.H. Exton 231

1. Introduction 231
2. The glucagon receptor 232

3.	Guanine nucleotide binding regulatory protein	233
4.	Adenylate cyclase catalytic subunit	235
5.	cAMP and cAMP-dependent protein kinase	236
6.	Substrates of cAMP-dependent protein kinase in liver	239
	6.1. Phosphorylase b kinase	239
	6.2. Glycogen synthase	241
	6.3. Pyruvate kinase	242
	6.4. 6-Phosphofructo 2-kinase/fructose 2,6-bisphosphatase	244
	6.5. Acetyl-CoA carboxylase, ATP-citrate lyase	245
7.	Effects of glucagon on cell calcium	245
8.	Synergistic interaction between glucagon and calcium-mobilizing agonists in liver	250
9.	Inhibitory action of phorbol esters on glucagon-induced calcium mobilization	252
10.	Other actions of glucagon	252
11.	Summary	256
	References	259

Chapter 13. Mechanism of action of growth hormone, by M. Wallis   265

1.	The growth hormone-prolactin family	265
2.	Growth hormone and the control of somatic growth	266
3.	Receptors for growth hormone	267
	3.1. Distribution of growth hormone receptors	267
	3.2. Heterogeneity of growth hormone receptors	268
	3.3. Structure and purification of growth hormone receptors	269
	3.4. Signal transduction following binding of growth hormone to its receptor	271
	3.5. Regulation of growth hormone receptor levels	271
4.	Somatomedins/IGFs and the actions of growth hormone	273
	4.1. The nature of somatomedins	273
	4.2. The actions of somatomedins	273
	4.3. Somatomedin-binding proteins	274
	4.4. Synthesis and secretion of somatomedins	274
	4.5. Regulation of somatomedin production by growth hormone	275
	4.6. Biochemical mechanisms involved in the action of growth hormone on somatomedin C production	276
	4.7. Somatomedin C and somatic growth	277
5.	Actions of growth hormone on production of other specific proteins	278
6.	Actions of growth hormone on protein metabolism	279
	6.1. Actions of growth hormone on protein synthesis in the liver	279
	6.2. Actions on muscle	279
7.	Actions of growth hormone on lipid and carbohydrate metabolism	280
	7.1. Lipid metabolism	281
	7.2. Carbohydrate metabolism	281
8.	Actions of growth hormone on cellular differentiation and proliferation	282
9.	Growth hormone and the control of lactation	283
10.	Applications of molecular biology to the study of the actions of growth hormone	283
	10.1. Protein engineering of growth hormone	283
	10.2. Transgenic mice	284
11.	Potentiation of the actions of growth hormone by monoclonal antibodies	284
12.	Growth hormone variants	286
	12.1. Naturally occurring variants	286

xiv

12.2. The multivalent nature of growth hormone	287
13. Conclusions	288
13.1. The multiple actions of growth hormone	288
13.2. The significance of somatomedin C/IGF-I	289
14. Addendum	289
Acknowledgements	290
References	290

Chapter 14. Mechanism of action of prolactin, by M. Wallis 295

1. Lactogenic hormones	295
2. The biological actions of prolactin	296
2.1. Actions on the mammary gland	296
2.2. Other actions in mammals	297
2.3. Actions in lower vertebrates	299
3. Receptors for prolactin	299
3.1. Characterization of receptors	299
3.2. Regulation of prolactin receptors	303
4. Biochemical mode of action of prolactin on the mammary gland	304
4.1. Actions on mammary gland differentiation and development	304
4.2. Effects on synthesis of milk proteins	306
4.3. Effects on other milk components	307
4.4. Second messengers in the actions of prolactin	307
5. Actions of prolactin on the pigeon crop sac	309
5.1. Synlactin and the actions of prolactin	309
6. Actions of prolactin on the immune system	311
6.1. Nb2 cell proliferation	311
6.2. Other tissues and cells of the immune system	313
7. Variants of prolactin	314
7.1. Fragments	314
7.2. Glycosylated prolactins	314
8. Prolactin and mammary cancer	314
9. Conclusions	315
References	315

Chapter 15. Structure and function of the receptor for insulin, by M.D. Houslay and M.J.O. Wakelam 321

1. Introduction	321
2. Insulin receptor structure	321
3. Cloning of the gene for the insulin receptor	324
4. Insulin receptor internalization	325
5. Insulin's stimulation of glucose transport	328
6. Insulin-like growth factors (IGFs)	329
7. Insulin receptor tyrosyl kinase activity	330
8. Insulin and its action on guanine nucleotide regulatory proteins	336
9. An intracellular 'mediator' of insulin's action	341
10. Concluding remarks	343
Acknowledgements	344
References	345

Chapter 16. A comparison of the structures of single polypeptide chain growth factor receptors that possess protein tyrosine kinase activity, by W.J. Gullick   349

1. Introduction 349
2. The EGF receptor and the c-erbB-2 protein 349
3. Platelet-derived growth factor receptor and colony-stimulating factor 1 receptor 354
4. Summary 358
References 359

Subject Index 361

CHAPTER 1

G proteins and transmembrane signalling

LUTZ BIRNBAUMER, JUAN CODINA, RAFAEL MATTERA, ATSUKO YATANI and ARTHUR M. BROWN

Departments of Cell Biology and Physiology and Molecular Biophysics, Baylor College of Medicine, Houston, TX 77030, U.S.A.

1. Introduction

G proteins are involved in the transduction of the signal generated by occupancy of cell membrane receptors by their specific ligands – neurotransmitters, hormones, para- and autocrine factors – into activation of membrane effector systems. They bind guanine nucleotides, share a common heterotrimeric subunit structure of the $\alpha\beta\gamma$-type, are activated by GTP and possess GTPase activity which confers to them a molecular clocking capacity. This clocking capacity impedes persistent activation of the G proteins and regulates the steady state activity level of effector functions.

Signal transducing G proteins were discovered during studies on the mechanism of hormonal activation of adenylyl cyclases. These studies led from the identification of a GTP regulatory step in adenylyl cyclase regulation to the purification and molecular characterization of G_s, the stimulatory regulatory component of adenylyl cyclase. Studies on the mode of action of rhodopsin in outer segments of retinal rod cells led from the identification of a GTP-dependent step in the photoactivation of cGMP-phosphodiesterase (cGMP-PDE) to the isolation and molecular characterization of a light-activated G protein, currently called transducin (T or G_t). The use of the ADP-ribosylating toxin of *Bordetella pertussis* (PTX, also called islet-activating protein or IAP) and, more recently, detailed studies on the mechanisms by which hormones and neurotransmitters regulate polyphosphoinositide hydrolysis and ion channel activity, have led to the identification of several additional G proteins. These G proteins either inhibit adenylyl cyclase (G_i), stimulate membrane-bound phospholipases (so-called G_ps), or activate K^+ channels (G_k). A list of seven to nine signal-transducing proteins can be made at this time. Some have been purified and cloned. The existence of other G proteins is inferred based on functional studies, but they have not yet been biochemically isolated. G proteins with still unknown function have been purified. A list of hormones and neurotransmitters which interact with receptors known to couple to G proteins is presented in Table I. The great variety of regulations mediated by G proteins points to their central role in cellular regulation.

In addition to factors, hormones, and neurotransmitters, known to act through receptors that couple to G proteins, Table I also lists effector systems that are or may be affected directly by activated G. Of these effector systems, positive and negative regulation of adenylyl cyclase, activation of phospholipases, activation of cGMP-PDE in photoreceptor cells, and activation of K^+ channels are well docu-

TABLE I
Examples of receptors acting on cells via G proteins

Receptor for	Membrane function/system affected	Effect	Coupling protein involved	Examples of target cell(s)/organ(s)
A. Neurotransmitters				
1. Adrenergic				
beta-1	AC	stimulation	G_s	heart, fat, symp. synapse
beta-2	AC	stimulation	G_s	liver, lung
alpha-1	PhL C	stimulation	G_{plc}	smooth muscle, liver
	PhL A_2	stimulation	G_{pla}	FRTL-1 cells
alpha-2	*a.* AC	inhibition	G_i	platelet, fat (human)
	b. Ca channel	closing	G_o (G_p?)	NG-108, symp. presynapse
2. Dopamine				
D-1	AC	stimulation	G_s	caudate nucleus
D-2	AC	inhibition	G_i	pituitary lactotrophs
3. Acetylcholine				
Muscarinic M_1	*a.* PhL C	stimulation	G_{plc}	pancreatic acinar cell
	b. K channel (M)	closing	?	CNS, Symp. ganglia
Muscarinic M_2	*a.* AC	inhibition	G_s	heart
	b. K channel	opening	G_k(G_i?)	heart, CNS
4. $GABA_B$	*a.* Ca channel	closing	G_o (G_p?)	neuroblastoma N1E
	b. K channel	opening	G_i (G_k?)	sympathetic ganglia
5. Adenosine				
A-1 or Ri	*a.* AC	inhibition	G_i	pituitary, CNS, heart
	b. K channel	opening	G_k (G_i?)	heart
A-2 or Ra	AC	stimulation	G_s	fat, kidney, CNS
6. Serotonin (5HT)				
S-1	PhL C	stimulation	G_{plc}	aplysia
S-1a	*a.* AC	inhibition	G_i	pyramidal cells
	b. K channel	opening	G_k (G_i?)	pyramidal cells
S-2	AC	stimulation	G_s	skeletal muscal

Examples of receptors acting on cells via G proteins

Receptor for	Membrane function/system affected	Effect	Coupling protein involved	Examples of target cell(s)/organ(s)
B. Peptide hormones				
1. Pituitary				
Adrenocorticotropin (ACTH)	*a.* AC	stimulation	G_s	fasciculata, glomerulosa
	b. Ca channel	opening	$G_s(?)$	glomerulosa
Opioid (u, k, d)	*a.* AC	inhibition	G_i	NG-108
	b. Ca channel	closing	G_o (G_p?)	NG-108
Luteinizing hormone (LH)	AC	stimulation	G_s	granulosa, luteal, Leydig
Follicle-stimulating hormone (FSH)	AC	stimulation	G_s	granulosa
Thyrotropin (TSH)	*a.* AC	stimulation	G_s	thyroid, FRTL-5
	b. phospholipase?	stimulation	$G_p(?)$	thyroid
Melanocyte-stimulating hormone (MSH)	AC	stimulation	G_s	melanocytes
2. Hypothalamic				
Corticotropin-releasing hormone (CRF)	AC	stimulation	G_s	corticotroph, hypothalamus
Growth hormone-releasing hormone (GRF)	AC	stimulation	G_s	somatotroph
Gonadotropin-releasing hormone (GnRH)	PhL A_2	stimulation	G_{pla}	gonadotroph
	PhL C	stimulation	G_{plc}	gonadotroph
Thyrotropin-releasing hormone (TRH)	PhL C	stimulation	G_{plc}	lactotroph, thyrotroph
Somatostatin (SST or SRIF)	*a.* AC	inhibition	G_i	pit. cells, endocr. pancr.
	b. K channel	opening	G_k (G_i?)	pit. cells, endocr. pancr.
	c. Ca channel	closing	?	pit. cells
3. Other hormones				
Glucagon	*a.* AC	stimulation	G_s	liver, fat, heart
	b. Ca pump	inhibition	G_s (?)	liver, heart (?)
	c. PhL C	stimulation	?	liver
Cholecystokinin (CCK)	PhL C	stimulation	G_{plc}	pancreatic acini
Secretin	AC	stimulation	G_s	pancreatic duct, fat
Vasoactive intestinal peptide (VIP)	AC	stimulation	G_s	pancreatic duct, CNS

TABLE I Contd.
Examples of receptors acting on cells via G proteins

Receptor for	Membrane function/system affected	Effect	Coupling protein involved	Examples of target cell(s)/organ(s)
Vasopressin				
VP-1 (vasopressor, glycogenolytic)	PhL C	stimulation	G_{plc}	smooth muscle, liver
	AC	inhibition	G_i	liver
VP-2 (antidiuretic)	AC	stimulation	G_s	distal and collecting tubule
C. Other regulatory factors				
Chemotactic (fMet-Leu-Phe or fMLP)	PhL C	stimulation	G_{plc}	neutrophils
Thrombin	a. PhL C	stimulation	G_{plc}	platelets, fibroblasts
	b. AC	inhibition	G_i	platelets
Bombesin	PhL C	stimulation	G_{plc}	fibroblasts
IgE	PhL C	stimulation	G_{plc}	mast cells
Bradykinin	a. PhL C	stimulation	G_{plc}	lung, fibroblasts, NG-108
	b. PhL A_2	stimulation	G_{pla}	fibroblasts, endothel. cells
	c. K channel	stimulation	G_k (G_i?)	NG-108
	d. AC	inhibition	G_i	NG-108
Angiontensin II	a. PhL C	stimulation	G_{plc}	liver, glomerulosa cells
	b. AC	inhibition	G_i	liver, glomerulosa cells
Light (Rhodopsin)	cGMP-PDE	stimulation	$Tr(G_{t-r})$	retinal rod cells (night)
Light (Rhodopsin)	?		$Tc(G_{t-c})$	retinal cone cells (color)
Histamine				
H-1	PhL C	stimulation	G-x	macrophages
H-2	AC	stimulation	G_s	heart
D. Prostanoids				
Prostaglandin E_1, E_2	AC	inhibition	G_i	fat, kidney
Prostacyclin (PGI_2, PGE_1, PGE_2)	AC	stimulation	G_s	luteal cells, endothel., kidney
Thromboxanes	PhL C	stimulation	G_{plc}	platelets
Platelet activating factor (PAF)	PhL C	stimulation	G_{plc}	platelets
E. Other				
Olfactory signals	AC phospholipases?	stimulation	G_s G_p?	olfactory cilia
Taste signals	AC phospholipases	stimulation	G_s G_p?	taste epithelium
Purinergic (ATP, ADP)	PhL C (PhtdChol)	stimulation	G_p	liver

AC, adenylyl cyclase; PhL C, unless denoted otherwise, phospholipase C with specificity for phosphatidylinositol bisphosphate; PhL A_2, phospholipase A_2 (substrate specificity unknown); PhtdChol, phosphatidylcholine.

mented as being regulated by G components. The identification of the other systems listed as G protein-regulated is more tentative, because direct cell-free reconstitution with the responsible pure G proteins has not yet been reported. These systems include possible negative regulation of phospholipase C and both positive and negative regulation of voltage-gated Ca^{2+} channels. The picture that is developing is one in which G proteins appear to constitute a complex, yet well coordinated intramembrane regulatory communications network, whereby a given stimulus may have pleiotropic effects. Functional characterization, purification, labeling with pertussis (PTX) and cholera (CTX) toxins, use of specific antibodies and molecular cloning are tools used to investigate signal transduction by G proteins. Each approach reveals a slightly different aspect of this process.

2. The G proteins identified by function and purification

2.1. G_s, the stimulatory regulatory component of adenylyl cyclase

The first evidences for a stimulatory role of GTP in regulation of adenylyl cyclase systems were published in 1971 [1,2]. By 1980 a separate component, responsible for mediation of hormonal stimulation of adenylyl cyclases, had been purified [3]. This component, initially referred to as G/F and N_s, is now called G_s. It is a heterotrimeric complex composed of: α subunits that migrate on SDS-PAGE at 42 and 52 kDa [3], β subunits of ca. 35 kDa [3], and γ subunits of 6–10 kDa [4] (For reviews see Refs. 5 and 6). Its α subunits are ADP-ribosylated by CTX [7], dissociate from the holocomplex after activation [8,9] and hydrolyze GTP [10]. The α subunits have been cloned in several laboratories [11–17] and their amino acid composition has been deduced from the cDNA nucleotide sequence. The amino acid sequence of one of two types of β subunits, called β_{36} [18], has also been deduced from its cDNA [19–21]. The amino acid sequence of the γ subunit is not yet known.

G_s is established to be the stimulatory regulatory component of adenylyl cyclase. This was first demonstrated by its ability to reconstitute the adenylyl cyclase system of cyc^- cells [3,22] concomitant with the reappearance of CTX labeling [23]. Cyc^- cells are derived from S49 murine lymphoma cells and lack G_s as indicated by lack of stimulation of adenylyl cyclase by NaF, GTP analogs and hormones (in spite of the presence of stimulatory receptors), by lack of substrate for CTX and by lack of mRNA encoding G_s-α subunits. Moreover, pure G_s also stimulates a 'resolved C' preparation [24], as well as fully purified C [25,26] of adenylyl cyclase, both in solution [27] and after reconstitution into phospholipid vesicles [28]. Thus, by all criteria the purified G_s is functional G_s.

The activation of G_s has been studied extensively both in native membranes and with purified G_s in solution. Non-hydrolyzable GTP analogs, but not GTP, activate soluble G_s. However, both the analogs and GTP elicit G_s activation in membranes, suggesting facilitation by receptors. Studies showed that activation by nucleotides entails a two-step process: G_s, under the combined influence of the GTP analog and Mg^{2+}, first changes conformation such that the nucleotide becomes tightly bound to the α subunit and then, in a temperature-dependent reaction, dissociates into α^G and $\beta\gamma$ complexes [8]. Isolated α^G complexes can reconstitute adenylyl cyclase regulation in cyc^- membranes, indicating that they are the effector molecules [9]. Even though GTP cannot substitute for GTP analogs (GTPγS or GMP-P(NH)P) to activate soluble G_s, it is assumed that activation of G_s in membranes also entails subunit dissociation, with receptors playing an obligatory 'helper' role in bringing about GTP (and Mg^{2+})-mediated formation of activated α^{GTP} complexes (α^{*GTP}). Studies with intact membranes showed that hormonal stimulation decreases the concentration of Mg^{2+} required for G_s activation by as much as 1000-times from 5–15 mM to 5–15 μM [29,30]. However, the exact mechanism by which a receptor facilitates activation of G_s by GTP is not known.

Pure G_s that has been incorporated into phospholipid vesicles exhibits a very low GTPase activity, ranging from 0.02 to 0.05 mol hydrolyzed per min per mol of G_s [31]. Co-incorporation into these vesicles of pure beta-adrenergic receptors increases this activity by a factor of 2–3 to 0.05–0.1 mol of GTP hydrolyzed per min per mol of G_s. Stimulation of the receptor with a beta-adrenergic agonist (isoproterenol) results in a further increase in GTP hydrolysis to rates of ca 1.0 mol of GTP hydrolyzed per min per mol of G_s [31,32].

The α and β subunits of G_s are water soluble; the γ subunits, on the other hand are strongly hydrophobic. Since the $\alpha\beta\gamma$ complex is hydrophobic, it is currently thought that G_s is a peripheral membrane protein anchored into the inner leaflet of the membrane bilayer through its γ subunit. The possibility exists that, upon activation, α^{*GTP} complexes could be released from the membrane. This led Rodbell to postulate functions for such 'programmable second messengers' [33].

Reconstitution studies, in which pure β-adrenergic receptors were incorporated into phospholipid vesicles either alone or with pure G_s, have shown that in the presence of G_s, up to 30% of the receptors are in a form with a high affinity for agonists. In the absence of G_s all the receptor molecules are in their low affinity form [31]. Further, in analogy to observations made in intact membranes, addition of a guanine nucleotide reverses the G_s effect. Thus, not only is G_s responsible for activation of the catalytic unit of adenylyl cyclase, through its α^{*GTP} form, but it also modulates the formation of high and low agonist affinity states of receptors. The high affinity state is being formed on interaction of nucleotide-free holo-G_s ($\alpha\beta\gamma$) with receptor. Figure 1 describes the regulatory cycle of G_s as it may occur under the influence of a hormone receptor. The scheme incorporates receptor-G_s interactions, the subunit dissociation reaction associated with G_s activation, as well as the interaction of G_s with the catalytic unit of the adenylyl cyclase system (E on the figure). The γ subunit is assumed to be the anchor for G_s when not dissociated; the effector E is presented as the 'anchor' for dissociated α^{*GTP}; and receptor R is pos-

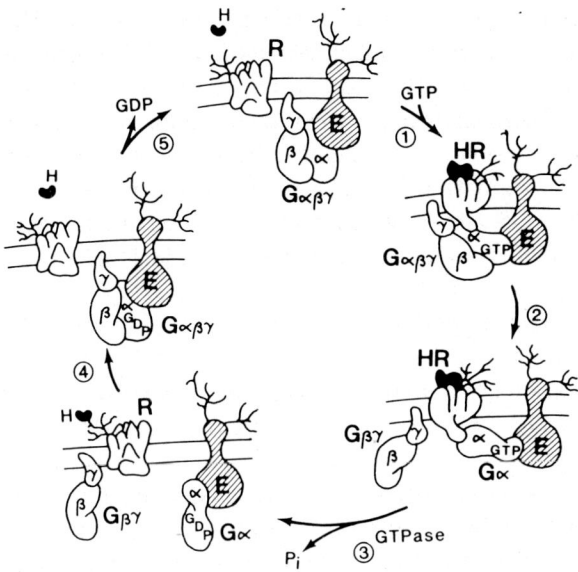

Fig. 1. Role of G protein in receptor-mediated regulation of effector function. The scheme is based on data from hormonal stimulation of adenylyl cyclase, but is applicable also to hormonal inhibition and, very likely, to G protein mediated regulation of any other effector function. R, receptor, is represented as a glycosylated transmembrane molecule having two conformations, one, unoccupied, with low affinity for hormone (H), and the other, occupied, with high affinity for both H and the α subunit of the signal transducing protein G. Under the influence of the HR complex the activation of G$\alpha\beta\gamma$ to Gα-GTP plus G$\beta\gamma$ is facilitated and the Gα-GTP complex reacts with and modulates the activity state of the effector E (reactions 1 and 2). The effector molecule E, like R, is represented as a glycosylated transmembrane protein. The signal transducing G protein is represented in its trimeric form, anchored to the inner leaflet of the plasma membrane through its γ subunit, and after activation in its dissociated forms as G$\beta\gamma$, still anchored to the membrane through γ, and Gα-GTP, which in this scheme is assumed to remain bound to the membrane complex through tight binding to the E. Reaction 3 (GTPase) is shown to convert Gα-GTP to Gα-GDP and to cause HR to dissociate, giving H plus low affinity R. However, separation of HR from G could have occurred also at the moment of Gα-GTP plus G$\beta\gamma$ formation. Reactions 4 and 5 lead to reassociation of the subunits of G to give G$\alpha\beta\gamma$ and dissociation of GDP. Although not observed with mammalian G$_s$, dissociation of GDP from the heterotrimer may require interaction with HR complex in the case of turkey G$_s$ and PTX sensitive G proteins, including transducin. Thus, the HR complex-G protein interaction may be part of reactions 5 and 6.

tulated as a 'catalyst' without which activation (dissociation) of G$_s$ would not occur. Deactivation of α^{*GTP} is shown to occur via conversion to α^{*GDP} + P$_i$. The receptor then separates from α_s^{GDP} and reverts to its low affinity form. This is followed by reassociation of $\beta\gamma$ to α^{GDP} and release of GDP to return to the starting point of the cycle. This cycle may need modifications if it is to be referred to G proteins other than G$_s$. One is the point of the cycle at which G proteins change receptors from high to low affinity. For example, with G$_s$, the high to low affinity transition is obtained with GDP at 10-times lower concentrations than with any other nucleotide

[34], but in other systems involving G_i or G_ps, GTP is equally as effective as GDP [35,36]. In this case, it is likely that receptors both change their affinity for agonists from high to low and separate from the system upon formation of the α^{*GTP}-effector complex. Another is the definition of which is slower: dissociation of GDP from $\alpha\beta\gamma$ or activation of $\alpha\beta\gamma$ by GTP to give α^{*GTP} plus $\beta\gamma$. With G_s activation is the slower – rate limiting – step [37–41]. With other G proteins, however, the rate of cycling appears to be limited by the dissociation of GDP [43,44]. Regardless, however, receptors act to accelerate both dissociation and activation [38,41,44,45].

2.2. Transducin (T), the light-activated GTPase

The first indication that phototransduction in the vertebrate retina involves a GTP-dependent step was reported in 1977 [46]. This led to the characterization of a G protein currently called transducin, or T. Like G_s, T is a heterotrimer of composition $\alpha\beta\gamma$ [47–50]. Of these, the β subunit is the same as β_{36} of G_s preparations [20], the α subunit (α_t) is distinct from α_s and is responsible for the coupling function of the protein, and the γ subunit is distinct from γ_s. The γ subunit is hydrophilic (in contrast to γ_s) and confers water solubility to the heterotrimer [51]. T is found in relatively high concentrations attached to the 'cytoplasmic' aspect of rhodopsin (Rho) containing disks of rod outer segments (ROS), in close proximity of cGMP-PDE. In contrast to G_s, T can be solubilized from ROS membranes without detergents under conditions that reflect its state of activity. At physiologic ionic strength, T associates in a Rho-dependent manner to membranes provided Rho is in its dark-adapted (inactive) state [47]. On lowering the ionic strength, T dissociates readily from dark adapted ROS membranes, but not from illuminated ROS membranes containing photoactivated rhodopsin (Rho*) [48]. However, addition of GTP results in dissociation of T from photoactivated (Rho*-containing) membranes [48]. Thus, in the absence of GTP, photoactivation stabilizes T on the membranes as Rho*-T complex, but in the presence of GTP, T interacts cyclically with photoactivated (Rho*-containing) membranes such that with each cycle 1 mol of GTP is hydrolyzed. As such, T functions as a light-activated GTPase [45–48]. In the presence of non-hydrolyzable GTP analogs, Rho*-T complexes undergo a dissociation reaction that results in release from the membrane both of free $\beta\gamma$ complexes and of free α subunits complexed with the non-hydrolyzable guanine nucleotide (α_t^{*G}). The latter activates a cGMP-phosphodiesterase in both illuminated and unilluminated ROS. This experiment demonstrates that α_t^{*G} represents the activated form of T and that T is the signal transducing protein mediating light-dependent activation of the cGMP-PDE [52]. This cGMP-PDE is itself an $\alpha\beta\gamma$ heterotrimer, of which the γ subunit inhibits the catalytic activity of the $\alpha\beta$ complex. Transducin-mediated activation of the ROS cGMP-PDE, in fact, entails release of $\alpha\beta_{pdc}$ from inhibition by γ_{pdc} through formation α_t^{*GTP}-γ_{pdc} [53,54] (for reviews see Refs. 55 and 56). Figure 2 depicts the cyclical activation/deactivation cycles thought to occur

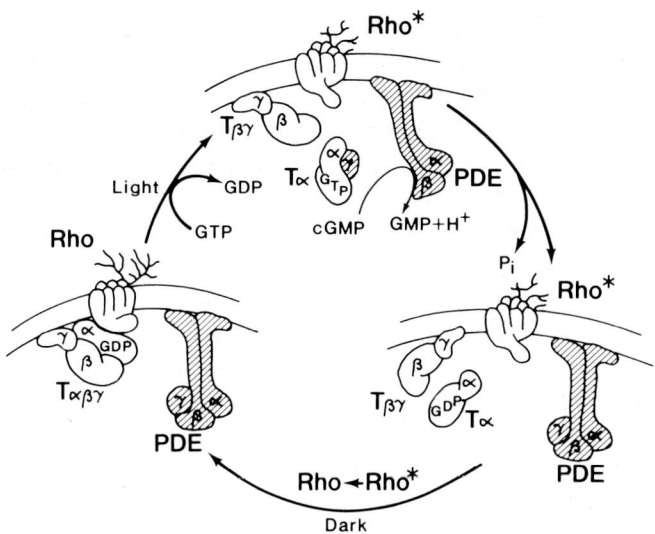

Fig. 2. Summary of regulatory GTPase cycle in photoactivation of cGMP-specific phosphodiesterase (PDE) in retinal rod cells. T, transducin (G_t); Rho, rhodopsin; Rho*, photoactivated Rho. PDE is represented as a heterotrimeric peripheral membrane protein, as is T. This regulatory cycle differs from that in Fig. 1 mainly in that the activation of PDE entails the dissociation of an inhibitory γ subunit (PDEγ) under the influence of activated Tα-GTP complex leading to formation of intermediary soluble Tα-GTP/PDEγ complex. This complex persists until GTP is hydrolyzed to GDP, at which moment the inhibited PDEαβγ heterotrimer reforms. Dark adapted – non-activated – Rho is then required for reassociation of Tα-GDP to Tβγ and release of GDP.

in ROS on photoactivation in the presence of GTP. The cycle differs from that shown in Fig. 1 for G_s in two important ways: (**1**) the transducing GTPase undergoes not only a subunit dissociation/reassociation cycle but also a membrane dissociation/association cycle and (**2**) the rate-limiting step in the cycle is the dissociation of GDP from transducin, as opposed to activation by GTP. Thus, in this case the primary function of the receptor appears to be the catalysis of GDP-GTP exchange. However, it cannot be excluded that Rho* also plays a role in promoting activation of T. For instance, Rho* may increase the character of GTP binding affinity of T, thereby converting the molecule to a form amenable to stabilization through subunit dissociation.

2.3. G_i, the inhibitory regulatory component of adenylyl cyclase

Inhibition of adenylyl cyclase by low (μM) concentrations of GTP was first reported in 1973–4 [57–59]. Further studies strongly suggested that hormones that attenuate adenylyl cyclase activity do so via a GTP-dependent step, akin to that intervening between stimulatory receptors and adenylyl cyclase (for reviews see Refs.

60, 61). The existence of a molecule coupling inhibitory hormonal receptors to adenylyl cyclase that is distinct from G_s was demonstrated in 1983 with the aid of PTX [62] and the cyc^- cell line [23]. The purification of PTX-substrates of subunit composition $\alpha\beta$, later revised to $\alpha\beta\gamma$ [5], was also reported in 1983 [63,64]. At that time, the only cellular function known to be blocked by PTX treatment was hormonal inhibition of adenylyl cyclase. The purified proteins, one from liver, with α subunit of 41 kDa [63], the other from human red blood cells (hRBCs), with α subunit of 40 kDa [64], were eventually named G_i (N_i).

G_i is thus functionally characterized as the GTP binding regulatory component that mediates hormonal inhibition of adenylyl cyclase [60,65–67]. Further, it can be activated by nonhydrolyzable GTP analogs [23,60,61] and, after ADP-ribosylation with PTX, loses its ability to interact with inhibitory receptors (R_i-type) (for review see Ref. 68) but not GTP analogs [23]. This uncoupling is correlated with ADP-ribosylation of a membrane component of M_r = 40–41 000 distinct from G_s [60]. PTX substrates are therefore good candidates for being G_i. Further, on hormonal stimulation, G_i displays also increased GTP hydrolysis, independent of and additive to stimulation of GTP hydrolysis by stimulatory hormones coupling to G_s [70], and increases its release of prebound GDP [45].

In contrast to G_s and T, purified 'G_i' has failed to work well in reconstitution assays designed to test for its ability to inhibit adenylyl cyclase in a manner predicted by the mode of action of G_s or T [69,71,72]. Thus, a definitive functional assay to confirm the identity of the purified proteins may be missing.

For discussion purposes, we shall refer to the functionally defined inhibitory G as G_i and the purified 40–41 000 Da PTX substrates, as 'G_i'.

2.4. G_o, a PTX substrate with an α subunit of M_r 39 000

Purification of GTP binding and hydrolyzing activity from bovine brain led to the isolation of a G protein with an α subunit of M_r 39 000, that is a substrate for PTX [73–75]. It co-purifies with another PTX substrate of much lower abundance, also an $\alpha\beta\gamma$ heterotrimer, but with an α subunit of M_r 41 000. The G protein with α of 39 kDa was termed G_o and its α subunit α_o or α_{39}. The subscript 'o' was meant to denote 'other', to distinguish it from the PTX substrate with α_{41}. The functional role of G_o is unknown. The protein with composition $\alpha_{41}\beta\gamma$ was termed G_i, because the GTP$_\gamma$S-activated α subunit inhibited adenylyl cyclase activity in G_s defective cyc^- cells [76]. Isolated G_o did not have this effect [75]. Thus, the cyc^- assay identifies $\alpha_{41}\beta\gamma$ of bovine brain but not $\alpha_{39}\beta\gamma$ as a possible G_i.

G_o and α_{41} 'G_i' have also been purified from rat [77] and porcine [78] brain and shown to be distinct from each other. Both bovine and rat brain G_o or 'G_i' were shown in reconstitution studies to interact with muscarinic receptors [79–81], by increasing the binding affinity for muscarinic agonists [80,81]. Platelet α_2-adrenergic receptors [82] stimulate the GTPase activity of the bovine G_o and 'G_i' proteins in an agonist specific manner. Since α_2-adrenergic receptors inhibit adenylyl cyclase in platelets, these studies identify the bovine G_o and 'G_i' from the bovine brain as G_i-type molecules. A similar study with porcine brain muscarinic receptors [80] and the rat G_o and 'G_i' proteins also showed stimulation of GTP hydrolysis by either G protein. However, since the porcine brain muscarinic receptor is of the M_1-type, which does not promote inhibition of adenylyl cyclase [83], the rat 'G_i' is not functionally identified as a G_i.

Isolated G_o and 'G_i' from rat brain also reconstitute with identical potency coupling of the chemotactic peptide (fMLP) receptor to a phosphoinositol bisphosphate-specific phospholipase C in PTX-treated membranes of HL-60 cells. This identifies both G proteins as stimulators of phospholipase C, i.e., as G_p-type molecules as opposed to G_i-type, suggesting that they have similar functions in cell regulation [84].

In contrast, porcine G_o (i.e., $\alpha_{39}\beta\gamma$), but not porcine 'G_i' ($\alpha_{41}\beta\gamma$), was shown to 'cure' opioid-ligand mediated closure of voltage-activated Ca^{2+} channels in cAMP-differentiated neuroblastoma × glioma hybrid cells that had been made refractory to ligands by pretreatment with PTX [85]. In this experiment, Ca^{2+} currents were recorded by the cell-attached broken-patch voltage-clamp technique and were injected into the cells through the patch pipette. This identifies G_o as a signal-transducing protein intervening in opioid receptor-mediated regulation of cellular Ca^{2+} channels, and differentiating it functionally from the PTX-sensitive G protein(s) that co-purified with it. However, it does not identify the effector molecule – adenylyl cyclase, phospholipase C or ion channel – coupled by G_o in these cells. Importantly, however, it ascribes dissimilar functions to G_o and 'G_i' in cell regulation.

The disparity in these conclusions may reside in differences between bovine 'G_i' and porcine 'G_i'; not detectable by simple SDS-PAGE migration. Indeed, a cDNA encoding bovine brain 'α_i' [86] and a cDNA encoding porcine brain 'α_i' [87] show significant differences in amino acid composition in spite of large stretches of identity (see below).

2.5. G_ps, the regulatory components of phospholipase (PhL) activity

As illustrated in Table I, many hormones act by stimulating membrane-bound phospholipases. The most commonly affected enzyme is a phospholipase C with specificity for phosphoinositides, i.e., a phosphoinositidase C (PIC) and, among these, the most relevant has specificity for phosphatidylinositol bisphosphate yielding inositol trisphosphate (IP_3) and diacylglycerol (DAG). IP_3 and DAG act as second messengers to mobilize Ca^{2+} from intracellular stores and activate the phospholipid- and Ca^{2+}-dependent protein kinase, respectively (protein kinase C) (for reviews see Refs. 87–90). A typical G_p-mediated response of this type occurs in neutrophils exposed to the chemoattractant peptide fMLP [91]. fMLP binds to specific membrane receptors which recognize proteolyzed fragments of bacterial proteins. The neutrophil response to fMLP is dependent on a G coupling proteins, as evidenced by the following findings. High affinity binding of fMLP to its receptor in isolated membranes is regulated by GTP and analogs [92] in much the same way as are hormone and neurotransmitter receptors that interact with G_s or G_i (cf. Refs. 1, 34, 35, 79). fMLP stimulates a low K_m GTPase in neutrophil membranes [93], causes release of IP_3 [94,95] and, in isolated membranes, stimulates PhL C activity in a GTP-dependent manner by decreasing the concentration of Ca^{2+} required for its activation [96]. Furthermore, non-hydrolyzable GTP analogs stimulate neutrophil membrane PhL C activity [96]. The argument that fMLP action depends on the intervention of a G protein is strengthened by the fact that effects of fMLP are blocked by pre-treatment of neutrophils with PTX under conditions that show ADP-ribosylation of an M_r 40–41 000 membrane component [97,98]. This functionally defined, but not yet biochemically identified, protein was named G_p (N_p), the 'p' standing for phospholipase [99,100] (for review see Ref. 101).

Just as neutrophils respond to fMLP [91], macrophages and mast cells exposed to IgE and other stimuli show activation of PhL C, formation of IP_3 and mobilization of Ca^{2+} [102]. This is followed by degranulation with release of lysosomal enzymes and histamine. These ligand-induced responses are blocked by PTX. Further, introduction of GTP$_\gamma$S into the cells causes degranulation that is blocked by neomycin, a substance that inhibits polyphosphoinositide hydrolysis by PhL C and interferes with the action of IP_3 to mobilize Ca^{2+} [100]. The data point to the existence of a PTX-sensitive G_p (PTX-sensitive G_p).

The response of mast cells includes release of arachidonic acid due to membrane PhL A_2 activation following a ligand-induced increase in cellular Ca^{2+}. PTX reduces the Ca^{2+}-mediated GTP$_\gamma$S-dependent release of this fatty acid in permeabilized cells [102,103]. This raised the possibility of a direct link, not only between receptors and PhL C, but also between receptors and PhL A_2. Existence of a G protein-mediated, PTX-sensitive, activation of PhL A_2 independent of G protein-mediated activation of PhL C was confirmed in studies first with fibroblasts [104] and then with FRTL thyroid cells [105]. Studies with the latter cells show that α_1-adrenergic receptors promote arachidonic acid release [105] and that this effect is mimicked in permeabilized cells by GTP$_\gamma$S and is not blocked by inhibition of PhL C with neomycin. Thus, at least two G_p proteins need to be defined: a G_p-stimulating PhL C (G_{plc}) and a G_p-stimulating PhL A_2 (G_{pla}). It is possible that rat brain G_o is PTX-sensitive G_{plc}.

In addition to PTX-sensitive G_ps, cells appear to have PTX insensitive G_ps, notably a PTX-insensitive G_{plc}. Functional G proteins of this type have been defined in several cells: in pituitary cells, responsive to TRH [106,107]; in liver, responsive to vasopressin, angiotensin II and α_1-adrenergic stimuli [108–110]; in 3T3-fibroblasts in response to various stimuli including bradykinin, thrombin and platelet-activating factor (PAF) [104]; in endothelial cells responsive to bradykinin [111]; and in FRTL-5 thyroid cells under stimulation by α_1-adrenergic ligands [105]. In all these systems IP_3 formation and/or the IP_3-mediated Ca^{2+} mobilization stimulated by these ligands are insensitive to PTX. Proper controls appear to have been done to establish the effectiveness of the PTX treatment, including the demonstration of simultaneous blockade of other PTX-sensitive pathways. Taking TRH-mediated effects on pituitary GH cells as an example, the cumulative evidence that a G protein is involved in signal transduction processes of this type is as follows: GTP regulates hormone binding [112], hormones stimulate GTP hydrolysis [113] and rapidly mobilize Ca^{2+} from intracellular stores [114,115], addition of $GTP_\gamma S$ to permeabilized cells mimics hormonal effects [116] and, as shown also in other systems [117–119], there is guanine nucleotide-dependent hormone-stimulated release of IP_3 from the isolated membranes [120].

Experiments with both fibroblasts [104] and FRTL-5 cells [105] indicate that a single ligand, presumably interacting with a single receptor, may activate in the same cell both a PTX-sensitive G_{pla} and a PTX-insensitive G_{plc}.

A guanine nucleotide-stimulation of phosphatidylcholine (PC) hydrolysis in isolated liver membranes by purinergic receptors (ATP, ADP) has also been observed [120] as has a phorbol ester-stimulated hydrolysis of PC to give diacylglycerol plus choline phosphate (CP) [122]. Interestingly, phorbol ester stimulation of PC hydrolysis also occurs in isolated membranes but depends on guanine nucleotide [120]. It appears therefore that one or more G proteins are involved in mediating PC hydrolysis in response to a class of receptors as well as in response to protein kinase C activation. The relation of the G protein mediating hydrolysis of phosphoinositides to the one (or both) involved in PC hydrolysis is not yet known. It has been a general observation that IP_3 formation is a rather transient response associated with intracellular Ca^{2+} mobilization while diacylglycerol formation is a more persistent response [123–125]. This suggests that the G protein acting on polyphosphoinositides may be the same as that acting on phosphatidylcholine.

Two laboratories reported activation of a cytosolic phospholipase C activity from platelets by guanine nucleotides [126,127] raising the possibility of existence of not only membrane bound G_ps but also of a cytosolic G_p.

Biochemical identification of these functionally distinct G proteins regulating phospholipid metabolism and generating second messengers in response to hormonal stimulation is an important goal of several laboratories.

2.6. G_k, the activator of 'ligand-gated' K^+ channels: mechanism of muscarinic regulation of atrial pacing

Muscarinic stimulation leads to activation of K^+ channels in sympathetic [128] and parasympathetic ganglia [129,130], as well as in central neurons [131,132]. Similarly, in atrial heart cells, acetylcholine combines with muscarinic receptors and opens a K^+ channel [133,134]. Increased K^+ causes the cells to hyperpolarize and become less excitable. Secretion is attenuated in neuronal cells and chronotropy is inhibited in atrial cells (for review see Ref. 135).

In heart, muscarinic receptors inhibit adenylyl cyclase, via activation of PTX-sensitive G_i (35,80,81,136,137). However, K^+ channel opening in response to muscarinic stimulation is not the result of decreased levels of cAMP [138,139]. Evidence obtained using patch-clamped cells [160] argues against involvement of any second messenger at all [141–143] in regulation of the K^+ channel. Moreover, experiments with 'inside-out' patches demonstrate unequivocally that K^+ channels couple directly to a receptor-regulated G protein [144,145]. We call this functionally identified G protein G_k.

Initial indications for direct receptor-G protein-K^+ channel coupling came from experiments which examined atrial muscarinic receptor regulation of K^+ channels present in a cell-attached patch [161]. It was found that K^+ channels in the patch are insensitive to acetylcholine (ACh) added to the bath, i.e., to the cell membrane outside the physically and electrically isolated patch. However, application of acetylcholine directly to the patch, using a specially constructed pipette opened the K^+ channels. If the effect of acetylcholine on the heart atrial K^+ channels were mediated by a change in the intracellular concentration of a second messenger, such as Ca^{2+} or cAMP, application of acetylcholine outside of the patch should have elicited a response. Moreover, the coupling mechanism was not addressed by this experiment.

The involvement of a G protein in the coupling of heart muscarinic receptors to K^+ channels was established in whole-cell broken-patch voltage-clamp experiments in which atrial heart cells were 'perfused' through a large patch pipette with medium containing or not GTP or the GTP analog, GMP-P(NH)P. In one set of experiments [142] it was established that acetylcholine was able to increase the K^+ currents only when the pipette/intracellular medium contained GTP. PTX treatment of the cells completely abolished this muscarinic response. In another set of experiments [143] using amphibian atrial cells, not only K^+ conductance but also Ca^{2+} conductance was measured. In these cells, acetylcholine induces an inward rectifying K^+ current and attenuates a β-adrenergic ligand (isoproterenol) induced slow inward Ca^{2+} current. The pipette/cytoplasm exchange was limited allowing detection of hormonal responses through the broken-patch pipette. Addition of GTP or of GMP-P(NH)P to the pipette had no effect unless hormones were added to the extracellular bathing fluid. In the presence of hormones, GMP-P(NH)P brought about agonist-induced, antagonist-resistant, persistent activation of the (acetylcholine-induced) inward rectifying K^+ channels. As before, this result indicates that muscarinic regulation of the K^+ channel is guanine nucleotide regulated. Further, the persistent nature of the response in the presence of GMP-P(NH)P indicates that the G protein intervening between receptor and K^+ channel resembles the G proteins that regulate adenylyl cyclase. The response of both G_s [147,148] and G_i [149] to GMP-P(NH)P is persistent. In agreement with the postulate that K^+ channel regulation is independent of cAMP [138,139], stimulation of the cell held with a GMP-P(NH)P-containing pipette with isoproterenol, which activates adenylyl cyclase, had no effect on induction of K$^+$ currents by ACh [143].

Direct coupling of a G protein to the heart K$^+$ channel activated by muscarinic ligands was demonstrated in a cell-free system using 'inside-out' patches. Addition to the bath (the cytoplasmic face of the patch) of GTP$_\gamma$S leads, after a lag, to permanent activation of K$^+$ channels [144]. Similarly, addition of purified PTX-sensitive G protein from human erythrocytes (referred to originally as 'G$_i$') or its α subunit complexed with GTP$_\gamma$ and free of $\beta\gamma$ subunits, also stimulates these channels, provided the G protein is preactivated by GTP$_\gamma$S [145]. These results defined the existence of a G$_k$ and identified it physically as an $\alpha\beta\gamma$ heterotrimer and a PTX substrate.

Neither non-activated 'G$_i$' nor non-activated or GTP$_\gamma$S-preactivated G$_s$ elicit K$^+$ channel opening under these conditions. Further, addition of complexes of α-GTP$_\gamma$S of G$_k$ resolved from $\beta\gamma$ subunits, but not of $\beta\gamma$ subunits, mimic the action of G$_k$ on atrial membrane patch K$^+$ channels [150]. This indicates that G$_k$ acts on K$^+$ channels via its α subunit as does G$_s$ acting on adenylyl cyclase and T acting on cGMP-specific PDE. Figure 3 summarizes evidence that led to identification of G$_k$ as the link between acetylcholine receptors and cardiac 'muscarinic' K$^+$ channels. Figure 4 presents key experiments that show direct activation of K$^+$ channels by G$_k$ and its α subunit.

Figure 5 summarizes the mechanism by which K$^+$ channels are activated by receptors such as the heart muscarinic acetylcholine channel (R$_k$-type) with involvement of G$_k$, as well as a view of how hormonal inhibition mediated by the PTX-sensitive G$_i$ may come about.

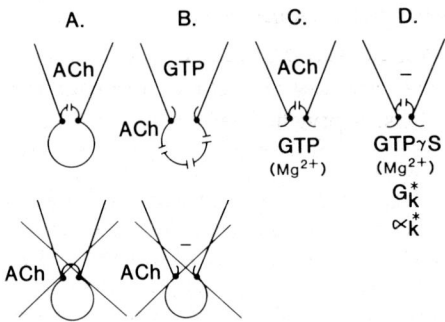

Fig. 3. Summary of results defining conditions that lead to opening of 'muscarinic' K$^+$ channels. *Panel A*. Experiments by Soejima and Noma [111] showed that opening of K channels in an isolated membrane patch occurs only by stimulation of a receptor located in the same patch (top) but not by stimulation of receptors outside of the patch (bottom). *B*. Experiments of Pfaffinger et al. [142] showed that acetylcholine (ACh) cannot lead to opening of K$^+$ channels unless GTP is supplied as a co-factor (top vs. bottom). *C*. Experiments by Kurachi et al. [144] and Yatani et al. [145] showed that addition of GTPγS to the inside face of 'inside-out' membrane patches leads to agonist-independent opening of K$^+$ channels. *D*. Activation of K$^+$ channels, independent of receptor occupancy, occurs on addition of GTPγS-activated G$_k$ (G$_k^*$) or its resolved subunit (α_k^*). For details see Fig. 4 and Refs. 144 and 150.

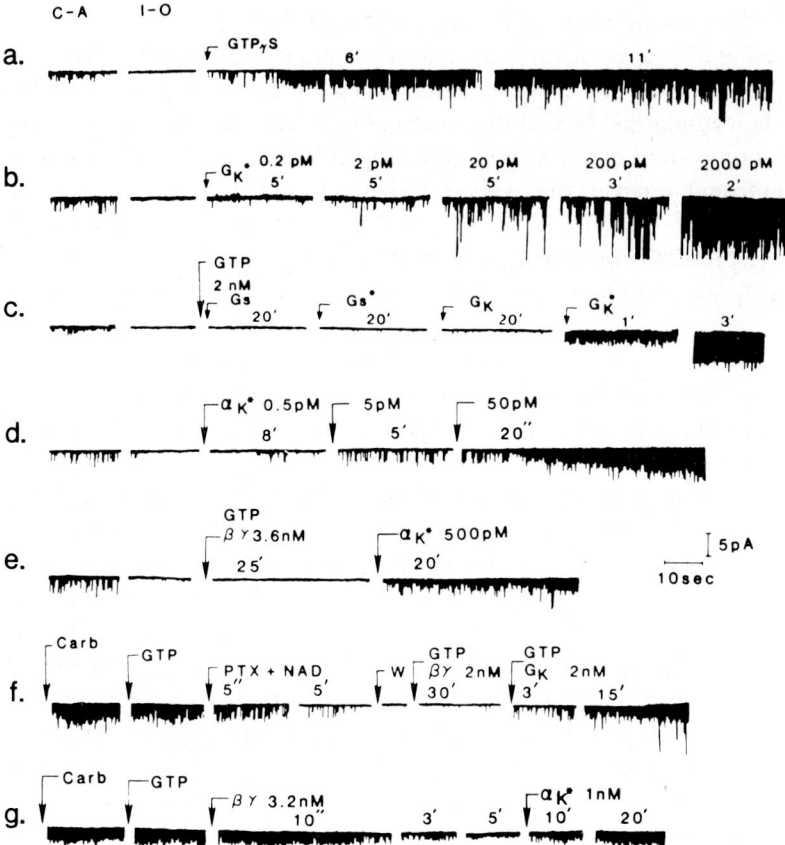

Fig. 4. Stimulation of opening of guinea pig atrial K$^+$ channels in isolated membrane patches. *Panel a*, effect of GTP$_\gamma$S; *b*, effect of increasing concentrations of G$_k^*$; *c*, specificity of effect of G$_k^*$; *d*, effect of α_k^* resolved from $\beta\gamma$ subunits; *e*, lack of effect of $\beta\gamma$ subunits; *f*, PTX sensitivity of acetylcholine receptor agonist (carbachol, CCh), lack of effect of $\beta\gamma$ on PTX-uncoupled system and recoupling of receptor-K$^+$ channel interaction with C$_k$ and GTP (W, wash); *g*, inhibitory effect of large excess of $\beta\gamma$ on receptor stimulated K$^+$ channel opening. Pipette buffer ('extracellular'): 140 mM KCl, 1.0 mM MgCl$_2$, 1.8 mM CaCl$_2$, 5 mM Hepes, pH 7.3, adjusted with NaOH. Bathing buffer ('intracellular'): 140 mM KCl, 2 mM MgCl$_2$, 1 mM EGTA, 2 mM ATP, 0.1 mM cAMP, and 5 mM Hepes, pH 7.3, adjusted with NaOH, plus additions (100 μM GTPγS (*a*); varying concentrations of G$_k^*$ (*b*); 100 μM GTP plus either G$_s$, G$_s^*$, G$_k$ or G$_k^*$, each at 2 nM (*c*); 0.5, 5 or 50 pM α_k^* (*d*); 3.6 nM $\beta\gamma$ plus 100 μM GTP followed by 500 pM α_k^* (*e*). 100 μM GTP followed in sequence by PTX (preactivated with DTT and AMP-P(NH)P) plus 1 mM NAD, 2 nM $\beta\gamma$ plus 100 μM GTP and 2 nM native G$_k$ plus 100 μM GTP (*f*); and 100 μM GTP followed by 3.2 nM $\beta\gamma$ in presence of GTP for 30 min and then 1 nM α_k^* (*g*). Note that patches used in panels *f* and *g* were held by pipettes containing 10 μM of the muscarinic receptor against carbachol (CCh). All G proteins and subunits were from human erythrocytes. G$_s$ and G$_k$, native non-activated G protein; G$_s^*$ and G$_k^*$, equimolar mixtures of activated α-GTPγS complexes plus $\beta\gamma$ dimers; α_k^*, α-GTPγS complex of G$_k$. Downward deflection, opening of K$^+$ channel(s). Patches were held in bathing buffer in a 100-μl chamber on a microscope stage. Additions were made at 25–30 min intervals, the first being 7–10 min after excision of the membrane patch from the cell. Numbers on top of records indicate time elapsed from last addition. CA, cell-attached, i.e., before patch excision; IO, inside-out, i.e., after patch excision. Holding potential, -80 mV except for panel *a* which was -90 mV. Each panel shows records obtained from a single membrane patch and is representative of at least three similar experiments.

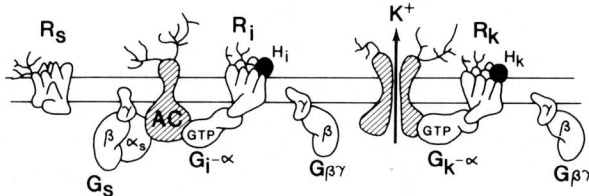

Fig. 5. Coupling of receptor-mediated inhibition of adenylyl cyclase and receptor-mediated activation of K^+ channel by G_i and G_k, respectively. G_i and G_k, coupling proteins responsible for inhibitory regulation of adenylyl cyclase and stimulatory regulation of K^+ channel, respectively. R_i and R_k, receptors of the type that inhibit adenylyl cyclase and/or activate the G_k-gated K^+ channels. For discussion of reasons leading to proposal that G_i acts on adenylyl cyclase primarily like G_k, G_s and T, i.e., through interaction of its α subunit with adenylyl cyclase, as opposed to primarily acting through its $\beta\gamma$ dimer to inhibit activation of G_s, see Ref. 235. Note that in heart muscarinic acetylcholine receptors play the role of both R_i- and R_k-type receptors. Likewise, somatostatin receptors (see Fig. 13) act on endocrine secretory cells to both inhibit adenylyl cyclase and stimulate K^+ channel opening. It is not known at present whether G_i and G_k are the same. As a consequence, the response of a cell in terms of inhibition of adenylyl cyclase and/or opening of K^+ channels is conditioned by the combination of: (i) whether or not R_i and R_k are the same – if different, then absence or presence of the receptor is decisive; (ii) whether or not G_i and G_k are the same – if different, absence of one or the other decides the type of response, even if receptors are the same; and (iii) presence or absence of the effector. Thus, even if $R_k = R_i$ and $G_i = G_k$, occurrence of a hyperpolarizing response still depends on expression of the gene encoding the G_k-gated K^+ channel. Simultaneous interaction of α subunits with receptor *and* effector is speculative.

ADP-ribosylation experiments show that up to five different PTX-sensitive G proteins may exist in heart [121]. One of them is G_i. It is possible that G_i is also G_k, that is, muscarinic receptors attenuate adenylyl cyclase activity and regulate K^+ channels by the same protein. However, muscarinic receptors may couple to more than one G protein, in which case K^+ channels and adenylyl cyclase would be regulated by different G proteins.

3. *G proteins detected by ADP-ribosylation*

The main question we are addressing is: how many G proteins are there? The answers gathered through functional analyses and direct purification were described above. Another looking glass is provided by the ADP-ribosylation of α subunits of G proteins with toxins, using $[^{32}P]NAD^+$ as substrate and analyzing their migration in electric fields by SDS-PAGE followed by autoradiography.

3.1. Labeling with CTX

CTX substrates, i.e., α subunits of G_s, when ADP-ribosylated and subjected to SDS-PAGE, migrate with apparent M_r between 42 000 and 52 000 [152–154]. In fact, most tissues contain at least two forms of α_s, one migrating as a relatively narrow band at 42–45 kDa while the other migrates as a broader band between 46 and 52 kDa. The M_r values reported for these forms of α_s are approximate, and vary depending on the reporting laboratory. The broader band is often reported as a doublet [154–156]. This raised the possibility that there exist not only two, but three types or forms of α_s subunits. In accord with predictions from ADP-ribosylation studies with CTX, G_s purified from rabbit liver is a mixture of at least two proteins: one having an α subunit migrating with an apparent M_r of 45 000 and the other with an α subunit that migrates with an M_r of ca. 52 000 [3]. α subunits of G_s prepared from human erythrocytes, in which cholera toxin ADP-ribosylates what appears as a single band of apparent M_r 42 000, migrates as a single band of M_r ca. 42 000 [29].

Thus, by CTX labeling there are at least two well defined types of α_s which we call α_{s1} (the smaller M_r 42–45 000 subunit) and α_{s2} (the larger M_r 50–52 000 subunit). Figure 6 illustrates autoradiograms of several membranes labeled with CTX.

Under special conditions other proteins can be labeled with CTX. Notable among these are bands at M_r ca. 39 000 in adipocyte membranes, which are observed only when the ADP-ribosylation is performed in the absence of GTP. These bands, la-

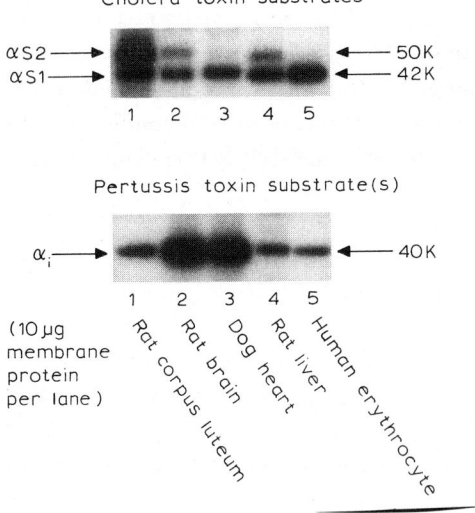

Fig. 6. Labelling of membranes from various sources with CTX and PTX. *Top panel*, 21-hour autoradiogram; *bottom panel*, 4-h autoradiogram. Note that, by purification, erythrocyte membranes contain approximately 5-times more Coomassie blue stainable PTX substrate than CTX substrate. SDS-PAGE was in 12.5% gels.

beled only partially by CTX, correspond to the α subunits of the PTX substrate(s), as evidenced by two-dimensional polyacrylamide gel isoelectric focusing/electrophoretic analysis [157]. Cloning experiments show that PTX substrates contain the CTX ADP-ribosylation site of α_s subunits (see below). Labeling of the β-subunit of eIF-2 by CTX has also been reported [154]. The significance of other bands of higher and lower M_r values than α subunits of G_s is not known. In ROS membranes, CTX labels T [158] at Arg^{174} [159].

3.2. Labeling with PTX

Like CTX, PTX also ADP-ribosylates T [160]. But, the ADP-ribosylated amino acid is Cys^{347}, four amino acids from the carboxyterminus of T [161].

Conditions for labeling with PTX differ markedly from those with CTX [162]. The abundance of PTX substrates in membranes is much greater than that of CTX substrates. Autoradiograms of labeling with PTX reveal fewer bands than those of with CTX and show ADP-ribosylation exclusively at M_r values that range from a maximum of 41 000 to a minimum of around 39 000. Under most circumstances the band at 41 000 (which is quantitatively minor compared to the others) is not observed, leaving only a cluster of bands – mostly fused – with M_r from 41 000 to 39 000 (Fig. 6).

It is currently not clear as to how many PTX substrates there are and, of these, how many may co-exist in any given tissue. Brain has been reported as containing three PTX substrates of M_r 39 000, 40 000 and 41 000 [73]. In heart and fat two PTX substrates were reported [163–165]. Human erythrocytes appear to have one PTX substrate of M_r 40 000 [29] which can be fractionated into two [151]. Prolonged exposures of autoradiograms of human erythrocyte, bovine brain, human neutrophil and rat liver PTX substrates reveal still a third substrate of apparent M_r = 43 000 [166,167]. As illustrated in Fig. 7, even more, up to five distinct proteins, can be visualized if membrane proteins are fractionated by SDS-PAGE in the presence of an urea gradient [151]. The functional difference between these PTX-labeled bands is not known at present.

cDNAs coding for PTX substrates have been cloned and found to have very similar carboxy termini:

```
              Ile         Lys Asn         Lys Asp           Phe
 -Val-Thr-Asp  -Ile-Ile-   -    -Asn-Leu-   -   C̄ȳs̄-Gly-Leu-   -COOH
              Val         Ala Glu         Arg Gly           Tyr
```

where the $\overline{Cys}$ is the ADP-ribosylated amino acid. The M_r 43 000 PTX substrate is not one of the CTX substrates, for α_s subunits have Tyr instead of Cys at the putative site of PTX ADP-ribosylation.

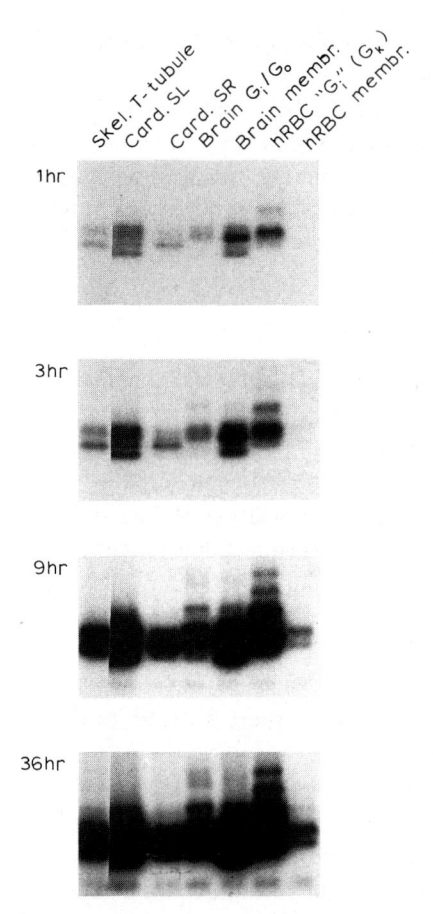

Fig. 7. Membranes and 'purified' proteins contain multiple PTX substrates. SDS-PAGE was in 8% gels, 15.5 cm long with a linear 4–8 M urea gradient. Only sections of the autoradiogram with molecules of M_r 35 000–45 000 are shown. Membrane samples, 1 µg protein/lane; brain G_i/G_o and hRBC 'G_i' (G_k), 20 ng/lane.

These PTX-labeling studies raise an obvious question: which PTX substrate mediates which coupling event?

4. G protein structure by cloning

Table II summarizes properties of G proteins, assuming that G_s is functionally an activator of adenylyl cyclase and Ca^{2+} channels (see below); G_i is functionally an inhibitor of adenylyl cyclase and activator of potassium channels (G_k), and G_o is functionally an activator of phospholipase C and an inhibitor of calcium channels [69]. Roles of $\beta\gamma$ complexes are also assigned. However, the mere existence of this rather large variety of functionally defined, biochemically isolated and/or toxin labeled G proteins unavoidably creates great difficulties in defining both how many G proteins there are and correctly ascribing functions to them. As shown in Fig. 7, even 'pure' protein preparations (see penultimate lane) are not pure.

Molecular cloning and expression of structurally defined molecules are expected to help clarify these problems. Major advances have been made during the last two years.

4.1. The α subunits

Purification of T from retinal rod outer segments and of G_o from brain, provided yields of these proteins that were sufficient for partial amino acid sequence analysis of their proteolytic fragments. This analysis revealed that α subunits of G proteins, while quite distinct from each other in general terms, are nevertheless similar. A partial sequence of 21 amino acids was determined to be common in bovine rod cell T and bovine brain G_o [168]. On the basis of this sequence, and other amino acid sequence analysis, four laboratories cloned cDNAs coding for transducin. The deduced amino acid structure of three of the cDNAs is the same [169–171]; the fourth differed [172]. Peptide-directed antibodies designed to distinguish between the two cloned forms localized one to rod cells (T-r) and the other to cone cells (T-c) [173].

Screening of a bovine brain cDNA library with a synthetic oligonucleotide mixture partially covering coding sequences of the region common to T and G_o, led to the isolation of a cDNA encoding G_s [11]. It was noted [168] that α subunits of G proteins are partially homologous to other G proteins including those of the family of *ras* oncogenes and bacterial elongation factor EFTu [174]. The similarities point to an origin in a common ancestral gene, especially those portions of the molecules now known to be involved in binding and hydrolyzing GTP [174,175].

At this time, the nucleotide sequences of the cDNAs encoding nine distinct α subunits have been published: T-r [169–171]; T-c [172]; G_s-1a, G_s-1b, G_s-2a and G_s-2b [12–17]; G_i-1 [10] and G_i-2 [14–16]; and G_o [14]. All have as identifying signature a common 18 amino acid identity box flanked on both sides by either Arg or Lys:

```
Arg                                                                          Arg
    -Leu-Leu-Leu-Leu-Gly-Ala-Gly-Glu-Ser-Gly-Lys-Ser-Thr-Ile-Val-Lys-Glu-Met-
Lys                                                                          Lys
```

The presence of such an identity box (*id* box) in any unknown cDNA may be used to identify it as the α subunit of a signal-transducing G protein. Figure 8 is a schematic representation of the various mRNAs encoding α subunits. Numerical data are given in Table III; Fig. 9 (from Ref. 176) shows the complete amino acid sequence of some of them; the more conserved sequences being enclosed by boxes. Based on the crystallographic structure obtained by Jurnak [175] for GDP-bonded

TABLE II
Signal transducing proteins: general subunit structure

G protein	Found in (mammals)	Subunit composition; effect of activation; rate limiting step	Properties of subunits	Function of protein	Regulated by	Effect of HR complexes
G_s (N_s)	all cells except spermatozoids	$\alpha_s\beta\gamma_G$ (polymorphic in α_s) dissociates into: $\alpha_s^{*G} + \beta\gamma_G$ rate limiting step in activation (mammals): $\alpha_s\beta\gamma_G \xrightarrow[Mg^{2+}]{GTP} \alpha_s^{*GTP} + \beta\gamma_G$	α_s: 380/394 a.a. M_r = 42–45 000 50–55 000 bind GTP hydrolyze GTP ADP-ribosylated by cholera toxin β: M_r = 35 000 γ_G: M_r = 5 000 $\beta\gamma_G$ is the same as in other Gs	stimulation of activity of catalytic unit of adenylyl cyclase opening of voltage-gated Ca channel (?)	H_sR_s complexes	acceleration of rate of activation by GTP decrease in susceptibility to deactivation by $\beta\gamma$
G_i (N_i) (G_k)	all cells (except in spermatozoids ?)	$\alpha_i\beta\gamma_G$ dissociates into: $\alpha_i^{*G} + \beta\gamma_G$ rate limiting step in activation (mammals): $\gamma_i^{GDP}\beta\gamma_G \xrightarrow[Mg^{2+}]{HR.GTP} \alpha_i^{*} + GTP\beta\gamma_G$	α_i: 355 a.a. M_r = 40–41 000 binds GTP hydrolyzes GTP ADP-ribosylated by pertussis toxin (IAP) β: M_r = 35 000 γ_G: M_r = 5 000 $\beta\gamma_G$ is the same as in other Gs	attenuation adenylyl cyclase activity opening of ligand-gated K channel	H_iR_i complexes and HR complexes that regulate excitability of membranes	opening of guanine nucleotide binding site and acceleration of activation by bound GTP
G_o (N_o)	nerve cells (brain, may also be in fat, heart and pituitary)	$\alpha_o\beta\gamma_G$ (other properties as for G_s and G_i)	α_o: M_r = 39 000 (other properties as for G_i)	similar to G_i, may be a G_p and/or a G_i closing of Ca channels involved in secretory events	H_iR_i-type complexes pre-synaptic receptors	same is in G_i

G_{plc} (PTX^+)	neutrophil macrophage fibroblast	assumed: $\alpha\beta\gamma_G$	α: ADP-ribosylated by PTX M_r ca. 40 000	activation of phospholipase C other: closing of voltage-gated K channels?	H_pR_p complexes	?
G_{plc} (PTX^-)	liver pituitary	assumed: $\alpha\beta\gamma_G$	α: not ADP-ribosylated by PTX	same as for G_{plc} (PTX^+)	H_pR_p complexes	?
G_{pla} (PTX^+)	fibroblasts thyroid	assumed: $\alpha\beta\gamma_G$	α: ADP-ribosylated by PTX M_r ca. 40 000	activation of phospholipase A_2	H_pR_p complexes	?
$\beta\gamma_G$ ('40 kDa protein')	all cells	$\beta\gamma_G$ complex of G_s, G_i, G_o and other $\alpha\beta\gamma_G$S; mixture of $\beta_{36}\gamma$ and $\beta_{35}\gamma$; $\beta_{35}\gamma$ abundant in human placenta	$M_r = 40\,000$	inhibition of G_s and G_i activation by unoccupied Rs mechanism: reversal of subunit dissociation		facilitate dissociation from holo-complex
T		$\alpha_T\beta\gamma_T$ dissociates into α_T^{*G} + $\beta\gamma_T$ rate limiting step in activation (rhodopsin dependent): $\alpha_T^{*GDP}\beta\gamma_T \xrightarrow{Rho^*} \alpha_T\beta\gamma$ + GDP	α_T: 350 a.a. $M_r = 39\,000$ binds GTP hydrolyzes GTP ADP-ribosylated by cholera and pertussis toxins β: $M_r = 35\,000$ γ_T: $M_r = 8500$ β in T is the same as β_{36} in G proteins; $\beta\gamma_T$ replaces functionally mixtures of $\beta_{35}\gamma_G/\beta_{36}\gamma_G$	activation of cGMP-specific phosphodiesterase	photon-activated rhodopsin (Rho*)	release of tightly bound GDP formed from GTP during the previous activation cycle, and concomitant facilitation of binding of new GTP and activation by it

a.a., amino acid; α_s, α_i, $\alpha_o = \alpha_{39}$, α_T, α subunits of the respective G proteins; γ_G and γ_T, γ subunits of G proteins and transducin, respectively; T, transducin; Rho*, photoactivated rhodopsin.

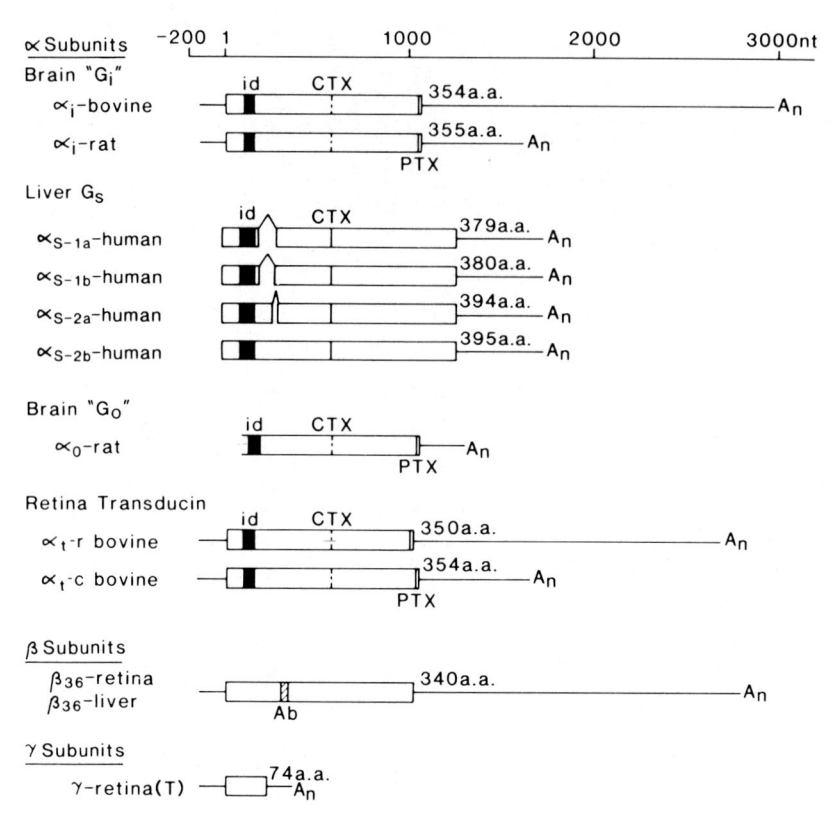

Fig. 8. Structure of mRNAs encoding subunits of G proteins as deduced from their respective cDNAs. *Open boxes*: open reading frames of mRNAs encoding the subunits; *lines*, 3′ and 5′ untranslated nucleotide sequences. The scale is in nucleotides. The lengths of the 5′ leader sequences preceding the open reading frames are approximate and may not be complete. α-*Subunits*: black boxes within open boxes represent the 'identity box' of the guanine nucleotide binding site (*id*). The locations of the arginine ADP-ribosylated by CTX and of the cysteine ADP-ribosylated by PTX are shown, as are those of the missing nucleotides of α_{s1a}, α_{s1b} and α_{s2a}. Note the large differences in sizes of 3′ untranslated regions of the mRNAs, even though they code for very similar proteins. β-*Subunits*: the location of the region against which a peptide-directed antibody has been made [18] is indicated by the *hatched box* within the open reading frame of the mRNA of β_{36}.

bacterial elongation factor (EFTu), and comparative amphipathic analysis of the amino acid composition of the α subunits shown in Fig. 9, Bourne and collaborators (Ref. 176) proposed a unifying model of the GTP binding site of α subunits and ascribed specific functions to different regions of the molecules. Figure 10 (from ref. 176) depicts this model.

Four forms of G_s-α subunits of 395, 394, 380 and 379 amino acids have been identified (Fig. 8 and Table III; Refs. 15, 177 and 17, and Mattera and Birnbaumer, unpublished). The differences appear to be due to alternate splicing of a common precursor mRNA resulting in insertion/deletion of a block of 45 and 3 nucleotides

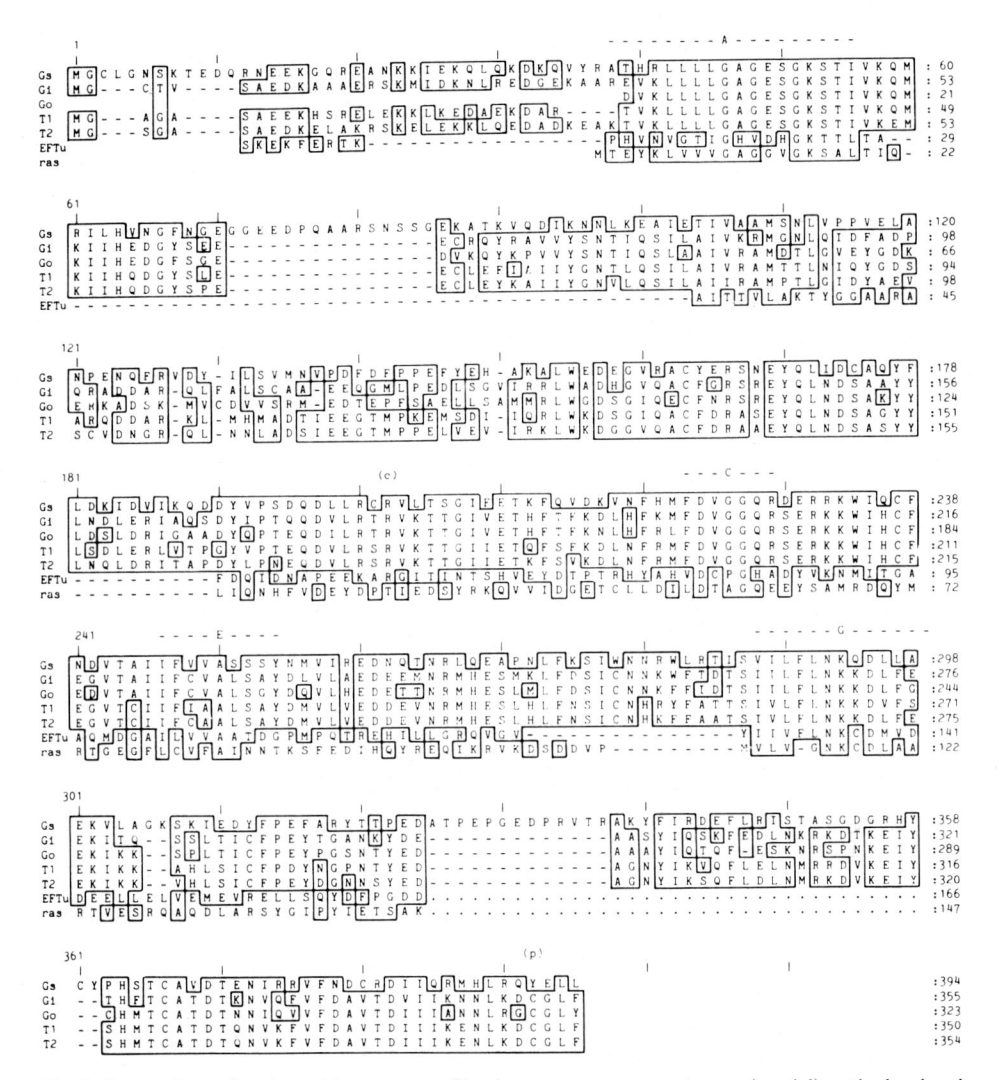

Fig. 9. Comparison of amino acid sequences of bovine α_{s2a}, mouse α_i, rat α_i, α_o (partial), retinal rod and cone cells α_t, bacterial elongation factor EFTu and c-Ha-ras, aligned to maximize matching. This leads to the definition of an α_{avg} of 396 amino acids (numbers above sequences). Boxes surround identical or conserved residues. (c) and (p), residues ADP-ribosylated by CTX and PTX, respectively. ---A---, ---C---, ---E--- and ---G---, sequences known to be intimately involved in guanine nucleotide binding by EFTu (From Refs. 175 and 176).

leading to presence or absence of 1, 15 or 16 (15 + 1) amino acids. We have named the shorter two α_{s1} a and b (α_{s1a} and α_{s1b}) and the longer two α_{s2} a and b (α_{s2a} and α_{s2b}) on the basis that type 1 and type 2, also called α_s-s (short) and α_s-l (long) [177], migrate on SDS-PAGE analysis as polypeptides with apparent M_r values of 42–45 000 and 50–52 000. This was shown by expression of the full length cDNAs of α_{s1b} and

TABLE III
Molecular parameters of G protein subunit cDNAs and the polypeptides encoded by them

Name of G protein	Source origin (ref.)	Toxin labelling		cDNA[a,c] (nucleotides)			Length amino acids	Calculated M_r	Apparent migration on SDS-PAGE
		CTX[b]	PTX	5'UT	ORF	3'UT			
A. α Subunits									
G_s-1a	human liver (15)	yes	no	117	1140	364	379	44108	42–45 kDa
	mouse macrophage (16)	yes	no	16	1132	345	377	43856	
G_s-1b	human liver (15)	yes	no	73	1140	364	379	44294	42–45 kDa
	bovine adrenal (177)	yes	no	nr	1140	364	380	44322	
G_s-2a[d]	human liver (15)	yes	no	68	1182	364	394	45664	49–52 kDa
	bovine brain/adrenal (12)	yes	no	2	1182	309	394	45692	
	bovine brain (13)	yes	no	52	1182	365	394	45706	
	rat brain (14)	yes	no	245	1182	290	394	45663	
G_s-2b	human brain (17)	yes	no	–	<1152>	121	incompl.	45751 (?)	49–52 kDa
G_i-a[e]	bovine brain (86)	(yes)	yes	116	1062	1921	354	40359	40–41 kDa

G_i-b	human monocyte (181)	(yes)	123	1065	514	355	40 456	40–41 kDa
	rat brain (14)	(yes)	48	1065	562	355	40 499	
	mouse macrophage (16)	(yes)	97	1965	153	355	40 482	
G_o	rat brain (14, 182)	(yes)	—	<1022>	115	incompl.	?	
T-r	bovine retinal rods (169–171)	yes	137	1050	1259	350	39 964	39 kDa
T-c[f]	bovine retinal cones (172)	yes	174	1062	429	354	40 100	?
B. β Subunit (β_{36})								
T-r	bovine retinal rods (19, 20)	—	138	1020	1664	340	37 375	36 kDa
G_sG_i	human liver (21)	—	280	1020	1788	340	identical	36 kDa
C. γ Subunit								
T-r	bovine retinal rod (186, 188)	—	67(58)[g]	222	145	74	8022	5–8 kDa

[a] 5'UT, 5' untranslated leader sequences; ORF, open reading frame; 3'UT, 3' untranslated sequences. If multiple forms reported from the same source, numbers are those of the longest.

[b] ARG^{201} for CTX and Cys at −4 amino acids from COOH end for PTX.

[c] If multiple forms reported from the same source, numbers are those of the longest.

[d] G_s-2a sequences differ in that human liver α_{s2a} is Ala^{18}, Asp^{139}, Ala^{188}; bovine adrenal is Gly^{18}, Asp^{139}, Asp^{188}; bovine brain α_{s2a} differs from bovine adrenal by one nucleotide having Ala^{18} instead of Gly^{18}; and rat α_{s2a} is Ala^{18}, Asn^{139}, Ala^{188}.

[e] Contain potential ADP-ribosylation site for CTX: QDxLRyRVzToGI, with x=L, V or I; y=C in α_s, T in all others; z=L in α_s, K in all others; and o=S in α_s, T in all others. Underscored R is ADP-ribosylated by CTX.

[f] Migration on SDS-PAGE not known.

[g] Different 5' untranslated (5'UT) leader sequences were reported.

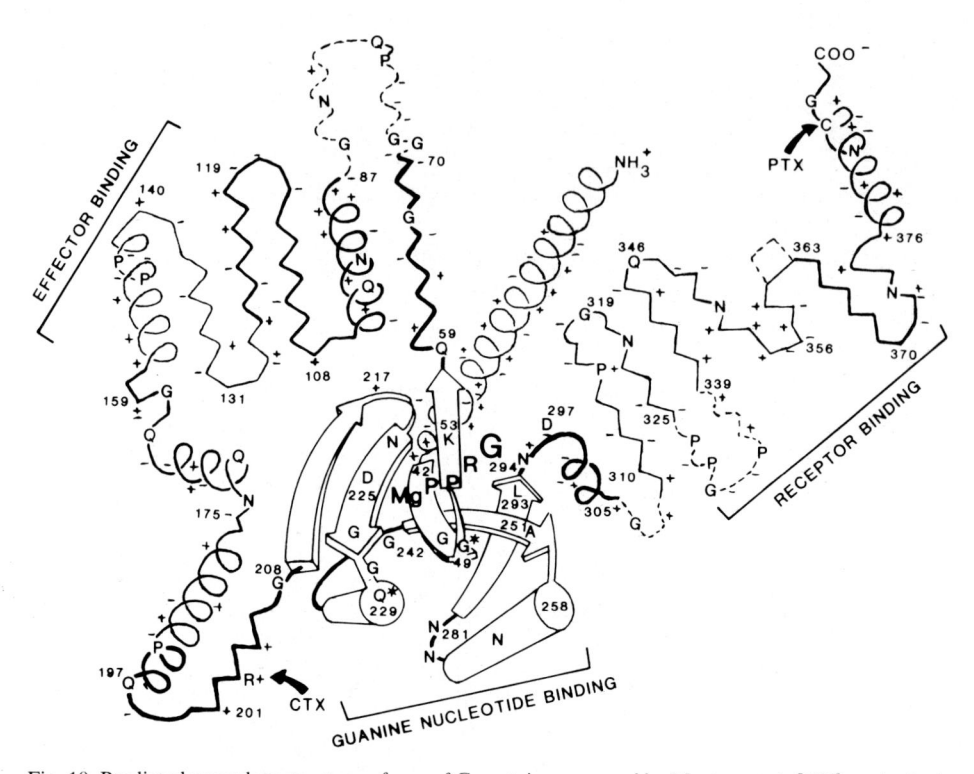

Fig. 10. Predicted secondary structure of α_{avg} of G proteins proposed by Masters et al. [176] on the basis of crystallography of EFTu [175] and amphipatic analysis of the primary amino acid sequence of five G-αs: α_s, α_i, α_o, α_t-r and α_t-c. Some of the amino acids present in at least four of the five G-α subunits are highlighted; numbers refer to numbering of α_{avg}. *Cylinders and ribbons*, guanine nucleotide binding region as deduced from X-ray crystallography; *thick lines*, regions in which the sequences of the five chains are >90% homologous; *lines with middle thickness*, regions in which the sequences are 70–89% homologous; *thin lines*, regions with less than 60% sequence similarity; *dashed lines*, sequences present only in α_s; residues 71–86 are present only in α_{s2a}. The molecule is shown with Mg-GDP bound to it. G^{42} and Q^{229} (marked by asterisks) correspond to the oncogenic mutation sites Gly^{12} and Glu^{61} of p21 *ras* proteins. (Adapted from Masters et al. [176]).

α_{s2a} in COS cells [177]. A model to account for the four forms of α_s as a consequence of alternate splicing reactions has been proposed [17]. G_s-α subunits are highly conserved between species (human vs. rat vs. bovine), differing only in two of three defined amino acids (Tables III and IV).

Earlier experiments have shown that cholate extracts from transformed lung fibroblasts having only the M_r 52 000 subunit of G_s are active in reconstituting hormone-, nucleotide- and fluoride-stimulated adenylyl cyclase activities in *cyc*$^-$ membranes [178]. Human erythrocyte G_s, which has only the M_r 42 000 α subunit(s) [22,179], also reconstitute(s) these functions. After partial separation, rabbit liver p52 G_s appeared to reconstitute hormone-stimulated activity better than p45 G_s

TABLE IV
Sequence differences between α subunits of G_s

A. α_{s1a} (M_r by SDS-PAGE migration: 42–45 000)

207																219	
AAT	GGA	GA-	...	...	...	...	...	...	...	...	...	...	...	...	-T	GAG	AAG
Asn	Gly	<u>Asp</u>	...	...	...	...	...	...	...	...	...	...	...	...	...	Glu	Lys
		70															73

B. α_{s1b} (M_r by SDS-PAGE migration: 42–45 000)

207																222	
		*												*	**		
AAT	GGA	GA-	...	...	...	...	...	...	...	...	...	...	...	-C	AGT	GAG	AAG
Asn	Gly	<u>Asp</u>	...	...	...	...	...	...	...	...	...	...	...	...	<u>Ser</u>	Glu	Lys
		70															74

C. α_{s2a} (M_r by SDS-PAGE migration: 48–52 000)

207																264	
	*	***	***	***	***	***	***	***	***	***	***	***	**		**		
AAT	GAG	GGC	GAA	GAG	GAC	CCG	CAG	GCT	GCA	AGG	AGC	AAC	AGC	GAT	GG-	-T	GAG	AAG
Asn	<u>Glu</u>	<u>Gly</u>	<u>Glu</u>	<u>Glu</u>	<u>Asp</u>	<u>Pro</u>	<u>Gln</u>	<u>Ala</u>	<u>Ala</u>	<u>Arg</u>	<u>Ser</u>	<u>Asn</u>	<u>Ser</u>	<u>Asp</u>	<u>Gly</u>		Glu	Lys
																		88

D. α_{s2b} (M_r by SDS-PAGE migration: 48–52 000)

207																267	
	*	***	***	***	***	***	***	***	***	***	***	***	***	***	**		
AAT	GAG	GGC	GAA	GAG	GAC	CCG	CAG	GCT	GCA	AGG	AGC	AAC	AGC	GAT	GGC	AGT	GAG	AAG
Asn	<u>Glu</u>	<u>Gly</u>	<u>Glu</u>	<u>Glu</u>	<u>Asp</u>	<u>Pro</u>	<u>Gln</u>	<u>Ala</u>	<u>Ala</u>	<u>Arg</u>	<u>Ser</u>	<u>Asn</u>	<u>Ser</u>	<u>Asp</u>	<u>Gly</u>	<u>Ser</u>	Glu	Lys
																		89

Differences in nucleotides are denoted by an asterisk. those in amino acids by underscoring. Note that type-1 (short) α_ss have Asp^{71}, while type-2 (long) α_ss have Glu^{71}. Type a differ from type b in that they lack a Ser.

[180]. The functional differences between the different forms of G_s may now be tested more specifically by using cloned proteins prepared by expression in bacteria or *cyc*⁻ cells.

Northern analysis of RNA from *cyc*⁻ S49 cells (Fig. 11) shows absence of α_s coding sequences, in agreement with the absence of α_s polypeptides in these cells. The levels of α_s mRNA in different tissues is variable. Very likely, α_s subunits are under developmental and hormonal control and experiments exploring this should be forthcoming shortly.

Two classes of cDNAs encoding for putative G_i-α subunits have been cloned [14,16,86,181]. They were identified by matching the deduced amino acid se-

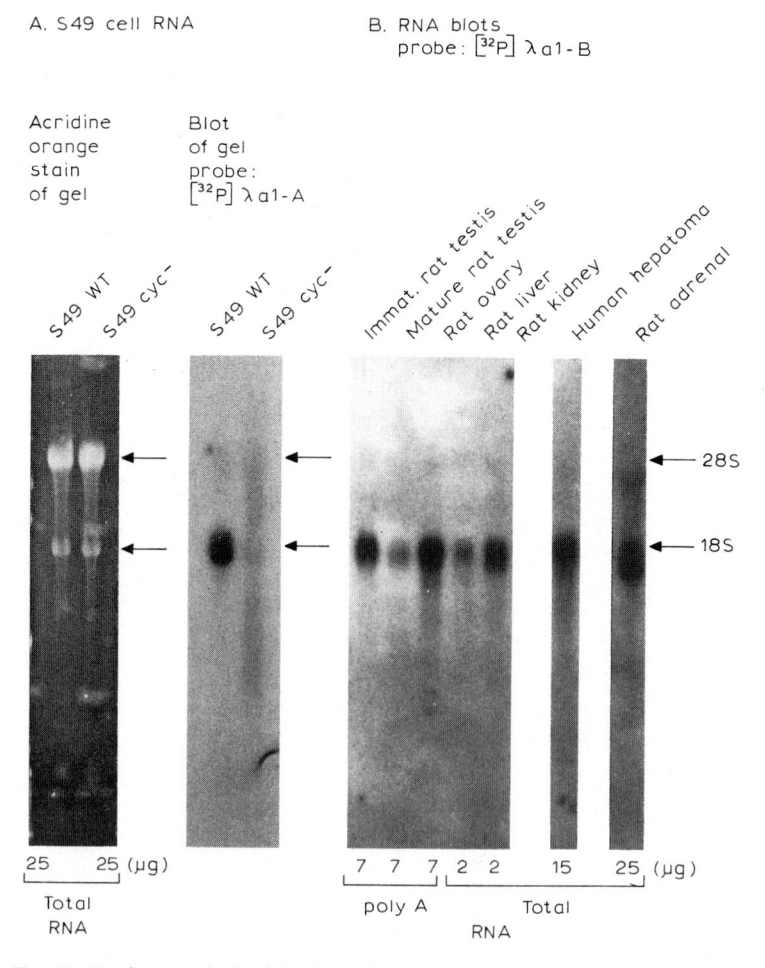

Fig. 11. Northern analysis of RNA fractions isolated from tissues and cell lines for presence of mRNA coding for α-subunits of G_s. Note absence of signal in S49 *cyc*⁻ cells and that levels of α_s mRNAs are not constant. Immature and mature rats were 10 and 180 days, respectively. (From Mattera et al. [15]).

quences to sequences obtained for several peptides isolated by tryptic digestion of purified α_i (PTX-sensitive p40–41) from bovine [86] and rat [14] brain. One (α_i-1) is present in bovine brain [86] and the other, α_i-2 is found in rat brain [14], mouse macrophages [16] and human neutrophils [179]. While encoding for two very similar proteins, the two-types of cDNAs differ sufficiently in the amino acids for which they code that they may represent functionally distinct polypeptides originating from non-allelic genes. Both contain identical PTX ADP-ribosylation sites. Figure 12 compares the amino acid differences between α_is as predicted from their cDNAs.

Partial G_o-α cDNAs have been reported as well [14–182]. The longest starts with the 9th amino acid of the G-α identity box, and codes for a PTX ADP-ribosylation site that differs by at least one amino acid from that of the G_i-αs (Fig. 8, Table III).

4.2. The β subunits

β-Subunits of G proteins migrate on SDS-PAGE gels as polypeptides of M_r 35–36 000. Close examination of the bands, especially if urea gradient gels are used, shows that the β subunits in G_s, G_i and G_o preparations migrate as a close doublet β_{35} and β_{36}, while that in T preparations migrates as a single band of 36 kDa at the β_{36} position. The two β subunits have not been separated biochemically. However, human placenta appears to express predominantly β_{35} [183]. βγ complexes derived from T appear to be functionally equivalent (exchangeable) with βγ complexes of brain, liver or erythrocyte origin ($G_{s/i}$-β). Amino acid composition, mono- and two-dimensional peptide analysis of β subunits derived from transducin, G_s, G_o and G_i from various sources (rabbit liver, bovine rod outer segments, bovine brain and human erythrocytes) revealed them to be very similar [4,22,184].

The cDNA encoding the β_{36} subunit of bovine T [19,20] and that of a human liver G_s/G_i [21] have been cloned. The two cDNAs encode for the identical polypeptide of 340 amino acids with a calculated M_r 37 375. Polyclonal antibodies raised against

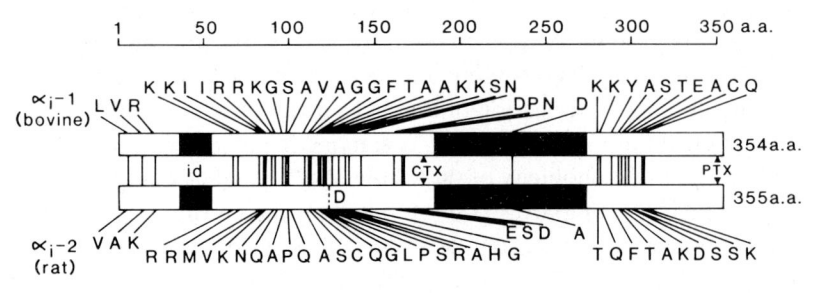

Fig. 12. Comparison of amino acid sequences of bovine and rat α_is as deduced from their respective cDNAs. The long boxes represent the two polypeptide chains, and the two black zones the stretches of amino acids involved in guanine nucleotide binding. The locations of amino acids that differ, identified above and below, are denoted by lines connecting the two polypeptide chains. Note that 75% of the differences are in the region proposed by Bourne and collaborators [176] as interacting with the effector.

a peptide deduced from this cDNA sequence immunoprecipitate β_{36} but not β_{35} [18]. Northern analysis of RNA from bovine retina, brain and liver showed the presence in each of these tissues of two mRNA species: one of about 1800 nucleotides, the other of ca. 3000 nucleotides [19]. The difference may lie in the presence of two AATAAA cleavage and polyadenylation signals in the 3' untranslated region of the mRNA. Cleavage at the signal proximal to the coding region would lead to the shorter mRNA while cleavage at the other would yield the longer mRNA. In addition, a probe to the 5' untranslated region of transducin β_{36} only hybridized to RNA extracted from retina, but not to liver or brain RNA [19]. Thus, there exists at least one tissue specific 5' untranslated leader sequences preceding the coding region of β_{36} mRNA. However, the situation may be more complex [20]. An antibody recognizing preferentially β_{35} has also been obtained [183]. The natural gene for neither β has been reported as yet.

4.3. The γ subunit: its role as a membrane anchor

The γ subunit of G_s is soluble in 10% trichloroacetic acid (TCA) and therefore was absent from SDS-PAGE gels analyzing TCA-precipitated material [3]. Its presence was only noted after analyzing G_s and G_i precipitated with acetone. This procedure was intended to remove the Lubrol PX used during isolation of the G proteins [4]. Two-dimensional peptide maps of γ subunits derived from G_s and G_i are indistinguishable, yet differ markedly from those derived from T [51]. Similarly, polyclonal antibodies raised against T cross-react well with T-γ, yet do not react with G_s/G_i [185]. The cDNAs of T-γ have been cloned by two laboratories and identified by either antibody screening plus cell free translation of hybrid selected RNA [186] or hybridization to oligonucleotide mixtures based on partial amino acid sequence analysis [187,188]. The cDNAs encode for a polypeptide of 74 amino acids. The polypeptide is very hydrophilic having 19 acidic and 11 basic amino acids. The cloning of a cDNA encoding $G_{s/i}$ γ has not yet been reported.

As pointed out earlier in this article, T differs from other G proteins in that it is a peripheral membrane protein. After activation by Rho* it seems to undergo subunit dissociation in which both its α subunit and its $\beta\gamma$ complex dissociate from the Rho-containing membranes. Purification of brain G-proteins has shown that free α subunits of G_o and G_i are also water soluble, remaining in solution in the absence of detergents [74]. The hydrophobicity of the whole $\alpha\beta\gamma$ G_o and G_i complexes was shown to be due to their $\beta\gamma$ complexes [189]. Indeed, purified α subunits associate with phospholipid vesicles only if $\beta\gamma$ complexes have been incorporated during vesicle formation [189]. Since the amino acid composition of T-β is equal to that of other G-βs, but their γs differ, it follows that the principal role of γ subunits should be to anchor non-T G proteins to the plasma membranes. This conclusion assumed, of course, that β subunits are not post-translationally modified in a tissue specific manner such that that they become water soluble in retinal photoreceptor cells and lipophilic in other tissues. Fatty acid acylation could be one such post-translational modification. It also assumes no fundamental functional differences between β_{36} (β of T) and β_{35}.

5. G protein mediation of receptor regulation of ion channels

5.1. Effects of inhibitory receptors on K^+ channels in tissues other than heart atria

Somatostatin (SST) antagonizes hormone-stimulated secretory events [190, 191]. This can be observed in clonal cell lines, such as rat pituitary GH cells of the GH_3 and GH_4C_1 type, which respond to TRH by secreting prolactin [192], or mouse pituitary AtT-20 cells, which secrete ACTH in response to a variety of hormones and neuromodulators, such as corticotropin-releasing factor (CRF), norepinephrine and vasopressin ACTH [193,194]. Similarly, neurosecretory events can be inhibited by opioid [195,196], α_2-adrenergic [197], $GABA_B$ [197] and a subclass ($5HT_{1a}$) of serotonin [197] receptors. Certain [198], but not all [199], muscarinic effects of acetylcholine are also of inhibitory nature. Inhibition due to occupation of α_2-adrenergic [200,201], opioid [202] and SST [194,203,204] receptors may be mediated by decreased adenylyl cyclase activity. Most of these effects are blocked by PTX (e.g., Refs. 106, 107, 205, 206), indicating the involvement of a PTX-sensitive G protein. In the case of adenylyl cyclase inhibition, this G protein is G_i. α_2-Adrenergic receptors can also stimulate GTP hydrolysis by human red blood cell 'G_i' as well as bovine brain 'G_i' and G_o [82]. Elevation of cAMP levels elicits secretory responses in pituitary cells, therefore the inhibitory effects of SST in these cells may be due to inhibition of adenylyl cyclase via G_i [203]. Because agonist binding to opioid [207], α_2-adrenergic [208,209] and $GABA_B$ [210] receptors is regulated by guanine nucleotides, it is also tempting to postulate that all neuronal inhibition is mediated by attenuation of adenylyl cyclase activity. Yet, even though these receptors do in many, and perhaps all, cases inhibit adenylyl cyclase activity, this does not completely account for their biological actions. This effect has most clearly been shown for the action of SST in GH cells.

In GH cells SST inhibits prolactin secretion induced by cAMP analogs and K^+-induced depolarization in addition to inhibiting adenylyl cyclase activity [192]. Similar data were also obtained with ACTH-secreting AtT-20 cells [203]. PTX blocks SST inhibition of ligand-induced and cAMP analog-induced secretions in both GH and AtT-20 cells [106,107]. Thus, it is clear that a G protein is involved in both the cAMP dependent and the cAMP-independent actions of SST. Quin-2 measurements showed that exposure of cells to SST lowers the intracellular Ca^{2+} levels. This effect is also PTX-sensitive [211–213]. The decline in intracellular Ca^{2+} was concluded to be secondary to SST hyperpolarizing the cells by increasing K^+ conductance [214,215]. Acetylcholine, like SST, causes inhibition of adenylyl cyclase, hyperpolarization and transient lowering of intracellular Ca^{2+} in GH cells [211–213,216,217]. The similarity of responses of GH cells to acetylcholine to those of heart cells suggested that both acetylcholine and SST might inhibit secretory responses in endocrine cells with an involvement of a G_k-protein akin to that found in atrial cells. Activation of G_k would cause hyperpolarization due to activation of a G protein-gated K^+ channel and reduce entry of Ca^{2+} and thereby reduce secretory activity. Patch-clamp experiments such as performed with atrial cell mem-

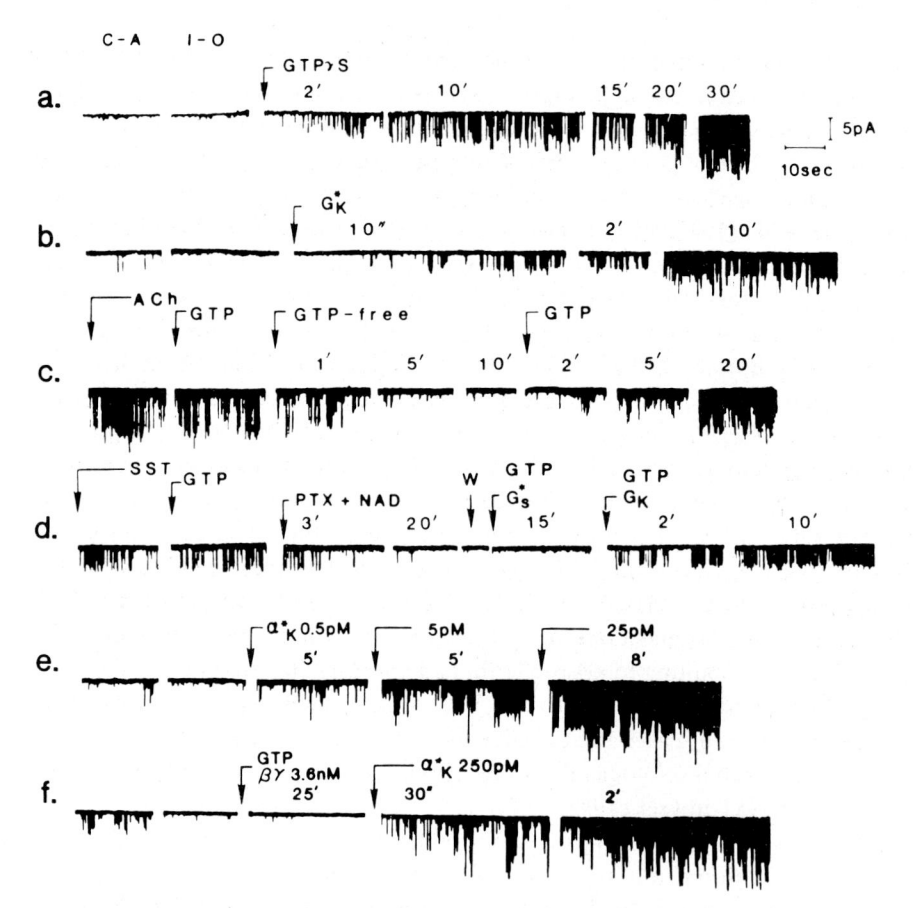

Fig. 13. Characteristics of activation of a G protein-gated K^+ channel as seen in endocrine (GH3) cells. Panel a, activation of the channel by 100 μM GTPγS; panel b, activation of the K^+ channel by 2 nM GTP-activated G_k (G_k^*); panel c, dependence of K^+ channel activation by receptors (10 μM acetylcholine, ACh) on GTP (when present 100 μM); panel d, sensitivity of GH_3 cell G_k to uncoupling by PTX, lack of effect of 2 nM GTPγS-activated G_s (G_s^*) and reconstitution of receptor (somatostatin, SST)-K^+ channel coupling by exogenously added native G_k (2 nM) in the presence of 100 μM GTP (W, wash); panel e, effect of α_k^*, i.e., α_k-GTPγS resolved from βγ dimers, added at 0.5, 5 and 25 pM; panel f, lack of effect of βγ dimers on K^+ conductance of an α_k^* responsive membrane. For rest of conditions see legend to Fig. 4 and Refs. 144, 145 and 232.

branes were performed with GH_3 cell membranes [146,147] and, as illustrated in Fig. 13, proved the existence in endocrine cells of both a G_k, activated by SST and acetylcholine receptors, and of a G_k-gated K^+ channel.

Very likely, inhibition of neuronal activities mediated by α_2-adrenergic, R_i-adenosine, $GABA_B$, opioid and M_2-muscarinic acetylcholine receptors [130,131,196,197,218], also has a biochemical basis in a direct coupling of G_k to the K^+ channel because stimulation of these receptors often induces a change in K^+ conductance resulting in hyperpolarization [130,131,196,218]. This conclusion received supporting evidence when it was shown that effects of $GABA_B$ and 5-HT on a slow inhibitory postsynaptic increase in K^+ conductance is inhibited by PTX [219]. G_k may also mediate part of the initial hyperpolarizing response of neuroblastoma × glioma cells to bradykinin [220,221]. This response is currently thought to be mediated exclusively by IP_3-stimulated Ca^{2+}-release and activation of Ca^{2+}-dependent, voltage-gated K^+ channels [221] (for comment see Ref. 222). If this is the case, it would require that bradykinin receptors couple to more than one G protein: a G_k to activate the 'G protein-gated' K^+ channels and a G_p (G_o?) to activate phospholipase C. Bradykinin seems to have the ability to interact potentially with multiple G proteins [220]. It both inhibits neuroblastoma × glioma cell adenylyl cyclase (a G_i function) and stimulates the GTPase activity of purified 'G_i' and G_o (a putative G_p [84]). Studies testing this hypothesis should soon be forthcoming. If bradykinin receptors do indeed interact with two G proteins in single cells they would join angiotensin II receptors of liver and adrenal glomerulosa cells which have very convincingly been shown to activate both a PTX sensitive G_i and a PTX insensitive G_p [110,223,224].

Although K^+ channel regulation could account for inhibitory responses of cells to what are here called 'inhibitory hormones' (e.g., SST, $\alpha_2$adrenergic, $5HT_{1a}$, $GABA_B$), voltage-clamp experiments testing the behavior of Ca^{2+} conductances indicate existence of a regulation of Ca^{2+} channels by inhibitory hormones that is independent of changes in K^+ conductances. The nature of the G protein mediating this effect is not known (see below).

5.2. Inhibitory regulation of voltage-gated Ca^{2+} channels: direct or indirect involvement of a G protein?

Several electrophysiological experiments indicate that a G protein intervenes between certain receptors and inhibition of Ca^{2+} channel conductance. First, measurements of voltage-gated Ca^{2+} currents in chick embryo dorsal root ganglia by the whole-cell broken-patch technique revealed that both norepinephrine (in this case acting as an α_2-adrenergic agonist) and GABA (acting on $GABA_B$ receptors) inhibit voltage-gated Ca^{2+} currents in a PTX-sensitive manner. Further injection of the GTP antagonist $GTP_\gamma S$ into cells blocks the effects of norepinephrine and GABA [206]. This indicates the participation of an intracellular GTP-binding site in the action of norepinephrine and GABA, which is distinct from the above mentioned G_k. Second, injection of porcine G_o (or its α subunit), but not porcine 'G_i', into PTX-treated neuroblastoma × glioma cells led to reconstitution of opioid receptor mediated inhibition of Ca^{2+} currents [85]. The ligand in this case was the enkephalin analog DADLE, interacting with δ-type opioid receptors. This result identified a specific function for G_o. Finally, whole-cell broken-patch voltage-clamp experiments with AtT-20 cells showed a PTX-sensitive inhibition of Ca^{2+} conductance [225].

The question arising from these experiments is whether the effect of these receptors is mediated via direct G protein-Ca^{2+} channel coupling (by G_k-, G_i- or G_o-type proteins) or via G protein stimulated formation of a second messenger, notably diacylglycerol leading to the activation of protein kinase C. Conclusive evidence to support either interpretation is missing. Involvement of protein kinase C is strongly indicated by the findings that phorbol esters inhibit Ca^{2+} influx through voltage gated Ca^{2+} channels into PC12 and RIN cells [226] as well as PTX-treated dorsal root ganglion cells [227]. Moreover, protein kinase C involvement cannot be ruled out in whole-cell broken-patch experiments using G_o [85]. In these experiments, the ratio of cell volume to pipette volume was such that dialysis of the cell was not complete. This was evidenced by the fact that agonist-induced effects, which by definition are GTP-dependent, were observed without adding GTP to the pipette medium. In addition, the solution in the pipette contained not only G proteins, but also ATP and Mg^{2+}. These ligands were included to stabilize the Ca^{2+} channels but might also activate the protein kinase C. However, lack of involvement of a second messenger between G protein and Ca^{2+} channel is suggested by experiments in which little or no neurotransmitter-mediated inhibition of Ca^{2+} currents could be measured in cell-attached patches which formed 'giga-seals' [228]. Future investigations into the mechanism of inhibitory regulation of Ca^{2+} channels and the possibility of a direct inhibitory effect of a G protein are needed.

5.3. Stimulatory regulation of Ca^{2+} channels: direct G protein coupling in spite of regulation by cAMP-dependent protein kinase

There is ample evidence indicating that Ca^{2+} channels are regulated by phosphorylation/dephosphorylation cycles involving cAMP-dependent protein kinase and cAMP levels. Indeed this is currently the principal mode by which ventricular contractile force is thought to be modulated positively by β-adrenergic receptors stimulating adenylyl cyclase and by M_2-type muscarinic receptors inhibiting adenylyl cyclase [138,139,145,229,230].

Yet, experiments on changes in intracellular Ca^{2+} levels in adrenal glomerulosa cells exposed to ACTH or Ang II strongly suggest that hormone receptors may also stimulate Ca^{2+} channels, and that regulation may be directly by a G protein (G_s?). Glomerulosa cells respond to both hormones by secreting aldosterone. However, the mechanism by which the hormones cause their effect differs markedly. Ang II-induced stimulus secretion involves activation of phospholipase C, formation of IP_3 and release of Ca^{2+} from intracellular stores to trigger increased aldosterone production. ACTH-induced secretion, on the other hand, is mediated by activation of adenylyl cyclase and formation of cAMP. cAMP triggers the same biochemical pathways as Ca^{2+} to produce aldosterone (for review see Ref. 231). ACTH also causes a rapid and transient influx of extracellular Ca^{2+} [232]. This influx is not caused by Ang II and is not mimicked by forskolin, i.e., by stimulation of adenylyl cyclase in a GTP-independent manner or by addition of cAMP analogs. The ACTH-induced Ca^{2+} influx is blocked by nifedipine, a Ca^{2+} channel blocker. This effect of ACTH does not appear to be the consequence of a second messenger since neither formation of cAMP nor activation of phospholipase C and its products mimics its action [233].

Since ACTH receptors stimulate adenylyl cyclase via G_s, the simplest explanation for the above data is that G_s may have the dual function of stimulating adenylyl cyclase and, albeit transiently, plasma membrane Ca^{2+} channels.

Thus, the possibility exists that Ca^{2+} channels are under dual control by stimulatory and inhibitory G proteins, as well as dual control by protein kinases. However, other explanations may apply. There may be more than one type of ACTH receptors on glomerulosa cells, one activating G_s, the other affecting Ca^{2+} channels activity by means independent of a G protein, such as activation of phospholipase A_2 activity and/or formation of cGMP. Future research in this area should clarify this question.

Figure 14 summarizes the type of G protein mediated regulation of effector functions known to exist (adenylyl cyclase, phospholipase, K^+ channels) or which may exist (Ca^{2+} channels) on the basis of current information.

6. Concluding remarks

Figure 15 schematizes the central regulatory role of G proteins in regulation of cellular metabolism in response to extracellular signals. The three most immediate questions that emerge from this scheme and the above discussions follow:
1. How many distinct effector systems interact with and are regulated by G proteins? Some were tentatively identified, but more may exist. Also, some of the systems that were proposed here as candidates may prove not to be.
2. How many different G proteins are there? What is the function of each?
To what extent are some simple isomers and others truly different coupling proteins?
3. Do receptors couple to more than one type of G protein in a given cell?, and hence, do G proteins contribute to the types of responses hormones can elicit from a given cell?

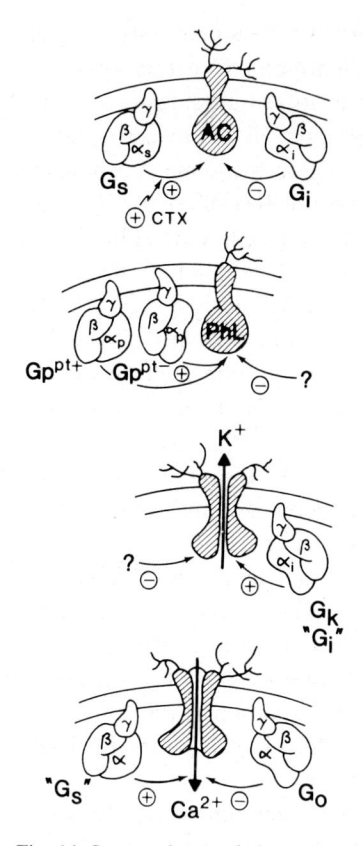

Fig. 14. Systems known (adenylyl cyclase, phospholipase, K$^+$ channel) and assumed (Ca^{2+} channel) to be under direct regulation of G proteins. The existence of at least two types of G$_p$, one PTX-sensitive (G$_p^{pt+}$) and the other PTX-insensitive (G$_p^{pt-}$), is denoted. The possibility of dual regulation of both, phospholipases and K$^+$ channels is raised. For further comments see text.

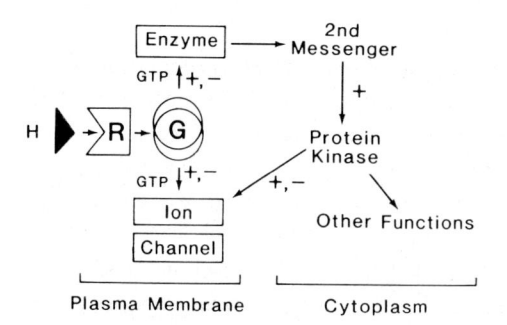

Fig. 15. Diagram of involvement of G proteins in receptor-mediated regulation of effector functions. G, signal transducing G proteins with structure $\alpha\beta\gamma$. G is shown as a double circle to denote the possibility that a single receptor may interact with more than one type of G protein (e.g., G$_i$ and G$_o$) and/or that a single G protein may interact with more than one effector system (e.g., G$_i$ = G$_k$).

There are other questions in need of answers as well. Among them, an intriguing and very fundamental question – which may be addressed through site-directed mutagenesis and controlled expression – relates to the importance of the GTPase function of G proteins. Why is pulsed activation important: is it because the effector system cannot sustain continued activity or is it because the continued activity of the effector systems would be too harmful for cellular survival?

It is clear that much has been learned since the original discovery of the effect of GTP on the binding of glucagon to liver membranes [1]. This knowledge was gained through the research of many, of which the contributions of only some were mentioned here [2–235]. We can now define a complex signal transducing machinery, the existence of which could not have been previously imagined. Surely, the constantly inquisitive mind of the scientific community will continue in its quest to uncover new facts and unveil novel features of cellular regulation which we cannot foresee. We should – and indeed can – look forward to these discoveries.

Acknowledgements

The preparation of this manuscript and original research reported in it were supported in part by grants from the National Institutes of Health to LB (HL-31164, DK-19318 and HD-09581), AMB (HL-36930 and HL-37044) and Baylor College of Medicine (Diabetes and Endocrinology Research Center Grant DK-27685). In addition, RM was supported by a Welch Foundation Grant (Q-952), and YA by an American Heart Association Grant (851159). Work with GH_3 cells was aided by a grant from the US-Spain Joint Committee for Scientific Research Grant (CCB8409-013).

We are especially thankful to Mrs Betty Kilday, without whose perseverance, endurance and diligent secretarial assistance the manuscript would not have been finished.

References

1. Rodbell, M., Krans, H.M.J., Pohl, S.L. and Birnbaumer, L. (1971) J. Biol. Chem. 246, 1872–1876.
2. Rodbell, M., Birnbaumer, L., Pohl, S.L. and Krans, H.M.J. (1971) J. Biol. Chem. 246, 1877–1882.
3. Northup, J.K., Sternweis, P.C., Smigel, M.D., Schleifer, L.S., Ross, E.M. and Gilman, A.G. (1980) Proc. Natl. Acad. Sci. U.S.A. 77, 6516–6520.
4. Hildebrandt, J.D., Codina, J., Risinger, R. and Birnbaumer, L. (1984) J. Biol. Chem. 259, 2039–2042.
5. Gilman, A.G. (1984) Cell 36, 577–579.
6. Birnbaumer, L., Hildebrandt, J.D., Codina, J., Mattera, R., Cerione, R.A., Hildebrandt, J.D., Sunyer, T., Rojas, F.J., Caron, M.G., Lefkowitz, R.J. and Iyengar, R. (1985) In: Molecular Mechanisms of Transmembrane Signalling (Cohen, P. and Houslay, M.D., eds.) pp. 131–182. Elsevier/North Holland Biomedical Press, Amsterdam.
7. Johnson, G.L., Kaslow, H.R. and Bourne, H.R. (1978) Proc. Natl. Acad. Sci. U.S.A. 75, 3113–3117.

8. Northup, J.K., Smigel, M.D., Sternweis, P.C. and Gilman, A.G. (1983) J. Biol. Chem. 258, 11369–11376.
9. Codina, J., Hildebrandt, J.D., Birnbaumer, L. and Sekura, R.D. (1984) J. Biol. Chem. 259, 11408–11418.
10. Brandt, D.R. and Ross, E.M. (1985) J. Biol. Chem. 260, 266–272.
11. Harris, B.A., Robishaw, J.D., Mumby, S.M. and Gilman, A.G. (1985) Science 229, 1274–1277.
12. Robishaw, J.D., Russell, D.W., Harris, B.A., Smigel, M.D. and Gilman, A.G. (1986) Proc. Natl. Acad. Sci. U.S.A. 83, 1251–1255.
13. Nukada, T., Tanabe, T., Takahashi, H., Noda, M., Hirose, T., Inayama, S. and Numa, S. (1986) FEBS Lett. 195, 220–224.
14. Itoh, H., Kozasa, T., Nagata, S., Nakamura, S., Katada, T., Ui, M., Iwai, S., Ohtsuka, E., Kawasaki, H., Suzuki, K. and Kaziro, Y. (1986) Proc. Natl. Acad. Sci. U.S.A. 83, 3776–3780.
15. Mattera, R., Codina, J., Crozat, A., Kidd, V., Woo, S.L.C. and Birnbaumer, L. (1986) FEBS Lett. 206, 36–42.
16. Sullivan, K.A., Liao, Y.-C., Alborzi, A., Beiderman, B., Chang, F.-H., Masters, S.B., Levinson, A.D. and Bourne, H.R. (1986) Proc. Natl. Acad. Sci. U.S.A. 83, 6687–6691.
17. Bray, P., Carter, A., Simons, C., Guo, V., Puckett, C., Kamholz, J., Spiegel, A. and Nirenberg, M. (1986) Proc. Natl. Acad. Sci. U.S.A. 83, 8893–8897.
18. Mumby, S.M., Kahn, R.A., Manning, D.R. and Gilman, A.G. (1986) Proc. Natl. Acad. Sci. U.S.A. 83, 265–269.
19. Sugimoto, K., Nukada, T., Takahashi, H., Noda, M., Minamino, N., Kangawa, K., Matsuo, H., Hirose, T., Inayama, S. and Numa, S. (1985) FEBS Lett. 191, 235–240.
20. Fong, H.K.W., Hurley, J.B., Hopkins, R.S., Miake-Lye, R., Johnson, M.S., Doolittle, R.F. and Simon, M.I. (1986) Proc. Natl. Acad. Sci. U.S.A. 83, 2162–2166.
21. Codina, J., Stengel, D., Woo, S.L.C. and Birnbaumer, L. (1986) FEBS Lett. 207, 187–192.
22. Codina, J., Hildebrandt, J.D., Sekura, R.D., Birnbaumer, M., Bryan, J., Manclark, C.R., Iyengar, R. and Birnbaumer, L. (1984) J. Biol. Chem. 259, 5871–5886.
23. Hildebrandt, J.D., Sekura, R.D., Codina, J., Iyengar, R., Manclark, C.R. and Birnbaumer, L. (1983) Nature 302, 706–709.
24. Strittmatter, S. and Neer, E.J. (1980) Proc. Natl. Acad. Sci. U.S.A. 77, 6344–6348.
25. Smigel, M.D. (1986) J. Biol. Chem. 261, 1976–1982.
26. May, D.C., Ross, E.M., Gilman, A.G. and Smigel, M.D. (1985) J. Biol. Chem. 260, 15829–15833.
27. Sunyer, T. and Birnbaumer, L., unpublished.
28. Cerione, R.A., Sibley, D.R., Codina, J., Benovic, J.L., Winslow, J., Neer, E.J., Birnbaumer, L., Caron, M.G. and Lefkowitz, R.J. (1984) J. Biol. Chem. 259, 9979–9982.
29. Iyengar, R. and Birnbaumer, L. (1982) J. Biol. Chem. 256, 11036–11041.
30. Iyengar, R. and Birnbaumer, L. (1982) Proc. Natl. Acad. Sci. U.S.A. 79, 5179–5183.
31. Cerione, R.A., Codina, J., Benovic, J.L., Lefkowitz, R.J., Birnbaumer, L. and Caron, M.G. (1984) Biochemistry 23, 4519–4525.
32. Cerione, R.A., Staniszewski, C., Benovic, J.L., Lefkowitz, R.J., Caron, M.C., Gierschick, P., Somers, R., Spiegel, A.L., Codina, J. and Birnbaumer, L. (1985) J. Biol. Chem. 260, 1493–1500.
33. Rodbell, M. (1985) Trends Biochem. Sci. 10, 461–463.
34. Rojas, F.J., Iyengar, R. and Birnbaumer, L. (1985) J. Biol. Chem. 260, 7829–7835.
35. Mattera, R., Pitts, B.J.R., Entman, M.S. and Birnbaumer, L. (1985) J. Biol. Chem. 260, 7410–7421.
36. Martin, M.W., Smith, M.M. and Harden, T.K. (1984) J. Pharmacol. Exp. Ther. 230, 424–430.
37. Birnbaumer, L., Swartz, T.L., Abramowitz, J., Mintz, P.W. and Iyengar, R. (1980) J. Biol. Chem. 255, 3542–3551.
38. Iyengar, R., Abramowitz, J., Riser, M. and Birnbaumer, L. (1980) J. Biol. Chem. 255, 3558–3564.
39. Iyengar, R. and Birnbaumer, L. (1981) J. Biol. Chem. 256, 11036–11041.
40. Iyengar, R. (1981) J. Biol. Chem. 256, 11042–11050.

41. Brandt, D.R. and Ross, E.M. (1986) J. Biol. Chem. 261, 1656–1664.
42. Ferguson, K.M., Higashijima, T., Smigel, M. and Gilman, A.G. (1986) J. Biol. Chem. 261, 7393–7399.
43. Higashijima, T., Ferguson, K.M., Sternweis, P.C., Smigel, M.D. and Gilman, A.G. (1987) J. Biol. Chem. 262, 762–766.
44. Lad, P.M., Nielsen, T.B., Preston, M.S. and Rodbell, M. (1980) J. Biol. Chem. 255, 988–995.
45. Murayama, T. and Ui, M. (1984) J. Biol. Chem. 259, 761–769.
46. Wheeler, G.L. and Bitensky, M.W. (1977) Proc. Natl. Acad. Sci. U.S.A. 74, 4238–4242.
47. Godchaux, W. III and Zimmerman, W.F. (1979) J. Biol. Chem. 254, 7874–7884.
48. Kühn, H. (1980) Nature 283, 587–589.
49. Bitensky, M.W., Wheeler, G.L., Yamazaki, A., Rasenick, M.M. and Stein, P.J. (1981) Curr. Top. Membrane Transport 15, 237–271.
50. Stryer, L., Hurley, J.B. and Fung, B.K. (1981) Curr. Top. Membrane Transport 15, 93–108.
51. Hildebrandt, J.D., Codina, J., Rosenthal, W., Birnbaumer, L., Neer, E.J., Yamazaki, A. and Bitensky, M.W. (1985) J. Biol. Chem. 260, 14867–14872.
52. Fung, B.K.-K., Hurley, J.B. and Stryer, L. (1981) Proc. Natl. Acad. Sci. U.S.A. 78, 152–156.
53. Yamazaki, A., Stein, P.J., Chernoff, N. and Bitensky, M. (1983) J. Biol. Chem. 258, 8188–8194.
54. Deterre, P., Bigay, J., Fourquet, F., Roberte, M. and Chabre, M. (1988) Proc. Natl. Acad. Sci. U.S.A. 85, 2424–2428.
55. Chabre, M. (1985) Ann. Rev. Biophys. Biophys. Chem. 14, 331–360.
56. Stryer, L. and Bourne, H.R. (1986) Ann. Rev. Cell Biol. 2, 391–419.
57. Birnbaumer, L. (1973) Biochim. Biophys. Acta (Reviews on Biomembranes) 300, 129–158.
58. Birnbaumer, L., Nakahara, T. and Yang, P.-Ch. (1974) J. Biol. Chem. 249, 7857–7866.
59. Yamamura, H., Lad, P.M. and Rodbell, M. (1977) J. Biol. Chem. 252, 7964–7966.
60. Rodbell, M. (1980) Nature 284, 17–22.
61. Cooper, D.M.F. (1982) FEBS Lett. 138, 157–163.
62. Katada, T. and Ui, M. (1982) Proc. Natl. Acad. Sci. U.S.A. 79, 3129–3133.
63. Bokoch, G.M., Katada, T., Northup, J.K., Hewlett, E.L. and Gilman, A.G. (1983) J. Biol. Chem. 258, 2072–2075.
64. Codina, J., Hildebrandt, J.D., Iyengar, R., Birnbaumer, L., Sekura, R.D. and Manclark, C.R. (1983) Proc. Natl. Acad. Sci. U.S.A. 80, 4276–4280.
65. Londos, C., Cooper, D.M.F., Schlegel, W. and Rodbell, M. (1978) Proc. Natl. Acad. Sci. U.S.A. 75, 5362–5366.
66. Jakobs, K.H., Saur, W. and Schultz, G. (1978) FEBS Lett. 85, 167–170.
67. Jakobs, K.H., Aktories, K. and Schultz, G. (1983) Nature 303, 177–178.
68. Ui, M. (1984) Trends Pharmacol. Sci. 5, 277–279.
69. Katada, T., Northup, J.K., Bokoch, G.M., Ui, M. and Gilman, A.G. (1984) J. Biol. Chem. 259, 3578–3858.
70. Aktories, K., Schultz, G. and Jakobs, K.H. (1982) FEBS Lett. 146, 65–68.
71. Katada, T., Bokoch, G.M., Smigel, M.D., Ui, M. and Gilman, A.G. (1984) J. Biol. Chem. 259, 3586–3595.
72. Katada, T., Bokoch, G.M., Northup, J.K., Ui, M. and Gilman, A.G. (1984) J. Biol. Chem. 259, 3586–3577.
73. Sternweis, P.C. and Robishaw, J.D. (1984) J. Biol. Chem. 259, 13806–13813.
74. Neer, E.J., Lok, J.M. and Wolf, L.G. (1984) J. Biol. Chem. 259, 14222–14229.
75. Milligan, G. and Klee, W.A. (1985) J. Biol. Chem. 260, 2057–2063.
76. Roof, D.J., Applebury, M.L. and Sternweis, P.C. (1985) J. Biol. Chem. 260, 16242–16249.
77. Katada, T., Oinuma, M. and Ui, M. (1986) J. Biol. Chem. 261, 8182–8191.
78. Rosenthal, W., Koesling, D., Rudolph, U., Kleuss, C., Pallast, M., Yajima, M. and Schultz, G. (1986) Eur. J. Biochem. 158, 255–263.
79. Florio, V.A. and Sternweis, P.C. (1985) J. Biol. Chem. 260, 3477–3483.

80. Haga, K., Haga, T., Ichiyama, A., Katada, T., Kurose, H. and Ui, M. (1985) Nature 316, 731–733.
81. Kurose, H., Katada, T., Haga, T., Haga, K., Ichiyama, A. and Ui, M. (1986) J. Biol. Chem. 261, 6423–6428.
82. Cerione, R.A., Regan, J.W., Nakata, H., Codina, J., Benovic, J.L., Gierschick, P., Somers, R.L., Spiegel, A.M., Birnbaumer, L., Lefkowitz, R.J. and Caron, M.G. (1986) J. Biol. Chem. 261, 3901–3909.
83. Kubo, T., Fukuda, K., Mikami, A., Maeda, A., Takahashi, H., Mishina, M., Haga, T., Haga, K., Ichiyama, A., Kangawa, K., Kojima, M., Matsuo, H., Hirose, T. and Numa, S. (1986) Nature 323, 411–416.
84. Kikuchi, A., Kozawa, O., Kaibuchi, K., Katada, T., Ui, M. and Takai, Y. (1986) J. Biol. Chem. 261, 11558–11562.
85. Hescheler, J., Rosenthal, W., Trautwein, W. and Schultz, G. (1987) Nature 325, 445–447.
86. Nukada, T., Tanabe, T., Takahashi, H., Noda, M., Haga, K., Haga, T., Ichiyama, A., Kangawa, K., Hiranaga, M., Matsuo, H. and Numa, S. (1986) FEBS Lett. 197, 305–310.
87. Michell, B. and Kirk, C. (1986) Nature 323, 112–113.
88. Downes, C.P. and Michell, R.H. (1985) In: Molecular Mechanisms of Transmembrane Signalling, (Cohen, P. and Houslay, M.D., eds.) pp. 3–56. Elsevier Science Publishers, Amsterdam.
89. Berridge, M.J. and Irvine, R.J. (1984) Nature 312, 315–321.
90. Nishizuka, Y. (1984) Nature 308, 693–698.
91. Snyderman, R. and Verghese, M.W. (1987) Rev. Infect. Dis. 9 (Suppl. 5), S562–S569.
92. Koo, C., Lefkowitz, R.J. and Snyderman, R. (1983) J. Clin. Invest. 72, 748–753.
93. Okajima, F., Katada, T. and Ui, M. (1985) J. Biol. Chem. 260, 6761–6768.
94. Volpi, M., Yassin, R., Naccache, P.H. and Sha'afi, R.I. (1983) Biochem. Biophys. Res. Commun. 112, 957–964.
95. Smith, C.D., Lane, B.C., Kusaka, I., Verghese, M.W. and Snyderman, R. (1985) J. Biol. Chem. 260, 5875–5878.
96. Smith, C.D., Cox, C.C. and Snyderman, R. (1986) Science 232, 97–100.
97. Bokoch, G.M. and Gilman, A.G. (1984) Cell 39, 301–308.
98. Okajima, F. and Ui, M. (1984) J. Biol. Chem. 259, 13863–13871.
99. Gomperts, B.D. (1983) Nature 306, 64–66.
100. Cockcroft, S. and Gomperts, B.D. (1985) Nature 314, 534–536.
101. Cockcroft, S. (1987) TIBS 12, 75–78.
102. Nakamura, T. and Ui, M. (1985) J. Biol. Chem. 260, 3584–3593.
103. Nakamura, T. and Ui, M. (1985) FEBS Lett. 173, 414–417.
104. Murayama, T. and Ui, M. (1985) J. Biol. Chem. 260, 7226–7233.
105. Burch, R.M., Luini, A. and Axelrod, J. (1986) Proc. Natl. Acad. Sci. U.S.A. 83, 7201–7205.
106. Koch, B.D., Dorflinger, L.J. and Schonbrunn, A. (1985) J. Biol. Chem. 260, 13138–13145.
107. Yajima, Y., Akita, Y. and Saito, T. (1986) J. Biol. Chem. 261, 2684–2689.
108. Williamson, J.R., Cooper, R.H., Joseph, S.K. and Thomas, A.P. (1985) Am. J. Physiol. 248 (Cell Physiol. 17), C203–C216.
109. Exton, J.H. (1985) J. Clin. Invest. 75, 1753–1757.
110. Pobiner, B.F., Hewlett, E.L. and Garrison, J.C. (1985) J. Biol. Chem. 260, 16200–16209.
111. Lambert, T.L., Kent, R.S. and Whorton, A.R. (1986) J. Biol. Chem. 261, 15288–15293.
112. Hinkle, P.M. and Kinsella, P.A. (1984) J. Biol. Chem. 259, 3345–3349.
113. Hinkle, P.M. and Phillips, W.J. (1984) Proc. Natl. Acad. Sci. U.S.A. 81, 6183–6187.
114. Albert, P.R. and Tashjian, A.H. (1984) J. Biol. Chem. 259, 5827–5832.
115. Gershengorn, M.C., Geras, E., Spina-Purello, V. and Rebecchi, M.J. (1984) J. Biol. Chem. 259, 10675–10681.
116. Martin, T.F.J., Lucas, D.O., Bajjaliah, S.M. and Kowalchyk, J.A. (1986) J. Biol. Chem. 261, 2918–2927.

117. Haslam, R.J. and Davidson, M.M.L. (1984) FEBS Lett. 174, 90–95.
118. Litosch, I., Wallis, C. and Fain, J.N. (1985) J. Biol. Chem. 260, 5464–5471.
119. Uhing, R.J., Prpic, V., Jiang, H. and Exton, J.H. (1986) J. Biol. Chem. 261, 2140–2146.
120. Straub, R.E. and Gershengorn, M.C. (1986) J. Biol. Chem. 261, 2712–2717.
121. Irving, H.R. and Exton, J.H. (1987) J. Biol. Chem. 262, 3440–3443.
122. Besterman, J.M., Duronio, V. and Cuatrecasas, P. (1986) Proc. Natl. Acad. Sci. U.S.A. 83, 6785–6789.
123. Thomas, A.P., Marks, J.S., Coll, K.E. and Williamson, J.R. (1983) J. Biol. Chem. 258, 5716–5725.
124. Griendling, K.K., Rittenhouse, S.E., Brock, T.A., Ekstein, L.S., Gimbrone, M.A., Jr. and Alexander, R.W. (1986) J. Biol. Chem. 261, 5901–5906.
125. Imai, A. and Gershengorn, M.C. (1986) Proc. Natl. Acad. Sci. U.S.A. 83, 8540–8544.
126. Baldassare, J.J. and Fisher, G.F. (1986) J. Biol. Chem. 261, 11942–11944.
127. Deckmyn, H., Tu, S.-M. and Majerus, P.W. (1986) J. Biol. Chem. 261, 16553–16558.
128. Eccles, R.M. and Libet, B. (1961) J. Physiol. (London) 157, 484–503.
129. Hartzell, H.C., Kuffler, S.W., Stickgold, R. and Yoshikami, D. (1977) J. Physiol. 271, 817–846.
130. Griffith, W.H. III, Gallagher, J.P. and Shinnick-Gallagher, P. (1981) Brain Res. 208, 446–451.
131. Hill-Smith, I. and Purves, R.D. (1978) J. Physiol. 279, 31–54.
132. Nakajima, Y., Nakajima, S., Leonard, R.J. and Yamagucchi, K. (1986) Proc. Natl. Acad. Sci. U.S.A. 83, 3022–3026.
133. Trautwein, W. and Dudel, J. (1958) Pflügers Archiv. 266, 324–334.
134. Giles, W. and Noble, S.J. (1976) J. Physiol. 261, 103–123.
135. Hartzell, H.C. (1981) Nature 291, 539–544.
136. Murad, F., Chi, Y.-M., Rall, T.W. and Sutherland, E.W. (1962) J. Biol. Chem. 237, 1233–1238.
137. Kurose, H. and Ui, M. (1983) J. Cyclic Nucleotide Protein Phosphor. Res. 9, 305–318.
138. Trautwein, W., Taniguchi, J. and Noma, A. (1982) Pflügers Archiv. 392, 307–314.
139. Nargeot, J., Nerbonne, J.M., Engels, J. and Lester, H.A. (1983) Proc. Natl. Acad. Sci. U.S.A. 80, 2395–2399.
140. Hamill, O.P., Marty, A., Neher, E., Sakmann, B. and Sigworth, F.J. (1981) Pflügers Arch. 391, 85–100.
141. Soejima, M. and Noma, A. (1984) Pflügers Arch. 400, 424–431.
142. Pfaffinger, P.J., Martin, J.M., Hunter, D.D., Nathanson, N.M. and Hille, B. (1985) Nature 317, 536–538.
143. Breitwieser, G.E. and Szabo, G. (1985) Nature 317, 538–540.
144. Kurachi, Y., Nakajima, T. and Sugimoto, T. (1986) Pflügers Arch. 407, 264–274.
145. Yatani, A., Codina, J., Brown, A.M. and Birnbaumer, L. (1986) Science 235, 207–211.
146. Yatani, A., Codina, J., Sekura, R.D., Birnbaumer, L. and Brown, A.M. (1987) Mol. Endocrinol. 1, 283–289.
147. Schramm, M. and Rodbell, M. (1975) J. Biol. Chem. 250, 2232–2237.
148. Iyengar, R. (1981) J. Biol. Chem. 256, 11042–11050.
149. Hildebrandt, J.D. and Birnbaumer, L. (1983) J. Biol. Chem. 258, 13141–13147.
150. Codina, J., Yatani, A., Grenet, D., Brown, A.M. and Birnbaumer, L. (1987) Science 236, 442–445.
151. Scherer, N.M., Toro, M.-J., Entman, M.L. and Birnbaumer, L. (1987) Arch. Biochem. Biophys. 259, 431–440.
152. Gill, D.M. and Meren, R. (1978) Proc. Natl. Acad. Sci. U.S.A. 75, 3050–3054.
153. Cassel, D. and Pfeuffer, T. (1978) Proc. Natl. Acad. Sci. U.S.A. 75, 2669–2673.
154. Cooper, D.M.F., Jagus, R., Somers, R.L. and Rodbell, M. (1981) Biochem. Biophys. Res. Commun. 101, 1179–1185.
155. Johnson, G.L., Kaslow, H.R. and Bourne, H.R. (1978) J. Biol. Chem. 253, 7120–7123.
156. Malbon, C.C. (1982) Biochim. Biophys. Acta 714, 429–434.
157. Owens, J.R., Frame, L.T., Ui, M. and Cooper, D.M.F. (1985) J. Biol. Chem. 260, 15946–15952.

158. Abood, M.E., Hurley, J.B., Pappone, M.-C., Bourne, H.R. and Stryer, L. (1982) J. Biol. Chem. 257, 10540–10543.
159. Van Dop, C., Tsubokawa, M., Bourne, H.R. and Ramachandran, J. (1984) J. Biol. Chem. 259, 696–699.
160. Van Dop, C., Yamanaka, G., Steinberg, F., Sekura, R.D., Manclark, C.R., Stryer, L. and Bourne, H.R. (1984) J. Biol. Chem. 259, 23–26.
161. West, R.E., Jr., Moss, J., Vaughan, M., Liu, T. and Liu, T.-Y. (1985) J. Biol. Chem. 260, 14428–14430.
162. Ribeiro Neto, F., Mattera, R., Grenet, D., Sekura, R.D., Birnbaumer, L. and Field, J.B. (1986) Mol. Endocrinol. 1, 472–481.
163. Malbon, C.C., Rapiejko, P.J. and Garciá-Sáinz, J.A. (1984) FEBS Lett. 176, 301–306.
164. Halvorsen, S.W. and Nathanson, N.M. (1984) Biochemistry 23, 5813–5321.
165. Malbon, C.C., Mangano, T.J. and Watkins, D.C. (1985) Biochem. Biophys. Res. Commun. 128, 809–815.
166. Hildebrandt, J.D., Codina, J., Tash, J., Kirchick, H.J., Lipschultz, l. and Birnbaumer, L. (1985) Endocrinology 116, 1357–1366.
167. Iyengar, R., Rich, K.A., Herberg, J.T., Grenet, D., Mumby, S.M. and Codina, J. (1987) J. Biol. Chem. 262, 9239–9245.
168. Hurley, J.B., Simon, M.I., Teplow, D.B., Robishaw, J.D. and Gilman, A.G. (1984) Science 226, 860–862.
169. Yatsunami, K. and Khorana, H.G. (1985) Proc. Natl. Acad. Sci. U.S.A. 82, 4316–4320.
170. Tanabe, T., Nukada, T., Nishikawa, Y., Sugimoto, K., Suzuki, H., Takahashi, H., Noda, M., Haga, T., Ichiyama, A., Kangawa, K., Minamino, N., Matsuo, H. and Numa, S. (1985) Nature 315, 242–245.
171. Medynski, D.C., Sullivan, K., Smith, D., Van Dop, C., Chang, F.-H., Fung, B.K.-K., Seeburg, P.H. and Bourne, H.R. (1985) Proc. Natl. Acad. Sci. U.S.A. 82, 4311–4315.
172. Lochrie, M.A., Hurley, J.B. and Simon, M.I. (1985) Science 228, 96–99.
173. Lerea, C.L., Somers, D.E., Hurley, J.B., Klock, I.B. and Bunt-Milan, A.H. (1986) Science 234, 77–80.
174. Halliday, K.R. (1983–84) Cyclic Nucleotide Protein Phosphoryl. Res. 9, 435–448.
175. Jurnak, F. (1985) Science 230, 32–36.
176. Masters, S.B., Stroud, R.M. and Bourne, H.R. (1986) Prot. Engineering 1, 47–54.
177. Robishaw, J.D., Smigel, M.D. and Gilman, A.G. (1986) J. Biol. Chem. 261, 9587–9590.
178. Kaslow, H.R., Cox, D., Groppi, V.E. and Bourne, H.R. (1981) Mol. Pharmacol. 19, 406–410.
179. Kaslow, H.R., Johnson, G.L., Brothers, V.M. and Bourne, H.R. (1980) J. Biol. Chem. 255, 3736–3741.
180. Sternweis, P.C., Northup, J.K., Smigel, M.D. and Gilman, A.G. (1981) J. Biol. Chem. 256, 11517–11527.
181. Didsbury, J.R., Ho, Y.-S. and Snyderman, R. (1987) FEBS Lett. 211, 160–164.
182. Angus, C.W., Van Meurs, K.P., Tsai, S.-C., Adamik, R., Miedel, M.C., Pan, Y.-C., Kung, H.-F., Moss, J. and Vaughan, M. (1986) Proc. Natl. Acad. Sci. U.S.A. 83, 5813–5816.
183. Evans, T., Fawzi, A., Fraser, E.D., Brown, M.L. and Northup, J.K. (1987) J. Biol. Chem. 262, 176–181.
184. Manning, D.R. and Gilman, A.G. (1983) J. Biol. Chem. 258, 7059–7063.
185. Gierschik, P., Codina, J., Simons, C., Birnbaumer, L. and Spiegel, A. (1985) Proc. Natl. Acad. Sci. U.S.A. 82, 721–727.
186. Yatsunami, K., Pandya, B.V., Oprian, D.D. and Khorana, H.G. (1985) Proc. Natl. Acad. Sci. U.S.A. 82, 1936–1940.
187. Van Dop, C., Medynski, D., Sullivan, K., Wu, A.M., Fung, B.K.-K. and Bourne, H.R. (1984) Biochem. Biophys. Res. Commun. 124, 250–255.

188. Hurley, J.B., Fong, H.K.W., Teplow, D.B., Dreyer, W.J. and Simon, M.I. (1984) Proc. Natl. Acad. Sci. U.S.A. 81, 6948–6952.
189. Sternweis, P.C. (1986) J. Biol. Chem. 261, 631–637.
190. Brazeau, P., Vale, W., Burgus, R., Ling, N., Butcher, M., Rivier, J. and Guillemin, R. (1973) Science 179, 77–79.
191. Guillemin, R. and Gerich, J.E. (1976) Annu. Rev. Med. 27, 379–388.
192. Schonbrunn, A. and Tashjian, A.H., Jr. (1978) J. Biol. Chem. 253, 6473–6483.
193. Richardson, U.I. and Schonbrunn, A. (1981) Endocrinology 108, 281–290.
194. Heisler, S., Reisine, T.D., Hook, V.Y.H. and Axelrod, J. (1982) Proc. Natl. Acad. Sci. U.S.A. 79, 6502–6506.
195. Jessel, T.M. and Iversen, L.L. (1987) Nature 268, 549–551.
196. Williams, J.T., Egan, T.M. and North, R.A. (1982) Nature 299, 74–77.
197. Dunlap, K. and Fischbach, G.D. (1978) Nature 276, 8837–8839.
198. Egan, T.M. and North, R.A. (1986) Nature 319, 405–407.
199. Adams, P.R., Brown, D.A. and Constanti, A. (1982) J. Physiol. (London) 332, 223–262.
200. Sabol, S.L. and Nirenberg, M. (1979) J. Biol. Chem. 254, 1913–1920.
201. Jakobs, K.H., Saur, W. and Schultz, G. (1976) J. Cyclic Nucleotide Res. 2, 381–392.
202. Sharma, S.K., Nirenberg, M. and Klee, W.A. (1975) Proc. Natl. Acad. Sci. U.S.A. 72, 590–594.
203. Reisine, T.D., Zhang, Y.-L. and Sekura, R.D. (1983) Biochem. Biophys. Res. Commun. 115, 794–799.
204. Koch, B.D. and Schonbrunn, A. (1984) Endocrinology 114, 1784–1790.
205. Reisine, T.D. (1986) J. Receptor Res. 4, 291–300.
206. Holz, G.G., Rane, S.G. and Dunlap, K. (1986) Nature 319, 670–672.
207. Blume, A.J. (1978) Proc. Natl. Acad. Sci. U.S.A. 75, 1713–1717.
208. Tsai, B.S. and Lefkowitz, R.J. (1979) Mol. Pharmacol. 16, 61–68.
209. Michel, T. and Lefkowitz, R.J. (1982) J. Biol. Chem. 257, 13557–13563.
210. Hill, D.R., Bowery, N.G. and Hudson, A.L. (1984) J. Neurochem. 42, 652–657.
211. Schlegel, W., Wuarin, F., Wollheim, C.B. and Zahnd, G.R. (1984) Cell Calcium 5, 223–236.
212. Schlegel, W., Wuarin, F., Zbaren, C., Wollheim, C.B. and Zahnd, G.R. (1985) FEBS Lett. 189, 27–32.
213. Schlegel, W., Wuarin, F., Zbaren, C. and Zahnd, G.R. (1985) Endocrinology 117, 976–981.
214. Barros, F., Katz, G.M., Kaczorowski, G.J., Vandlen, R.L. and Reuben, J.P. (1985) Proc. Natl. Acad. Sci. U.S.A. 82, 1108–1112.
215. Yamashita, N., Shibuya, N. and Ogata, E. (1986) Proc. Natl. Acad. Sci. U.S.A. 83, 6198–6202.
216. Wojcikiewicz, R.J.H., Dobson, P.R.M. and Brown, B.L. (1984) Biochim. Biophys. Acta 805, 25–29.
217. Wojcikiewicz, R.J.H., Dobson, R.M., Irons, L.I., Robinson, A. and Brown, B.L. (1984) Biochem. J. 224, 339–342.
218. Newberry, N.R. and Nicoll, R.A. (1984) Nature 308, 450–452.
219. Andrade, R., Malenka, R.C. and Nicoll, R.A. (1986) Science 234, 1261–1265.
220. Higashida, H., Streaty, R.A., Klee, W. and Nirenberg, M. (1986) Proc. Natl. Acad. Sci. U.S.A. 83, 942–946.
221. Higashida, H. and Brown, D.A. (1986) Nature 323, 333–335.
222. Berridge, M. (1986) Nature (News and Views) 323, 294–295.
223. Woodcock, E.A. and McLeod, J.K. (1986) Endocrinology 119, 1697–1702.
224. Hausdorff, W.P., Sekura, R.D., Aguilera, G. and Catt, K.J. (1987) Endocrinology 120, 1668–1678.
225. Lewis, D.L., Weight, F.F. and Luini, A. (1986) Proc. Natl. Acad. Sci. U.S.A. 83, 9035–9039.
226. Di Virgilio, F., Possan, T., Wollheim, C.B., Vicentini, L.M. and Meldolesi, J. (1986) J. Biol. Chem. 261, 32–35.
227. Rane, S.O. and Dunlap, K. (1986) Proc. Natl. Acad. Sci. U.S.A. 83, 184–188.
228. Forscher, P., Oxford, G.S. and Schulz, D. (1986) J. Physiol. 379, 131–144.

229. Flockerzi, V., Oeken, H.J., Hofmann, F., Pelzer, D., Cavalie, A. and Trautwein, W. (1986) Nature 323, 66–68.
230. Kameyama, M., Hescheler, J., Hofmann, F. and Trautwein, W. (1986) Pflügers Arch. 407, 121–128.
231. Barrett, P.Q. and Rasmussen, H. (1984) Physiol. Rev. 64, 938–984.
232. Kojima, I., Kojima, K. and Rasmussen, H. (1985) J. Biol. Chem. 260, 4248–4256.
233. Kojima, I. and Ogata, E. (1986) J. Biol. Chem. 261, 9832–9838.
234. Codina, J., Grenet, D., Yatani, A., Birnbaumer, L. and Brown, A.M. (1987) FEBS Lett. 216, 104–106.
235. Birnbaumer, L. (1987) Trends Pharmacol. Sci. 8, 209–211.

CHAPTER 2

Inositol phospholipids and cellular signalling

GRAEME R. GUY and CHRISTOPHER J. KIRK

Biochemistry Department, University of Birmingham, P.O. Box 363, Birmingham B15 2TT, England

1. Introduction

The multicellular organism is composed of a number of different cell types that have differentiated from a limited number of stem cells. Communication between cells is required to control their growth and division and to co-ordinate their diverse activities. Co-ordination in an organism is controlled by molecules, often peptides, that can be secreted by specialised cells into the bloodstream (endocrine factors), by neighbouring cells (paracrine factors) or by the cell itself (autocrine factors). Many extracellular signalling molecules cannot cross the lipophilic membrane bilayer so they bind to receptors which are specialised transmembrane proteins or glycoproteins that act as molecular antennae for external messages. The incoming signal, thought to be initiated by a conformational change in the receptor protein, is transduced into a second messenger which amplifies the signal by further modifying various intracellular targets. One class of receptors mediates its response through formation of cAMP which is catalysed by adenylate cyclase, a protein present on the cytoplasmic side of the membrane. The coupling of adenylate cyclase to the receptor is further regulated by intramembrane GTP-binding transducing proteins.

Another class of receptors mediate intracellular responses through Ca^{2+} mobilisation. A specific class of phospholipids, the phosophoinositides play an important role in signal transduction from receptors at the plasma membrane. The cellular responses that use phosphoinositide hydrolysis and Ca^{2+} mobilisation are diverse, they include general metabolism, secretion, contraction, phototransduction and proliferation. An imbalance of the second messenger system in the proliferation cascade may be responsible for normal cells becoming cancerous.

In the early 1950s, several hormones were shown to enhance the incorporation of $^{32}P_i$ into the anionic membrane phospholipid, phosphatidylinositol (for review see Ref. 1). In 1975, Michell [2] noted that agonists which influence inositol lipid metabolism also provoke Ca^{2+} mobilisation in stimulated cells. It was first thought that agonists induced inositol lipid turnover by stimulating the hydrolysis of phosphatidylinositol. However, it is now clear that the primary receptor mediated event is the hydrolysis of one or both of the 'polyphosphoinositides' which are phosphorylated derivatives of phosphatidylinositol.

2. Inositol phospholipids

In animal cells there are three major myo-inositol-containing phospholipids (Fig. 1); phosphatidylinositol [1-(3-*sn*-phosphatidyl)-D-myo-inositol) (PtdIns)], which usually accounts for over 90% of the total inositol lipid, phosphatidylinositol-4-phosphate (PtdIns 4-P) and phosphatidylinositol 4,5-bisphosphate (PtdIns 4,5-P_2). The polyphosphorylated inositides generally constitute 1–10% of total inositol lipids, which are themselves 6–10% of total phospholipids. The polyphosphoinositides are considered to be located on the cytoplasmic side of the plasma membrane, which places them in an ideal position to play a role in signal transduction. PtdIns, on the other hand, is found in all cellular membranes. In some cells there is evidence for separate hormone sensitive and hormone insensitive pools of the inositol lipids [3].

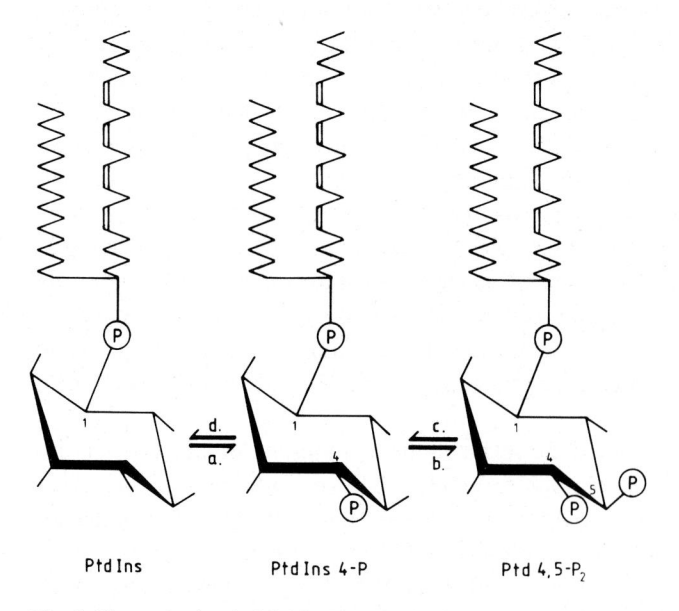

Fig. 1. The major inositol lipids. Phosphatidylinositol (PtdIns), the major membrane inositol phospholipid, is phosphorylated to phosphatidylinositol 4-phosphate (PtdIns 4-P) by a phosphatidylinositol kinase (a). PtdIns 4-P is further phosphorylated to phosphatidylinositol 4,5-bisphosphate (PtdIns 4,5-P_2) by a phosphatidylinositol 4-phosphate kinase (b). PtdIns 4,5-P_2 is converted back to PtdIns 4-P by phosphatidylinositol 4,5-bisphosphate phosphomonoesterase (c) and then to PtdIns by phosphatidylinositol 4-phosphate monoesterase (d). The pathway of phosphorylation and dephosphorylation constitutes a futile cycle and is only interrupted by agonist-induced hydrolysis of PtdIns 4,5-P_2.

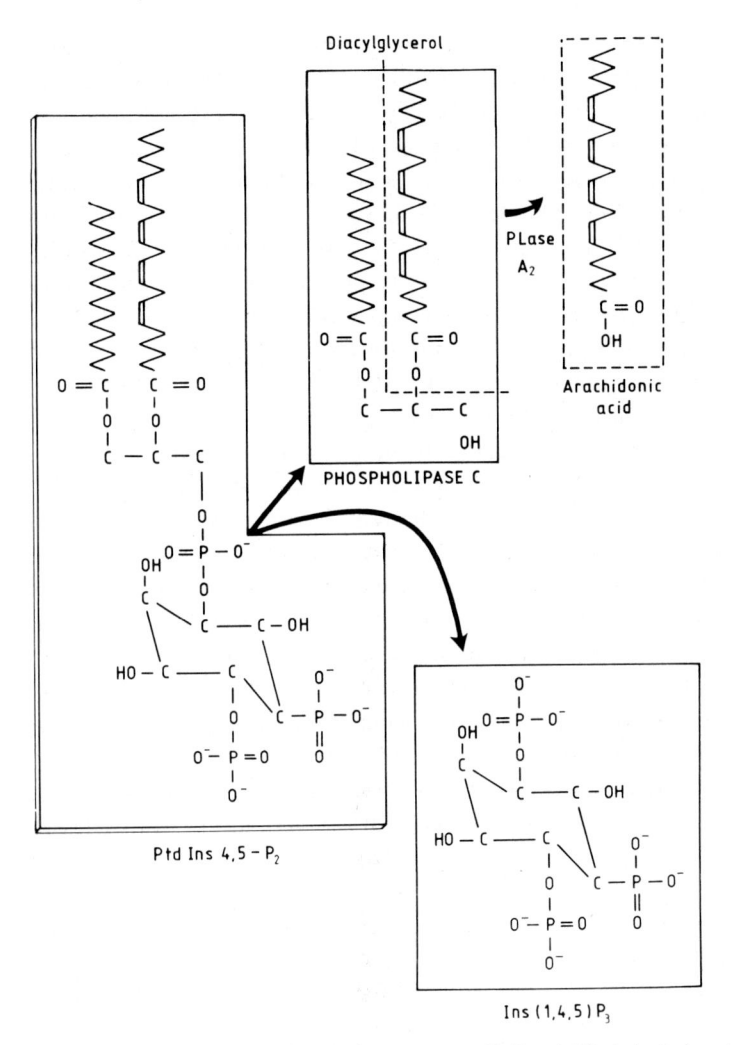

Fig. 2. PtdIns 4,5-P$_2$-derived second messengers. PtdIns 4,5-P$_2$ is hydrolysed when a phospholipase C (PtdIns 4,5-P$_2$ phosphodiesterase) is activated following the binding of specific agonists to their surface receptor proteins. The PtdIns 4,5-P$_2$ is cleaved to yield diacylglycerol (DG), which is a co-activator of protein kinase C and other enzymes, and Ins(1,4,5)P$_3$, which is capable of releasing Ca^{2+} from intracellular stores. The DG often contains arachidonic acid, which is the source of the prostanoids, which are also capable of controlling diverse cellular functions. The arachidonic acid is cleaved from the parent lipid or from DG by specific phospholipase A$_2$ enzymes.

PtdIns 4,5-P$_2$ is formed by a two-stage phosphorylation of PtdIns, which is first phosphorylated at the 4-position of its inositol headgroup by a specific kinase to form PtdIns 4-P; this is in turn phosphorylated at the 5-position to give PtdIns 4,5-P$_2$, the substrate for receptor-mediated lipid hydrolysis. As well as the two kinases responsible for the stepwise phosphorylation of PtdIns there are corresponding phosphomonoesterase enzymes which convert PtdIns 4,5-P_2 back to PtdIns. The reactions which add and remove phosphates from the 4 and 5 positions on the inositol headgroup constitute a futile cycle (Figs. 1 and 4). Such futile cycles are metabolically expensive and are operative in non-stimulated cells. It is only when the receptor is occupied by a specific agonist that PtdIns 4,5-P_2 is diverted out of the futile cycle towards a phosphodiesterase (phospholipase C) enzyme, by which it is cleaved into *sn*-1,2-diacylglycerol and $InsP_3$ (Fig. 2). The possible role of polyphosphoinositides in cell signalling was first described by Durell et al. [4] then later emphasised with the finding that Ca^{2+} mobilising agonists, such as vasopressin in liver, caused a very rapid breakdown of PtdIns 4,5-P_2 [5]. Since then, PtdIns 4,5-P_2 hydrolysis has been reported in a number of other tissues as a primary biochemical event following agonist binding to specific receptors [6] (Table I).

Quantitation of the loss of lipid from the polyphosphoinositides was difficult because of the low levels of these lipids and the presence of non-activated pools leading to a high background on which to detect changes. The analysis, in stimulated cells, of the inositol phosphate products of lipid hydrolysis offered a more sensitive means of analysing which of the phosphoinositides were hydrolysed. A useful tool for this analysis was the anti-manic cation, lithium, which inhibits Ins-P phosphatase, resulting in an increased concentration of InsP and a decreased concentration of inositol in the brain [7]. Lithium amplifies the agonist-stimulated accumulation of InsP in many tissues, including blowfly salivary gland, rat brain slices and rat parotid gland [6,8]. Increased levels of $InsP_2$ and $InsP_3$ were also observed and these occurred at earlier times following receptor activation. Hence the accumulation of these higher inositol phosphates was consistent with the notion that PtdIns 4,5-P_2 hydrolysis is the primary event following receptor stimulation. It is now clear that the receptor mediated hydrolysis of PtdIns 4,5-P_2 and the accumulation of Ins(1,4,5)P_3 and diacylglycerol (DG) is at the hub of the second messenger system that elevates intracellular Ca^{2+}.

3. Role of GTP-binding proteins in receptor-response coupling

The manner in which receptors are coupled to the PtdIns 4,5-P_2 phosphodiesterase is not certain. In a number of cell types the receptor and phosphodiesterase may be coupled by GTP-binding proteins in an analogous manner to the systems operating through cAMP as a second messenger [6,9]. Most of the evidence for the role of GTP-binding proteins comes from the activation of phosphoinositide hydrolysing systems by GTP or its non-hydrolysable analogues GTPγS or GppNHp and/or the inhibition of such systems by pertussis or cholera toxins, both of which ADP-ribosylate certain GTP-binding proteins [10].

The *ras* gene product, p21, which is a membrane-associated protein capable of binding and hydrolysing GTP, may function like the postulated GTP-binding pro-

TABLE I

Summary of the agonists that induce the hydrolysis of PtdIns 4,5-P_2, and some examples of target tissues (from Refs. 6, 8 and 13)

Stimulus (receptor)	Target tissue
Acetylcholine (muscarinic)	Smooth muscle Brain and other neural tissues Pancreas Parotid Adrenal medulla
Noradrenaline (α_1 adrenergic)	Smooth muscle Brain Hepatocytes Brown adipocytes Parotid gland
Histamine (H_1)	Brain Chromaffin cells
Serotonin (H_2)	Blowfly salivary gland Brain Platelets
Vasopressin (V_1)	Smooth muscle Sympathetic ganglion Hepatocytes
Angiotensin II	Smooth muscle Hepatocytes Kidney Anterior pituitary cells
Bradykinin	Neuroblastoma hybrid lines
Substance P	Smooth muscle Parotid
f-Methionyl-leucyl-phenylalanine	Neutrophils Leucocytes
Thrombin	Platelets
Thyrotropin releasing hormone	GH_3 pituitary cells
Glucose	Pancreatic islets
Platelet activating factor	Platelets Hepatocytes
Light	Photoreceptors
Cross-linking antigens and plant lectins	B-lymphocytes (anti-Ig) T-lymphocytes (T3/Ti crosslinking) Mast cells (anti-IgE)
Sperm	Sea urchin eggs
ATP	Ehrlich ascites tumour cells

tein to link receptors to PtdIns 4,5-P_2 phosphodiesterase [11]. The normal *ras* gene can both bind and hydrolyse GTP, but on activation by a point mutation at codon 12, the resulting oncogenic protein can still bind GTP, although its ability to hydrolyse this nucleotide is severely impaired [12]. If *ras* plays a part in this receptor mechanism, by analogy with the adenylate cyclase system, the loss of GTPase activity might mean that the oncogenic protein would continue to activate the formation of Ins(1,4,5)P_3 and DG in an uncontrolled way, independently of agonists.

4. Products of phosphatidylinositol 4,5-bisphosphate hydrolysis and their roles as second messengers in the cell

4.1. Inositol trisphosphate and calcium mobilisation

Cytosolic Ca^{2+} may be mobilised from either intracellular stores or from the extracellular compartment or both. The mobilisation of calcium from intracellular stores is an important source of activator Ca^{2+} following stimulation of many different cells, including oocytes, Swiss 3T3 cells, insulin-secreting β-cells, liver and smooth muscle (for reviews see Refs. 6,8,13). It was first demonstrated in 1983 that the addition of InsP_3 (1,4,5-isomer) to 'permeabilised' pancreatic acinar cells resulted in the release of Ca^{2+} from these cells [14]. Since then, similar results have been demonstrated in other permeabilised preparations as well as in the microsomal fraction of exocrine pancreas, rat insulinoma cells and liver [6,8,13]. The release of Ca^{2+} is rapid and occurs at less than micromolar concentrations of Ins(1,4,5)P_3. Specificity of the 1,4,5 isomer of InsP_3 for Ca^{2+} release has been demonstrated in pancreas and Swiss mouse 3T3 cells. InsP and InsP_2 were unable to release Ca^{2+}, while Ins(1,3,4)P_3 was only effective at much higher concentrations [6]. The release of Ca^{2+} by Ins(1,4,5)P_3 is transient due to the further metabolism of this messenger (see below).

4.2. Diacylglycerol mobilisation and the activation of protein kinase C

A rapid and usually transient accumulation of DG associated with stimulated phosphoinositide hydrolysis has been demonstrated in platelets, pancreas, mast cells, 3T3 cells, pituitary cells, GH3 cells and liver [6,8,13].

Interest in DG as a product of phospholipid degradation was stimulated by the discovery in the early 1980s of a phospholipid-dependent DG-stimulated and Ca^{2+}-activated protein kinase, termed protein kinase C (PKC) [15]. This kinase is ubiquitous in animal cells and is found in some plant cells. Activation of this enzyme by DG occurs by a shift in affinity for Ca^{2+} from the mM to the sub-μM range, which exists in stimulated cells. Further interest in the DG-activated PKC stimulatory pathway was aroused by the observation that this kinase is the intracellular receptor

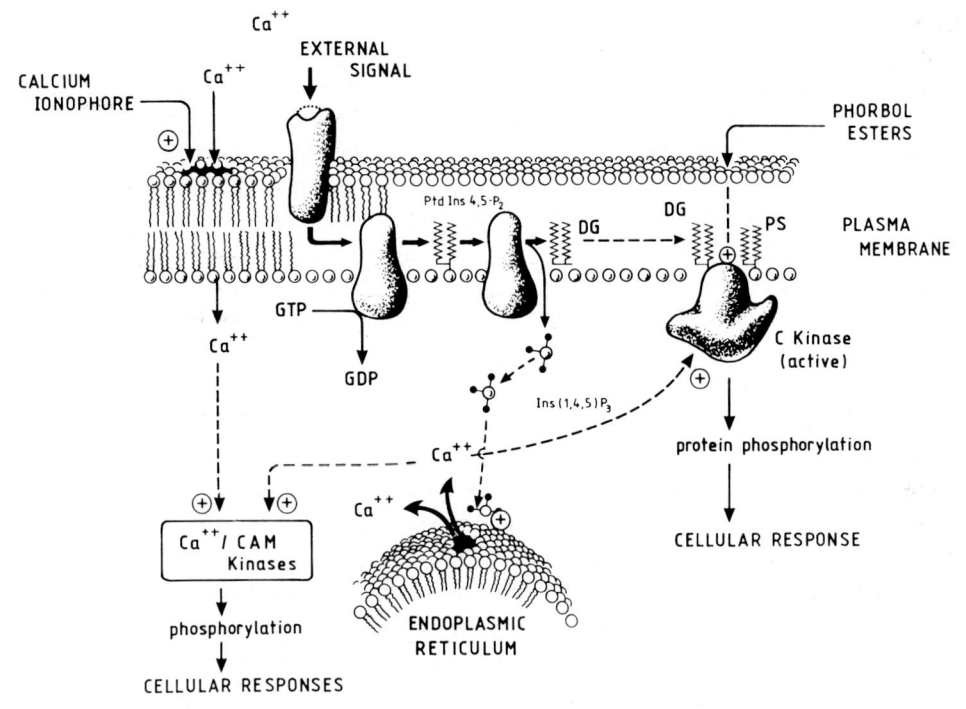

Fig. 3. The two second messenger pathways associated with receptor mediated PtdIns 4,5-P_2 hydrolysis. The external signal, i.e. an agonist binding to its receptor, is transduced through the membrane by a mechanism involving a GTP-binding protein, to activate a phospholipase C (PtdIns 4,5-P_2 phosphodiesterase). The activated phospholipase C cleaves the PtdIns 4,5-P_2 to generate diacylglycerol (DG) and Ins(1,4,5)P_3. The former is hydrophobic and stays in the plane of the membrane, activating protein kinase C. The latter is released into the cell interior and liberates Ca^{2+} from intracellular stores. The activated protein kinase C is capable of phosphorylating, and altering the activity of a number of substrate proteins, thus provoking some cellular response. The liberated Ca^{2+} is thought to activate, via calmodulin kinases, a distinct set of cellular substrates, some of which may overlap with those activated by protein kinase C. Ca^{2+} is also a co-stimulator of protein kinase C. The summation of these respective activations is postulated to synergistically activate a focal target and this can be mimicked in certain systems by the co-addition of calcium ionophores, which raise cytosolic Ca^{2+}, and phorbol esters or diacylglycerol analogues, which activate protein kinase C. (Redrawn from Berridge [23].)

for the tumour-promoting phorbol esters [16]. Protein kinase C is a mainly cytosolic enzyme but during DG or phorbol ester activation it appears to become membrane-associated. Nishizuka [17] postulated that the products of PtdIns 4,5-P_2 hydrolysis, Ins(1,4,5)P_3 and DG respectively activate separate pathways that result in a synergistic activation of later events, such as secretion or proliferation (Fig. 3).

A number of endogenous substrate proteins have been identified for protein kinase C in various cell types. These include nuclear proteins, histones, high mobility group proteins, enzymes such as tyrosine hydroxylase and guanylate cyclase, ribosomal S6 protein, and various membrane proteins important in growth such as the epidermal growth factor receptor, transferrin receptor, the glucose transporter, T200 glycoprotein, class 1 HLA receptors and the interleukin 2 receptor [18–20]. The subset of proteins phosphorylated by PKC has been shown to be mainly different from those stimulated by the ionophore-induced elevation of intracellular Ca^{2+}. However, a combination of the two subsets of phosphorylated proteins is seen when hepatocytes are stimulated by vasopressin, which induces the hydrolysis of PtdIns 4,5-P_2 [21]. Another interesting substrate for PKC is a 40 kDa protein that is phosphorylated when platelets are stimulated by TPA, other cell-permeant diacylglycerol analogues, or by thrombin, all of which induce PtdIns 4,5-P_2 hydrolysis [22]. The hydrolysis of PtdIns 4,5-P_2 is postulated to result in a synergistic effector response by activating PKC, and possibly other DG activated enzymes, and by parallel activation of calcium-dependent kinases, phosphatases and proteases. The physiological consequences of receptor-mediated PtdIns 4,5-P_2 hydrolysis can be mimicked by the co-addition of calcium ionophores, such as ionomycin and A23187, and TPA or other cell-permeant, diacylglycerol analogues. Furthermore, co-addition of TPA and calcium ionophores often leads to synergistic effects on a number of physiological functions. These have been observed in lymphocytes, hepatocytes, adrenal glomerulosa cells, exocrine pancreas, mast cells, adrenocortical cells, pancreatic islets, neutrophils, GH_4 cells, ileum and parotid cells [6,8,13].

5. Metabolism of the hydrolysis products of PtdIns 4,5-P_2

5.1. Inositol trisphosphate

The intracellular levels of DG and Ins(1,4,5)P_3 are determined by a balance between their rate of formation and by their removal. In order for a signalling system to function effectively, it is just as important to control the inactivation of second messengers as it is to control their production. The Ins(1,4,5)P_3 in stimulated cells is degraded to free inositol which is incorporated again into PtdIns. The cycle begins with an inositol trisphosphatase which attenuates the second messenger activity of the Ins(1,4,5)P_3 by removing phosphate from the 5-position to yield inositol 1,4-bisphosphate (Ins(1,4)P_2). This is then hydrolysed to inositol 1-phosphate (Ins1-P) and/or Ins4P, suggesting that either the bisphosphatase is not specific or there are two separate enzymes with different specificity. The inositol monophosphates are then converted to free inositol by inositol 1- and/or 4-phosphatases [6,8,13].

A cyclic InsP_3 has also been found in several tissues. This isomer is an extremely labile compound and can be rapidly metabolised to InsP_3 by phosphodiesterases [24]. It is not known whether it has any second messenger function but it appears to be present in only very small amounts. Inositol phosphates containing four, five and six phosphates have been isolated from stimulated tissues [25–27]. There is likely

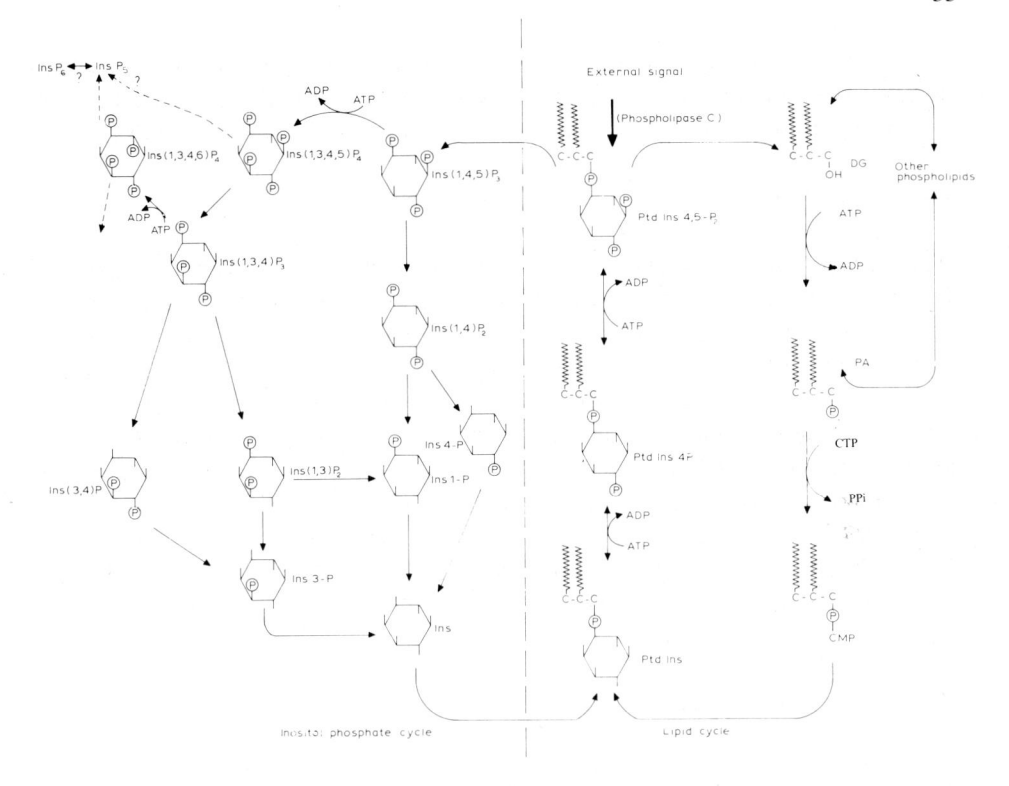

Fig. 4. Metabolism of inositol lipids and inositol phosphates in stimulated cells. In the lipid cycle, PtdIns is converted through PtdIns 4-P to Ptd 4,5-P_2 which is hydrolysed upon receptor stimulation to yield diacylglycerol (DG) and Ins(1,4,5)P_3. DG is phosphorylated via diacylglycerol kinase to give phosphatidic acid (PA). PA is activated by CTP to form CDP-diacylglycerol which can combine with free inositol (Ins) to re-form PtdIns. The other product of the PtdIns 4,5-P_2 hydrolysis, Ins(1,4,5)P_3, may be dephosphorylated to various InsP_2 and InsP isomers which ultimately provide free inositol (Ins) for resynthesis of PtdIns. Ins(1,4,5)P_3 may be phosphorylated in steps to give an array of InsP_4, InsP_5 or InsP_6 isomers.

to be a complex series of kinase and phosphatase pathways for these compounds. Evidence indicates that all of the inositol tris, tetrakis, pentakis and hexakis-phosphates derive initially from Ins(1,4,5)P_3. For instance, the Ins(1,3,4,5)tetrakisphosphate found in activated cells is formed by specific phosphorylation of Ins(1,4,5)trisphosphate in the 3-position [28]. The other InsP_3 isomer, Ins(1,3,4)P_3, arises in a degradative step catalysed by a phosphatase that specifically removes a phosphate from the 5-position of the inositol headgroup [25]. Recent evidence suggests that this Ins(1,3,4)P_3 may itself be phosphorylated to Ins(1,3,4,6)P_4 [29]. It is likely that the number of inositol phosphates and their associated anabolic and catabolic pathways will continue to increase and some of these may be found to have functions as physiological second messengers. They are formed at the expense of a large amount of energy and they are good candidates for intracellular modulators with their unique arrays of mono-ester phosphates. The presently recognised metabolic pathways for these compounds are shown in Fig. 4.

5.2. Diacylglycerol

Diacylglycerol, like InsP$_3$, appears transiently in stimulated cells. The control of DG accumulation is important as it is a co-activator of a number of enzymes. As well as protein kinase C, phosphatidate phosphohydrolase, diacylglycerol kinase and p36 have been reported to be activated by DG [30,31]. DG and phosphatidic acid are also intermediates in the synthetic and degradative pathways of the other major, membrane phospholipids but the accumulated evidence from radioactive tracer studies suggests that DG formed during PtdIns 4,5-P$_2$ hydrolysis is directed back to resynthesise PtdIns [8]. This must occur as a result of physical or chemical compartmentation and is instigated by the phosphorylation of DG to PA by DG-kinase. PA is converted back to PtdIns by the sequential actions of CTP-DG cytidyl transferase and PtdIns-synthetase.

6. Fertilisation, proliferation and oncogenes

6.1. Role of inositol lipid degradation

Most of the stimulus response systems that utilise increased phosphoinositide hydrolysis as a messenger system exhibit only a very short delay, e.g. muscle contraction, light transduction and secretion. Cell proliferation, the creation of two cells from one, is the culmination of a vast number of biochemical events which are needed to duplicate all of the cell's components. The chain of events instigated by the binding of growth factors to their receptors can be divided arbitrarily into four major parts (Fig. 5): (a) agonist binding to receptor, (b) membrane transduction and genesis of second messenger(s), (c) cytoplasmic modifications by second messengers, and (d) specific gene expression and synthesis of active product.

The involvement of calcium in the initiation of proliferation was suggested by the observation in lymphocytes that A23187 could mimic the effects of phytohaemagglutinin [8]. Several laboratories have also shown that a transient elevation of Ca^{2+} is essential at some stage during the cell cycle [32]. Early observations regarding the involvement of phosphoinositides in cell growth were reported by Diringer and co-workers [33]. Since then stimulated inositol phospholipid metabolism has been demonstrated in the proliferative response of a number of cell types as an effect of growth factors [6,8]. A similar response is seen also during the fertilisation process in several invertebrates. The large increase in polyphosphoinositide metabolism in sea urchin eggs following fertilisation has been implicated in the mobilisation of intracellular Ca^{2+}. Direct evidence for the involvement of $Ins(1,4,5)P_3$ is provided by the observation that this isomer, when micro-injected, is specific in releasing Ca^{2+} from intracellular sources and triggering other early events that normally follow fertilisation of sea urchin or starfish oocytes by sperm [6]. A variety of growth factors such as platelet-derived growth factor (PDGF), bombesin or vasopressin also result in a rapid, increased formation of diacylglycerol and $Ins(1,4,5)P_3$ in Swiss 3T3 cells [6,8,13].

It is an advantage when investigating the role of various biochemical pathways in proliferative studies to use a primary cell type and to mimic as closely as possible the physiological situation. Cell lines, while offering the convenience of quantities of material, have the disadvantage of biochemical lesions superimposed during immortalisation or transformation. Lymphocytes are a good model to study proliferative pathways and modern techniques of cell separation, and the use of monoclonal antibodies and immunofluorescent labeling techniques has enabled the purification of homogeneous populations of both T and B lymphocytes.

The stimulated turnover of inositol phospholipids in lymphocytes, following the crosslinking of cell-surface receptors by plant lectins, was observed by several groups in the 1960s [1]. More recently it was demonstrated that PtdIns 4,5-P_2 is rapidly hydrolysed following crosslinking of surface immunoglobulins on B lymphocytes [34] or the T_i/T_3 antigen complex on T lymphocytes [35]. The analysis of intracellular

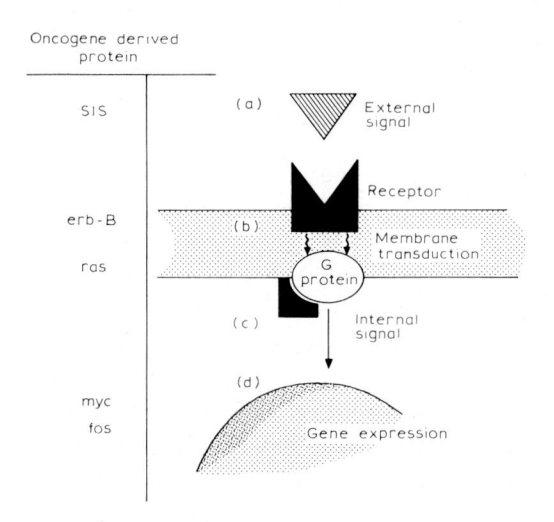

Fig. 5. Components of the proliferation signalling cascade and the role of oncogene products. The cascade of signal transduction involves a number of arbitrary signal transduction stages each of which contains key proteins that are potential sites for oncogenic protein transformation. The response is activated by an external signal (a) binding to receptor (b). The signal is then transferred across the cytoplasm, resulting in the expression of a specific subset of proliferative genes (d). Each of the oncogene products listed has been shown to have at least an indirect relationship to the inositol phospholipid component of the signalling cascade.

Ca^{2+} levels, with one of the intracellular calcium-chelating, fluorescent probes, quin-2, fura-2 or indo-1, demonstrates that, in both cases, there is a rapid rise in intracellular Ca^{2+} as this ion is released from intracellular stores. Analysis of stimulated B lymphocytes, using the probe fura-2, indicates that if Ca^{2+} in the external medium is removed the intracellular Ca^{2+} level returns to basal levels in 5 to 7 minutes, but if there is Ca^{2+} present in the external media a sustained increase in intracellular Ca^{2+} is detected [36]. Such analysis suggests the opening of a plasma membrane calcium channel but the nature of the channel or mechanism of its opening are not presently known. It is possible that the opening of this channel could be stimulated by one of the inositol phosphates.

The increase in phosphoinositide hydrolysis appears to occur during the G_0 and early G_1 phases of the cell cycle. Cells will not tolerate a sustained increase in intracellular Ca^{2+} and actively extrude the ion. Recent evidence in B lymphocytes, using calcium probes and image intensifiers, shows that when surface immunoglobulin is crosslinked the rise of intracellular Ca^{2+} occurs in repeated short bursts [37].

6.2. Influence of ionophores and synthetic stimulators of protein kinase C

In both T and B lymphocytes the co-addition of calcium ionophore (A21387 or ionomycin) and a PKC activating agent (TPA or cell permeant diacylglycerols) produces a synergistic proliferative response at doses that are submitogenic when added alone [38,39]. This indicates that the two pathways normally activated by the hydrolytic products of PtdIns 4,5-P_2 are capable of inducing proliferation or the expression and secretion of other factors that are necessary for the completion of the cell cycle and their corresponding receptors.

In the physiological situation, a cell of a particular lineage will be subject to control by a combination of exogenous factors and by the expression of a succession of appropriate receptors on the cell surface. These factors have been arbitrarily assigned as competence (primary stage) and progression (secondary stage) factors. The competence stage, which ends in the initiation of DNA synthesis, is controlled by a group of factors such as PDGF. On the other hand, factors like epidermal growth factor, insulin and the somatomedins direct growth at the progression stage. In the lymphocyte systems the accumulated evidence suggests that the stimulated hydrolysis of PtdIns 4,5-P_2 is an early event induced by competence factors, and subsequent activation of the bifurcating pathway results in the increased expression of secondary factors and possibly their receptors that maintain the proliferation cascade.

6.3. Oncogenes

The enzymatic reactions associated with the generation of these multiple activating pathways based on the inositol phospholipids may represent the site of action of certain oncogene products, and a detailed knowledge of these will aid in the understanding of oncogenesis. The *sis* proto-oncogene encodes for PDGF [40], which is shown to stimulate PtdIns 4,5-P_2 hydrolysis. The *v-erb* B gene codes for a protein which shows homology to part of the EGF receptor [41]. EGF also stimulates phosphoinositide turnover and an increase in cytosolic [Ca^{2+}] in fibroblasts [42]. The possible perturbation of the proteins discussed above and that of the GTP-binding *ras* gene product relate to the proliferation cascade as shown in Fig. 5. Ultimately the signals for proliferation, that are received at the cell membrane, must be transmitted to the nucleus and converted into gene expression (stage (d) Fig. 5). There are strong indications that cellular oncogene counterparts may be expressed as a result of the incoming signal. Several groups have shown that PDGF, as well as other designated competence factors, and tumour promotors activate the *myc* and *fos* genes [43]. The proteins translated from these genes are located in the nucleus where they may be involved in controlling the expression of other genes.

7. *Release of arachidonic acid (for review see Refs. 8 and 13)*

Arachidonate metabolites (prostaglandins, leukotrienes and thromboxanes) are potent regulators of various physiological responses. The levels of these compounds are regulated by the level of free arachidonic acid, which may be liberated from esterified lipids. It is believed that PtdIns in all mammalian tissues is rich in arachidonate and this has been demonstrated in a number of tissues. The polyphosphoinositides have also been shown to contain a large amount of arachidonate, mainly in the 2 position on the diacylglycerol backbone (see Fig. 2). Although some of the other phospholipids, PtdCho and PtdEt, have been shown to contain varying amounts of arachidonate, there is an increasing list of tissues where a release of free arachidonic acid occurs in parallel with the phosphoinositide response (Table II).

7.1. *Mechanisms of arachidonate liberation*

During hormonal stimulation of PtdIns 4,5-P_2 hydrolysis there appears to be a preferential degradation of molecules, such as diacylglycerol and phosphatidic acid, which contain arachidonate in the 2-position. Two separate pathways have been proposed for the release of arachidonic acid from these two products of the phosphoinositide response. The first proposes that diacylglycerol is the source of the liberated arachidonate and that diacylglycerol lipase acts on the DG released by hydrolysis of phosphoinositides. The second suggests that a phosphatidic acid-specific phospholipase A_2 is responsible for cleaving the arachidonic acid from phosphatidate.

The release of arachidonic acid from these lipids by the synchronised activation of phospholipase C/phospholipase A_2 could occur in complexes of proteins that are

TABLE II

Summary of stimuli that induce both PtdIns 4,5-P_2 hydrolysis and the release of arachidonate, and some examples of target tissues (from Refs. 6 and 8)

Stimulus	Target tissue
Thrombin	Platelets
Muscarinic cholinergic	Pancreas Vascular smooth muscle
α_1-Adrenergic	Spleen Kidney Brain Iris smooth muscle
Serotonin	Blowfly salivary gland
Vasopressin	Renal mesangial cells
Angiotensin II	Kidney
Bradykinin	Kidney BALB/3T3 cells
Substance P	Iris smooth muscle
f-Methionyl-leucyl-phenylalanine	Neutrophils
Caerulin	Pancreas
Platelet activating factor	Platelets Iris smooth muscle

closely controlled by Ca^{2+} levels and DG availability. As well as the controlled release of arachidonic acid that is initiated by calcium mobilising receptors there is a separate pathway that releases arachidonate from various phospholipids by a relatively less specific phospholipase A_2.

8. Summary

In the past few years our understanding of a new second messenger system involving inositol phospholipids has developed rapidly. It is clear the action of many hormones and neurotransmitters depends on the hydrolysis of membrane phosphoinositides. Agonists induce the cleavage of PtdIns 4,5-P_2, resulting in the formation of diacylglycerol which activates protein kinase C, and inositol trisphosphate which mobilises Ca^{2+}. Both of these important second messengers are inactivated by a succession of closely regulated steps. The level of these signals will always depend upon the balance between their rates of formation and degradation. The two products of PtdIns $4,5\text{-}P_2$ hydrolysis may also be responsible for releasing arachidonic acid, which is the immediate precursor for a whole subset of cellular messengers. While the number of tissues and agonists shown to use phosphoinositide hydrolysis as a mechanism for generating second messengers grows and the Ca^{2+} mobilising capacity of $Ins(1,4,5)P_3$ has been solidly established, there are many aspects of this signalling system that remain to be resolved. For example: what is the nature of the G-protein(s) that couple receptors to the phosphodiesterase (phospholipase C)? What is the nature and location of the $Ins(1,4,5)P_3$ receptor? What is the significance of the $InsP_4$ pathway and the other more highly phosphorylated isomers? How is the entry of Ca^{2+} to the cell regulated and how is this signalling system integrated with the other second messenger systems, such as those for cyclic nucleotides and tyrosine phosphorylation? One of the most exciting areas of future research will be the elucidation of the exact contribution of the inositide-generated signalling molecules to the proliferation cascade and the capacity for oncogenic disregulation that leads to malignant transformation.

References

1. Michell, R.H. (1982) Cell Calcium 3, 285–294.
2. Michell, R.H. (1975) Biochim. Biophys. Acta 415, 81–147.
3. Monaco, M.E. and Woods, D. (1983) J. Biol. Chem. 256, 15125–15129.
4. Durell, J., Garland, J.T. and Friedel, R.O. (1969) Science 165, 862–866.
5. Michell, R.H., Kirk, C.J., Jones, L.M., Downes, C.P. and Creba, J.A. (1981) Phil. Trans. R. Soc. Lond. 296, 123–137.
6. Berridge, M.J. and Irvine, R.F. (1984) Nature 315–324.
7. Alison, J.H. and Stewart, M.A. (1971) Nature 233, 267–268.
8. Chandra Sekar, M. and Hokin, L.E. (1986) J. Memb. Biol. 89, 192–210.
9. Goodhardt, M., Ferry, N., Geynet, P. and Hanoune, J. (1982) J. Biol. Chem. 257, 11577–11583.
10. Haslam, R.J. and Davidson, M.M.L. (1984) FEBS Lett. 174, 90–95.
11. Wakelam, M.J.O., Murphy, G.J., Hornby, V.R. and Houslay, M.D. (1986) Nature 323, 68–71.
12. Finkel, T., Der, C.J. and Cooper, G.M. (1984) Cell 37, 151–158.
13. Abdel-Latif, A.A. (1986) Pharmacol. Rev. 38, 227–272.
14. Streb, H., Irvine, R.F., Berridge, M.J. and Schulz, I. (1983) Nature 306, 67–69.
15. Takai, Y., Kishimoto, A., Inoue, M. and Nishizuka, Y. (1977) J. Biol. Chem. 252, 7603–7609.
16. Castagna, M., Takai, Y., Kaibuchi, K., Sano, K., Kikkawa, U. and Nishizuka, Y. (1982) J. Biol. Chem. 257, 11404–11414.
17. Nishizuka, Y. (1984) Nature 308, 693–698.
18. Schwantke, N., Le Bouffant, F., Doree, M. and Le Peuch, C.J. (1985) Biochimie 67, 1103–1110.
19. Shackelford, D.A. and Trowbridge, I.S. (1986) J. Biol. Chem. 261, 8334–8341.
20. Nishizuka, Y. (1986) Science 233, 305–312.
21. Garrison, J.C., Johnson, D.E. and Capanile, C.P. (1984) J. Biol. Chem. 259, 3283–3292.

22. Kaibuchi, K., Takai, Y., Sawamura, M., Hoshijima, M., Fujikura, T. and Nishizuka, Y. (1983) J. Biol. Chem. 258, 6701–6704.
23. Berridge, M.J. (1985) Sci. Am. 253, 124–135.
24. Wilson, D.B., Bross, T.E., Sherman, W.R., Berger, R.A. and Majerus, P.W. (1985) Proc. Natl. Acad. Sci. USA 82, 4013–4017.
25. Batty, I.R., Nahorsky, S.R. and Irvine, R.F. (1985) Biochem. J. 232, 211–215.
26. Morgan, R.O., Chang, J.P. and Catt, K.J. (1986) J. Biol. Chem. 262, 1166–1171.
27. Heslop, J.P., Irvine, R.F., Tashjian, A.H. and Berridge, M.J. (1985) J. Exp. Biol. 119, 395–401.
28. Irvine, R.F., Letcher, R.F., Heslop, J.P. and Berridge, M.J. (1986) Nature 320, 631–634.
29. Shears, S.B., Parry, J.B., Tang, E.K.Y., Irvine, R.F., Michell, R.H. and Kirk, C.J. (1987) Biochem. J. 246, 139–147.
30. Glenney, J.R. (1985) FEBS Lett. 192, 79–82.
31. Besterman, J.M., Pollenz, R.S., Booker, E.L. and Cuatrecasas, P. (1986) Proc. Natl. Acad. Sci. USA 83, 9378–9382.
32. Berridge, M.J. (1984) Biochem. J. 220, 345–360.
33. Diringer, H. and Koch, M.A. (1974) Hoppe-Zeyler's Z. Phys. Chemie. 355, 93–97.
34. Bjisterbosch, M.K., Meade, J.C., Turner, G.A. and Klaus, G.G.B. (1985) Cell 41, 999–1005.
35. Imboden, J. and Stobo, J. (1985) J. Exp. Med. 161, 446–456.
36. Guy, G.R., Michell, R.H. and Gordon, J. In: Inositol Lipids Function and Metabolism. (Michell, R.H., Downes, P.T. and Drummond, A., eds.) Academic Press, New York, in press.
37. Tsien, R.Y. and Peonie, M. (1986) Trends Biol. Sci. 11, 450–456.
38. Truneh, A., Alberts, F., Golstein, P. and Schmitt-Verhulst, A. (1985) Nature 313, 318–320.
39. Guy, G.R., Bunce, C.M., Gordon, J., Michell, R.H. and Brown, G. (1985) Scand. J. Immunol. 22, 591–596.
40. Waterfield, M.D., Scrace, G.T., Whittle, N., Stroobant, P., Johnsson, A., Watteson, A., Westermark, B., Helder, C.-H., Huang, J.B. and Deuel, T.F. (1983) Nature 304, 35–39.
41. Downward, J., Yarden, Y., Mayes, E., Scrace, G., Totty, N., Stockwell, P., Ullrich, A., Schlessinger, J. and Waterfield, M.D. (1982) Nature 307, 521–527.
42. Moolenaar, W.H., Tertoolen, L.G.J. and de Laat, S.W. (1984) J. Biol. Chem. 259, 8066–8069.
43. Moore, J.P., Todd, J.A., Hesketh, T.R. and Metcalf, J.C. (1986) J. Biol. Chem. 261, 8158–8186.

CHAPTER 3

The role of calcium binding proteins in signal transduction

NAVIN C. KHANNA, MASAAKI TOKUDA and
DAVID M. WAISMAN

Cell Regulation Group, Department of Medical Biochemistry, The University of Calgary, Calgary, Alberta, Canada, T2N 4N1

1. Introduction

Hormone action at cellular levels begins with association of the hormone and its specific receptor. Hormones can be classified by the location of the receptor and by the nature of the signal used to mediate hormone action within the cell. Steroid hormones are lipophilic and thus can cross the plasma membrane to interact with their intracellular receptors. On the other hand, polypeptide and catecholamine hormones bind to the plasma membrane of the target cells. These hormones communicate with intracellular metabolic process through a limited number of intermediary molecules, so called second messengers (the hormone itself is the first messenger), which are generated as a consequence of the ligand-receptor interaction. Cyclic AMP, cyclic GMP and Ca^{2+} are the classical messenger molecules for the majority of hormones. Several hormones, many of which were previously thought to affect cAMP, appear to use calcium or phosphatidylinositide metabolites (or both) as intracellular messengers. Table 1 lists the polypeptide hormones and catecholamines that act through cAMP or calcium. These intracellular messengers are known to regulate many cellular processes including metabolism, secretion, contraction, differentiation, proliferation and phototransduction. When the cell is quiescent the intracellular messenger concentration is below the binding affinity of the receptor protein; when the cell is activated by the extracellular messenger the concentration of the intracellular messenger rises to a level such that binding with the intracellular receptor is possible. As a consequence of the binding event the intracellular receptor can bind to and influence its intracellular target proteins. In the case of the cyclic AMP second messenger system, it is probable that all the actions of cyclic AMP are mediated through the action of a single intracellular receptor protein, the cyclic AMP dependent protein kinase. Similarly, it is also suspected that the second messenger

TABLE I
Mediators of polypeptide and catecholamine hormone action

Second messenger is cyclic AMP	Second messenger is Ca^{2+} and/or phosphoinositides
ACTH	ACTH
Calcitonin	HCG
HCG	FSH
CRF	GnRH
GHRF	LH
LH	PTH
PTH	TRH
TSH	TSH
Prostaglandin E_1	Vasopressin
Vasopressin	α-Adrenergic agents
Glucagon	PDGF[a]
β-Adrenergic agents	EGF[a]
$α_2$-Adrenergic agents	Insulin[a]
Angiotensin II	
Insulin	
Muscarinic agents	
Opiates	
Oxytocin	
Somatostatin	

[a]Stimulates protein tyrosine kinases.

action of cGMP is mediated through the cyclic GMP dependent protein kinase. Therefore, the phosphorylation of target proteins appears to be the mechanism of signal transduction by cAMP and cGMP.

The calcium messenger system is more complex than either the cyclic AMP or cyclic GMP second messenger systems. This is reflected in part by the many intracellular receptors, i.e., calcium binding proteins that exist for calcium. As will be discussed, the calcium binding proteins utilize both protein phosphorylation mechanisms as well as direct protein-protein interaction to transduce the calcium signal. When the extracellular messenger interacts with receptor on the cell surface, an increase in the concentration of the intracellular messenger occurs; this increase is detected by one or more specific receptor proteins. The receptor proteins have a high affinity for and a narrow concentration range over which they bind the intracellular messenger.

Unfortunately, at the present time the physiological function of the majority of calcium binding proteins is unknown, and before the complexity of the calcium messenger system can be completely elucidated, the function of these proteins must be clarified. Accordingly, the purpose of the review is to briefly document the various cellular calcium binding proteins and where possible discuss the role of these proteins in the signal transduction process. Since our knowledge of the role of calcium binding proteins in signal transduction is incomplete, we have discussed many diverse cellular systems, including hormonally activated systems, in order to attempt to integrate the available information into a general framework. This basic framework will provide a generalized model for an understanding of the role of calcium binding proteins in the signal transduction process.

2. The calcium transient

The role of calcium as an intracellular messenger was initially suggested by Heilbrunn and Wiercinski [1] in 1947. These investigators showed that the calcium ion was the only physiological substance that, when microinjected into a living muscle, could induce localized constrictions at the site of injection. In 1970, Rasmussen [2] presented evidence based on analysis of a number of hormonally activated systems, that calcium served as an intracellular messenger in signal transduction. An abundance of evidence (reviewed in Ref. 3) now supports the hypothesis that, in many systems, cellular activation initiated by direct interaction of hormones, growth factors or neurotransmitters with plasma membrane receptors results in an increase in intracellular free calcium.

The total intracellular calcium concentration has been estimated to be about 1.0 mM, while the cytosol has been estimated to contain about 50–100 μM calcium [4,5]. Measurement of the free intracellular calcium $[Ca^{2+}]_i$ within cells suggests that it is very low, of the order of 10–100 nM. This means that the vast majority of intracellular calcium is bound. Considering that the extracellular calcium concentration is about 1 mM (compared to 10 nM $[Ca^{2+}]_i$) and that the interior of the cell is negatively charged there is a large electrochemical gradient driving calcium ions into the cell, however the relative impermeability of the plasma membrane to calcium ions prevents large movements of calcium ions across the plasma membrane.

The calcium transient is controlled by the uptake and release of calcium across the three major membrane systems which border the cytoplasm, namely the plasma membrane, the inner mitochondrial membrane and the endoplasmic reticulum. Each membrane possesses distinctive transport mechanisms which act in concert to regulate and modulate intracellular calcium. Therefore, the transient changes in $[Ca^{2+}]_i$ are set against a background of intracellular buffering and any change in calcium signal reflects a balance between mechanisms that tend to remove or bind Ca^{2+}. There are two mechanisms by which the calcium signal can be activated (**1**) by a release of calcium ions from intracellular stores, namely the endoplasmic reticulum [6,7] and (**2**) by movement of calcium ions down their electrochemical gradient from the extracellular space through the plasma membrane into the cell (Fig. 1). Activation of the calcium signal by the first mechanism begins with the occupation of extracellular plasma membrane receptors by such agonists as hormones, neurotransmitters or growth factors. As a result of agonist-receptor interaction a plasma

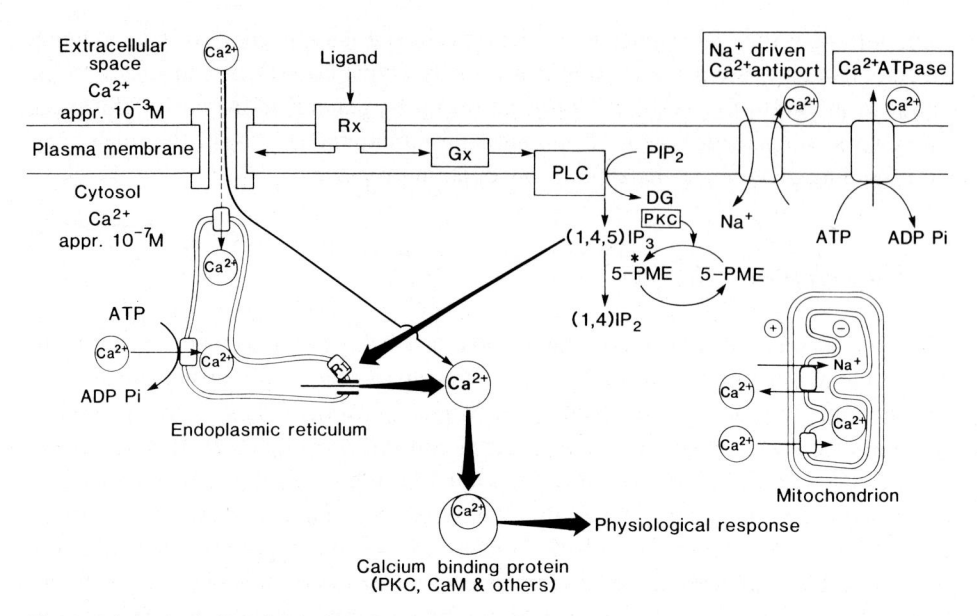

Fig. 1. A hypothetical model depicting the role of calcium in cellular activation. The binding of agonist to its receptor (Rx) results in the activation of phospholipase (PLC). Coupling the receptor to phospholipase C is a guanine nucleotide-dependent regulatory protein (Gx). The PLC catalyzes the breakdown of PIP_2 to diacylglycerol (DG) and $(1,4,5)IP_3$ (or its cyclic derivative c-IP_3). The $(1,4,5)IP_3$ interacts with a specific receptor on the endoplasmic reticulum (RI) which causes the release of Ca^{2+} from the lumen of the endoplasmic reticulum thereby generating an increase in cytosolic Ca^{2+}, the calcium transient. The generation of the calcium transient activates calcium binding proteins which by interacting with target proteins produce the physiological response. DG, an important component of PIP_2 hydrolysis, activates protein kinase C (PKC). Activation of PKC results in the phosphorylation and activation of the $(1,4,5)IP_3$ 5-phosphomonoesterase (5-PME) to 5-*PME, an enzyme which dephosphorylates and inactivates $(1,4,5)IP_3$. Lowering of the $(1,4,5)IP_3$ concentration results in a lowering of the cytosolic Ca^{2+}, to levels slightly above resting levels, but at a sufficient concentration to maintain the cell in the activated state. Extracellular calcium participates in this model of receptor-mediated phosphoinositide-induced Ca^{2+} mobilization by refilling of the intracellular hormone-sensitive stores of the endoplasmic reticulum. The Ca^{2+} influx pathway responsible for filling the endoplasmic reticulum Ca^{2+} stores are agonist independent. Distinct from this Ca^{2+} influx pathway are the receptor operated Ca^{2+} channels which are thought to represent an agonist regulated mechanism for activating Ca^{2+} influx. In many excitable cells, Ca^{2+} channels exist which allow Ca^{2+} influx in response to membrane depolarization, the depolarization operated Ca^{2+} channels. The calcium transient is buffered by, and after removal of agonist, terminated by the calcium extrusion proteins of the endoplasmic reticulum, the plasma membrane, and under certain conditions, the mitochondria. 5-*PME represents the phosphorylated form of 5-PME.

membrane phospholipase C (PLC) is activated to hydrolyze a unique lipid, phosphatidylinositol 4,5-biphosphate (PIP_2) to inositol 1,4,5,-trisphosphate (IP_3) and diacylglycerol (DG). The IP_3 is released into the cytosol where it functions to release Ca^{2+} from the endoplasmic reticulum (Fig. 1).

As a result of the action of IP_3, $[Ca^{2+}]_i$ is increased, i.e., the calcium signal is generated. The other product of PIP_2 hydrolysis, DG, has been shown to activate a protein kinase (protein kinase C) [8]. The DG/protein kinase C pathway appears to function in the modulation of the calcium signal. Protein kinase can phosphorylate and activate the 5-phosphomonoesterase thereby lowering the IP_3 concentration and subsequently decreasing the intracellular free calcium concentration.

A second mechanism for increasing $[Ca^{2+}]_i$ involves the movement of Ca^{2+} from the extracellular space, across the plasma membrane into the cell (Fig. 1). Two distinct types of Ca^{2+} channels are thought to regulate the entry of Ca^{2+} into the cell [9]. The Ca^{2+} channels that open in response to membrane depolarization are described as depolarization-operated Ca^{2+} channels (DOC) [10,11], whereas those which are controlled by receptor activation are called receptor-operated Ca^{2+} channels [12,13]. It is interesting to note that in some cells, e.g., smooth muscle, Ca^{2+} mobilization is thought to occur by both types of channels.

3. Calcium binding proteins and signal transduction

It was not until 1969 that the first calcium receptor protein, troponin-C was isolated from muscle by Ebashi [14]. Shortly afterward, another calcium receptor, calmodulin [15] was identified and it became apparent that unlike the cAMP messenger system, more than one receptor protein existed for calcium.

In order to fulfill the role of cytosolic calcium receptor, a calcium binding protein must demonstrate high affinity and specificity for calcium. In the resting state, the cytosolic free calcium is thought to be 50–100 μM (reviewed in Ref. 4). During cellular activation, the calcium concentration may rise to 10 μM. This means that in order to be physiologically relevant to a second messenger role, calcium binding must occur within a range of 50 nM to 10 μM. Since intracellular magnesium is about 2 mM, it is also important that calcium receptor proteins bind calcium with specificity, i.e., high affinity calcium binding in the presence of millimolar magnesium. The activation of calcium binding proteins by changes in the amplitude of the calcium signal is called amplitude modulation [16]. It is important to point out that besides responding to changes in the amplitude of the calcium signal, calcium binding proteins can be activated by changes in their sensitivity to the calcium. For example, calmodulin has been shown to increase its affinity for calcium after binding to many of its target enzymes [16]. Similarly, the binding of diacylglycerol to protein kinase C dramatically increases the affinity of this enzyme for calcium [8]. Activation of calcium binding proteins, independent of changes in intracellular free calcium, is called sensitivity modulation [16].

A calcium binding protein could play a role in transduction of the calcium signal in one or more of three possible ways. First, it could serve as a calcium buffer protein; for instance in response to cellular activation it could release bound calcium and potentiate the rise in intracellular free calcium. Alternatively, a calcium buffer protein could be important for termination of the calcium signal by binding large quantities of calcium (e.g., calsequestrin). The existence of calcium buffer proteins has been suggested from a comparison of the total cytosolic calcium concentration (100 μM) with the cytosolic free calcium concentration (10–100 nM). The ability of the cytoplasm to buffer Ca^{2+} means that changes in free calcium can be localized within cells. For example, it has been shown that when calcium is injected into cells the resultant increase in intracellular free calcium is highly localized to the site of injection and does not freely diffuse in the cytosol. This means that changes in intracellular free calcium are spatially restricted in the cell and can vary from one region of the cell to another.

Secondly, a calcium binding protein could function as a calcium dependent regulatory protein. A typical example is calmodulin (for a review see Ref. 17); at resting $[Ca^{2+}]_i$, this protein is inactive but during activation of the calcium signal the protein binds calcium and activates several target enzymes and proteins.

A calcium regulated enzyme or protein could be considered as the third type of calcium binding protein. The calcium regulated enzyme is contrasted to the target protein of the calcium dependent regulatory protein in that the calcium regulated enzymes are capable of binding calcium directly and do not require additional proteins to confer calcium sensitivity. An example of this type of calcium receptor is the protein kinase C [8].

Temporally, two distinct types of calcium signals exist. The first type is charac-

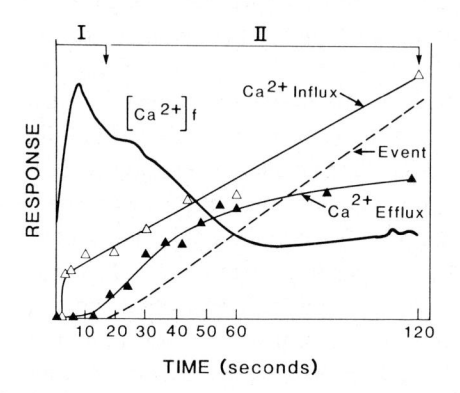

Fig. 2. The calcium transient. A diagrammatic presentation of the calcium transient of a typical cell during a sustained response. Addition of agonist produces a rapid increase in cytosolic free calcium (phase I) which returns to a concentration slightly above the resting cytosolic free calcium (phase II). Phase I represents the receptor-mediated phosphoinositide-induced Ca^{2+} mobilization from the endoplasmic reticulum. Phase II of the calcium transient represents the sustained phase which is dependent upon extracellular calcium. Although the cytoplasmic Ca^{2+} concentration is only slightly elevated from resting levels during phase II, the cell remains in the activated state. During phase II both calcium influx and efflux mechanisms are activated and therefore calcium cycles across the plasma membrane. The physiological response of the cell (event) appears to be activated during phase II. Adapted from Ref. 149.

terized by a rapid rise and decline of the calcium signal to resting concentrations. This is most commonly exhibited by cells such as skeletal muscle, which respond to external stimulation both rapidly and transiently. The second type of calcium signal is presented in Fig. 2, and is commonly exhibited by cells which exhibit a sustained response to external stimuli. Interestingly, in the case of sustained hormonal activation, the calcium transient is not maintained at the peak concentration but is actually maintained at a final concentration only slightly above resting intracellular free calcium. It is also important to point out that the calcium transient is not uniform throughout the cell. Both the magnitude and kinetics of changes in intracellular free calcium have been shown to be dramatically different in different regions of the cell.

In conclusion, cellular signal transduction can actually be divided into two separate components: **(1)** the generation of an intracellular signal from an external event and **(2)** transduction of the intracellular signal into the physiological response of the cell. In the first component, various agonists such as hormones, growth factors or neurotransmitters bind to specific receptors located on the extracellular face of the plasma membrane and, as a result of this interaction, a complex series of biochemical events are set into motion which result in generation of an intracellular signal. This intracellular signal can take the form of an increase in the intracellular free calcium ion concentration. In the second component of cellular signal transduction, the physiological response of the cell is activated as a consequence of the conversion of the calcium signal into the physiological response of the cell. This involves a class of proteins called the calcium binding proteins which when activated by the calcium signal can act on target proteins through protein phosphorylation mechanisms or directly via protein-protein interaction.

4. Calcium binding proteins: structure and function

Calcium binding proteins can be broadly divided into three major groups: the extracellular calcium binding proteins, the membrane bound calcium binding proteins and the intracellular calcium binding proteins. The extracellular calcium binding proteins are not involved in the process of signal transduction and the binding of calcium to these proteins may play a role in catalysis or be important for protection against autocatalytic degradation of the protein. In addition, calcium binding to extracellular proteins may function to 'bridge' the protein to solid surfaces such as membranes, or hydroxylapatite of bone. The members of the second group of calcium binding proteins, the membranous calcium binding proteins, have two distinct functions; some of these proteins are involved in calcium dependent cell-cell adhesion while others are involved in the maintenance of calcium homeostasis. The membranous calcium binding proteins involved in calcium homeostasis are important components of the transmembrane signal transduction process since they serve to modulate the intracellular calcium signal. The third class of calcium binding proteins, the intracellular calcium binding proteins are central to the signal transduction process since these proteins are responsible for the transduction of the calcium signal into the cellular response.

4.1. Extracellular calcium binding proteins

Unlike the intracellular calcium binding proteins, the extracellular calcium binding proteins are surrounded by a stable, non-fluctuating environment (the extracellular calcium concentration is about 1 mM) and, therefore, these proteins tend to exhibit low affinity calcium binding. These calcium binding proteins represent a large family of proteins, therefore only representatives of this family are discussed below. These calcium binding proteins can be grouped into three different classes (Table I). The first class of extracellular calcium binding proteins include such proteins as trypsin or thermolysin, and require calcium for stability and protection against autocatalytic degradation. Other members of this class of proteins such as phospholipase A_2, require calcium for catalysis. In contrast, the second class of extracellular calcium binding proteins uses the post-translationally acquired amino acid, gamma-carboxyglutamic acid (Gla) to confer upon these proteins functionally important interaction potential with calcium and acidic phospholipid or hydroxylapitite. For example, the binding of calcium by osteocalcin (a major calcium binding protein of bone) results in an increased α-helical content and enhanced adsorption of the protein to hydroxylapatite. Similarly, the binding of calcium by prothrombin or several blood clotting protease factors is important for the binding of the protein to phospholipid. The formation of Gla residues requires vitamin K, and inhibition of this process by dicoumarol or warfarin produces inhibition of blood clotting as well as abnormal bone formation (Table II).

The third class of extracellular calcium binding proteins are the salivary acidic proline-rich proteins. The major proteins have been named salivary proteins A and C. These proteins bind to and inhibit the formation of hydroxylapatite and it has been suggested that they function to aid in the maintenance of the integrity of the teeth. Calcium is bound by the salivary proteins by a series of negatively charged phosphoserine residues. Treatment with phosphatases to selectively remove the phosphoryl-moieties from phosphoserine dramatically reduces calcium binding affinity.

4.2. Membranous calcium binding proteins

The first class of membranous calcium binding proteins are represented by those proteins involved in the generation and modulation of the calcium signal (Table III). The calcium signal is regulated by the uptake and release of calcium across the three major membrane systems which bound the cytoplasm (the plasma membrane, the

TABLE II
Extracellular calcium binding proteins

Protein	M_r	$pK_d(Ca^{2+})$	nmol/mol	Ca^{2+} requirement	Refs.
Trypsin	24 000	3.2	1	Stability	18
Thermolysin	34 000	6.0 (2)	4	Activity & stability	19
		4.7 (2)			
Phospholipase A_2	14 000	3.0	1	Activity	20
DNAase I	31 000	4.8	2	Activity	21
α-Amylase	50 000	6.0	1	Activity	22
Osteocalcin	5 500	3.0	3	Structure & binding	23
Prothrombin	70 000	3.0	6–15	Binding	24
Factor X	55 000	3.5	12	Binding	24
Factor IX	55 000	4.0	2	Binding	24
		3.0	10		
Salivary protein A	11 000	4.3	2	Binding	25,26
		3.0	4		
Salivary protein C	15 000	4.4	1	Binding	25,26
		3.0	4		

inner mitochondrial membrane and the endoplasmic reticulum).

The observation that mitochondria can take up extremely large amounts of calcium in an energy dependent manner has for many years focused attention on these organelles as the major intracellular store controlling cytosolic calcium (with the exception of muscle where the sarcoplasmic reticulum was considered to be the key organelle).

The inner membrane of the mitochondria contains two important calcium binding proteins; a calcium importer that moves calcium from the cytoplasm to the mi-

TABLE III
Membraneous calcium binding proteins

Protein	Organelle	Tissue	$K_m(Ca^{2+})$ μM	V_{max} of transport nmol Ca^{2+}/mg protein
Ca^{2+}-ATPase[a]	Plasma membrane	Heart	0.5	0.5
Na^+/Ca^{2+} exchanger[a]	Plasma membrane	Heart	2–20	15–30
Ca^{2+}-ATPase[a]	Sarcoplasmic reticulum	Heart	0.1-0.5	20–30
Na^+/Ca^{2+} exchanger[a]	Mitochondria	Heart	13	0.2–0.3
Ca^{2+} uniporter[a]	Mitochondria	Heart	15–30	< 0.5
Cadherin-E[b]	Plasma membrane	Epithelium	–	–
Cadherin-N[b]	Plasma membrane	Neuronal and muscle	–	–
Thrombospondin[c]	Plasma membrane	Platelets, etc.	100 μM	12 mol Ca^{2+}/mol

[a]Adapted from reference 36; [b]adapted from reference 41,43; [c]adapted from reference 42

tochondrial matrix and whose conductance rate increases as the cube of the cytosolic free calcium, and a $Ca^{2+}/3Na^+$ antiporter which transports calcium from the mitochondrial matrix to the cytoplasm. The activity of this efflux pathway is dependent on the amount of calcium stored in the mitochondria, the mitochondrial matrix load. At matrix loads <10 nmol Ca^{2+}/mg, the activity of the efflux pathway increases in proportion to the mitochondrial matrix free calcium concentration [27,28]. Under these conditions, the influx and efflux mechanisms do not buffer cytosolic calcium but rather serve to regulate the mitochondrial matrix free calcium. This means that increases in the cytosolic free calcium will stimulate the influx pathway and as mitochondrial matrix free calcium increases the activity of the efflux mechanism will increase. Eventually a new mitochondria matrix free calcium concentration will be reached where influx equals efflux.

The ability of the mitochondria to buffer matrix free calcium is important because of the presence of three calcium regulated enzymes, pyruvate dehydrogenase, NAD-linked isocitrate dehydrogenase and 2-oxoglutarate dehydrogenase [29–31]. Half-maximal activation of these enzymes occurs from 0.4 to 0.5 μM calcium. At higher mitochondrial matrix calcium level (>10 nmol/mg) the calcium efflux pathway is thought to become saturated and therefore no longer dependent on mitochondrial matrix free calcium. Under these conditions, and at about 1 μM cytosolic free calcium, efflux and influx are balanced. This suggests that the mitochondria may buffer the upper end of the calcium signal, i.e., when the calcium signal reaches > 1 μM the mitochondria will respond, with net uptake of cytosolic calcium and, therefore, resist any further increase in cytosolic free calcium. The reports that liver or heart mitochondria, in situ, may contain only about 1 nmol Ca^{2+}/mg mitochondrial protein [32] has suggested that mitochondria may not be important regulators of cytosolic calcium. It has been suggested, however, that hormones or other agents might regulate either the influx or efflux pathway of the mitochondria and thereby modulate the calcium signal [33].

Electron-probing of fast frozen tissue section [32] has shown that the endoplasmic reticulum contains most of the intracellular stores of calcium. The general consensus reached only in recent years has suggested that the endoplasmic reticular system is the primary intracellular controller of cytosolic free calcium, and it is from the endoplasmic reticulum that the calcium signal originates. Two important membranous calcium proteins participate in the handling of cytoplasmic calcium by the endoplasmic reticulum, an electrogenic, ATP dependent calcium pump (moving calcium into the lumen of the endoplasmic reticulum) and a calcium channel (releasing calcium into the cytoplasm) which in many cells is activated by IP_3 [34]. The mechanism by which occupancy of this receptor leads to the release of calcium is largely unknown. Of the calcium stored in the endoplasmic reticulum only about 50% is susceptible to release in response to IP_3. While the endoplasmic reticulum is the central organelle for the IP_3 dependent release of calcium into the cytoplasm, this organelle also functions to terminate the calcium signal. Experiments with permeabilized cells (the plasma membrane of the cell is made 'leaky' by detergents or electrical shocks) have suggested that the endoplasmic reticulum will buffer the cytoplasmic calcium to a resting level of 0.1–0.2 μM [35].

The plasma membrane possesses three membranous calcium binding proteins which are involved in the modulation of the calcium signal; a calcium pump activated by calmodulin (Section 4.3.1) which uses the energy of ATP to remove calcium from the cytoplasm to the extracellular space, an electrogenic Na^+/Ca^{2+} exchanger which is also believed to extrude calcium from the cell and one or more calcium binding proteins that serve as membranous calcium channels and allow calcium influx into the cytoplasm (the calcium channels may be opened by membrane depolarization, i.e., depolarization operated calcium channels or DOC, or by hormonal activation, i.e., receptor operated calcium channels or ROC). It has been suggested that the calcium pump and Na^+/Ca^{2+} exchanger are synergistic in extruding calcium from the cell. The Na^+/Ca^{2+} exchanger possessing a low affinity for calcium but a high rate of calcium transport might be primarily responsible for the initial lowering of elevated cytosolic free calcium. In contrast, the calcium pump with a higher affinity for calcium but lower rate of transport might be more effective at lower cytosolic free calcium [36]. In heart cells it has been suggested that the plasma membrane handles only a minor portion of the total calcium that participates in the functional cycle of the cell [37]. In most cells the plasma membrane appears to be responsible for the receptor linked breakdown of PIP_2 to IP_3 which provides the intracellular signal for mobilization of calcium from the endoplasmic reticulum. As shown in Fig. 2 the agonist induced calcium transient consists of two components, a rapid component which consists of a rapid rise (phase I) and fall of intracellular free calcium to moderately elevated levels and a prolonged phase of moderately elevated intracellular free calcium (phase II). Extracellular calcium appears to be important for the refilling of endoplasmic reticular calcium stores (which are responsible for generation of IP_3 induced rapid component of the calcium transient), and for the maintenance of the prolonged phase of the calcium transient (in the absence of extracellular calcium this component is not observed). Current experimental evidence [37–39] suggests that two separate types of calcium channels may be involved in the generation of the calcium transient; an agonist insensitive calcium channel which allows the refilling of the endoplasmic reticulum by extracellular calcium and an agonist activated calcium channel (ROC) which is responsible for the prolonged phase of the calcium transient. In addition to these two types of calcium channels many cells (e.g., heart, skeletal and smooth muscle) may utilize DOC channels. As the name suggests these channels open transiently when the membrane is depolarized. In the case of heart muscle it is thought that calcium influx through DOC is responsible for the release of calcium from the sarcoplasmic reticulum, and contraction in cardiac muscle is dependent on both calcium influx through the DOC as well as from calcium released from the sarcoplasmic reticulum. In contrast, aortic smooth muscles appear to have both DOC and ROC [40].

It is therefore apparent that the generation of the calcium transient involves the participation of calcium influx, calcium efflux and IP_3 generated release of endoplasmic reticular calcium (Figs 1,2).

The second class of membranous calcium binding proteins (Table III) are the calcium dependent cell-cell adhesion molecules [41–43]. Cell adhesion molecules were first identified by means of immunologically based adhesion assays in which specific antibodies capable of blocking cell adhesion in vivo were used to purify cell surface molecules as cell adhesion molecules. Later these proteins were shown to be involved in calcium dependent or calcium independent cell adhesion. Two homologous cell surface intrinsic membrane glycoproteins, called cadherins, have been suggested to be involved in calcium dependent adhesion. They have been detected in most mammalian and avian tissues, and have been divided into two subclasses. The E type of cadherin (M_r 124 000) is distributed exclusively in epithelial cells of various tissues while the N type of cadherin (M_r 127 000) is distributed in neuronal and muscle tissue. Interestingly, cells which express E-cadherin do not cross-adhere with cells expressing N-cadherin.

A 420 000 Da calcium binding glycoprotein called thrombospondin [42] has also been identified as a member of the adhesive proteins that mediate cell-to-cell and cell-to-matrix interaction. Physical studies have suggested that this protein binds 12 mol Ca^{2+}/mol protein with a $K_d(Ca^{2+})$ of 100 μM. Unlike the cadherins this protein appears to be secreted from a variety of cells and then becomes associated with the plasma membrane. Thrombospondin possesses binding domains for fibrinogen, fibronectin, laminin and type V collagen. Sequence studies have suggested that the carboxy terminal of the molecule possesses a cell surface binding site.

4.3. Intracellular calcium binding proteins

The main function of the intracellular calcium binding proteins is to modulate cellular events in response to the calcium signal. Analysis of the sequence of many of the intracellular calcium binding proteins has suggested the existence of two distinct families; the 'EF domain' family and the annexin fold family. For completeness, we have grouped the remainder of the intracellular calcium binding proteins into a miscellaneous category.

4.3.1. The 'EF' domain family
In 1973, R.H. Kretsinger [58] and his colleagues reported the first crystal structure for an intracellular calcium binding protein, parvalbumin. This protein was shown to consist of six homologous regions of α-helix (A through F) with a non-helical N terminus region and five loops of β turns between the six regions of α-helix (Fig. 3). One atom of calcium was shown to be coordinated by six amino acids of the twelve amino acid residue long loop, between helix C and helix D (called the CD loop) and helix E and helix F (the EF loop). In contrast to the EF loop the AB loop

of parvalbumin does not bind calcium. The calcium binding unit consisting of helix E, EF loop and helix F is referred to as the 'EF' domain. The homology between the three EF domains of parvalbumin suggested that these domains evolved from a primordial gene.

Subsequently, amino acid sequence and crystallographic data has suggested that a variety of calcium binding proteins consist of homologous EF domains (Table IV). By gene duplication and deletion the separate genes for the six-domain calbindin-28K, the four domain calmodulin and troponin C, the three domain parvalbumin and oncomodulin and the two-domain S-100 are thought to have arisen. In several cases (for example, the AB loop of parvalbumin) the EF loops have lost their ability to bind calcium. Furthermore, the specificity of the EF loops for calcium has changed during evolution. Some EF loops have been shown to bind calcium specifically (called Ca^{2+} sites) while other EF loops bind calcium and magnesium competitively (called Ca^{2+}/Mg^{2+} sites). It has been suggested that the Ca^{2+}/Mg^{2+} sites of parvalbumin, troponin C and myosin light chain are occupied by Mg^{2+} under physiological conditions in the resting muscle cell. This means that before calcium can bind to these sites magnesium must first dissociate. The dissociation of magnesium is very slow compared to that of calcium and as a result calculations have suggested that the percent saturation of the Ca^{2+}/Mg^{2+} sites with calcium changes very little in response to the calcium transient [59]. This result has suggested that Ca^{2+}/Mg^{2+} sites may not respond to the calcium transient under physiological conditions. It is also interesting to note that the interaction of some intracellular calcium binding proteins with their target protein may alter their affinity for calcium. For example, the binding of calmodulin to the calcium pump reduces the $K_d(Ca^{2+})$ of calmodulin from 8 μM to 0.1 μM. Analysis of the calcium binding properties of free calmodulin and the calmodulin-troponin I complex have suggested that the interaction of calmodulin with troponin I results in an increase in the affinity of calcium binding but not of the total amount of calcium bound [60].

As shown in Table III, the function of many of the EF domain calcium binding proteins is at present unknown. Only troponin C and calmodulin have established functions: troponin C is involved in the calcium dependent regulation of skeletal and cardiac muscle contraction (Section 5.1), while calmodulin has been shown to regulate a wide variety of proteins and enzymes (Table V and Section 5). One of the most striking aspects of calmodulin is its ability to interact with a wide variety of target enzymes. The purpose of this interaction is to confer calcium sensitivity to an enzyme which would otherwise not be calcium sensitive. It has become clear that the mechanism of action of calmodulin correlates with the calcium-induced exposure of hydrophobic patches on its surface [61], and these hydrophobic domains are thought to represent recognition domains for target enzymes. Recent data [62] has suggested that the interaction of calmodulin with myosin light chain kinase may involve a different functional domain than the binding domain between calmodulin and calcineurin, calmodulin-dependent multifunctional protein kinase and phosphorylase kinase.

TABLE IV
The EF domain family

Protein	M_r (kDa)	IpH	[b,c]App·K_d (Ca^{2+}) (μM)	n(Ca^{2+}) (mol/mol)	Number of EF domains	Major tissue	Function	Ref.
Parvalbumin	12	$4.2 \to 5.0$[d]	$0.1 \to 0.4$[d]	2	3	White muscle	Unknown	44–46
Skeletal troponin C	18	4.2	0.1	4	4	Skeletal muscle	Regulation of muscle contraction	47–48
Cardiac troponin C	18	4.2	0.1	3	4	Cardiac muscle	Regulation of muscle contraction	47–48
S-100a	10	4.2	500	2	2	Glial cells	Unknown	
S-100b	10	4.4	500	2	2	Glial cells	Unknown	
Calmodulin	17	4.2	7.0	4	4	Widely distributed	Multifunctional regulatory protein	17,50, 51
Calcineurin B	19	4.8	$1 \to 5.0$	4	4	Brain	Protein phosphatase	52
Oncomodulin	12	3.9	≈1.0	2	2	Hepatoma	Unknown	53
Sarcoplasmic calcium binding protein	20	$4.8 \to 5.0$[d]	0.15	3	4	Invertebrate muscle	Unknown	46,54
Calbindin-10k	9	4.8	0.7	2	2	Duodenum	Unknown	55
Calbindin-28k	28	4.2	0.5	4	6	Cerebellum	Unknown	56
Calpain light chain	30	5.8	nd	nd	4	Widely distributed	Unknown	57
Sorcin	22	5.7	nd	nd	4	Drug resistant cells	Drug resistance (?)	58
Myosin light chain	18.5	4.4	30	1	4	Muscle	Myosin regulation	47

[a]Derived from SDS-PAGE; [b]determined in the presence of 150 mM KCl and 3 mM MgCl$_2$; [c]where appropriate, K_d values are for the oligomer; [d]represents species variation. nd, not determined

TABLE V
Calmodulin regulated enzymes

Cyclic nucleotide phosphodiesterase
Adenylate cyclase
Multifunctional calmodulin-dependent protein kinases
Myosin light chain kinase
Calcineurin
Phosphorylase kinase
NAD kinase
Calcium-transport ATPase

4.3.2. The annexin-fold family

Over the last few years several laboratories interested in the role of calcium in secretion and in the control of the membrane cytoskeleton have identified calcium binding proteins which appear to form a group distinct from the EF domain family of calcium binding proteins (Table VI). That a calcium binding protein might be involved in secretion was first suggested by Creutz [77] who purified an adrenal medullary protein, called synexin, that caused calcium-dependent aggregation of isolated chromaffin granules. Based on this work, Creutz suggested that synexin could be an intracellular receptor for calcium in the process of exocytosis, acting to promote close association of granules both with other granules as well as with plasma membranes prior to secretion. As a result of this discovery, several laboratories set out to identify the proteins which could associate in a calcium dependent manner with membranes. These experiments were accomplished by preparation of membrane fractions and analysis of the cytosol for proteins which bind in the presence of calcium and are released in the absence of calcium (in the presence of excess EGTA). The membrane preparations used for these experiments have included chromaffin granule membranes [79–80] (the proteins that bound in a calcium dependent manner were called chromobindins) and electric organ membranes from the ray *Torpedo marmorata* [81] (the single calcium binding protein isolated was called calelectrin); other preparations used include detergent extracted, insoluble, actin-rich cytoskeletons of lymphocytes [73], intestinal epithelial cells [66] (the major membrane associated proteins were called proteins I–III), smooth muscle membranes [82], liver [72] (one of the membrane associated proteins was named endonexin) and bovine lung membranes [70]. Independently, Dedman [74,88] used a phenothiazine Sepharose column (a hydrophobic resin) to purify several proteins from smooth muscle homogenates that bound to this column in a calcium dependent manner. Recent studies have confirmed that the various experimental approaches discussed above have yielded a common set of calcium binding proteins. Typically, calcium binding proteins of M_r 68 000, 36 000, 35 000, 34 000 and 32 500 have been reported. The main confusion in the literature has been over the relationship of the M_r 36 000 and 35 000 proteins both to each other and to other pre-

TABLE VI
The annexin fold family

Name	$M_r^a \times 10^3$	IpH	K_d Ca^{2+} μM	n(Ca^{2+}) mol/mol	Function	Refs.
Calpactin II Lipocortin I	38.0	6.9	10.0^b	2.0	Actin binding, phospholipase A$_2$ inhibitor	63–65
Protein I Calpactin I Lipocortin-85	α_2 36.0 β_2 11.2	7.4	5.0^b	4.0	Actin binding, phospholipase A$_2$ inhibitor	66–69
Calpactin I Lipocortin II	36.0	7.6	5.0^b	2.0	Actin binding, phospholipase A$_2$ inhibitor	66,68,70,71
Protein III p68	68.0	5.8	1.5	1.0	Unknown	72–74
Protein II Endonexin	32.5	5.5	25.0	1.0	Unknown	75,66,74
Calelectrin	34.0	5.5	50.0	nd	Unknown	76,74
Synexin I	47.0	6.3	200	nd	Unknown	77
Synexin II	56.0	6.4	nd	nd	Unknown	78

^{a}Determined by SDS-PAGE; ^{b}binding reported in the presence of phosphatidylserine: nd, not determined.

viously described proteins. Recently, Gerke and Weber [66] have identified one of their intestinal calcium binding proteins, protein I (composed of two 36 kDa subunits and two 10 kDa subunits), as p36, the major substrate of the protein tyrosine kinase of Rous sarcoma virus pp60 v-src [84–85]. Recent work has suggested that many cells contain both the oligomeric and monomeric forms of the M_r 36 000 protein. Huang et al. [86] demonstrated the existence in human placenta of two distinct M_r 36 000–35 000 proteins that were potent inhibitors of phospholipase A$_2$. Peptide and sequence analysis identified one of these proteins as p35, a protein originally reported by Fava and Cohen [63] as a major in vivo substrate of the epidermal growth factor receptor protein tyrosine kinase. This protein was also shown to be identical to lipocortin, a protein identified by a variety of laboratories to be involved in the anti-inflammatory effects of steroids (reviewed in Ref. 87). The second phospholipase A$_2$ inhibitory protein identified by Huang [86] was identified as p36, the major cellular substrate of pp60src. Analysis of the complete sequence of p35 and p36 has shown that these proteins are about 50% homologous, and each protein contains four regions of internal homology, each about seventy amino acids long. Within each of the four homologous regions a highly conserved region of seventeen amino acids was identified. Based on the partial sequence analysis of p36, p35, and the M_r 34 000, 32 000 and 68 000 proteins, Giesow [83] has reported that these proteins all contain a 17 amino-acid consensus sequence which was conserved and present in multiple copies. This consensus sequence has been named the annexin fold.

The exact function of the annexin fold family is at present unclear. All of these proteins appear to show calcium-dependent binding to phosphatidylethanolamine or phosphatidylinositol liposomes. In addition, they can promote fusion of liposomes, and because of this property, it has been suggested that these proteins might mediate calcium dependent exocytosis. P36 and p35 have also been shown to bind to F-actin and spectrin [65,66]. Recently, Khanna et al. [70] have reported a procedure for the simultaneous purification of p35, p36 oligomer and p36 monomer from bovine lung, and identified all three proteins as substrates of protein kinase C. Furthermore, the work of Huang et al. [86] and Khanna et al. [69] has suggested that all three proteins are inhibitors of phospholipase A_2. Further experiments will be required to clarify the function of these proteins.

4.3.3. Miscellaneous calcium binding proteins

A unique approach to the identification of intracellular calcium binding proteins has been reported by Waisman and co-workers [89–94]. These investigations have used the chelex-100 competitive calcium binding assay to identify the complete spectrum of intracellular high affinity calcium binding proteins. Their procedure involved the chromatography of the $100\,000 \times g$ supernatants of bovine brain, heart and liver on ion exchange columns (DEAE-cellulose) followed by analysis of the resultant fractions for calcium binding activity.

The peaks of calcium binding activity eluted from the ion exchange columns were subjected to further purification and the calcium binding proteins responsible for the calcium binding activity peaks have been purified and characterized. This approach has identified several novel calcium binding proteins of M_r 63 000, 48 000 and 27 000 (Table VII). These proteins have been shown to be present in tissues at high concentrations (> 100 mg/kg tissue) and to bind calcium under physiological conditions. The function of these proteins is currently unknown.

TABLE VII
Miscellaneous calcium binding proteins

Name	$M_r \times 10^{-3}$	IpH	K_d (Ca^{2+}) μM	n(Ca^{2+}) mol/mol	Function	References
Calregulin	63.0	4.65	0.05	1	Unknown	89–91
CAB-27	27.0	4.8	0.2	2	Unknown	92–94
CAB-48	48.0	4.7	15.0	1	Unknown	93,94
Villin	95.0	nd	5.0	3	Actin-binding	95
Profilin	15.0	9.1	0.1		Actin-binding	96
Gelsolin	90.0	6.1	1.0	2	Actin-binding	97
Calsequestrin	65.0	nd	1300	50	Calcium sequestration	98

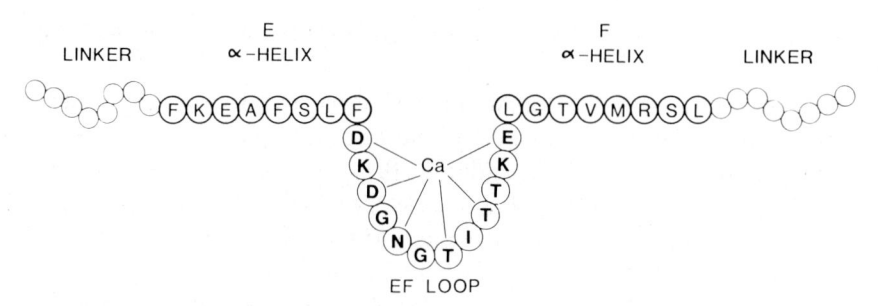

Fig. 3. The EF domain. The EF primordial gene is regarded as consisting of coding sequences for one domain: an α-helix of about eight amino acids (E), a calcium binding loop of about twelve amino acids (EF loop) and another region of α-helix (F). The EF domains are joined by linker sequences. Calcium coordination is accomplished by oxygen atoms of six amino acids of EF loop. Shown in this figure is the structure of the EF domain I of calmodulin.

Another approach to the study of intracellular calcium binding proteins has involved the analysis of the cytosol for proteins which demonstrate a calcium dependent interaction with actin [97]. In most non-muscle cells, the motile system, composed of actin and myosin, does not make any permanent structure but displays repeated cycles of assembly and disassembly. Among the various proteins that bind to actin, one group has been isolated from non-muscle source that acts specifically on the polymer state of actin in a calcium sensitive manner. Calcium binding proteins of this group include gelsolin from macrophages and villin from the microvilli of the intestinal brush border. Profilin, isolated from a variety of cells, predominantly binds in a calcium dependent manner to a single actin monomer, and impairs the ability of actin to assemble into filaments.

The last calcium binding protein to be discussed in the miscellaneous category is calsequestrin. This protein was isolated from the lumen of the sarcoplasmic reticulum and because of the high binding capacity, has been suggested to be a major site of calcium binding in the interior of the sarcoplasmic reticulum.

5. Calcium binding proteins and cellular function

There are a large number of cellular functions which have been postulated or proven to be dependent on, or regulated by, calcium. For example, contraction of muscle myofibrils, events dependent on microtubules and/or microfilaments (cell division, motility of cilia and flagella), activation of the egg at fertilization, cell adhesion, cell communication via gap junctions, cell proliferation and differentiation have all been suggested to be activated as a consequence of changes in intracellular free calcium. Calcium is also involved in many secretory mechanisms, such as intracellular transport, discharge of neurotransmitters from nerve endings, release of histamine from mast cells, secretion of hormones from endocrine glands, release of proteins and mucus from exocrine glands and regulation of electrolyte and water transport by secretory and absorptive epithelia. Calcium also has a profound effect on cellular metabolism as illustrated by the large number of enzymes that require calcium for their activity. In the following section, we will discuss some of these calcium regulated events and where appropriate, the role of calcium binding proteins in these processes.

5.1. Muscle contraction

Calcium is known to be an important key regulator for contractile activities of both skeletal and smooth muscles. There are two general mechanisms of regulation of muscle contraction: actin based and myosin based.

5.1.1. Actin based regulation (skeletal and cardiac muscles)

Calcium regulates muscle contraction by an allosteric mechanism which is mediated by troponin, tropomyosin and F-actin system. Troponin is composed of three components, troponin I, troponin T and the calcium binding component, troponin C.

In skeletal muscle in the relaxed state, the sarcoplasm has a high $MgATP^{2-}$-concentration, but the concentration of calcium is below the threshold required for initiation of contraction. The myosin head, under resting conditions is unable to react with actin of the thin filaments because in the absence of calcium the tropomyosin molecule masks the myosin binding site on G-actin monomer or holds it in a conformation that is unreactive, through the action of TN-1 subunit of troponin. One tropomyosin molecule inhibits the myosin binding activity of seven G-actin monomers.

When free calcium is released into the sarcoplasm by the incoming nerve signal, calcium is immediately bound to the calcium binding site of troponin, troponin C. Binding of calcium by troponin C results in a conformational change in the troponin complex. The myosin binding site on the G-actin monomer becomes exposed as a consequence of this conformational change and combines with energized myosin head to form the force generating complex. The myosin head is now believed to undergo an energy yielding conformational change, so that the cross bridge changes its angular relationship to the axis of the heavy filaments, causing the thin filaments to be moved along the thick filaments and therefore producing contraction.

Relaxation of the muscle is brought about by removal of the ionic calcium from the sarcoplasm. This calcium is transported across the membrane of the sarcoplasmic reticulum, in an energy requiring process. In addition to the calcium pumping ATPase, the sarcoplasmic reticulum also contains a calcium binding protein called calsequestrin (Section 4.3.3). Some of the calcium segregated by the sarcoplasmic reticulum is apparently bound to this protein within the lumen of the sarcoplasmic reticulum. As sequestration of calcium ions into sarcoplasmic reticulum proceeds, more calcium ions dissociate from their binding sites on troponin C, resulting in a restoration of actomyosin-inhibitory property of troponin I and, accordingly, a return of the thick and thin filaments to the resting state (relaxation).

5.1.2. Myosin based regulation (smooth muscle)

Smooth muscles are not regulated by the troponin system. Rather it is the phosphorylation of the myosin light chains which appears to be the event which activates smooth muscle contraction.

In the resting state of smooth muscle, intracellular free calcium (10–100 nM) is at too low a concentration to support binding to the intracellular receptor calmodulin. However after activation by an appropriate stimulus, the intracellular free calcium increases to a new concentration such that calcium binding to calmodulin is promoted. The resultant calcium-calmodulin complex interacts with the catalytic subunit of myosin light chain kinase (MLCK) converting it from an inactive to an active state. The activated kinase catalyses the phosphorylation of serine 19 on each of the two 20 000 Da light chains of myosin. In this phosphorylated state, actin-myosin interaction is promoted and the muscle contracts presumably by a cross-bridge cycle mechanism analogous to that of skeletal muscle [99]. Thus, the phosphorylation of the myosin light chains (a calcium-calmodulin dependent phenomenon) commences the attachment-detachment contraction cycle of smooth muscle.

Relaxation follows the drop in cytosolic calcium below the activation threshold, which initiates release of calcium from calmodulin, whereupon calmodulin dissociates from the kinase catalytic subunit. This results in loss of myosin light chain kinase activity. Under these conditions, the dephosphorylation of myosin light chains will be the predominant reaction. Consequently, actin and myosin will no longer interact and the muscle relaxes.

More recently, considerable evidence has accumulated, which has shown a correlation between myosin light chain phosphorylation and smooth muscle contraction in vivo. However, there are many reports which suggest a weak correlation between these two events. For example, it has been shown that after prolonged stimulation smooth muscle can maintain steady-state tension, even though myosin light chain phosphorylation had decreased to low levels [100]. Maintenance of tension at low myosin phosphorylation is not consistent within the phosphorylation theory and thus alternate regulatory mechanisms have been implicated [101]. Some evidence suggests that once myosin phosphorylation is achieved, it is possible to dephosphorylate the myosin light chains in the presence of calcium and to maintain force development by activation of another calcium control mechanism.

Apart from the phosphorylation theory, other regulatory mechanisms have also been suggested for smooth muscle contraction. A thin-filament protein that has been proposed as a regulatory component is caldesmon [102]. Purified caldesmon is a potent inhibitor of actin-tropomyosin interaction with myosin. The mechanisms by which calcium removes this inhibition are controversial. Furthermore, phosphorylation of caldesmon by a caldesmon kinase in vitro has also been implicated in this regulatory process [103]. However, extent of caldesmon phosphorylation in vivo is unknown, and it has not yet been directly demonstrated that phosphorylation of caldesmon can alter its regulatory properties [104].

5.2. Metabolism

The regulatory role of calcium ions in intermediary metabolism is well documented. Calcium has been shown to be involved in activation or inhibition of specific enzyme systems [105]. For example, it activates cyclic nucleotide phosphodiesterase, phosphofructokinase, fructose 1:6 biphosphatase, glycerol phosphate dehydrogenase, pyruvate dehydrogenase phosphatase and pyruvate dehydrogenase kinase. Calcium ions inhibit pyruvate kinase, pyruvate carboxylase, Na^+/K^+-ATPase and adenylate cyclase.

Phosphorylase kinase is one of the best characterized enzyme systems to illustrate the role of calcium ions in regulation of intermediary metabolism. Phosphorylase kinase is composed of four different subunits termed α (M_r 145000), β (M_r 128000), γ (M_r 45000) and δ (M_r 17000) and has the structure $(\alpha\beta\gamma\delta)_4$ [106]. Only one of its four subunits actually catalyses the phosphorylation reaction: the other three subunits are regulatory and enable the enzyme complex to be activated both by calcium and cyclic AMP. The γ subunit carries the catalytic activity; the δ subunit is the calcium binding protein calmodulin and is responsible for the calcium dependence of the enzyme. The α and β subunits are the targets for cyclic-AMP mediated regulation, both being phosphorylated by the cyclic-AMP dependent protein kinase. Calmodulin appears to interact with phosphorylase kinase in a different manner from other enzymes, since it is an integral component of the enzyme. Phosphorylase kinase has an absolute requirement for calcium, and is inactive in its absence.

In the presence of calcium, the enzyme can interact with either a second molecule of calmodulin (termed the δ'-subunit) or with skeletal muscle troponin C leading to further enhancement of the activity. It appears that regulation of different forms of phosphorylase kinase in vivo is achieved through the interaction of the enzyme with these two calcium binding proteins [106]. Troponin C is the dominant calcium-dependent regulator of the dephosphorylated form of the enzyme (phosphorylase kinase b) at calcium ion concentrations in the micromolar range, i.e., below 3 μM the activity of phosphorylase kinase b was almost completely dependent on troponin C. The effect is specific for skeletal muscle troponin C as cardiac troponin C or parvalbumin were not effective activators of the enzyme. In the absence of troponin C, 23 μM calcium is required for half maximal stimulation of the enzyme. Thus troponin C rather than calmodulin is the physiological activator of phosphorylase kinase b – providing an attractive mechanism for coupling glycogenolysis and muscle contraction, since the same calcium binding protein (troponin C) activates both processes. In contrast the phosphorylated form of the enzyme (phosphorylase kinase *a*) is only slightly activated by troponin C. Phosphorylase kinase *a* is 15-fold more active than the *b*-form. In addition, its requirement for calcium is also 15-fold lower than the dephosphorylated form. Calmodulin is therefore the dominant calcium-dependent regulator of phosphorylase kinase in its phosphorylated form, i.e., in its hormonally, activated state.

Since phosphorylase kinase not only activates phosphorylase, but also phosphorylates glycogen synthase thereby decreasing its activity, the regulation of phosphorylase kinase by calcium may also provide a mechanism for co-ordinating the rates of glycogenolysis and glycogen synthesis during muscle contraction.

5.3. Secretion and exocytosis

Calcium is an important intracellular messenger in stimulus-secretion coupling. It has been shown to be required for neurotransmitter release, endocrine and exocrine secretion, mast cell degranulation and activation of ionic mechanisms that regulate fluid secretion [107,108]. Secretion and exocytosis depends on functioning microfilament and microtubule systems to provide the motile force and pathway for exocytosis. Calcium ions play a central role in the control of regulated pathway of the exocytotic secretion, as a majority of secretory cells show a dependency on extracellular calcium. Most secretagogues either directly or indirectly raise cytosolic free calcium levels which in turn affects granule biogenesis, contractile events, gel/sol transition in intracellular matrix and membrane fusion events occurring during exocytosis. A number of instances of calcium independent secretion have been reported, where exocytosis can be activated without an increase in calcium, or even where a decrease in cytosolic free calcium occurs [109–111]. However, in many of these cases calcium probably remains an obligatory component of the effector system which is under modulation by another second messenger system. Cyclic nucleotides, phosphatidylinositol metabolites, intracellular pH and arachidonic acid metabolites interact with calcium signal by affecting one another's concentrations in cell and by acting both synergistically and antagonistically on effector systems coupled to these messengers [112–113]. As the system generating these signals are often activated sequentially and transiently, a complex relationship is seen between the secretory response and cytosolic calcium during sustained secretion.

There is indirect evidence to show the involvement of calmodulin in stimulus-secretion coupling. Calmodulin can activate myosin light chain kinase and influence microtubule [114]. The microtubule system consists of tubulin dimers in a polymerized-depolymerized equilibrium, and calcium-calmodulin promotes the depolymerization of microtubules. Following the activation of secretion microtubules possibly play a part in conferring directionality to the movement of the granules to their site of secretion. It has been postulated that microtubule associated proteins (MAPs) and granule associated proteins provide sites of anchorage of actin, and movement is generated by the myosin ATPase [115]. Actin or actin-like proteins readily associate with secretory granules and with microtubules via the MAPs. The latter interaction is possibly regulated by cAMP and calcium dependent phosphorylation of MAP [116].

Considerable attention has been paid to actin polymerization and calcium dependent gel-sol transitions as a regulatory component in the secretory process. One of the earliest proteins described in this context in mast cells and platelets is the calcium dependent actin binding protein, gelsolin (Section 4.3.3). This protein is widely distributed among secretory tissues [117]. In the presence of micromolar calcium, gelsolin induces a rapid gel-sol transition of actin filamentous network. Given that a number of the actin binding proteins found in association with isolated secretory granules can be regulated by calcium-calmodulin, it is not surprising that calmodulin has also been identified as a calcium dependent associable protein with a variety of secretory granule types [117].

There are many soluble proteins that bind to purified secretory vesicles or to other membranes in a calcium dependent manner in vitro [77] (Section 4.3.2). Using immobilized chromaffin granule membranes, a series of 22 cytosolic proteins termed chromobindins have been isolated [77]. They are further classified according to the dependence on calcium, ATP and phospholipid. Among these have been identified calmodulin, synexin, protein kinase C, a 37 kDa protein, kinase C substrate (chromobindin IX), caldesmon and synhybin. Similar experiments have localized the bovine calelectrins (67, 36 and 32.5 kDa) to chromaffin granules where their calcium and phospholipid-dependent association can, like synexin, result in granule aggregation in vivo [75,118]. The calelectrins appear to be a subset of chromobindins and overlap with a series of chicken gizzard peptides, the calcimedins, the phospholipase A_2 inhibitors, lipocortin I and II, or calpactins [66–71] (Section 4.3.2). The 67 kDa calelectrin may be identical to synhybin, 67 kDa calcimedins. The 36 kDa calelectrin appears to correspond to the 36 kDa subunit of calpactin I (p36). The 32.5 kDa calelectrin may be equivalent to chromobindin IV (endonexin). Another isotype of 35 kDa protein which is not in calelectrin series includes chromobindin IX, calpactin II or lipocortin I. A seventeen amino acid consensus sequence [83] deduced by Geisow et al. appears to be present in all these hydrophobic proteins and has been implicated for their calcium and phospholipid binding properties. Saris et al. [119] have speculated that all these proteins may represent only four distinct gene products.

The homology of p36 with lipocortin II and its ability to inhibit pancreatic phospholipase A_2 suggests a possible but unproven role as inhibitor of intracellular phospholipase A_2 [86]. Phospholipase A_2 activity provides one of the many means of arachidonate production which would have important consequences to the activation of secretory response and membrane fusion events. The observation that the 40 kDa protein kinase C substrate, whose phosphorylation state changes dramatically upon platelet activation, may partially cross-react with a polyclonal antibody raised to rat renal lipocortin, is therefore interesting [120]. An alternate suggestion is that the 40 kDa platelet protein is an inositol 1,4,5-trisphosphate 5′ phosphomonoesterase which is activated by phosphorylation by protein kinase C [121].

An important role for protein kinase C in coupling secretagogue stimulation to exocytosis in a wide variety of secretory cells has been suggested from investigation of the broad class of secretagogues whose activity is coupled to receptor-mediated phosphoinositide turnover [38]. A general model for these secretagogues envisages receptor coupling to phospholipase C through a novel GTP binding protein with receptor occupation resulting in diacylglycerol and inositol 1,4,5-triphosphate may then mobilize calcium from intracellular stores. Diacylglycerol may exert direct effects on the fusogenicity of the membranes in which it is generated and indirect effects through its action as an allosteric regulator of protein kinase C and as a precursor of arachidonic acid. Receptor occupancy may not be an essential component in eliciting the response of phosphoinoside turnover, since in sea urchin eggs phospholipase C activation, which is thought to precede the cortical granule release reaction, is either activated directly by the sperm or by a coupled calcium transient following fertilization [112].

5.4. Egg fertilization and maturation

There is considerable evidence that calcium acts as a primary trigger for egg maturation and fertilization in several phyla. Calcium regulation has been demonstrated or suggested for numerous specific events in fertilization, including sperm motility, the acrosome reaction, sperm-egg binding and fusion and metabolic activation of egg, etc. However, very little is known concerning the mechanisms whereby calcium exerts its effects.

One of the best studied systems concerning the role of calcium in egg fertilization is the sea urchin egg [123]. Upon fertilization, two ion fluxes occur in short prereplicative phase which are obligatory for the subsequent initiation of processes, including DNA synthesis [124]. The first is an internal surge of calcium ions at about 0–3 s post fertilization, and second is the efflux of hydrogen ion coupled to an influx of Na^+ resulting in the alkalization of egg cytoplasm [125].

Activation of eggs and subsequent DNA synthesis by artificially raising the internal calcium ion concentration has been accomplished by a variety of methods. Simply incubating eggs in high calcium sea water, or wounding the egg in sea water containing calcium stimulates egg division [126]. However, extracellular calcium does not appear necessary, since calcium ionophore A23187 can induce a transient internal calcium surge in calcium free sea water, and also cause a subsequent increase in DNA synthesis [125]. The prevention of this fertilization-linked intracellular calcium surge by injection of calcium chelators has been shown to block subsequent egg development [127]. Whether the calcium surge alone is an adequate signal for succeeding initiation of DNA synthesis remains unclear. The efflux of H^+ is another prerequisite for egg development [125] and whether calcium is linked to this efflux remains unresolved.

The cortical granule exocytosis, occurring after fertilization has been suggested to be responsible for the late block to polyspermy. This cortical reaction is mediated by a large rise in the concentration of free calcium in the cytosol. The importance of calcium in triggering the cortical reaction can be demonstrated directly in plasma membrane isolated from sea urchin eggs with cortical granules still attached to their cytoplasmic surfaces. When small amounts of calcium are added to such preparations exocytosis occurs within seconds. The protease and other enzymes released in cortical reaction alter the structure of the egg coat in such a way that additional sperm cannot penetrate. In mammalian eggs, the cortical reaction acts in a similar way to prevent polyspermy, causing a glycoprotein in the zona pellucida to become altered so that it can no longer bind sperm or activate them to undergo an acrosomal reaction [128].

While there is substantial evidence that calcium is involved in processes regulating in egg maturation and egg activation, very little evidence exists demonstrating the activity of specific calcium binding proteins in these events. Calmodulin has been shown to play a role in the activation of NAD kinase after fertilization of sea urchin eggs [129]. This kinase is one of the first enzymes activated after fertilization and may provide a link between rapid increase in egg calcium, activation of egg metabolism and DNA synthesis [129]. The possibility of calmodulin being involved in the initiation of DNA synthesis is strengthened by studies in other cells which have shown increases in this protein during liver regeneration [130] and the ability of a calmodulin antagonist to inhibit cell proliferation [131]. Studies using PKC agonists and antagonists have suggested the involvement of this pathway in egg activation [131], but such involvement has not been unambiguously demonstrated.

5.5. Cell growth and proliferation

Studies on cell proliferation in vivo has provided indirect evidence that calcium may be an important mediator of cell growth. Most of the data linking calcium to the regulation of cell growth have been obtained from studies of cells in culture, using defined serum free medium and purified growth factors [132]. Growth factors initiate their action by binding to specific cell-surface receptors. Immediate consequences of receptor activation include a transient rise in free cytosolic calcium, a sustained increase in pH and tyrosine specific protein phosphorylation [133]. The rapid rise in calcium signal is due either to release from internal stores, via the IP_3 pathway (Section 2) (platelet derived growth factor), or influx of external calcium via a voltage independent channel in plasma membrane (epidermal growth factor) [34].

Whatever the mechanism of growth factor induced transient rise in intracellular calcium (internal release or influx), the key question is the physiological role of calcium transient in the action of growth factors. Of particular relevance of this question is the recent finding that calcium ionophores mimic EGF and PDGF in rapidly inducing the expression of c-fos and c-myc proto-oncogenes in responsive cells [135]. Interestingly, calcium mobilizing neurohormones such as bradykinin and histamine also induce c-fos expression and cell division in their target cells [134]. More recently, it has been demonstrated that agents or conditions that affect voltage-dependent calcium channels, also influence the control of c-fos expression [136].

Calcium sensitive, phospholipid-dependent protein kinase C also plays a crucial role in cell growth and proliferation. The enzyme serves as a receptor for the tumor promoters and is involved in signal transduction for a variety of growth factors and mitogens. Platelet derived growth factor (PDGF) markedly activates protein kinase C in intact, quiescent Swiss 3T3 cells [137]. Further, the desensitization of the protein kinase C pathway by phorbol esters reduces the stimulation of DNA synthesis by low concentrations of PDGF in these cells [138]. Protein kinase C has also been implicated in transmodulation of epidermal growth factor (EGF) receptor. Addition of PDGF inhibits the binding of ^{125}I-labeled EGF to specific receptors in Swiss 3T3 cells. The modulation of EGF binding is rapid in onset and results from a decrease in the apparent affinity of EGF receptor population for EGF [139]. Since PDGF does not bind to EGF receptor [139], the decrease in the affinity of EGF receptors occurs through an indirect mechanism termed 'transmodulation' [140].

Considerable evidence implicates protein kinase C in the transmodulation of the EGF receptor. Phorbol esters [141] cause a rapid and striking decrease in the apparent affinity of this receptor without changing the total number of sites. The transmodulation induced by these agents is prevented by prolonged treatment of the cells with phorbol esters, which desensitizes the protein kinase C pathway [142]. Further, protein kinase C phosphorylates the EGF receptor of human epidermal carcinoma A431 cells at a specific threonine residue, located nine amino acids from the transmembrane domain of this receptor [143]. PDGF stimulates the phosphorylation of EGF receptor in human fibroblasts at an identical site [143]. Thus, the transmodulation of EGF receptors may result from the covalent modification of the EGF receptor catalysed by protein kinase C. Tyrosine specific protein kinases including the growth factor receptor tyrosine kinases and the retroviral tyrosine protein kinases have been shown to phosphorylate in vivo two related but distinct calcium binding proteins, p36 and p35 [144]. P36 is phosphorylated on tyrosine in cells transformed by many oncogenic retroviruses [145]. P35 has been identified as the major protein phosphorylated in membrane vesicles from A431 cells treated with EGF [145]. However, the role of p36 phosphorylation in oncogenic transformation or in growth-factor stimulated cell proliferation has not been clearly defined. Genetic and biochemical analyses of v-src mutants of Rous sarcoma virus (RSV) have provided evidence that phosphorylation of p36 is not sufficient to elicit morphological transformations [146,147], but that it may be necessary for tumorigenicity [148]. Taken together, these findings indicate that calcium could function as a second messenger that mediates either directly or indirectly (through calcium binding proteins), the early transcriptional effects of growth factors, and they provide a framework for future studies on the role of calcium binding proteins in growth control.

References

1. Heilbrunn, L.V. and Wiercenski, F.J. (1947) J. Cell Comp. Physiol. 29, 15.
2. Rasmussen, H. (1970) Science 170, 404.
3. Rasmussen, H. (1981) Calcium and c-AMP as synarchic messengers. John Wiley and Sons, New York.
4. Borle, A.B. (1981) Rev. Physiol. Biochem. Pharmacol. 90, 13–153.
5. Baker, P.F. (1986) Calcium and the Cell. (Ciba Foundation) Symposium 122, pp. 1–4. Wiley, Chichester.
6. Berridge, M.J. (1984) Biochem. J. 220, 345–360.
7. Irvine, R.F. (1986) Br. Med. Bull. 42, 369–374.
8. Nishizuka, Y. (1984) Nature (London) 308, 693–697.
9. Bolton, T.B. (1979) Physiol. Rev. 59, 606–658.
10. Curtis, B.M. and Catterall, W.A. (1983) J. Biol. Chem. 258, 7280–7283.
11. Reuter, H. (1986) Calcium and The Cell. (Ciba Foundation) Symposium 122, pp. 5–22. Wiley, Chichester.
12. Chiu, A.T., McCall, D.E. and Timmermans, P.M. (1986) Eur. J. Pharmacol. 127, 1–8.
13. Poggioli, J., Mauger, J.-P., Guesdon, F. and Claret, M. (1985) J. Biol. Chem. 260, 3289–3294.
14. Ebashi, S., Endo, M. and Ohtsuki, I. (1969) Q. Rev. Biophys. 2, 351.
15. Cheung, W.Y. (1970) Biochem. Biophys. Res. Commun. 38, 533.
16. Rasmussen, H. and Waisman, D.M. (1983) In: Reviews of Physiology, Biochemistry and Pharmacology. Volume 95, pp. 111–148. (Adrian, R.H. et al., eds.) Springer-Verlag, Berlin, New York.
17. Klee, C.B., Crouch, T.H. and Richman, P. (1980) Ann. Rev. Biochem. 49, 489.
18. Bode, W. and Schwager, P. (1975) FEBS Lett. 56, 139–143.
19. Fisher, E.H. and Stein, E.A. (1960) The Enzymes 4, 313–343.
20. Hanahan, D.J. (1971) The Enzymes, 5, 71–85.
21. Price, P.A. (1975) J. Biol. Chem. 250, 1981–1986.
22. Fisher, E.H. and Stein, E.A. (1960) The Enzymes 4, 313–343.
23. Hauschka, P.V. (1986) Haemostasis 16, 258–272.
24. Furie, B.C., Borowski, M., Keyt, B. and Furie, B. (1982) Calcium Cell Function 2, 217–238.
25. Bennick, A., McLaughlin, A.C., Grey, A.A. and Madapallimattam, (1981) J. Biol. Chem. 256, 4741–4746.
26. Braunlin, W.H., Vogel, H.J., Drakenberg, T., Bennick, A. (1986) Biochem. 25, 6712–6716.
27. Nicholls, D.G. (1986) Br. Med. Bull. 42, 353–358.
28. Hansford, R.G. (1985) Rev. Physiol. Biochem. Pharmacol. 102, 1–72.
29. Denton, R.M., McCormack, J.G. and Edgell, N.J. (1980) Biochem. J. 190, 107–117.
30. Denton, R.M., Randle, P.J. and Martin, B.R. (1972) Biochem. J. 128, 161–163.
31. Denton, R.M., Richards, D.A. and Chin, J.G. (1978) Biochem. J. 176, 899–906.
32. Somlyo, A.P., Bond, M. and Somlyo, A.V. (1985) Nature 314, 622–625.
33. Compton, M. (1985) Membrane Transport, Vol. 3, pp. 249–286. (Martonosi, A.N., ed.) Plenum Press, New York.
34. Muallem, S., Schoeffield, M., Pandol, S. and Sachs, G. (1985) Proc. Natl. Acad. Sci. U.S.A. 82, 4433–4437.
35. Streb, H. and Schultz, I. (1983) Am. J. Physiol. 245, G347–G357.

36. Carafoli, E. (1985) J. Mol. Cell. Cardiol. 17, 203–212.
37. Joseph, S.K., Coll, K.E., Thomas, A.P., Rubin, R. and Williamson, J.R. (1985) J. Biol. Chem. 260, 12508–12515.
38. Putney, J.W. (1987) Am. J. Physiol. 252, G149–G157.
39. Altin, J.G. and Bygrave, F.L. (1987) Biochem. J. 242, 43–50.
40. Chiu, A.T., McCall, D.E. and Timmermans, P.B. (1986) Eur. J. Pharmacol. 127, 1–8.
41. Obrink, B. (1986) Exp. Cell. Res. 163, 1–21.
42. Lawler, J. and Hynes, R.O. (1986) J. Cell. Biol. 103, 1635–1648.
43. Shirayoshi, Y., Hatta, K., Hosoda, M., Tsunasawa, S., Sakiyama, F. and Takeichi, M. (1986) EMBO J. 5, 2485–2488.
44. Moeschler, H.J., Schaer, J.-J. and Cox, J.A. (1980) FEBS Lett. 111, 73–78.
45. Heizmann, C.W. (1984) Experentia 40, 910–920.
46. Wnuk, W., Cox, J.A. and Stein, E.A. (1982) Calcium and Cell Function, Vol. II, pp. 243–273. (Cheung, W.Y., ed.) Academic Press, New York.
47. Holroyde, M.J., Potter, J.D. and Solaro, R.J. (1979) J. Biol. Chem. 254, 6478–6482.
48. Potter, J.D. and Gergely, J. (1975) J. Biol. Chem. 250, 4628–4633.
49. Baudier, J., Glasser, N. and Gerard, D. (1986) J. Biol. Chem. 261, 8192–8203.
50. Means, A.R., Tash, J.S. and Chafoulos, J.G. (1982) Physiol. Rev. 62, 1–39.
51. Klee, C.B. and Vanaman, T.C. (1982) Adv. Protein Chem. 35, 213–312.
52. Tallant, E.A. and Cheung, W.Y. (1986) Calcium Cell Function VI, 72–104.
53. MacManus, J.P., Szabo, A.G. and Williams, R.E. (1984) Biochem. J. 220, 261–268.
54. Takagi, T., Konishi, K. and Cox, J.A. (1986) Biochemistry 25, 3585–3592.
55. Bryant, D.T. and Andrews, P. (1984) Biochem. J. 219, 287–292.
56. Thomasset, M., Parkes, C.O. and Cuisinier-Gleizes, P. (1982) Am. J. Physiol. 243, E483–E488.
57. Sakihama, T., Kakidani, H., Zenita, K. and Murachi, T. (1985) Proc. Natl. Acad. Sci. U.S.A. 82, 6075–6079.
58. Kretsinger, R.H. and Nockolds, C.E. (1973) J. Biol. Chem. 248, 3313–3321.
59. Robertson, S.P., Johnson, J.D. and Potter, J.D. (1981) Biophys. J. 34, 559–569.
60. Olwin, B.B. and Storm, D.R. (1985) Biochemistry 24, 8081–8086.
61. LaPorte, D.C., Wierman, B.M. and Storm, D.R. (1980) Biochemistry 19, 3814–3819.
62. Putkey, J.A., Draetta, G.F., Slaughter, G.R., Klee, C.B., Cohen, P., Stull, J.T. and Means, A.R. (1986) 261, 9896–9903.
63. Fava, R.A. and Cohen, S. (1984) J. Biol. Chem. 259, 2635–3645.
64. De, B.K., Misono, K.S., Lukas, T.J., Mroczkowski, B. and Cohen, S. (1986) J. Biol. Chem. 261, 13784–13792.
65. Glenney, J.R., Tack, B. and Powell, M.A. (1987) J. Cell Biol. 104, 503–511.
66. Gerke, V. and Weber, K. (1985) EMBO J. 4, 2917–2920.
67. Glenney, J.R. (1985) FEBS Lett. 192, 79–82.
68. Glenney, J.R. (1986) J. Biol. Chem. 261, 7247–7252.
69. Khanna, N.C., Hee-Cheong, M., Severson, D.L., Tokuda, M., Chong, S.M. and Waisman, D.M. (1986) Biochem. Biophys. Res. Commun. 139, 455–460.
70. Khanna, N.C., Tokuda, M. and Waisman, D.M. (1986) Biochem. Biophys. Res. Commun. 141, 547–554.
71. Wallner, B.P., Matatliano, R.J., Hession, C., Cate, R.L., Tizard, R., Sinclair, L.K., Foeller, C., Chow, E.P., Browning, J.L., Ramachandran, K.L. and Pepinsky, R.B. (1986) Nature 320, 77–81.
72. Giesow, M.J., Childs, J., Dash, B., Harris, A., Panayotou, G., Sudhof, T. and Walker, J.H. (1984) EMBO J. 3, 2969–2974.
73. Owens, R.J., Gallagher, C.J. and Crumpton, M.J. (1984) EMBO J. 3, 945–952.
74. Moore, P.M. and Dedman, J.R. (1982) J. Biol. Chem. 257, 9663–9667.
75. Sudhof, T.C., Ebbecke, M., Walker, J.H., Fritsche, U. and Boustead, C. (1984) Biochemistry 23, 1103–1109.

76. Sudhof, T.C., Walker, J.H. and Fritsche, U. (1985) J. Neurochem. 44, 1302–1307.
77. Creutz, C.E., Pazles, C.J. and Pollard, H.B. (1978) J. Biol. Chem. 253, 2858–2866.
78. Odenwald, W.F. and Morris, S.J. (1983) Biochem. Biophys. Res. Commun. 112, 147–154.
79. Creutz, C.E. (1981) Biochem. Biophys. Res. Commun. 103, 1395–1400.
80. Geisow, M.J. and Burgoyne, R.D. (1982) J. Neurochem. 38, 1735–1741.
81. Walker, J.H. (1982) J. Neurochem. 39, 815–823.
82. Raeymaekers, L., Wuytack, F. and Casteels, R. (1985) Biochem. Biophys. Res. Commun. 132, 526–532.
83. Giesow, M.J., Fritsche, U., Hexam, J.M., Dash, B. and Johnson, T. (1986) Nature 320, 874–876.
84. Radke, K., Gilmore, T. and Martin, G.S. (1980) Cell 21, 821–828.
85. Erikson, E. and Erickson, R.L. (1980) Cell 21, 829–836.
86. Huang, K.S., Walner, B.P., Mattalioano, R.J., Tizzard, R., Burne, C., Frey, A., Hession, C., McGray, P., Sinclair, L.K., Chow, E.P., Browning, J.L., Ramachandran, K.L., Tang, J., Smart, J.E. and Pepinsky, R.B. (1986) Cell 46, 191–199.
87. Flower, R., Wood, J.N. and Parente, L. (1984) Adv. Inflammation Res. 7, 61–69.
88. Dedman, J.R. (1986) Cell Calcium 7, 297–307.
89. Waisman, D.M., Salimath, B. and Anderson, J. (1985) J. Biol. Chem. 260, 1652–1660.
90. Khanna, N.C. and Waisman, D.M. (1986) Biochemistry 25, 1078–1982.
91. Khanna, N.C., Tokuda, M. and Waisman, D.M. (1987) Methods Enzymol. 139, 36–50.
92. Waisman, D.M., Muranyi, J. and Ahmed, M. (1983) FEBS Lett. 164, 80–84.
93. Waisman, D.M., Tokuda, M., Morys, S.J., Buckland, L. and Clark, T. (1985) Biochem. Biophys. Res. Commun. 128, 1138–1144.
94. Tokuda, M., Khanna, N.C. and Waisman, D.M. (1987) Methods Enzymol. 139, 68–79.
95. Hesterberg, L.K. and Weber, J. (1983) J. Biol. Chem. 258, 365–369.
96. Carlsson, L., Mystrom, L.E., Sundkrivst, I., Markey, F. and Lindberg, U. (1977) J. Mol. Biol. 115, 465–483.
97. Yin, H.L. and Stossel, T.P. (1979) Nature 281, 583–586.
98. MacLennan, D.H. and Wong, P.T.S. (1971) Proc. Natl. Acad. Sci. U.S.A. 1231–1235.
99. Harrington, W.F. (1979) In: The Proteins Vol. 4, pp. 245–409. (Neurath, H. and Hill, R.L., eds.).
100. Hoar, P.E., Pato, M.E. and Kerrick, W.G.L. (1985) J. Biol. Chem. 260, 8760–8764.
101. Kerrick, W.G.L. and Hoar, P.E. (1986) In: Calcium and the cell. (Ciba Foundation) Symposium 122. pp. 183–196, Wiley, Chichester.
102. Dabrowska, R., Goch, A., Galazkiewicz, B. and Osinska, H. (1985) Biochim. Biophys. Acta 842, 70–75.
103. Ngai, P.K. and Walsh, M.P. (1984) J. Biol. Chem. 259, 13656–13659.
104. Marston, S.B. (1986) Biochem. J. 237, 605–607.
105. Case, R.M. (1980) Cell Calcium 1, 1–5.
106. Picton, C., Klee, C.B. and Cohen, P. (1981) Cell Calcium 2, 281–294.
107. Rubin, R.P. (1982) Fed. Proc. 41, 2181–2187.
108. Baker, P.F. and Knight, D.E. (1986) Br. Med. Bull. 42, 399–404.
109. Muff, R. and Fischer, J.A. (1986) FEBS Lett. 194, 215–218.
110. Frey, E.A., Kebabian, J.W. and Guild, S. (1986) Mol. Pharmacol. 29, 461–467.
111. Gomperts, B.D. (1986) Trends Biochem. Sci. 11, 290–292.
112. Rasmussen, H. (1986) N. Engl. J. Med. 314, 1094–1101.
113. Rasmussen, H. (1986) N. Engl. J. Med. 314, 1164–1170.
114. Daniel, J.F. (1985) In: Calcium in biological systems (Rubin, R.P., Weiss, G.B. and Putney, J.W. Jr., eds.) pp. 165–171. Plenum Press, New York.
115. Kakiuchi, S. and Sobue, K. (1983) Trends Biochem. Sci. 8, 59–62.
116. Umekawa, M., Kaka, M., Ingaki, M., Onishi, H., Wakabayashi, and Hidaka, M. (1985) J. Biol. Chem. 260, 9833–9837.

117. Pollard, T.D. and Cooper, J.A. (1986) Ann. Rev. Biochem. 55, 987–1036.
118. Geisow, M.J. and Burgoyne, R.D. (1984) J. Neurochem. 38, 1735–1741.
119. Saris, C.J.M., Tack, B.F., Kristensen, T., Glenney, J.R. and Hunter, T. (1986) Cell 46, 201–212.
120. Taqui, L., Rothhut, B., Shaw, A.W., Fradin, A., Vragaftig, B.B. and Russo-Marie, F. (1986) Nature 321, 177–180.
121. Connolly, T.M., Lawing, W.J. and Majerus, P.W. (1986) Cell 46, 951–958.
122. Whitaker, M. and Aitchison, M. (1985) FEBS Lett. 182, 119–124.
123. Metz, C.B. and Monroy, A. (1985) Biology of Fertilization, Vol. 3, Academic Press, New York.
124. Jaffe, F.F. (1980) Ann. NY Acad. Sci. 339, 86–101.
125. Epel, D. (1980) Ann. NY Acad. Sci. 339, 74–85.
126. Tyler, A. (1941) Biol. Rev. 16, 291–305.
127. Zucker, R.S. and Steinhardt, R.A. (1978) Biochim. Biophys. Acta 541, 459–466.
128. Epel, O., Patton, C., Wallace, R.W. and Cheung, W.Y. (1981) Cell 23, 543–549.
129. MacMannus, J.P., Braceland, B.M., Rixon, R.H., Whitfeild, J.F. and Morris, H.P. (1981) FEBS Lett. 133, 99–103.
130. Hidaka, H., Susaki, S., Tanaka, T., Endo, T., Ohno, S., Fijii, T. and Nagata, T. (1981) Proc. Natl. Acad. Sci. U.S.A. 78, 4354–4558.
131. Eckberg, W.R. (1986) Cell Calcium 7, 365–375.
132. Metcalfe, J.C., Moore, J.P., Smith, G.A. and Hesketh, T.R. (1986) Br. Med. Bull. 42, 405–412.
133. Rozengurt, E. (1986) Science 234, 161–166.
134. Moolenaar, W.H., Defize, L.H.K. and Laat, D.S.W. (1986) In: Calcium and the cell. (Ciba Foundation) Symposium 122, pp. 212–231. Wiley, Chichester.
135. Bravo, R., Buckhardt, J., Curran, T. and Muller, R. (1986) EMBO J. 4, 1193–1198.
136. Morgan, J.I. and Curran, T. (1986) 322, 552–555.
137. Rodriguez, A.P. and Rozengurt, E. (1985) EMBO J. 4, 71–75.
138. Rozengurt, E. (1986) Science 234, 161–166.
140. Rozengurt, E. and Collins, M. (1983) J. Pathol. 141, 309–331.
141. Shoyab, M., DeLarco, J.E. and Todardo, G.J. (1979) Nature 279, 387–391.
142. Smith, J.W.S. and Rozengurt, E. (1985) J. Cell Physiol. 124, 81–86.
143. Lin, C.R., Chen, W.S., Lazar, C.S., Carpenter, C.D., Gill, G.N., Evans, R.M. and Rosenfeld, M.G. (1986) Cell 44, 839–848.
144. Cooper, J.A. and Hunter, T. (1983) Curr. Top. Microbiol. Immunol. 107, 125–161.
145. Fava, R.A. and Cohen, S. (1985) J. Biol. Chem. 260, 8233–8236.
146. Kamps, M.P., Buss, J.E. and Sefton, B.M. (1985) Proc. Natl. Acad. Sci. U.S.A. 82, 4625–4628.
147. Kamps, M.P., Buss, J.E. and Sefton, B.M. (1986) Cell 45, 105–112.
148. Nawrocki, J.F., Lau, A.F. and Faras, A.J. (1984) Mol. Cell. Biol. 4, 212–215.
149. Korachak, H.M., Vienne, K., Rutherford, L.E., Wilkenfeld, C., Finkelstein, M.C. and Weissmann, G. (1984) J. Biol. Chem. 259, 4076–4082.

CHAPTER 4

Mechanism of action of Ca^{2+}-dependent hormones

HOWARD RASMUSSEN and PAULA Q. BARRETT

Departments of Cell Biology, Physiology and Internal Medicine, Yale University School of Medicine, New Haven, CT 06510, U.S.A.

1. Introduction

Calcium ion, Ca^{2+}, is a major and nearly universal intracellular messenger throughout the animal kingdom. This messenger operates in the mammalian cell under considerable constraint because Ca^{2+} is also a cellular toxin. Normally a 5000–10 000-fold Ca^{2+} concentration gradient exists across the plasma membrane of most cells. Hence, in terms of the $[Ca^{2+}]$ in the extracellular fluids (circa 1000 μM), all cells of a multicellular organism exist in a hostile environment. Yet, such cells survive and function normally for decades employing changes in intracellular Ca^{2+} concentration, $[Ca^{2+}]_i$, in the range of 0.1 to 2 μM, as a means of coupling a variety of stimuli to a particular cellular response. Moreover, when changes in $[Ca^{2+}]_i$ are employed, these same cells are able to maintain Ca^{2+} homeostasis and prevent dangerous accumulations of intracellular Ca^{2+}. From these considerations, it is clear that an understanding of the cellular metabolism of Ca^{2+} is a prerequisite to an understanding of the messenger function of Ca^{2+} [1–3].

A second important aspect of Ca^{2+} messenger function is its complexity. There are multiple mechanisms by which a Ca^{2+} messenger can be generated or terminated. Messenger generation can result from the turnover of polyphosphoinosi-

---

Abbreviations: $[Ca^{2+}]_i$, intracellular Ca^{2+} concentration as measured by a Ca^{2+} indicator such as aequorin; $[Ca^{2+}]_c$, Ca^{2+} concentration in the bulk cytosol (hypothetical value); $[Ca^{2+}]_{sm}$, Ca^{2+} concentration in submembrane domain just beneath the plasma membrane (a hypothetical value); PI, phosphatidylinositol; PIP_2, phosphatidylinositol 4,5-bisphosphate; PIP, phosphatidylinositol 4-phosphate; $Ins1,4,5,P_3$, inositol 1,4,5,-trisphosphate; $Ins1,3,4,P_3$, inositol 1,3,4-trisphosphate; $Ins1,3,4,5P_4$, inositol 1,3,4,5-tetrakisphosphate; $Ins1,4P_2$, inositol 1,4-bisphosphate; CaM, calmodulin; C-kinase, protein kinase C; $[cAMP]_c$, cAMP concentration in the bulk cytosol; $[cAMP]_{sm}$, cAMP concentration in submembrane domain just beneath the plasma membrane.

tides, the activation of the cAMP messenger system, and the activation of receptor-operated or potential-dependent Ca^{2+} channels in the plasma membrane. In many cases, in addition to cAMP serving to generate a Ca^{2+} signal, cAMP and Ca^{2+} can also serve as interrelated or synarchic messengers in the regulation of a specific cellular response [4]. Messenger termination can result from the activation of Ca^{2+} pumps in cellular membranes as a result of changes in cAMP or cGMP concentrations, and/or by autoregulatory mechanisms including the activation of the plasma membrane Ca^{2+} pump by either Ca^{2+}-calmodulin or protein kinase C.

The most common mechanism by which Ca^{2+}-dependent hormones act is via a receptor-linked phospholipase C which catalyses the hydrolysis of polyphosphatidylinositol, resulting in the production of two intracellular messengers: inositol 1,4,5-trisphosphate and diacylglycerol. The many cell responses which are mediated by this system are discussed in the chapter by Guy and Kirk (Chapter 2).

The mechanisms by which the Ca^{2+} message leads to a change in cell function are also complex. The effects of Ca^{2+} on targets in specific subcellular domains involve different molecular mechanisms of action. The action of Ca^{2+} can be achieved via either amplitude or sensitivity modulation [4]: this means that Ca^{2+} can serve as a messenger either by an increase in the amplitude (concentration) of the message, or by a change in the affinity (sensitivity) of the Ca^{2+} receptor unit for the message. In the latter circumstance, Ca^{2+} serves as an intracellular messenger even though the Ca^{2+} concentration does not change.

The actions of Ca^{2+} can also be expressed in different spatial domains within the cell, and these domains may change during a sustained cellular response. Hence, the concept of the messenger function of Ca^{2+} goes well beyond the simple classic view that hormone-receptor interaction leads to an increase in intracellular Ca^{2+} concentration $[Ca^{2+}]_i$ (as measured with the photoprotein aequorin), which acts as a messenger throughout the cell and is elevated as long as hormone is present. Rather, it is now apparent that even during sustained cellular responses, a transient rise in $[Ca^{2+}]_i$ is seen, and is associated with changing temporal patterns of PI and/or cyclic nucleotide metabolism and response element activity. These various aspects of Ca^{2+} messenger function will be considered in this chapter.

2. Cellular calcium metabolism

An appreciation of Ca^{2+} messenger function requires an understanding of cellular Ca^{2+} metabolism [1–4]. Our present view of cellular Ca^{2+} metabolism is represented schematically in Fig. 1. There are three membranes and four cellular compartments to consider. The membranes are the plasma membrane, the inner mitochondrial membrane, and the membrane of the endoplasmic reticulum (or a component of it). The compartments are the extracellular, the cytosolic, the mitochondrial matrix and the endoplasmic reticulum. The Ca^{2+} concentrations in these

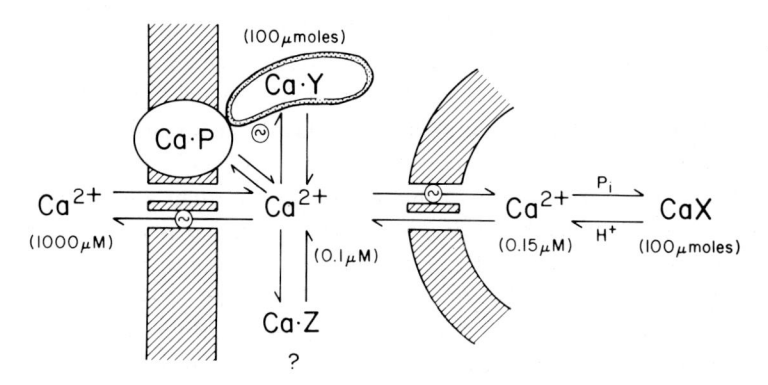

Fig. 1. A schematic representation of cellular calcium metabolism in which the plasma membrane (striped area on left), the endoplasmic reticulum (Ca·Y), and the inner mitochondrial membrane (striped area on right) are represented as separating the extracellular Ca^{2+} concentration (left), the endoplasmic pool of Ca^{2+} (Ca·Y) and the intramitochondrial pool of ionized (Ca^{2+}) and non-ionized calcium (CaX) from the cytosolic calcium Ca^{2+} (center). A possible plasma membrane pool of calcium (Ca·P) is also represented, as is a possible buffer pool in the cell cytosol (Ca·Z). The approximate concentrations and sizes of the respective pools are given. Also depicted are the pump (—⊖—) leak (—) systems that operate across the plasma and mitochondrial membrane. See text for discussion.

compartments are: (a) extracellular (mammalian), 1000 µM; (b) cytosolic, circa 0.05–0.150 µM; (c) mitochondrial matrix, circa 0.1–0.2 µM, but a larger component of non-ionic calcium phosphate exists which may be equivalent to 50–100 µmol/l cell H_2O; and (d) endoplasmic reticulum; the ionic concentration is unknown but most of the Ca^{2+} is in a non-ionic form (possibly protein-bound) and is equivalent to 50–100 µmol/l cell H_2O. In each membrane there are both passive Ca^{2+} channels through which Ca^{2+} flows, and energy-dependent Ca^{2+} 'pump' pathways. These channels and pumps are organized in such a way that the pumps in each membrane are always oriented so that they move Ca^{2+} away from the cytosolic compartment; e.g., the pump in the plasma membrane moves Ca^{2+} from the cytosol, a region of low $[Ca^{2+}]$, to the ECF, one of high $[Ca^{2+}]$. Hence, Ca^{2+} enters the cytosol via passive leaks in each membrane and leaves via energy-dependent pumps in each membrane.

2.1. Plasma membrane

In spite of the large $[Ca^{2+}]$ gradient across the plasma membrane and the necessity of maintaining the $[Ca^{2+}]_i$ at a low and constant value at all times, the energy expenditure needed to maintain this gradient represents less than 1 percent of the basal energy consumption of the cell. This low energy consumption is due to the fact that the plasma membrane has a very low permeability to Ca^{2+}, and the Ca^{2+} that enters the cell across this membrane does so solely via specific channels.

There are specific Ca^{2+} channels in the plasma membrane which allow Ca^{2+} to enter the cell [5,6]. At least three types of voltage-sensitive Ca^{2+} channel have been identified in neuronal and non-neuronal cells: T, N, and L channels. The T channels give rise to small and transient increases in Ca^{2+} current as a result of small depolarizing steps (from negative holding potentials). Such channels from neural tissues are insensitive to inhibition by dihydropyridine Ca^{2+} channel antagonists, but similar channels from non-neuronal cells demonstrate some sensitivity to such inhibitors. The second type of channel, the N channel, is opened by stronger depolarizing steps; it inactivates less readily than does the T channel, and is insensitive to inhibition by dihydropyridines. This type of channel appears to be closely linked to neurotransmitter release. The third or L channel is also activated by strong depolarization, is available at depolarized potentials and is sensitive to inhibition by dihydropyridines. Similar L channels are found in both neural and non-neural cells.

Less well-defined but particularly important in terms of the function of non-neuronal cells are so-called receptor-operated channels [6,7]. By definition these are channels in the plasma membrane which open in response to hormone-receptor interaction without a change in membrane potential. The mechanism of their opening may either be by a direct coupling of receptor (possibly via a G protein) with the channel, or by an indirect coupling via the generation of an intracellular messenger such as cAMP or the putative messenger, inositol 1,3,4,5-tetrakisphosphate.

In some cases, it is clear that the receptor-operated channel and the voltage-operated channel are one and the same entity, i.e., channel activity is controlled by several different mechanisms. For example, in cardiac myocytes, cAMP by catalysing the phosphorylation of L-type channels makes them responsive to changes in membrane potential. In hepatocytes a receptor-operated channel is opened by the action of angiotensin II. Angiotensin II acts synergistically with cAMP, generated by the action of high concentrations of glucagon, to increase Ca^{2+} flux through these 'receptor-operated' channels. The mechanism by which fluxes of Ca^{2+} through these receptor-operated channels is controlled is not established. Two possibilities are currently under investigation: (a) either inositol 1,3,4,5-tetrakisphosphate or inositol 1,3,4-trisphosphate serve to regulate channel activity; or (b) there is a direct coupling of receptor to channel, possibly involving a G protein-linked function. In other cases, there is more direct evidence that a hormone-receptor interaction directly regulates channel activity. For example, in adrenal glomerulosa cells, ACTH activates one class of Ca^{2+} channels directly, and angiotensin II another class via one of the mechanisms discussed above, but each class of channel is also voltage-operated.

When a typical cell is activated by a hormone, or other extracellular messenger, a common event is a 2–4-fold increase in Ca^{2+} influx rate through Ca^{2+} channels in the plasma membrane [1]. Such an increase in Ca^{2+} influx rate leads to only a transient increase in cellular Ca^{2+} concentration $[Ca^{2+}]_i$, because of the existence of an elegant autoregulatory system in the plasma membrane.

There are two energy-dependent mechanisms by which Ca^{2+} is pumped back out of the cell: an Na^+-Ca^{2+} exchanger which uses the Na^+ gradient across the membrane (maintained by the activity of the Na^+ pump) to drive Ca^{2+} efflux, and a $2H^+$-Ca^{2+}-ATPase [3,8]. The $3Na^+$-Ca^{2+} exchanger appears to operate as a high-capacity low-affinity system to move relatively large amounts of Ca^{2+} out of the cell rapidly. The $2H^+$-Ca^{2+}-ATPase is a lower-capacity high-affinity system which is crucial for long-term maintenance of cellular Ca^{2+} homeostasis. This pump is regulated by a variety of intracellular signals so that Ca^{2+} homeostasis is maintained in various states of cell activity. When, for example, an increase in extracellular K^+ concentration leads to a sustained increase in Ca^{2+} influx via voltage-dependent channels, the $[Ca^{2+}]_c$ rises transiently and then falls to a value only slightly greater than basal, and total cell Ca^{2+} rises transiently but then remains stable because of a compensatory increase in Ca^{2+} efflux rate. A major means by which this increase in Ca^{2+} efflux is achieved is via the Ca^{2+}-dependent activation of the Ca^{2+} pump by the intracellular Ca^{2+} receptor protein, calmodulin (CaM) [9]. An increase in the Ca^{2+} concentration in the cell cytosol, $[Ca^{2+}]_c$, leads to an increase in the content of $Ca_4 \cdot CaM$, which associates with the pump, bringing about its activation by both increasing its V_{max} and lowering K_m for Ca^{2+}. In many cells, additional controls of pump activity may also operate. These include: (a) cAMP-dependent activation of the pump; (b) cGMP-dependent activation of the pump; and/or (c) activation of the pump by protein kinase C.

In operational terms the existence of these autoregulatory systems means that when a hormone acts to bring about a sustained increase in Ca^{2+} influx rate, there follows a sustained increase in efflux rate. Hence, during the sustained phase of many cellular responses, there is a sustained increase in the rate of Ca^{2+} cycling across the plasma membrane with little or no increase in intracellular free $[Ca^{2+}]_i$, as measured, for example, with the photoprotein aequorin. Nevertheless, as discussed below, this cycling of Ca^{2+} across the plasma membrane acts as a messenger during the sustained phase of the response probably by changing the $[Ca^{2+}]$ in a subdomain of the cell just beneath the plasma membrane $[Ca^{2+}]_{sm}$.

2.2. Endoplasmic reticulum

It has become increasingly clear that there is a non-mitochondrial, intracellular calcium pool which plays an important role in cell activation in a large number of non-muscle cells as well as in smooth and skeletal muscle. This pool is relatively enormous in skeletal muscle, provides the bulk of the Ca^{2+} needed to regulate skeletal muscle contraction, and is located in a distinct organelle, the sarcoplasmic reticulum. The pool is smaller in non-muscle and in smooth muscle cells, and its location less obvious [10,11]. To fill the pool requires ATP, i.e., uptake of Ca^{2+} into the pool is driven by a distinct Ca^{2+}-ATPase, an enzyme which purifies with the microsomal (or endoplasmic reticulum) fraction of the cell. The release of Ca^{2+} from this pool can be induced by inositol 1,4,5-trisphosphate. The interaction of Ca^{2+}-linked hormones with their cell surface receptors leads to the hydrolysis of polyphosphatidylinositides and the generation of inositol 1,4,5-trisphosphate; there is a rapid release of Ca^{2+} from this trigger pool leading to a rapid, but transient, increase in $[Ca^{2+}]_i$. This is almost always associated with a rapid net loss of Ca^{2+} from the cell. Even though the rise in $Ins1,4,5P_3$ concentration is transient, this trigger pool does not refill in the presence of a sustained increase in hormone concentration, but rapidly refills upon hormone removal or inhibition of hormone action. Nonetheless, when hormone action is terminated, the $[Ca^{2+}]_i$ changes only minimally. These kinetic characteristics, along with the structural evidence of subsurface cisternae in many cells (called a primitive sarcoplasmic reticulum in smooth muscle cells), suggest that the trigger pool from which Ca^{2+} is mobilized by inositol 1,4,5-trisphosphate is located in a membrane-enclosed compartment distinct from, but geographically close to, the endoplasmic surface of the plasma membrane (see Fig. 3). Its major function is to serve as a source of rapidly releasable Ca^{2+} which can bring about a transient 5–10-fold increase in cytosolic Ca^{2+} concentration, $[Ca^{2+}]_c$. Such an increase can serve either to elicit a brief cellular response or to initiate a sustained cellular response. It is noteworthy that the release of approximately 30–40 μmol of Ca^{2+} (per liter of cell H_2O) is needed to raise the $[Ca^{2+}]_i$ from 0.1 to 1.0 μM in 10 s. Even so, the rise in $[Ca^{2+}]_i$ is over in 30–60 s because the bulk of the Ca^{2+} is pumped out of the cell – another indication of the close geographic localization of the membrane system containing the trigger pool with the plasma membrane.

2.3. Mitochondrial matrix

The uptake of Ca^{2+} by the mitochondria occurs via a Ca^{2+} uniporter in the inner membrane and is driven by the proton-electrochemical gradient across the membrane [12]. If Ca^{2+} uptake occurs in the presence of acetate (and absence of phosphate), then both Ca^{2+} and acetate$^-$ accumulate, and the mitochondria swell because both are osmotically active. However, if phosphate is now added, the mitochondria shrink without losing Ca^{2+}; rather they exchange acetate for phosphate, resulting in the formation of a non-ionic complex of calcium-phosphate within the mitochondrial matrix which is in rapid exchange with a small pool of Ca^{2+} within the matrix space, $[Ca^{2+}]_{mm}$. The $[Ca^{2+}]_{mm}$ is approximately the same as the $[Ca^{2+}]_c$ in the resting cell (0.1–0.2 μM) and represents less than 1% of total mitochondrial calcium. The $[Ca^{2+}]_{mm}$ is in rapid exchange with the $[Ca^{2+}]_c$. Upon activation of the cell by a Ca^{2+}-dependent hormone, the $[Ca^{2+}]_{mm}$ rises as the $[Ca^{2+}]_c$ rises. This leads to the activation of three Ca^{2+}-sensitive mitochondrial dehydrogenases, pyruvate, NAD-linked isocitrate and α-ketoglutarate dehydrogenase [13]. These enzymes remain activated when the $[Ca^{2+}]_c$ falls. It is not clear how this sustained activation is achieved. Either a second non-Ca^{2+}-dependent mechanism is responsible, or more likely the $[Ca^{2+}]_{mm}$ remains high, even though the $[Ca^{2+}]_c$ falls back close to its basal value, because of a sustained increase in Ca^{2+} influx and efflux across the mitochondrial membrane (a sustained Ca^{2+} cycling) due to a decrease in the K_m for Ca^{2+} of the mitochondrial uniporter.

The mitochondrial Ca^{2+} pool plays a second role in cellular Ca^{2+} homeostasis by serving as a sink for Ca^{2+} during times of excessive Ca^{2+} uptake by the cell. Under this circumstance, the non-ionic calcium pool in the matrix space can increase 10-fold or more, thereby protecting the cell from Ca^{2+} intoxication. This mechanism provides a temporary device by which the cell can protect itself, but in the long term only by regulating Ca^{2+} fluxes across the plasma membrane can the cell maintain Ca^{2+} homeostasis [14].

3. Mechanisms of Ca^{2+} messenger generation

In discussing the mechanisms of Ca^{2+} messenger generation, it is first necessary to define what constitutes a Ca^{2+} messenger. Within the context of the original second messenger model, a classic second messenger is generated at the plasma membrane and diffuses throughout the cell cytosol. It interacts, therein, with specific receptor proteins resulting in the transmission of the message from cell surface to cell interior. The second messenger model, with minor modifications, has been applied to the Ca^{2+} messenger system. In this view, an increase in Ca^{2+} concentration arising as a result of the release of Ca^{2+} from the intracellular trigger pool as well as an increase in influx leads to an increase in $[Ca^{2+}]_c$ which interacts with Ca^{2+} receptor proteins, e.g., calmodulin, to initiate and sustain a cellular response. However, measurement of the time course of change in $[Ca^{2+}]_i$ following agonist action demonstrates that the $[Ca^{2+}]_i$ rises transiently after agonist action, and then falls to a value close to its original basal value in spite of the fact that there is a sustained increase in rate of Ca^{2+} cycling across the plasma membrane. Furthermore, it is clear that this increased cycling of Ca^{2+} plays a messenger role even though it is not associated with a rise in $[Ca^{2+}]_c$. For example, inhibition of cycling leads to an inhibition of the sustained response. Hence, Ca^{2+} appears to serve a messenger function in a highly restricted subcellular domain, $[Ca^{2+}]_{sm}$, as well as throughout the total cytosolic domain.

4. Messenger calcium

There are at least four distinct mechanisms by which messenger Ca^{2+} can be generated (Fig. 2). The most widely employed is that mediated by the receptor-linked activation of the inositol lipid pathway [15]. In this case, mobilization of Ca^{2+} from

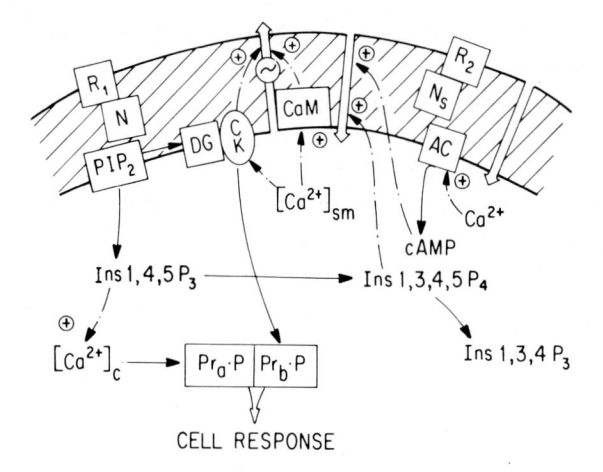

Fig. 2. A schematic representation of some of the mechanisms by which Ca^{2+} fluxes across the plasma membrane are regulated. In the plasma membrane (the striped area) there are both influx ($\Rightarrow$) and energy-dependent ($\Rightarrow$) efflux pathways. Two mechanisms by which Ca^{2+} influx can be increased are via the actions of the intracellular messengers inositol 1,3,4,5-tetrakisphosphate, and cAMP generated via activation of specific classes of surface receptors (R_1 and R_2) linked to specific N proteins which activate either phosphatidylinositol 4,5-bisphosphate (PIP_2) hydrolysis or adenylate cyclase (AC). Additionally, influx can be increased either by a direct receptor-coupled event or by a membrane depolarization (not shown). A rise in the Ca^{2+} concentration in the domain just beneath the plasma membrane, $[Ca^{2+}]_{sm}$, can lead to an activation of the Ca^{2+} pump either via a direct calmodulin (CaM)-dependent mechanism, or indirectly via the activation of protein kinase C (CK). Additionally, in some cells, an increase in cGMP concentration also increases Ca^{2+} efflux (not shown), and in still others cAMP may stimulate Ca^{2+} efflux.

the trigger pool and increased plasma membrane Ca^{2+} cycling are involved in Ca^{2+} messenger generation. In the other three cases, only an increase in plasma membrane Ca^{2+} cycling is involved. Such an increase can be brought about by: (a) the direct activation of a receptor-operated Ca^{2+} channel; (b) the activation of a voltage-dependent Ca^{2+} channel; (c) the activation of a Ca^{2+} channel secondary to the receptor-mediated activation of adenylate cyclase; or (d) possibly by an increase in either inositol 1,4,5-trisphosphate or inositol 1,3,4,5-tetrakisphosphate.

These mechanisms can produce either a brief response such as a transient neurosecretory response due to a brief depolarization of the presynaptic membrane and a transient influx of Ca^{2+}, or a more sustained cellular response, such as is seen when peptide hormones act on target cells. Only the latter will be considered in the present discussion.

4.1. Coordinated changes in PI and Ca^{2+}-metabolism

Receptor-mediated stimulation of polyphosphoinositide hydrolysis is discussed in detail by Guy and Kirk in Chapter 2 of this volume (see also Refs. 15–17). These

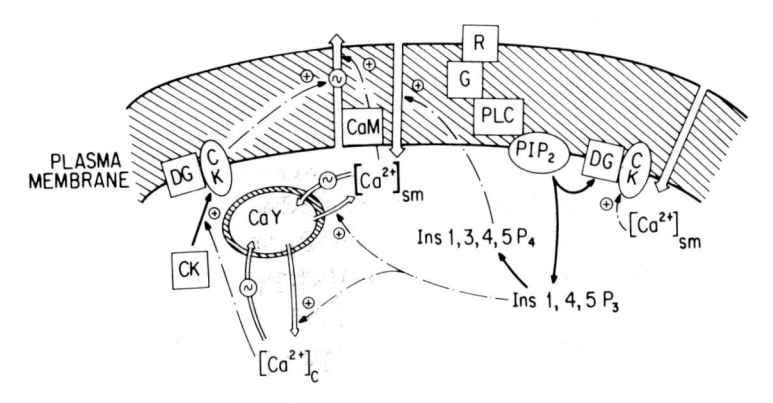

Fig. 3. The interrelated changes in the metabolism of phosphatidylinositol 4,5-bisphosphate (PIP_2) and Ca^{2+} during activation of the cell by a typical Ca^{2+}-dependent hormone. R, receptor; G, guanine regulatory protein; PLC, phospholipase C; DG, diacylglycerol; CK, protein kinase C; $[Ca^{2+}]_{sm}$, the Ca^{2+} concentration in a cellular domain just beneath the plasma membrane (striped area); $Ins1,4,5P_3$, inositol 1,4,5-trisphosphate; $Ins1,3,4,5P_4$, inositol 1,3,4,5-tetrakisphosphate; $[Ca^{2+}]_c$, cytosolic Ca^{2+} concentration; CaM, calmodulin; arrows ($\Rightarrow$), fluxes of Ca^{2+} across membranes; ⊚⇒, energy-dependent fluxes; CaY, a calcium pool in specialized compartment of the endoplasmic reticulum. See text for discussion.

changes in polyphosphoinositide metabolism are closely linked to alterations in Ca^{2+} metabolism (Fig. 3). Hormone-receptor interaction leads to a stimulation of phospholipase C activity, the hydrolysis of PIP_2, and the production of inositol 1,4,5-trisphosphate ($Ins1,4,5P_3$) and diacylglycerol (DG). The inositol $1,4,5P_3$ is converted successively to inositol $1,3,4,5P_4$, inositol $1,3,4P_3$, inositol $1,4P_2$, inositol 1P and/or inositol 4P and free inositol. The two inositol phosphates which have been shown to have messenger functions are $Ins1,4,5P_3$ and possibly $Ins1,3,4,5P_4$. Upon activation of the cell the $Ins1,4,5P_3$ concentration rises in seconds, reaches a peak in 5–15 s, and then falls to a value only slightly greater than its basal value. In contrast, the $Ins1,3,4,5P_4$ concentration rises rapidly to a sustained plateau. The rise in $Ins1,4,5P_3$ triggers the release of Ca^{2+} from the trigger pool of Ca^{2+} stored in a membrane compartment located just beneath the plasma membrane. As a result, the $[Ca^{2+}]_c$ and the $[Ca^{2+}]_{sm}$ both rise. The rise in $[Ca^{2+}]_{sm}$ is associated with a net loss of Ca^{2+} from the cell because of the CaM-dependent activation of the plasma membrane Ca^{2+} pump.

The trigger pool of Ca^{2+} does not refill as long as agonist is present because the concentration of $Ins1,4,5P_3$ remains high enough to stimulate a continued efflux of Ca^{2+} from this pool. Since in the steady state Ca^{2+} influx into the trigger pool is maintained, there is an increased rate of Ca^{2+} cycling across the membrane of this intracellular system.

This model describes a system in which Ca^{2+} serves a messenger function during both phases of a cellular response, but in which the Ca^{2+} pools involved and the molecular targets of Ca^{2+} action differ. During the initial phase of the response, it is the rise in $[Ca^{2+}]_c$ which is the messenger, and CaM-dependent enzymes, including protein kinases, are the molecular targets. During the sustained phase the increase in $[Ca^{2+}]_{sm}$ is the messenger, and the molecular targets are the plasma membrane Ca^{2+} pump, the plasma membrane-associated enzyme, protein kinase C, and other Ca^{2+}-sensitive enzymes associated with the plasma membrane.

The existence of these temporally distinct phases of cellular response means that there is a mechanism by which cells recognize both the strength and the duration of their exposure to a particular extracellular messenger.

4.2. Smooth muscle contraction

The contraction of tracheal smooth muscle in response to carbacholamine [18–22] can be taken as an example of a system in which a temporal sequence of changes in the metabolism of inositol lipids, in Ca^{2+} metabolism, and in protein phosphorylation underlies cellular response (Fig. 4). Upon activation of the cell there is a transient rise in $[Ca^{2+}]_c$ lasting 1–1.5 min which is followed by a transient rise in the content of MLC·P peaking in 3–5 min, and then declining to near basal values in 20–25 min. The development of force occurs simultaneously with the phosphoryl-

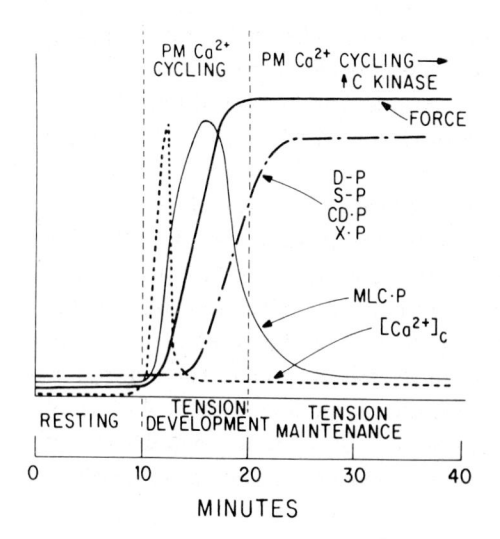

Fig. 4. The temporal sequence of events when a resting strip of tracheal smooth muscle is activated by carbacholamine addition at 10 min. There is a transient rise in $[Ca^{2+}]_c$ (---) followed by a transient increase in the content (—) of phosphorylated myosin light chains (MLC·P) which lead in turn to the initiation of force development (—). Increased force is sustained even though the content of MLC·P declines. Preceding the sustained phase of force maintenance, there is an increase in the phosphorylation of desmin (D-P), synemin (S-P), caldesmon (CD-P) and a number of low molecular weight cytosolic proteins (X-P). These remain phosphorylated throughout the sustained phase of the response during which there is a sustained increase in Ca^{2+} cycling across the plasma membrane which regulates the activity of the membrane-associated protein kinase C.

ation of the light chains of myosin, but is then sustained even though the extent of myosin light chain phosphorylation declines. With a time delay of 5–10 min, there is a progressive increase in the extent of phosphorylation of caldesmon, of the intermediate filament proteins desmin and synemin, and of several low molecular weight cytosolic proteins. The increase in the extent of phosphorylation of these late-phase proteins reaches a plateau and is then sustained, as is the increase in tension. The phosphorylation of these proteins is due to the increase in Ca^{2+} cycling across the plasma membrane regulating the activity of protein kinase C.

If the postulate is accepted that the major means by which intracellular messengers bring about changes in cell function is via the activation of specific protein kinases, and thus the phosphorylation of cell specific proteins, then this type of result indicates that a different subset of cellular proteins is phosphorylated during the initial and sustained phase of the response. Further, this difference implies that the two phases of the response are mediated by different molecular mechanisms.

The other feature of smooth muscle contraction is its reversal by agents which increase the intracellular concentrations of either cGMP or cAMP. An example of particular interest is that of atrial natriuretic peptide, a recently discovered hormone secreted from the heart [23]. This peptide acts by stimulating a receptor-linked, plasma membrane-associated guanylate cyclase, causing, thereby, an increase in [cGMP]. This rise in [cGMP] may act by at least two mechanisms. The first is the stimulation of Ca^{2+} efflux by an activation of the plasma membrane Ca^{2+} pump; and the second is the activation of one or more phosphoprotein phosphatases which catalyse the dephosphorylation of such proteins as the phosphorylated form of myosin light chain.

4.3. Coordinate changes in cAMP and Ca^{2+} metabolism

Although many sustained cellular responses are mediated by the PI-linked Ca^{2+} messenger system, other such responses occur in which PIP_2 hydrolysis and its related changes in Ca^{2+} metabolism do not occur. Rather, cellular response depends upon the coordinate activities of two intracellular messengers, Ca^{2+} and cAMP. Two examples will suffice to illustrate some of the possible relationships between the two messenger systems: K^+-dependent regulation of aldosterone secretion, and the glucagon-dependent regulation of hepatic cell intermediary metabolism.

4.3.1. K^+-mediated aldosterone secretion
An increase in the extracellular K^+ concentration from 2 to 8 mM (the limits of the physiological range) leads to a 4–5-fold increase in the rate of aldosterone secretion by the adrenal glomerulosa cell [24,25]. This increase in [K^+] leads to a depolarization of the plasma membrane of the cell. This depolarization causes the opening of two types of voltage-dependent Ca^{2+} channel in this membrane, and thus to a 4-fold increase in the rate of Ca^{2+} influx. The total cell calcium and the $[Ca^{2+}]_i$ both

rise initially. The peak rise in $[Ca^{2+}]_i$ occurs in 15–30 s, and then declines slowly so that in 1–1.5 min it is back close to, but slightly above, its basal value even though Ca^{2+} influx rate remains 4-fold greater than its basal value. Yet during the sustained activation, total cell calcium does not increase further indicating that there is a compensatory increase in Ca^{2+} efflux rate so that during the sustained response there is an increase in the cycling of Ca^{2+} across the plasma membrane, and presumably, therefore, an increase in $[Ca^{2+}]_{sm}$.

Thus, during the sustained phase of the K^+-mediated aldosterone secretory response, the $[Ca^{2+}]_i$ is only slightly greater than basal, but Ca^{2+} cycling across the plasma membrane is increased 4-fold. Hence, it is very likely that the increase in $[Ca^{2+}]_i$ reflects largely an increase in $[Ca^{2+}]_{sm}$. This Ca^{2+} cycling plays a messenger function during the sustained phase of the response. Its inhibition leads to a prompt inhibition of secretion. However, a rise in $[K^+]$ does not lead to a change in PI turnover or to the activation of protein kinase C. Therefore, Ca^{2+} cycling must control aldosterone secretion by mechanisms that differ when K^+ or angiotensin II acts. Equally important is the observation that an increase in Ca^{2+} cycling alone, induced, for example, by addition of a Ca^{2+} ionophore such as A23187, is not sufficient to induce a sustained response. Addition of A23187 in a concentration sufficient to cause a 4-fold increase in Ca^{2+} influx rate induces a transient rather than a sustained aldosterone secretory response. Therefore, K^+-dependent stimulation of aldosterone secretion must involve an additional messenger. This messenger is cAMP. The K^+-dependent opening of one of the two classes of Ca^{2+} channels in the plasma membrane is linked to the Ca^{2+}-CaM-dependent activation of adenylate cyclase. Addition of A23187 does not lead to an activation of cyclase even though it increases Ca^{2+} influx rate to the same extent as K^+. The activation of the cyclase by K^+ is sustained. However, despite a sustained 4-fold increase in cyclase activity, the [cAMP] (after a transient rise) falls to a value approximately 1.5-fold above basal.

A working model which describes the operation of this control system stresses a temporal pattern of events in which the sustained response is mediated by a rapid cycling of Ca^{2+} across the plasma membrane linked to a high rate of cAMP turnover. The cycling of Ca^{2+} is associated with an increase in $[Ca^{2+}]_{sm}$ but not $[Ca^{2+}]_c$. There is also a rapid synthesis and hydrolysis of cAMP associated with an increase in $[cAMP]_{sm}$ but not $[cAMP]_c$. The mechanisms by which these coordinate changes in $[Ca^{2+}]_{sm}$ and $[cAMP]_{sm}$ regulate the sustained phase of aldosterone secretion remain to be elucidated. On the other hand, a number of the possibilities by which they regulate each other's concentration and turnover can be surmised. The increase in $[Ca^{2+}]_{sm}$ stimulates not only adenylate cyclase but also phosphodiesterase so that both synthesis and hydrolysis of cAMP are controlled by Ca^{2+}. Conversely, cAMP may well play a role along with Ca^{2+} in regulating the efflux of Ca^{2+} via the Ca^{2+} pump.

4.3.2. Control of hepatic metabolism by glucagon

Glucagon has a number of effects on various aspects of hepatocyte metabolism. Many of these are covered in other parts of this treatise. The focus of our present attention is the identity of the intracellular messengers which mediate the effects of glucagon. In considering this question, it is first necessary to recognize that different mechanisms operate when cells are exposed to low or high concentrations of glucagon.

When hepatocytes are exposed to a low concentration of glucagon (10^{-12} M), the hormone interacts with one class of receptors linked to phospholipase C and the hydrolysis of PIP_2 [26]. These events lead to an increase in $Ins1,4,5P_3$, and presumably to both a mobilization of intracellular Ca^{2+} and an increase in Ca^{2+} influx rate. Hence, low concentrations of glucagon act exclusively via the PI system.

At a higher glucagon concentration (10^{-9} M) the hormone acts via a second class of receptors which are coupled to adenylate cyclase via a stimulatory G protein. The resulting rise in [cAMP] leads to three important effects, all of which are thought to result from activation of the cAMP-dependent protein kinase. It stimulates the phosphorylation of key glycogenolytic and gluconeogenic enzymes; it inhibits the actions of glucagon on the first class of receptors, so that the effects of glucagon on PI metabolism are not expressed; and it stimulates the influx of Ca^{2+} across the plasma membrane of the hepatocyte. In terms of this latter effect, high concentrations of glucagon acting via the rise in [cAMP] act synergistically with hormones such as angiotensin II, which activates the PI system, to stimulate Ca^{2+} influx [27]. This action of cAMP is thought to be brought about by the cAMP-dependent phosphorylation of Ca^{2+} channels. Of note is the fact that the rise in [cAMP] induced by high concentrations of glucagon inhibits the effect of glucagon on PIP_2 turnover, but inhibits only partially the effects of angiotensin II, phenylephrine or vasopressin on PIP_2 turnover.

The point to be made from these findings is that in the hepatocyte, there is only one common feature of the actions of the different hormones which regulate glycogenolysis: it is an increase in $[Ca^{2+}]_{sm}$ due to an increase in Ca^{2+} cycling across the plasma membrane. This increase in $[Ca^{2+}]_{sm}$ may be brought about by one of two mechanisms: activation of PIP_2 turnover and the associated increase in $Ins1,3,4,5P_4$ (or directly via a receptor-operated channel), or by activation of adenylate cyclase and the rise in cAMP. Hence, even in this classic cAMP-dependent system, a critically important messenger, particularly during the sustained phase of the response, is an increase in Ca^{2+} influx and Ca^{2+} cycling across the plasma membrane of the hepatocyte leading to an increase in $[Ca^{2+}]_{sm}$. How this messenger participates in the regulation of the sustained changes in energy metabolism and glucose production in these cells has not yet been adequately addressed.

5. Synarchic regulation

The examples of K^+-mediated aldosterone secretion and the control of hepatic metabolism by glucagon are two of the many relationships which exist between the Ca^{2+} and cAMP messenger systems. What has emerged from the analysis of the control of specific cellular responses of particular differentiated tissues is the widespread use of both messenger systems to regulate the cell's response to environmental stimuli. Ca^{2+} and cAMP commonly serve as synarchic messengers in the regulation of cell function [4]. However, their interrelationships do not follow a stereotyped pattern but can take a number of forms. One can recognize at least five major patterns in their relationship: coordinate control in which the same hormone activates both systems and the intracellular messengers Ca^{2+} and cAMP act in a coordinate fashion to mediate cellular response; hierarchical control in which different concentrations of the same hormone or different hormones activate the separate messenger systems via different receptor classes in such a way that activation of the second class of receptors and the generation of an additional second messenger lead to an amplification of the response elicited by activation of the first class of receptors; redundant control in which activation of either pathway leads to a cellular response providing thereby the same tissue response to a number of different stimuli; antagonistic control in which activation of one pathway leads to an inhibition of the response mediated by the other; and sequential control in which activation of one pathway leads to the generation of one messenger which now serves to bring about the generation of the other.

This concept and examples of these different patterns are presented in detail elsewhere [4], but there is one particular relationship which appears to be quite common: that in which the concentration of cAMP determines the responsiveness of a particular system to activation by the Ca^{2+} messenger system, i.e., cAMP determines the sensitivity of the response elements within the cell to a given Ca^{2+} message.

5.1. Regulation of insulin secretion by CCK and glucose

A specific example of this type of control system is seen in the case of the interrelated roles of cholecystokinin 8-sulfate (CCK8S) and glucose in the regulation of insulin secretion from the beta cell of the Islets of Langerhans (Fig. 5). Each of these extracellular messengers stimulates insulin secretion under the appropriate conditions, but each does so by different mechanisms, allowing for a considerable range of responsiveness [28,29].

CCK8S acts in the beta cell as a typical PI-linked agonist. Binding of CCK8S to its receptor leads to a breakdown of phosphatidylinositol 4,5-bisphosphate, with the generation of inositol 1,4,5-trisphosphate and diacylglycerol. The rise in inositol 1,4,5-trisphosphate concentration leads to the mobilization of an intracellular trig-

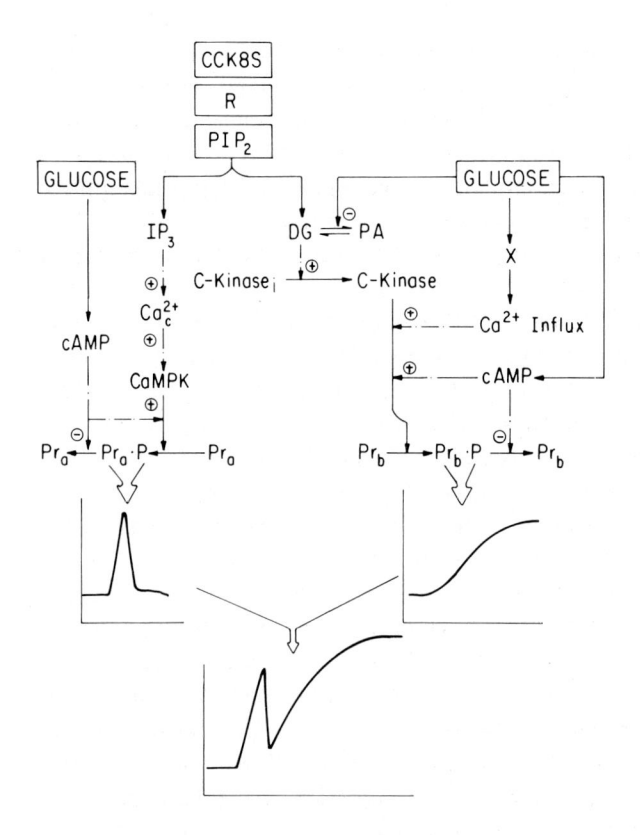

Fig. 5. The interrelated actions of cholecystokinin 8-sulfate (CCK8S) and glucose on insulin secretion from the beta cells of the Islets of Langerhans. When CCK8S interacts with its receptor (R), there is an increased hydrolysis of phosphatidylinositol 4,5-bisphosphate (PIP_2) with the production of two intracellular messengers, inositol 1,4,5-trisphosphate (IP_3) and diacylglycerol (DG). These messengers activate separate branches of the Ca^{2+} messenger system when the plasma glucose is >7 mM: the calmodulin (CaM) branch responsible for the initial phase of insulin, and the C-kinase branch responsible for the sustained phase. When glucose concentration is low (e.g., 2 mM) then CCK8S still stimulates PIP_2 hydrolysis and a rise in the cytosolic calcium ion concentration, Ca^{2+}_c, but these changes will not lead to an increase in either phase of insulin secretion. The first phase response and the phosphorylation of a specific subset of cellular proteins, P_{ra}, by CaM-dependent protein kinases can be restored by increasing the cAMP concentration either by raising glucose concentration or by stimulating adenylate cyclase activity with forskolin (not shown); but under these circumstances, CCK8S will not evoke a second phase response. A second phase response to CCK8S depends upon three effects of glucose: an increase in plasma membrane Ca^{2+} influx which serves to regulate the activity of the Ca^{2+}-sensitive form of C-kinase; and an alteration in the synthesis and/or metabolism of DG so that the DG content of the cell rises and C-kinase is converted from its Ca^{2+}-insensitive to its Ca^{2+}-sensitive form, and an increase in cAMP concentration which determines whether or not an activation of protein kinase C will lead to an increase in the extent of phosphorylation of a second subset of cellular proteins, P_{rb}. It is not yet clear how an increase in cAMP concentration determines whether the messenger generated during either phase of response is read out or not. Either an increase in cAMP concentration leads to a sensitization of the kinases to activation by Ca^{2+}, or it leads to an inhibition of phosphoprotein phosphatase(s).

ger pool of Ca^{2+}. These events occur in beta cells exposed to either 2 or 7 mM glucose, but lead to an increase in insulin secretion only when the glucose concentration is 7 mM or above. Hence, the same change in Ca^{2+} messenger concentration leads to quite different insulin secretory responses depending on the concentration of glucose; low glucose – no response; moderate to high glucose – a typical biphasic pattern of insulin secretion. It is postulated that the first phase of secretion is mediated via the calmodulin branch, and the second phase by the C-kinase branch of the Ca^{2+} messenger system.

Three of the known effects of glucose on islet cell function are:(a) to increase the rate of Ca^{2+} influx via an ATP-dependent inhibition of K^+ efflux leading to a depolarization of the plasma membrane and the opening of voltage-dependent Ca^{2+} channels; (b) to increase the cAMP content of the beta cells; and (c) to activate PI turnover to a modest degree in a Ca^{2+}-dependent manner. Using specific drugs, two of these effects can be produced pharmacologically in beta cells: forskolin activates adenylate cyclase and causes an increase in [cAMP]; and tolbutamide acts on the K^+ channel to cause membrane depolarization and an increased rate of Ca^{2+} influx.

If beta cells are incubated in media containing 2 mM glucose and then treated with forskolin and/or tolbutamide, there is a small transient increase in insulin secretion. The subsequent addition of CCK8S leads to a very marked first phase of insulin secretion, but causes no sustained increase or second phase of insulin secretion. These results mean that an increase in cAMP alters the Ca^{2+} sensitivity of the response elements underlying the first phase of secretion. These elements, presumed to be Ca^{2+}-calmodulin-dependent processes including CaM-dependent protein kinases, become more sensitive to activation by Ca^{2+} either because cAMP acts to enhance the sensitivity of CaM-dependent kinases to Ca^{2+}, or because cAMP inhibits, by an unknown mechanism, the activity of phosphoprotein phosphatases.

What is of particular interest is the fact that even though diacylglycerol (DG) is produced in response to CCK8S action, there is no second, C-kinase-dependent secretion of insulin when glucose concentration is low even if forskolin is present. However, there is such a secretion, if the glucose concentration is above 7 mM. This means that an effect of glucose in addition to an increase in Ca^{2+} influx and/or an increase in cAMP concentration allows the DG generated by CCK8S receptor interaction to produce an increase in second or sustained phase insulin secretion. The exact mechanism by which glucose modulates this second phase secretion is not yet worked out, but the evidence suggests that glucose has an independent effect on DG metabolism so that when the medium glucose is low, any DG generated in response to CCK8S action does not accumulate, but when the glucose concentration is higher it does accumulate.

The lesson is that in the case of hormones which act via the PI system, the two phases of the typical response can be modulated separately. The system need not behave in a single stereotyped fashion (see also Ref. 30).

6. Integration of extracellular messenger inputs

The example just discussed of the interrelated action of CCK8S and glucose in the regulation of insulin secretion is just one of many instances in which a particular response is controlled by multiple extracellular messengers. In fact, such an arrangement is the rule rather than the exception.

One particularly instructive example of the mechanism by which the actions of multiple extracellular messengers are integrated to achieve a modulated cellular response is seen in the case of the control of aldosterone secretion by extracellular K^+, ACTH and angiotensin II (AII). The effect of K^+ on this process has been discussed above and the actions of the two hormones are discussed elsewhere (Chapters 10 and 11), so the focus of this discussion is upon their interrelationships. The most important fact about the action of all three agonists is that they stimulate Ca^{2+} influx. In each case an increase in agonist concentration leads to a sustained increase in Ca^{2+} influx rate, but the mechanisms by which these changes are brought about differ: K^+ acts by increasing the flux of Ca^{2+} through two voltage-dependent Ca^{2+} channels; AII acts either by a direct receptor coupling to a Ca^{2+} channel or by increasing the concentration of inositol 1,3,4,5-tetrakisphosphate which acts on one class of 'receptor-operated' Ca^{2+} channels; and ACTH acts via a second class of receptor-operated channel. However, the two classes of voltage-dependent channels and the two types of receptor-operated channels are equivalent, i.e., there are apparently only two classes of channels each of which is voltage-dependent, but one of which is, in addition, receptor-operated via an ACTH receptor, and the other via the AII receptor. Furthermore, the flux of Ca^{2+} through these two different channels is linked to different elements in the plasma membrane: adenylate cyclase (ACTH) and protein kinase C (angiotensin II).

The flux of Ca^{2+} through the ACTH receptor-operated channel leads to the CaM-dependent activation of adenylate cyclase. Hence, either a low concentration of ACTH and/or an increase in K^+ concentration leads both to an influx of Ca^{2+} through this channel and to the activation of adenylate cyclase. Higher concentrations of ACTH acting via a second type of receptor activate adenylate cyclase via the classic G_s protein mechanism. These two ACTH-dependent inputs activate adenylate cyclase in a synergistic fashion.

The flux of Ca^{2+} through the AII-'receptor-operated' channel is increased only when PIP_2 turnover is stimulated and C-kinase activated. An increase in flux through this channel leads to a stimulation of C-kinase activity but not to a stimulation of adenylate cyclase.

The integrated cellular response of the adrenal glomerulosa cell is determined by the activities of three different protein kinases: cAMP-dependent protein kinase, CaM-dependent protein kinases, and C-kinase. A sustained response can be produced by the combined activation either of the cAMP- and CaM-dependent kinases, or of the CaM-dependent kinases and C-kinase. Activation of each type of kinase leads to the phosphorylation of a specific subset of intracellular proteins, a few of which may be common substrates for more than one of the kinases, but some of which are distinct substrates for the particular kinase.

This arrangement means that Ca^{2+} influx is the key messenger in determining the magnitude of a cellular response mediated by any of the three extracellular messengers. This can be most dramatically illustrated by analysing the response of these cells to either peptide hormone when the cells are incubated in 2 mM K^+. Under these conditions, the plasma membrane is hyperpolarized, and addition of either ACTH or AII leads to no increase in Ca^{2+} influx rate; i.e., the specific channels are no longer responsive to the hormonal messengers. In the case of ACTH action, this means that an increase in neither intracellular Ca^{2+} nor cAMP concentration occurs (because the cyclase is Ca^{2+}-dependent); the peptide hormone is ineffective in eliciting an aldosterone secretory response. In the case of AII action, this means that Ca^{2+} mobilization from the trigger pool will occur and be enough to induce a transient aldosterone secretory response, but that no increase in Ca^{2+} influx through the other 'receptor-operated' channel will occur, and hence no sustained aldosterone secretory response will be observed because protein kinase C will not be active.

The conclusions from this analysis are (a) that Ca^{2+} influx rate (and hence Ca^{2+} cycling) is the prime determinant of the magnitude of a sustained cellular response in cells whose response is controlled via the Ca^{2+} messenger system; and (b) this is true whether the particular response is mediated via the synarchic interaction of Ca^{2+} and cAMP, or by the coordinate, Ca^{2+}-dependent regulation of CaM-dependent protein kinases and protein kinase C [31].

References

1. Borle, A.B. (1981) Rev. Physiol. Biochem. Physiol. 90, 13–153.
2. Campbell, A.K. (1983) Intracellular Calcium. John Wiley and Sons, New York.
3. Carafoli, E. (1985) J. Mol. Cell Cardiol. 17, 203–212.
4. Rasmussen, H. (1981) Calcium and cAMP as Synarchic Messengers. John Wiley and Sons, New York.
5. Norwycky, M.C., Fox, R.W. and Tsien, R.W. (1985) Nature 316, 440–446.
6. Miller, R.J. (1987) Science 235, 46–52.
7. Bolton, T.B. (1979) Physiol. Rev. 59, 606–718.
8. Smallwood, J., Waisman, D.M., Lafreniere, D. and Rasmussen, H. (1983) J. Biol. Chem. 258, 11092–11097.
9. Vincenzi, F.F. and Larsen, F.L. (1980) Fed. Proc. 39, 2427–2431.
10. Somlyo, A.P. (1985) Circulation Res. 57, 497–506.
11. Putney, J.W., Jr. (1986) Cell Calcium 7, 1–12.
12. Akerman, K.E.O. and Nicholls, D.G. (1983) Rev. Physiol. Biochem. Pharmacol. 95, 149–201.
13. Denton, R.M. and McCormick, J.G. (1986) Cell Calcium 7, 377–386.
14. Rasmussen, H., Kojima, K., Zawalich, W. and Apfeldorf, W, (1984) Adv. Cyclic Nucl. Prot. Phosphorylation Res. 18, 159–193.
15. Berridge, M.J. and Irvine, R.F. (1984) Nature 312, 315–321.
16. Rasmussen, H. and Barrett, P.Q. (1984) Physiol. Rev. 64, 938–984.

17. Rasmussen, H. (1986) N. Engl. J. Med. 314, 1094–1101; 1164–1170.
18. Silver, P.J. and Stull, J.T. (1984) Mol. Pharmacol. 25, 267–274.
19. Park, S. and Rasmussen, H. (1985) Proc. Natl. Acad. Sci. USA 82, 8835–8839.
20. Park, S. and Rasmussen, H. (1986) J. Biol. Chem. 261, 15723–15739.
21. Morgan, J.P. and Morgan, K.G. (1984) J. Physiol. 351, 155–167.
22. Takuwa, Y., Takuwa, N. and Rasmussen, H. (1986) J. Biol. Chem. 261, 14670–14675.
23. Ballermann, B.J. and Brenner, B.M. (1985) J. Clin. Invest. 76, 2041–2048.
24. Kojima, I., Kojima, K. and Rasmussen, H. (1985) Biochem. J. 228, 69–76.
25. Hyatt, P.J., Tait, J.F. and Tait, S.A.S. (1986) Proc. Soc. Lond. B227, 21–42.
26. Wakelam, M.J.O., Murphy, G.J., Hruby, V.J. and Houslay, M.D. (1986) Nature 323, 68–71.
27. Mauger, J.-P., Poggioli, J. and Claret, M. (1985) J. Biol. Chem. 260, 11635–11642.
28. Zawalich, W., Cote, S.B. and Diaz, V.A. (1986) Endocrinology 119, 616–621.
29. Zawalich, W., Takuwa, N., Diaz, V.A. and Rasmussen, H. (1987) Diabetes 36, 426–433.
30. Abedl-Latif, A.A. (1986) Pharmacol. Rev. 38, 227–272.
31. Alkon, D.L. and Rasmussen, H. (1988) Science 239, 998–1005.

B.A. Cooke, R.J.B. King and H.J. van der Molen (eds.)
Hormones and their Actions, Part II

CHAPTER 5

Mechanism of action of pituitary hormone releasing and inhibiting factors

CARL DENEF

Laboratory of Cell Pharmacology, Faculty of Medicine, University of Leuven, Campus Gasthuisberg, B-3000 Leuven, Belgium

Releasing and inhibiting factors control pituitary hormone secretion through binding to cell surface receptors and require intracellular second messengers to couple the extracellular signal for secretion to the intracellular secretory machinery. It is becoming increasingly clear that there is no universal and single messenger for all these releasing factors. In addition to the classical adenylate cyclase-cAMP system, various other intracellular messengers have been identified: the Ca^{2+}-calmodulin system, the inositol polyphosphate–diacylglycerol–protein kinase C system and arachidonic acid derivatives, and the relative importance of each of them appears to differ according to the particular hormone-secreting cell type and releasing factor. This chapter focuses on intracellular messengers underlying the stimulus-secretion coupling of thyrotropin releasing hormone (TRH) and dopamine (DA) on prolactin (PRL) and thyrotropin (TSH) secretion, of luteinizing hormone (LH) releasing hormone (LHRH) on LH secretion, of vasoactive intestinal peptide (VIP) on PRL secretion, of corticotropin (ACTH) releasing factor (CRF) and vasopressin on ACTH and β-endorphin secretion and of growth hormone (GH) releasing factor (GRF) and somatostatin (SRIF) on GH secretion.

Abbreviations: ACTH, corticotropin; CRF, corticotropin releasing factor; GH, growth hormone; GRF, growth hormone releasing factor; HETE, hydroxyeicosatetraenoic acid; IBMX, isobutylmethylxanthine; IP_3, inositol triphosphate; LH, luteinizing hormone; LHRH, luteinizing hormone releasing hormone; NDGA, nordihydroguaiaretic acid; PGE_2, prostaglandin E_2; PI, phosphatidylinositol; PIP, phosphatidylinositolphosphate; PIP_2, phosphatidylinositolbiphosphate; PRL, prolactin; SRIF, somatostatin; TRH, thyrotropin releasing hormone; TSH, thyrotropin; VIP, vasoactive intestinal peptide; 5,6-EET, 5,6-epoxyeicosatrienoic acid.

1. The adenylate cyclase–cAMP system

1.1. TRH

It was originally suggested that activation of adenylate cyclase is one of the early events in the stimulus-secretion coupling of TRH in lactotrophs as well as thyrotrophs. Indirect evidence for this suggestion was obtained from PRL-secreting rat pituitary tumor cells (GH_3 cells) in which TRH was found to rapidly increase intracellular cAMP levels [1–4]. TRH is capable of stimulating a cAMP-dependent protein kinase [2], and cyclic nucleotide analogues [1,2], phosphodiesterase inhibitors and forskolin, an activator of adenylate cyclase, were found to stimulate PRL release [5]. TSH release was also found to be stimulated by agents which increase intracellular levels of cAMP such as cAMP analogues [6], inhibitors of cyclic nucleotide phosphodiesterase [6] as well as by cholera toxin [6]. A number of studies have also shown that TRH increases cAMP content in anterior pituitary explants [6] and cell cultures [6] as well as in a population of anterior pituitary cells enriched in thyrotrophs by gradient sedimentation [6]. However, several of the above findings could not be confirmed by others [6], and several investigators have argued against a primary role of cAMP in mediating TRH stimulation of TSH as well as PRL release. Kotani et al. found that in vivo TRH stimulated TSH release at a much lower dose than that necessary to increase cellular cAMP content [7]. TSH secretion in vivo is blocked by thyroid hormones despite the latter raised pituitary cAMP levels. In addition, Sundberg et al. did not find an increase of TSH release by cholera toxin [8].

Gershengorn et al. [9] found in thyrotropic tumor cells in culture that cAMP analogues, phosphodiesterase inhibitors and cholera toxin caused an increase in TSH release that was additive to that of TRH, that there was no correlation between TSH release and cAMP levels and that there was no change in binding of cAMP to protein kinase, suggesting that cAMP is not the physiological mediator of TRH-induced TSH release.

1.2. VIP

In contrast to TRH, there is general agreement among authors that VIP is a potent and efficient agonist of adenylate cyclase activity in the pituitary and that cAMP functions as a second messenger system for increasing PRL release. VIP stimulates PRL release in tumor as well as in normal lactotrophs [10,11] with an ED_{50} of 30–50 nM [12]. In perifusion systems with high time resolution, release from tumoral GH_4C_1 cells occurred with a delay of 60 s [12]. VIP activated adenylate cyclase with a latency of less than 30 s (the shortest time interval taken), clearly anteceding the onset of PRL secretion [12]. The dose-response curve of cAMP accumulation is superimposable on that for PRL secretion, and effective concentrations of VIP correspond to those found in hypophyseal portal blood [12]. VIP is also considerably more potent than TRH in stimulating cAMP accumulation [12]. Moreover, the time response and pattern of secretion strongly differs among the two peptides [12], the onset of VIP response being slower than that of TRH. These data, taken together with the fact that agents which increase intracellular levels of cAMP stimulate PRL release, clearly demonstrate the second messenger role of cAMP in VIP stimulated PRL secretion.

1.3. DA

Although earlier studies failed to find consistent effects of DA on cAMP levels or adenylate cyclase activity in anterior pituitary cells [13], intact pituitary gland [14,15] or homogenates [13,16], a functional connection between the two is now supported by many experimental approaches. DA and DA agonists inhibit cAMP levels in cultured rat pituitary cells at concentrations in the nanomolar range, comparable to those which inhibit PRL release [17–21]. DA also inhibits cAMP accumulation stimulated by VIP or TRH [20]. Inhibition is also seen in human prolactinoma cells [22].

Inhibition of intracellular cAMP levels is a fast response as it is seen within 1 min of exposure to DA (the shortest exposure time explored) in incubations performed at room temperature, suggesting an even faster response at 37°C [17]. PRL response to DA, measured in a perifusion system of anterior pituitary cell aggregates, also kept at room temperature, showed a lag time of 3 min, showing that inhibition of cAMP levels antecedes that of PRL release [17]. The effects of DA on cAMP levels and PRL release could be blocked by different DA receptor antagonists with roughly similar rank orders of potency [17]. Of most importance is the finding not only that cAMP levels were much higher in cell populations highly enriched ($\approx$ 70–75%) in lactotrophs obtained by unit gravity sedimentation from adult female rats but also that inhibition of cAMP accumulation was much more pronounced in these enriched populations than in populations containing less lactotrophs [17]. In lactotroph-poor populations DA failed to affect cAMP accumulation [17]. Moreover, in lactotroph-rich populations DA is capable of inhibiting basal PRL release even in the absence of the phosphodiesterase inhibitor IBMX [17]. Another argument for a causal relationship between DA, cAMP and PRL release comes from experiments with pertussis toxin, a protein exotoxin produced by *Bordetella pertussis* which modifies the response of a variety of cells to several hormones mainly by blocking inhibitory hormone action at the level of the GTP-dependent inhibitory regulatory subunit of adenylate cyclase. The toxin strongly reduces the ability of DA to inhibit PRL release from anterior pituitary cells in vitro as well as the inhibition by DA of cAMP accumulation [23]. Also in favor of a role of cAMP in DA action is the fact that agents which increase cAMP levels such as IBMX, theophylline and cAMP analogues increase PRL release and partially reverse DA inhibition of PRL release (Ref. 17 and references cited therein) although some authors failed to see such effects (see Ref. 17). The elevation of cAMP in rat anterior pituitary and GH_3 cells and PRL release by cholera toxin are also attenuated by DA [24], although in sheep anterior pituitary cells DA failed to affect cholera toxin-elevated cAMP levels [25]. Forskolin stimulation of the adenylate cyclase enzyme strongly increases intracellular levels of cAMP and PRL release in GH_3 cells, normal pituitary cells, and partially purified pituitary PRL-secreting tumor cells induced by estrogen treatment [26–29]. However, although DA only partially reduced cAMP levels induced by forskolin, it completely blocked the rise of PRL release by forskolin [28,29]. Thus, under certain conditions, DA seems to retain its full potency of inhibiting PRL release even when cAMP levels are elevated. All the above discrepancies, however, cannot be taken as convincing evidence against the role of cAMP in DA action, since the observed partial inhibition of cAMP levels might have represented the fall in lactotrophs, the remainder being the elevated levels in other pituitary cell types not influenced by DA. In particular, folliculostellate cells appear to have an extremely high rate of cAMP accumulation [30,31] and these cells may even be present in pituitary tumors [32].

cAMP also seems to mediate stimulation in PRL gene transcription [33,34] whereas inhibition of PRL gene transcription by DA agonists may be mediated by inhibition of cAMP formation.

1.4. LHRH

As reviewed in more detail elsewhere [35], several investigations have led to the contention that the cAMP messenger system is not coupled to LHRH-evoked gonadotropin release. Although it was originally reported that LHRH increases cAMP levels in hemipituitaries, the increase does not occur before but rather later than LH release. Maximally effective doses of cAMP analogues are able to stimulate LH release but not with the same efficacy as LHRH. Cholera toxin, which increases cAMP levels, does not stimulate LH release from cultured pituitary cells. In cultured cells, LHRH was not able to enhance cAMP levels. However, in mixed cell populations in which gonadotrophs represent a low percentage, cAMP accumulation might be obscured by background cAMP levels, particularly those in lactotrophs which are highly active secretors in culture. In cultured populations which were enriched in gonadotrophs, LHRH did stimulate cAMP accumulation but there was no correlation in response time with LH. During the first 30 min of LHRH stimulation there was LH release but there was no effect on cAMP levels [36]. In later studies evidence was found that cAMP is not a second messenger for release but may take part in intracellular processing of LH and in regulating the tonic availability of LH for release. In primary cultures, agents which increase intracellular levels of cAMP (cholera toxin, forskolin and cAMP analogues) amplified basal and LHRH stimulated LH release after a lag period of 1–4 h [37]. Pertussis toxin had a similar effect [38]. In neither case was there enhanced LH synthesis. Others have shown evidence for a delayed effect of cAMP on glycosylation of LH [39,40].

1.5. CRF

The binding of the 41-residue peptide CRF to its receptor stimulates adenylate cyclase activity in an Mg^{2+}- and GTP-dependent manner [41,42]. In normal pituitary cells as well as in tumoral cells intracellular levels of cAMP rise in response to physiological concentrations of CRF and this enhances the activity of cAMP-dependent protein kinase I [42,43]. As far as tested in ACTH-secreting tumor cells, CRF increases cAMP levels within 30 s, reaching a maximum at 5 min [44] and this rise antecedes that of ACTH release. cAMP analogues and other agents which increase the intracellular levels of cAMP mimic the effect of CRF in stimulating ACTH and β-endorphin release [45]. CRF also induces proopiomelanocortin mRNA and this effect also seems to depend on cAMP although the effect requires 48 h to be established [46]. Thus, all experimental data today are in agreement with a second messenger role for cAMP in the ACTH (and β-endorphin) response to CRF.

1.6. Vasopressin

Vasopressin is now established as a physiological releasing factor of ACTH. Although it stimulates ACTH secretion on its own, the peptide's peculiar activity seems to consist of potentiating the releasing activity of CRF. When applied alone, vasopressin does not act through activation of cAMP accumulation [47]. However, in the presence of CRF, it potentiates CRF-induced cAMP formation most probably through both inhibition of phosphodiesterase and activation of cAMP formation [48]. The latter effect on cAMP formation seems to involve activation of protein kinase C through another second messenger system (see below), which in turn would phosphorylate one of the regulatory components of the adenylate cyclase complex [48].

1.7. GRF and SRIF

These peptides appear to act in concert in regulating GH secretion, GRF being stimulatory and SRIF inhibitory [49].

There is general agreement that GRF uses cAMP as a second messenger system. It has been known for many years that cAMP derivatives stimulate GH in pituitary in in vitro preparations [50,51]. Forskolin and cholera toxin have a similar effect [52]. GRF stimulates the accumulation of intracellular cAMP in cultured pituitary cells as early as 1 min after its addition [53]. In further support of the messenger role of cAMP is the finding that the addition of various agents increasing intracellular cAMP levels to a maximally effective dose of GRF does not further increase

GH release [52], that dexamethasone, which increases the GH response to GRF, also increases the stimulation of cAMP accumulation by GRF but not basal levels [54] and that the ED_{50} of increasing GH (1.6×10^{-12} M) closely matches the ED_{50} of increasing cAMP levels (6×10^{-12} M). The increase of cAMP levels is caused by a stimulation of the adenylate cyclase in a GTP-dependent manner [55].

The GH release inhibitory action of SRIF at least in part involves inhibition of adenylate cyclase [55,56]. Most likely this occurs through coupling with the inhibitory regulatory G protein of the enzyme, since the effect is GTP-dependent and is blocked by pertussis toxin. However, in sheep cells SRIF inhibits GRF-stimulated GH release without any effect on cAMP levels whereas in rats GRF stimulation of GH is completely inhibited by SRIF but cAMP accumulation only in part [57]. An additional site of action of SRIF appears to be at the level of GRF-stimulated Ca^{2+} mobilization (see below).

2. The Ca^{2+} messenger system

The secretion of all pituitary hormones requires extracellular ionic calcium (Ca^{2+}) and many investigations have shown that stimulation of secretion requires an increase of the concentration of free Ca^{2+} in the cytoplasm ($[Ca^{2+}]_i$).

2.1. TRH

There is considerable evidence for a primary messenger role of Ca^{2+} influx in TRH stimulation of TSH and PRL release. High extracellular $[K^+]$, Ca^{2+} ionophores such as A23187 and Ca^{2+} channel agonists such as BAY 8644 and maitotoxin, which all increase cytosolic Ca^{2+} concentration through Ca^{2+} influx from the extracellular compartment, reproducibly provoke TSH and PRL release [6,58,59], whereas removal of Ca^{2+} from the medium in which isolated pituitary cells are incubated or addition of Ca^{2+} channel blockers such as Ca^{2+}, verapamil or D600 depresses secretion [6,60–63]. However, the fact that experimental manipulation of Ca^{2+} influx affects release is by itself insufficient evidence for a primary messenger role of extracellular Ca^{2+} influx and does not rule out the possibility that the rise in $[Ca^{2+}]_i$ is due to intracellular Ca mobilization. Indeed, Ca^{2+} channel blockers also deplete the non-mitochondrial intracellular Ca-pool from which Ca^{2+} can be released to raise $[Ca^{2+}]_i$ [64]. Exposing cells to medium without Ca^{2+} causes a rapid decrease in $[Ca^{2+}]_i$ [65]. Finally, increasing $[Ca^{2+}]_i$ results in net efflux of Ca^{2+} [66].

There is also ample evidence that intracellular Ca^{2+} fluxes are involved in TRH action. However, the relative importance of stimulus-evoked increase of extra- and intracellular Ca^{2+} seems to be different for PRL and TSH release. Geras et al. reported that TRH increased Ca^{2+} uptake in thyrotropic tumor cells (TtT cells) but not in PRL-secreting GH_3 cells [67] whereas verapamil had no effect on TRH-stimulated Ca^{2+} efflux and PRL release from GH_3 cells but depressed TRH-stimulated Ca^{2+} efflux and TSH release from perifused TtT cells [67,68]. Similar results were obtained when normal pituitary cells were used, suggesting that TRH stimulation of lactotrophs involves only mobilization of intracellular Ca^{2+} whereas TRH action in thyrotrophs involves both intracellular Ca^{2+} fluxes and extracellular Ca^{2+} influx, the latter probably serving to support further TSH secretion [68]. However, more recent work has demonstrated a biphasic pattern of intracellular Ca^{2+} fluxes upon TRH stimulation of PRL releasing GH_3 cells [69–73]. This was possible by introducing more refined techniques in which intracellular levels of Ca^{2+} can be accurately determined in individual cells by using fluorescent tetracarboxylate dyes such as Quin-2 [73]. These substances bind Ca^{2+} with consequent changes in fluorescent emission intensity. They can be loaded into cells without disrupting cellular membranes as membrane-permeant esters that are cleaved by cytoplasmic esterases, in this way entrapping the dye intracellularly. Quin-2 is highly specific for Ca^{2+} binding (not affected by Mg^{2+}) and Ca^{2+} binding is not influenced by pH or $[Na^+]$, at least within the physiological range [73]. Digital image analysis with microscope spectrofluorometry allows accurate measurement of intracellular Ca^{2+} in individual cells [73]. Using this technique it has been shown in GH_3 cells that TRH raises intracellular Ca^{2+} within 10 s, followed by a decline towards resting levels over 1.5 min and then a sustained elevation [72]. The first rapid increase was only weakly affected, if at all, by decreasing extracellular Ca^{2+} or by blocking Ca^{2+} influx with nifedipine and verapamil, whereas the second phase was strongly inhibited by these manipulations, clearly suggesting that TRH mobilizes both cellular and extracellular Ca^{2+}, but in a time-dependent fashion [72]. This biphasic response corresponds temporarily with the pattern of PRL secretion from GH_3 cells induced by TRH in a perifusion system [74]. Whether or not these data can be extrapolated to normal pituitary lactotrophs remains to be studied and such studies are mandatory as GH_3 cells differ widely from normal cells. They lack DA receptors, have much lower levels of intracellular PRL stores and also secrete GH. In this respect it is noteworthy that tumoral PRL cells seem to have larger intracellular Ca stores and are less dependent on extracellular Ca^{2+} than normal lactotrophs [27].

2.2. VIP

Using the Quin-2 method for measuring $[Ca^{2+}]_i$, VIP stimulated $[Ca^{2+}]_i$ [12]. However, the latency (20–30 s) was considerably longer than that for TRH and seemed to be slightly longer than the time course of VIP-stimulated cAMP accumulation, suggesting that cAMP may by the messenger for the rise of $[Ca^{2+}]_i$. This is further supported by the finding that forskolin, which stimulates cAMP accumulation, also increases $[Ca^{2+}]_i$, the latency being shorter than in the case of the increment of $[Ca^{2+}]_i$ induced by VIP [12].

2.3. DA

As already mentioned, PRL secretion is a Ca^{2+}-dependent process and experimental data show that a change of $[Ca^{2+}]_i$ is involved in the inhibition of PRL secretion by DA. In partially purified lactotrophs DA decreases $[Ca^{2+}]_i$ and prevents the rise of $[Ca^{2+}]_i$ induced by TRH [75]. Furthermore, PRL release by Ca^{2+} ionophores or Ca^{2+} entry agonists but not by K^+ can be inhibited by DA [76–79]. However, as DA does not affect uptake of Ca^{2+} in either the presence or the absence of A23187 [78], it is likely that DA inhibits $[Ca^{2+}]_i$ at a step after Ca^{2+} uptake or at a step of intracellular Ca^{2+} mobilization. There appears to be a close interrelationship between Ca^{2+} and the cAMP-dependent signaling system. The ionophore A23187 and maitotoxin stimulate cAMP accumulation in cultured rat anterior pituitary cells or hemipituitaries [79,80] although no such effect was seen in cultured ovine pituitary cells [77]. This effect seems to be mediated by calmodulin, an intracellular target of $[Ca^{2+}]_i$, as stimulation of cAMP by A23187 is inhibited by W7, a selective blocker of Ca^{2+}-calmodulin action [80], whereas addition of calmodulin to pituitary homogenates activates adenylate cyclase in the presence of Ca^{2+} [79]. Furthermore, DA is capable of inhibiting the stimulation of cAMP accumulation and PRL release by A23187 [80] or maitotoxin [79]. The complementarity of the Ca^{2+} (+ calmodulin) and cAMP messenger systems is also supported by the findings of Delbeke and Dannies [29]. These authors found that the inhibition of PRL release by DA is only partially overcome by agents mobilizing Ca^{2+} and by forskolin or cAMP analogue. Combining these agents, however, caused a complete reversal of DA inhibition. In further support of this interrelationship are the data reported by Judd et al. [81] that DA is unable to inhibit PRL release in 7315a PRL tumor cells, despite the fact that it does inhibit cAMP accumulation, unless intracellular $[Ca^{2+}]_i$ was raised by maitotoxin or A23187. Thus it seems that inhibition of PRL release is mediated by inhibition of cAMP accumulation but that at some step sufficient $[Ca^{2+}]_i$ has to be mobilized.

2.4. LHRH

There is convincing evidence that elevation of $[Ca^{2+}]_i$ is both necessary and sufficient for enhanced LH release when gonadotrophs are exposed to LHRH [82]. Removal of Ca^{2+} from the incubation medium either by leaving the ion out or by chelating Ca^{2+} with EGTA blocks the release action of LHRH in cultured pituitary cells. Ca^{2+} entry blockers also block LH release although the various Ca^{2+} blockers markedly differ in order of potency when compared with other tissues [83]. Elevation of $[Ca^{2+}]_i$ by using ionophores such as A23187 or ionomycin elicits LH release and only when there is extracellular Ca^{2+} available [81]. Other Ca^{2+} ionophores, such as X537A, do not require extracellular Ca^{2+} to raise LH release but appear to mobilize Ca^{2+} from intracellular stores [81]. Using Quin-2, it has been clearly shown that LHRH enhances $[Ca^{2+}]_i$ with a latency of only 6–8 s [84]. As far as the sites from which Ca^{2+} is mobilized are concerned Conn's group has presented evidence that influx of Ca^{2+} from the extracellular milieu is sufficient for maintaining LH release once started. There was a fall in secretion in less than 2 min after introducing D600 or EGTA to perifused cells under continuous stimulation by LHRH. Moreover the fall in LH release upon removal of LHRH followed a similar time course [85]. Comparable results were obtained by Borgis et al. [86]. Furthermore, preventing intracellular Ca^{2+} mobilization by Dantrolene or TMB-8 was shown not to alter LHRH-stimulated LH release [86]. However, addition of D600 or EGTA or Ca^{2+} withdrawal from the perifusion medium is less effective in blocking the initial response to LHRH than the sustained release phase [87,88]. It should also be noted that Hopkins and Walker, using porcine instead of rat pituitary cell cultures, found that although LHRH stimulates Ca^{2+} influx, inhibition of Ca^{2+} influx by La^{3+} or verapamil did not inhibit the initial phase of LH release [89]. Thus, the relative importance of extra- and intracellular sources of Ca mobilization does not seem to be firmly established. Most recently, however, additional evidence has been presented that the first-phase response to LHRH is based not on influx of extracellular Ca^{2+} but on mobilization of intracellular Ca to the cytosol. Using the Quin-2 method in an enriched population of gonadotrophs, it was found that the rapid rise (within 6–8 s) of $[Ca^{2+}]_i$ induced by LHRH was not abolished in Ca^{2+}-free medium containing EGTA but that this rise was terminated within 2 min [84]. Readdition of Ca^{2+} to the medium provoked a second slower rise of $[Ca^{2+}]_i$ provided LHRH was also added and this resembled the second-phase response of Ca^{2+} to LHRH [84]. The results indicated that the initial response to LHRH depends on intracellular Ca^{2+} whereas sustained release also requires influx of extracellular Ca^{2+}. The primary intracellular target for mobilized Ca^{2+} appears to be calmodulin. During LHRH stimulation calmodulin is redistributed from the cytosol to the plasma membrane and more precisely in the LHRH receptor patch [82]. On the other hand, calmodulin inhibitors strongly inhibit LHRH-stimulated LHRH release [82]. Which calmodulin-dependent cellular function is then activated to lead to secretion remains unknown.

2.5. CRF

As measured by the fluorescent indicator Quin-2 in AtT-20 ACTH-secreting tumor cells, CRF increases $[Ca^{2+}]_i$ although to a smaller extent than other secretagogues do in other cell types [90]. In AtT-20 cells this increase in $[Ca^{2+}]_i$ seems to be dependent on cAMP, since cAMP analogues and forskolin also increase $[Ca^{2+}]_i$. As the effect was blocked by EGTA-chelation of Ca^{2+} or by the Ca^{2+} entry blocker nifedipine, the source of cAMP-induced rise of $[Ca^{2+}]_i$ seems to be extracellular Ca^{2+} [90]. These findings are at least in part consistent with findings using cultured normal pituitary cells. In those cells omission of Ca^{2+} from the incubation medium or addition of nifedipine reduced the ACTH response to CRF [91]. However, others found no effect of the Ca^{2+} channel blocker, verapamil, on CRF action [92]. Moreover, the same authors found that removal of Ca^{2+} from the medium reduced CRF-stimulated ACTH release without affecting cAMP levels. ACTH release induced by 8-bromo-cAMP was also reduced in Ca^{2+}-free medium, whereas the Ca^{2+} ionophore A23187 did not affect ACTH release [92]. All these data suggest that in normal cells an increase of $[Ca^{2+}]_i$ from an intracellular source is required for CRF stimulation of ACTH release but that the role of Ca^{2+} is distal to enhanced cAMP formation. The source of Ca^{2+} mobilization by CRF and the relationship of Ca^{2+} to cAMP require further study. The effect of CRF on proopiomelanocortin mRNA levels also requires Ca^{2+} [46]. Moreover, there is evidence that Ca^{2+} acts in concert with cAMP although the exact interrelationship and the sequence of events remain an open question.

2.6. Vasopressin

Vasopressin when applied alone may use Ca^{2+} from an intracellular source as a primary messenger system for enhancing ACTH release [93]. When acting in the presence of CRF, this system may synergistically interact with the CRF-stimulated cAMP messenger system, which may be another mechanism underlying vasopressin potentiation of ACTH release by CRF. Experimental elevation of $[Ca^{2+}]_i$ caused by high K^+ is indeed synergistic with the stimulation of ACTH release caused by elevating cAMP levels by forskolin [94].

2.7. GRF and SRIF

Using the Quin-2 method it has been shown that GRF rapidly increases $[Ca^{2+}]_i$ in cultured pituitary cells. The response time was 15–20 s, which was slower than the response times for TRH and LHRH measured in the same study [84]. The Ca^{2+} response was specific, as it was only seen in a population enriched in somatotrophs but not in populations enriched in other cell types [84]. GRF stimulates $^{45}Ca^{2+}$ efflux in $^{45}Ca^{2+}$ prelabeled cells in a dose-dependent manner and with a response time closely matching the response time of GH release [95]. These data suggest an early action of GRF on $[Ca^{2+}]_i$ fluxes but do not distinguish whether mobilization is from intra- or extracellular stores. Anyhow, the participation of Ca^{2+} and its target calmodulin in stimulus-secretion coupling of GRF is also suggested by the findings that Ca^{2+}-channel blockers and calmodulin antagonists block GRF and Ca^{2+} ionophore-stimulated GH release (see Ref. 50). Moreover, the Ca^{2+}-calmodulin and the adenylate cyclase system seem to be interdependent but data about the exact sequence of events remain controversial (see Ref. 50). One group of authors found no effect of Ca^{2+} channel blockers or Ca^{2+}-ionophores on cAMP levels but found that cAMP-stimulated GH release was blocked by the Ca^{2+}-channel blocker, suggesting that Ca^{2+} mobilization is distal to stimulation of cAMP accumulation. Others found that Ca^{2+}-ionophores did stimulate cAMP and GH release and that calmodulin blockers inhibited this effect. Still others found that another calmodulin antagonist increased GRF stimulation of GH release and cAMP levels.

As far as SRIF is concerned, there is no unanimity about the messenger role of Ca^{2+} in inhibition of GH release. As reviewed in Ref. 95, in static incubations of normal rat pituitary using non-physiological (μM) concentrations of SRIF, $^{45}Ca^{2+}$ uptake is inhibited by SRIF but this effect is not seen in bovine pituitary cells. GH release stimulated by the Ca^{2+}-ionophore A23187 is inhibited by SRIF but $^{45}Ca^{2+}$ efflux is not. In more recent work, however, Ca^{2+} fluxes were measured in perifused pituitary cells [95]. In this study SRIF was shown to inhibit fractional $^{45}Ca^{2+}$ efflux from prelabeled cells with response times closely parallel to inhibition of GH release. Moreover, the effects were seen at physiological concentrations of SRIF (0.1–1 nM).

3. The inositol polyphosphate–diacylglycerol–protein kinase C system

The hydrolysis of phosphatidylinositol 4,5-biphosphate (PIP_2) by phospholipase C at the inner leaflet of the plasma membrane into inositol 1,4,5-triphosphate (IP_3) and 1,2-diacylglycerol is believed to be a major second messenger system of many hormones acting through plasma membrane receptors [97]. IP_3 appears to mobilize sequestered Ca from an intracellular pool which is not mitochondrial but most probably the endoplasmic reticulum. This causes elevation of free $[Ca^{2+}]_i$. 1,2-Diacylglycerol activates protein kinase C, a phospholipid-dependent protein kinase in the plasma membrane. The latter activation is based on an increase of the affinity of protein kinase C for Ca^{2+}. Enhanced $[Ca^{2+}]_i$ and phosphorylation catalysed by protein kinase C are thought to alter secretory activity and other cellular functions. The coupling between the plasma membrane receptor and phospholipase C may be a GTP-dependent G protein [98].

3.1. TRH

Within a few seconds after exposure of GH_3 cells to TRH, enhanced hydrolysis of PIP_2 by a phospholipase C into IP_3 and 1,2-diacylglycerol is observed [99–101]. TRH receptor occupancy closely correlates with the rate of PIP_2 hydrolysis [99] and there is also a close correlation between the concentration of TRH receptors and TRH stimulation of PIP_2 hydrolysis [102]. It has also been shown that IP_3 formation can be rapidly switched off by displacing TRH from its receptor [102]. Based on all these data it has been proposed that the TRH receptor, when occupied by its ligand TRH, tightly couples to phospholipase C, resulting in activation of the enzyme. In permeabilized GH_3 cells, exogenous IP_3 is capable of directly mobilizing Ca^{2+} from a non-mitochondrial pool [103]. The extremely short latency of IP_3 formation and consequent Ca^{2+} mobilization strongly suggests that this is the sequence of events triggering PRL release in GH_3 cells, at least as far as the first release phase (burst release) is concerned. Mobilization of Ca^{2+} occurs in coincidence with IP_3 formation, i.e. within 5 s IP_3 formation is detectable [99] and within 10 s $[Ca^{2+}]_i$ reaches maximal levels (see earlier). The activation of protein kinase C by 1,2-diacylglycerol results in increased protein phosphorylation in GH_3 cells [104], initiating in this way changes in cellular activities. Bypassing the TRH receptor coupling mechanism by activation of protein kinase C by phorbol esters which mimic the action of 1,2-diacylglycerol also results in PRL release [105] and this has raised the question about the precise individual role of IP_3 and 1,2-diacylglycerol in PRL secretion. In GH_3 cells, the early burst of PRL secretion in response to TRH seems to require only a rise of $[Ca^{2+}]_i$, presumably through IP_3, irrespective of elevation of 1,2-diacylglycerol [106]. Subsequent sustained release occurs by the coordinated effects of raised $[Ca^{2+}]_i$ and 1,2-diacylglycerol activated protein kinase C [106]. There is also evidence that the latter system under certain conditions may function as a negative feedback on $[Ca^{2+}]_i$ and secretion [107]. The increase of IP_3 and $[Ca^{2+}]_i$ caused by TRH is antagonized by phorbol esters given 5 min in advance [107]. These compounds mimic 1,2-diacylglycerol activation of protein kinase C [97]. A similar fall of $[Ca^{2+}]_i$ was observed when TRH-stimulated GH_3 cells were exposed to exogenous phospholipase C, a condition which increases the formation of 1,2-diacylglycerol and protein kinase C activity [107].

Stimulation of IP_3 formation by TRH has also been shown in normal pituitary cells in culture.

3.2. VIP

In contrast to TRH, VIP does not seem to increase PIP_2 hydrolysis [108], a finding suggesting that Ca^{2+} mobilization by IP_3 is not involved in VIP-stimulated PRL secretion. This is consistent with the notion that cAMP is the messenger for this effect.

3.3. DA

The incorporation of labeled phosphate in PI to generate PIP_2 in normal as well as in 7315a PRL-secreting tumor cells appears to be inhibited by DA [109,110]. Whether this effect is related to the hydrolysis of PIP_2 into IP_3 and consequent Ca^{2+} mobilization from endoplasmic reticulum remains to be shown.

3.4. LHRH

There is general agreement that LHRH stimulates the incorporation of labeled phosphate into the phosphoinositides PI, PIP and PIP_2 [111–115]. This has also been shown using enriched populations of rat gonadotrophs in culture [112]. LHRH also increases IP_3 and diacylglycerol levels presumably derived from hydrolysis of PIP_2 by phospholipase C in these preparations [115, 116]. The response time of IP_3 formation is extremely fast, the shortest time reported being 15 s [116]. The formation of IP_2 and IP_1 is considerably slower [116], suggesting PIP_2 is the principal substrate for phospholipase C. The formation of IP_3 is not transient, it is sustained but rapidly switched off when LHRH is displaced from its receptor by an LHRH receptor blocker [116]. LHRH also increases diacylglycerol levels and translocation of protein kinase C from cytosol to the plasma membrane [117]. It has also been shown that signal transduction from the LHRH receptor to phospholipase C in gonadotrophs requires GTP and occurs most probably through a GTP-binding protein (G-protein) [114]. Ca^{2+} ionophores do not affect PI turnover in pituitary cells [115], excluding the possibility that formation of IP_3 is secondary to $[Ca^{2+}]_i$ mobilization induced by LHRH. Taken together these data are consistent with a second messenger role of IP_3 in LHRH-stimulated LH release. This second messenger role may consist of $[Ca^{2+}]_i$ mobilization by IP_3 but definite proof has still to be given. There are also indications for a specific role of 1,2-diacylglycerol. This lipid together with Ca^{2+} activates protein kinase C. Phorbol esters, which can substitute for diacylglycerol in activating protein kinase C, stimulate LH release (see Ref. 82). Synthetic diacylglycerols with different chain lengths showed an order of potency in activating protein kinase C equal to their order of potency in stimulating LH release [82]. Inactive analogues were also ineffective on LH release. Another finding was that the action of diacylglycerols on LH release does not require extracellular Ca^{2+} and is not blocked by calmodulin blockers or D600 [82]. However, there is synergism between the Ca^{2+} ionophore A23187 and diacylglycerol [82]. Taken together these data suggest that diacylglycerol may be a second messenger amplifying the Ca^{2+} messenger system and thus the LHRH-stimulated release process. The exact biochemical processes underlying this action remain unknown.

3.5. CRF and vasopressin

Indirect evidence suggests that CRF does not or only weakly stimulates PIP_2 hydrolysis into IP_3 and 1,2-diacylglycerol. In cultured normal pituitary cells the phorbol esters and the synthetic diacylglycerol dioctanoylglycerol caused a dose-dependent increase of ACTH release to levels equal to those by CRF but they were additive to CRF when given with a maximally effective CRF concentration [118]. Stimulation of endogenous diacylglycerol formation by adding phospholipase C had the same effect(s). However, inhibition of phospholipase C by retinal reduced CRF-stimulated ACTH release [118,119], suggesting the presence of a tonic activity of diacylglycerol formation.

In contrast, vasopressin does stimulate diacylglycerol formation from phospholipid [118,119]. Through this messenger system, vasopressin seems to potentiate

CRF. Activation of protein kinase C by phorbol esters or synthetic diacylglycerols of CRF-stimulated cells mimics the synergistic action of vasopressin and CRF [48]. As mentioned above, this may be mediated by phosphorylation by protein kinase C of one of the regulatory components of CRF-stimulated adenylate cyclase [48].

3.6. GRF and SRIF

Today only limited data are available as to the involvement of IP_3 and diacylglycerol as possible second messenger systems in GRF and SRIF action. GRF stimulates PI turnover, and phospholipase C, phorbol esters and diacylglycerols stimulate GH release from cultured pituitary cells; SRIF is inhibitory (see Ref. 50). However, GRF and cAMP analogues enhance the maximal stimulatory effect of phorbol esters on GH release, suggesting that the principal mechanism mediating GRF's stimulation of GH release is not through the diacylglycerol–protein kinase activation system (see Ref. 50).

4. *Arachidonic acid derivatives*

4.1. TRH

Much of the cellular PI contains the polyunsaturated fatty acid arachidonic acid at position 2. A further possible link between receptor occupancy, PI metabolism and TRH stimulation of PRL secretion is the finding that 1,2-diacylglycerol can supply arachidonic acid from which cyclooxygenase products (prostaglandins, thromboxane and prostacyclin), lipoxygenase products (leucotrienes) and epoxygenase products can be synthesized. Prostaglandins and thromboxanes probably have no important effects on basal PRL secretion in vitro [120]. In addition to phospholipase A_2, which cleaves arachidonic acid from membrane phospholipids to generate arachidonic acid, diacylglycerol lipase appears to be involved in generating arachidonic acid and this polyunsaturated fatty acid seems to play a role in basal as well as TRH-stimulated PRL release [121,122]. In primary cultures of normal anterior pituitary cells, the selective inhibitor of diacylglycerol lipase activity RHC 80207 produces a concentration-dependent reduction of PRL release and abolishes PRL release by TRH [121]. On the other hand agents that increase intracellular arachidonate concentration, such as phospholipase A_2, phorbol myristate acetate, mellitin, a phospholipase A_2 stimulator, or arachidonic acid itself stimulate PRL release in vitro [121,122], whereas phospholipase A_2 inhibitors inhibit PRL release from GH_3 cells, hemipituitary glands and dispersed anterior pituitary cells [121,122]. To determine whether arachidonic acid itself or one of its lipoxygenase, epoxygenase or cyclooxygenase derivatives is involved, the influence of inhibitors of these different pathways on TRH-stimulated PRL release was studied using normal anterior pituitary cells [123,124]. Nordihydroguaiaretic acid (NDGA) and BW755C, both lipoxygenase inhibitors, inhibited stimulated and basal PRL release. Indomethacin, a cyclooxygenase inhibitor, had no significant effect. Lipoxygenase but not cyclooxygenase inhibitors also blocked PRL release stimulated by exogenous arachidonic acid or phospholipase A_2 [123,124]. The lipoxygenase product 5-hydroxyeicosatetraenoic acid (5-HETE) significantly increased PRL release from primary cultures [125] and overcame the inhibition of PRL release by lipoxygenase inhibitors [125]. It has been suggested that leukotriene C_4 may be an important component [126]. The stimulation effect of the Ca^{2+} channel agonist maitotoxin also seems to require conversion of arachidonic acid to lipoxygenase product(s) [126], suggesting an interrelationship between the Ca^{2+} and arachidonic acid derivative messenger systems. Evidence for an opposite relationship, i.e. mobilization of Ca^{2+} by arachidonic acid and PRL release, has also been given [127].

As far as has been studied in mouse TSH-secreting tumor cells, TRH appears to increase arachidonic acid liberation and exogenously added arachidonic acid increases TSH secretion [128]. As is the case with PRL release, there is good evidence that cyclooxygenase products are not directly involved in the stimulus-secretion coupling in thyrotrophs. However, the TSH response to TSH seems to be potentiated by prostaglandin E_2, suggesting a local positive feedback role for this prostanoid [129].

4.2. VIP

Only little information is available as to the role of arachidonic acid derivatives in VIP-stimulated PRL release. As in the case of TRH, agents that inhibit arachidonic acid generation blunt the PRL response to VIP [122].

4.3. DA

There is no evidence that DA has an inhibitory action on the generation of arachidonic acid derivatives. DA inhibition of stimulated PRL release seems to occur at some step after stimulation of release by 5-HETE or other lipoxygenase products, as DA completely blocks 5-HETE-stimulated PRL release [125]. Also the finding that forskolin- and cAMP-induced PRL release is blocked by inhibitors of arachidonic acid generation [122] suggests that the cAMP messenger system and, hence, inhibition by DA act at a step subsequent to the action of arachidonic acid derivatives. There is also evidence that inhibition of PRL release through the Ca^{2+}-calmodulin messenger system operates at a subsequent step, as stimulation of PRL release by maitotoxin, a Ca^{2+}-channel agonist, is blocked by the lipoxygenase inhibitor BW755C [126].

4.4. LHRH

In in vitro anterior pituitary assay systems prostaglandins do not stimulate basal LH release and cyclooxygenase inhibitors do not inhibit LHRH-stimulated LH release [130], indicating that cyclooxygenase products derived from arachidonic acid do not function as second messenger systems in LHRH action. However, various observations suggest that substances generated in the lipoxygenase pathway and perhaps even more importantly in the recently discovered epoxygenase pathway might play a role. Stimulation of gonadotrophs by LHRH provokes arachidonic acid release [131], whereas lipoxygenase inhibitors such as NDGA, ETYA and BW755C blunt LHRH-stimulated LH release [132]. Arachidonic acid stimulates LH release but so do the lipoxygenase product 5-HETE and leucotriene C_4 [131,133]. Since lipoxygenase activity has been demonstrated mainly in enriched populations of gonadotrophs and appears to be very low in lactotroph- and somatotroph-enriched populations [134], a role of oxygenated arachidonic acid derivatives in LH release seems most likely. Of the different substances identified, 12-HETE and 15-HETE did not elicit LH release, 5-HETE did but only at high concentrations [132] and this substance was not detectably formed in gonadotrophs [134]. In a study in which the concentration of NDGA was used not in excess of that inhibiting lipoxygenase, NDGA did not attenuate LHRH-induced LH release [135]. It should be noted, however, that leucotriene C_4 stimulates LH release and this at femtomole concentrations [133]. Other candidates for a second messenger role are products generated by the NADPH cytochrome P-450-dependent epoxygenase pathway. Inhibition of this pathway by eicosatetraenoic acid blocked LHRH-stimulated LH release, whereas addition of several epoxygenated arachidonic acid metabolites, particularly 5,6-epoxyicosatrienoic acid (5,6-EET), potently (nanomolar concentration) stimulate LH release with an efficacy even higher than LHRH [135]. Furthermore, arachidonic acid appears to be metabolized to 5,6-EET in pituitary microsomes (see Ref. 135). Although it has not been shown in which cell type this occurs, it has been speculated that epoxygenated products might increase the membrane permeability for Ca^{2+} [136].

4.5. CRF and vasopressin

There are data indicating that arachidonic acid derivatives may function as a messenger system in ACTH release. A pulse of CRF as short as 2 min causes a 2-fold rise of arachidonic acid release in cultured rat anterior pituitary cells [137]. Exogenously added or endogenously generated (by phospholipase A_2 or mellitin) arachidonic acid stimulates ACTH release [137]. Dose-response curves were parallel to those of CRF [137]. Addition of phospholipase C, but not of phospholipase D, to cultured pituitary cells also stimulates ACTH release, possibly through 1,2-diacylglycerol from which arachidonic acid can be released [137]. However, diacylglycerol might also act by activating protein kinase C and in this way, by phosphorylating one of the regulatory elements of the adenylate cyclase system, stimulate via cAMP formation ACTH release [48]. Inhibitors of lipoxygenase inhibit CRF-stimulated ACTH release whereas cyclooxygenase inhibitors are stimulatory [138,139], suggesting that a lipoxygenase product is responsible for the stimulatory effects of arachidonic acid. However, the lipoxygenase inhibitors used also inhibit the epoxygenase pathway and it has been shown in AtT-20 cells that the selective epoxygenase inhibitor SKF525A is able to completely block CRF- or cAMP-stimulated ACTH release [140], suggesting that arachidonic acid derivatives formed via the epoxygenase pathway or both the lipoxygenase and epoxygenase pathway are involved in stimulation of ACTH release. Epoxygenase products are indeed formed from arachidonic acid in AtT-20 cells (see Ref. 140).

Some data have shed light on the interrelationship between the arachidonic acid derivatives and the other messenger systems. The lipoxy- and epoxygenase inhibitors that block CRF-induced ACTH release also block cAMP-stimulated release [140] but CRF stimulation of cAMP accumulation is not blocked by the above inhibitors [140]. These inhibitors also do not block ACTH secretion stimulated by the Ca^{2+} ionophore A23187 [140]. Taken together, the data suggest a causal relationship between the arachidonic acid derivative(s) and CRF-induced rise of $[Ca^{2+}]_i$ but at a step subsequent to activation of the adenylate cyclase. Other work has suggested that the intracellular signal for arachidonic acid generation in response to CRF may be cAMP or protein kinase C, as 8-bromo-cAMP and phorbol esters (which stimulate protein kinase C) also generate arachidonic acid [137].

The potentiation of the ACTH response to CRF by vasopressin also seems to be based on the arachidonic acid derivative system, as vasopressin-stimulated ACTH secretion is inhibited by lipoxygenase inhibitors [141].

4.6. GRF and SRIF

As reviewed in more detail in Ref. 50, the basal release of GH is strongly stimulated by prostaglandin E_2 (PGE_2), the effect being detectable within 1 min. This effect is inhibited by SRIF. PGE_2 also stimulates cAMP levels in anterior pituitary, an effect also inhibited by SRIF. Since GRF stimulates arachidonic acid release as well as the formation of PGE_2 in vitro and cyclooxygenase inhibitors inhibit GRF-induced GH release, a messenger role for PGE_2 seems possible. However, PGE_2 and GRF show additive effects on GH release even when both are used at a maximal dose. This finding does not support a second messenger role for prostaglandins in GH release, although an interaction between this system and the adenylate cyclase system may be an important modulatory mechanism.

GRF-stimulated GH release is also inhibited by lipoxygenase inhibitors, suggesting participation of some lipoxygenase product(s). However, in this area of research more work is also required to attribute a specific messenger role to these arachidonic acid metabolites.

5. Concluding remarks

Although remarkable progress has been made in our understanding of the mechanism of action of releasing factors many questions remain unresolved. How is the receptor activation signal transduced to Ca^{2+} mobilization, phospholipid turnover and arachidonic acid metabolism? What is the sequence of biochemical changes induced by second messengers? In several cases the exact interrelationship between the different second-messenger systems needs to be defined. A general and unitary mechanism for the various releasing factors does not seem to exist. In those experiments in which the total pituitary cell or even enriched populations are used, it remains uncertain whether molecules such as lipoxygenase products are true second messengers or are generated in other cell types and act as *inter*cellular rather than *intra*cellular messengers. Can findings obtained using tumoral cell lines be extrapolated to normal pituitary cells? Many of the conclusions on the involvement of Ca^{2+}, IP_3 and arachidonic acid metabolites are based on the use of enzyme inhibitors and since many of these drugs are lipophilic substances they may have non-specific membrane effects, implying that certain models need additional experimental evidence. Finally, much remains to be done to settle the question of which Ca^{2+} channels, voltage-dependent or receptor-activated, are involved when Ca^{2+} influx occurs.

References

1. Dannies, P.S., Gautvik, K.M. and Tashjian, A.H., Jr. (1976) Endocrinology 98, 1147–1159.
2. Gautvik, K.M., Walaas, E. and Walaas, O. (1977) Biochem. J. 162, 379–386.
3. Gautvik, K.M. and Lystad, E. (1981) Eur. J. Biochem. 116, 235–242.
4. Labrie, F., Borgeat, P., Lemay, A., Lemaire, S., Barden, N., Drouin, J., Lemaire, I., Jolicoeur, P. and Belanger, A. (1975) Adv. Cyclic Nucleotide Res. 5, 787–801.
5. Delbeke, D., Scammel, J.G. and Dannies, P.S. (1984) Endocrinology 114, 1433–1440.
6. Morley, J.E. (1981) Endocr. Rev. 2, 396–436.
7. Kotani, M., Kobayashi, R., Harada, A., Tsukui, T. and Yamada, T. (1978) Arch. Int. Pharmacodyn. Ther. 234, 145–155.
8. Sundberg, D.K., Fawcett, C.P., Illner, P. and McCann, S.M. (1975) Proc. Soc. Exp. Biol. Med. 148, 54–59.
9. Gershengorn, M.C., Rebecchi, M.J., Geras, E. and Arevalo, C.O. (1980) Endocrinology 107, 665–670.
10. Rosselin, G., Maletti, M., Besson, J. and Rostene, W. (1982) Mol. Cell. Endocrinol. 27, 243–262.
11. Guild, S. and Drummond, A.H. (1983) Biochem. J. 216, 551–557.
12. Bjoro, T., Ostberg, B.C., Sand, O., Gordeladze, J., Iversen, J.-G., Torjesen, P.A., Gautvik, K.M. and Haug, E. (1987) Mol. Cell. Endocrinol. 49, 119–128.
13. Thorner, M.O., Hackett, J.T., Murad, F. and MacLeod, R.M. (1980) Neuroendocrinology 31, 390–402.
14. Zor, U., Kaneko, T., Schneider, H.P.G., McCann, S.M., Lowe, I.P., Bloom, G., Borland, B. and Field, J.B. (1969) Proc. Natl. Acad. Sci. USA 63, 918–925.
15. Schmidt, J. and Hill, L.E. (1977) Life Sci. 20, 789–798.

16. Kimura, H., Calbro, M.A. and MacLeod, R.M. (1976) Fed. Proc. 35, 305.
17. Swennen, L. and Denef, C. (1982) Endocrinology 111, 398–405.
18. Tam, J.W. and Dannies, P.S. (1981) Endocrinology 109, 403–408.
19. Schettini, G., Cronin, M.J. and MacLeod, R.M. (1983) Endocrinology 112, 1801–1807.
20. Barnes, G.D., Brown, B.L., Gard, T.C., Atkinson, D. and Ekins, R.P. (1978) Mol. Cell. Endocrinol. 12, 273–284.
21. Enjalbert, A. and Bockart, J. (1982) Mol. Pharmacol. 23, 576–584.
22. DeCamilli, R., Macconi, D. and Spada, A. (1979) Nature (London) 278, 252–254.
23. Cronin, M.J., Myers, G.A., MacLeod, R.M. and Hewlett, E.L. (1983) Am. J. Physiol. 244, E499–E504.
24. Cronin, M.J. and Thorner, M.O. (1982) J. Cyclic Nucleotide Res. 8, 267–275.
25. Ray, K.P. and Wallis, M. (1982) Mol. Cell. Endocrinol. 27, 139–155.
26. Cronin, M.J., Myers, G.A., MacLeod, R.M. and Hewlett, E.L. (1983) J. Cyclic Nucleotide Res. 9, 245–258.
27. Delbeke, D., Scammell, J.G. and Dannies, P.S. (1984) Endocrinology 114, 1433–1440.
28. Delbeke, D., Scammell, J.G., Martinez-Camps, A. and Dannies, P.S. (1986) Endocrinology 118, 1271–1277.
29. Delbeke, D. and Dannies, P.S. (1985) Endocrinology 117, 439–446.
30. Swennen, L., Baes, M., Schramme, C. and Denef, C. (1985) Neuroendocrinology 40, 78–83.
31. Baes, M., Allaerts, W. and Denef, C. (1987) Endocrinology 120, 685–691.
32. Lauriola, L., Cocchia, D., Sentinelli, S., Maggiano, N., Maira, G. and Michetti, F. (1984) Virchows Arch. (Cell. Pathol.) 47, 189–197.
33. Maurer, R.A. (1982) Endocrinology 110, 1957–1963.
34. Maurer, R.A. (1981) Nature 294, 94–97.
35. Conn, P.M., Marian, J., McMillian, M. and Rogers, D. (1980) Cell Calcium 1, 7–20.
36. Benoist, L., Le Dafniet, M., Rotsztejn, W., Besson, J. and Duval, J. (1981) Acta Endocrinol. 97, 329–337.
37. Cronin, M.J., Evans, W.S., Hewlett, E.L. and Thorner, M.O. (1984) Am. J. Physiol. 246, E44–E51.
38. Cronin, M.J., Evans, W.S., Hewlett, E.L., Rogol, A.D. and Thorner, M.O. (1983) Neuroendocrinology 37, 161–163.
39. Liu, T.-C., Wang, P.S. and Jackson, G.L. (1981) Am. J. Physiol. 241, E14–E21.
40. Liu, T.-C. and Jackson, G.L. (1981) Am. J. Physiol. 241, E6–E13.
41. Holmes, M.C., Antoni, F.A., Szentendrei, T. (1984) Neuroendocrinology 39, 162–169.
42. Aguilera, G., Harwood, J.P., Wilson, J.X., Morell, J., Brown, J.H. and Catt, K.J. (1983) J. Biol. Chem. 258, 8039–8045.
43. Labrie, F., Veilleux, R., Lafevre, G., Coy, D.H., Sueiras-Diaz, J., Schally, A.V. (1982) Science 216, 1007–1008.
44. Litvin, Y., PasMantier, R., Fleischer, N. and Erlichman, J. (1984) J. Biol. Chem. 259, 10296–10302.
45. Axelrod, J. and Reisine, T.D. (1985) Science 224, 452–459.
46. Loeffler, J.P., Kley, N., Pittius, C.W. and Höllt, V. (1986) Endocrinology 119, 2840–2847.
47. Aguilera, G., Harwood, J.P., Wilson, J.X., Morell, J., Brown, J.H. and Catt, K.J. (1983) J. Biol. Chem. 258, 8039–8045.
48. Abou-Samra, A.B., Harwood, J.P., Manganiello, V.C., Catt, K.J. and Aguilera, G. (1987) J. Biol. Chem. 262, 1129–1136.
49. Frohman, L.A. and Jansson, J.O. (1986) Endocr. Rev. 7, 223–253.
50. MacLeod, R.M. and Lehmeyer, J.E. (1970) Proc. Natl. Acad. Sci. USA 67, 1172–1179.
51. Vale, W., Brazeau, P., Rivier, C., Brown, M., Boss, B., Rivier, J., Burgus, R., Ling, N. and Guillemin, R. (1975) Recent Prog. Horm. Res. 31, 365–397.
52. Brazeau, P., Ling, N., Esh, F., Bohlen, P., Mougin, C. and Guillemin, R. (1982) Biochem. Biophys. Res. Commun. 109, 588–594.

53. Michel, D., Lefèvre, G. and Labrie, F. (1983) Mol. Cell. Endocrinol. 33, 255–264.
54. Michel, D., Lefèvre, G. and Labrie, F. (1984) Life Sci. 35, 597–602.
55. Harwood, J.P., Grewe, C. and Aguilera, G. (1984) Mol. Cell. Endocrinol. 37, 277–284.
56. Cronin, M.J., Rogol, A.D., Myers, G.A. and Hewlett, E.L. (1983) Endocrinology 113, 209–215.
57. Law, G.J., Ray, K.P. and Wallis, M. (1984) FEBS Lett. 179, 12–16.
58. Enyeart, J.J. and Hinkle, P.M. (1984) Biochem. Biophys. Res. Commun. 122, 991–996.
59. Login, I.S., Mudd, A.M., Cronin, M.J., Koike, K., Schettini, G., Yasumoto, T. and MacLeod, R.M. (1985) Endocrinology 116, 622–627.
60. Snowdowne, K.W. and Borle, A.B. (1984) Am. J. Physiol. 246, E198–E201.
61. Struthers, A.D., Millar, J.A., Beastall, G.H., McIntosh, W.B. and Reid, J.L. (1984) J. Clin. Endocrinol. Metab. 56, 401–422.
62. Stachura, M.E. (1983) Proc. Soc. Exp. Biol. Med. 173, 109–117.
63. Drust, D.S. and Martin, T.F.J. (1982) J. Biol. Chem. 257, 7566–7573.
64. Rasmussen, H. and Goodman, D.B.P. (1977) Physiol. Rev. 57, 421–509.
65. Moriarty, C.M. and Leuschen, M.P. (1981) Am. J. Physiol. 240, E705–E711.
66. Gershengorn, M.C. (1977) J. Biol. Chem. 255, 1801–1803.
67. Geras, E., Rebecchi, M.J. and Gershengorn, M.C. (1982) Endocrinology 110, 901–906.
68. Geras, E.J. and Gershengorn, M.C. (1982) Am. J. Physiol. 242, E109–E114.
69. Tan, K.-N. and Tashjian, A.H., Jr. (1981) J. Biol. Chem. 256, 8994–9002.
70. Tan, K.-N. and Tashjian, A.H., Jr. (1984) J. Biol. Chem. 259, 427–434.
71. Schlegel, W. and Wollheim, C.B. (1984) J. Cell Biol. 99, 83–87.
72. Gershengorn, M.C. and Thaw, C. (1985) Endocrinology 116, 591–596.
73. Kruskal, B.A., Keith, C.H. and Maxfield, F.R. (1984) J. Cell Biol. 99, 1167–1172.
74. Aizawa, T. and Hinkle, P.M. (1985) Endocrinology 116, 73–82.
75. Shofield, J.G. (1983) FEBS Lett. 159, 79–82.
76. Merritt, J.E. and Brown, B.L. (1984) Life Sci. 35, 707–711.
77. Ray, K.P. and Wallis, M. (1982) Mol. Cell. Endocrinol. 28, 691–703.
78. Tam, S.W. and Dannies, P.S. (1980) J. Biol. Chem. 255, 6595–6599.
79. Schettini, G., Hewlett, E.L., Cronin, M.J., Koike, K. and Yasumoto, T. (1986) Neuroendocrinology 44, 1–7.
80. Schettini, G., Cronin, M.J. and MacLeod, R.M. (1983) Endocrinology 112, 1801–1807.
81. Judd, A.M., Koike, K., Schettini, G., Login, I.S., Hewlett, E.L., Yasumoto, T. and MacLeod, R.M. (1985) Endocrinology 117, 1215–1221.
82. Conn, P.M. (1986) Endocr. Rev. 7, 3–10.
83. Conn, P.M., Rogers, D.C. and Seay, S.G. (1983) Endocrinology 113, 1592–1595.
84. Limor, R., Ayalon, D., Capponi, A.M., Childs, G.V. and Naor, Z. (1987) Endocrinology 120, 497–503.
85. Bates, M.D. and Conn, P.M. (1984) Endocrinology 115, 1380–1385.
86. Borges, J.L.C., Scott, D., Kaiser, D.L., Evans, W.S. and Thorner, M.O. (1983) Endocrinology 113, 557–562.
87. Stern, J.E. and Conn, P.M. (1981) Am. J. Physiol. 240, E504–E509.
88. Zolman, J.C. and Valenta, L.J. (1981) Acta Endocrinol. 97, 26–32.
89. Hopkins, C.R. and Walker, A.M. (1978) Mol. Cell. Endocrinol. 12, 189–208.
90. Luini, A., Lewis, D., Guild, S., Corda, D. and Axelrod, J. (1985) Proc. Natl. Acad. Sci. USA 82, 8034–8038.
91. Sobel, D.O. (1986) Peptides 7, 443–448.
92. Giguere, V., Lefevre, G. and Labrie, F. (1982) Life Sci. 31, 3057–3062.
93. Raymond, V., Leung, P.C.K., Veilleux, R. and Labrie, F. (1985) FEBS Lett. 182, 196–200.
94. Guild, S., Itoh, Y., Kebabian, J.W., Luini, A. and Reisini, T. (1986) Endocrinology 118, 268–279.
95. Login, I.S., Judd, A.M. and MacLeod, R.M. (1986) Endocrinology 118, 239–243.

96. Login, I.S. and Judd, A.M. (1986) Endocrinology 119, 1703–1707.
97. Abdel-Latif, A.A. (1986) Pharmacol. Rev. 38, 227–247.
98. Spiegel, A.M. (1987) Mol. Cell. Endocrinol. 49, 1–16.
99. Martin, T.F.J. (1983) J. Biol. Chem. 258, 14816–14822.
100. Rebecchi, M.J., Kolesnick, R.N. and Gershengorn, M.C. (1983) J. Biol. Chem. 258, 227–234.
101. Macphee, C.H. and Drummond, A.H. (1984) Mol. Pharmacol. 25, 193–200.
102. Gershengorn, M.C. and Paul, M.E. (1986) Endocrinology 119, 836–839.
103. Gershengorn, M.C., Geras, E., Spina-Purello, V. and Rebecchi, M.J. (1984) J. Biol. Chem. 259, 10675–10681.
104. Drust, D.S. and Martin, T.F.J. (1982) J. Biol. Chem. 257, 7566–7573.
105. Negro-Vilar, A. and Lapetina, E.G. (1985) Endocrinology 117, 1559–1564.
106. Kolesnick, R.N. and Gershengorn, M.C. (1986) Endocrinology 119, 2461–2466.
107. Drummond, A.H. (1985) Nature 315, 752–755.
108. Sutton, C.A. and Martin, T.F.J. (1982) Endocrinology 110, 1273–1280.
109. Canonico, P.G., Valdenegro, C.A. and MacLeod, R.M. (1983) Endocrinology 113, 7–14.
110. Judd, A.M., Canonico, P.L. and MacLeod, R.M. (1984) Mol. Cell. Endocrinol. 36, 221–228.
111. Snyder, G.D. and Bleasdale, J.E. (1982) Mol. Endocrinol. 28, 55–63.
112. Raymond, V., Leung, P.C.K., Veilleux, R., Lefèvre, G. and Labrie, F. (1984) Mol. Cell. Endocrinol. 36, 157–164.
113. Harris, C.E., Staley, D. and Conn, P.M. (1985) Mol. Pharmacol. 27, 532–536.
114. Andrews, W.V., Staley, D.D., Huckle, W.R. and Conn, P.M. (1986) Endocrinology 119, 2537–2546.
115. Andrews, W.V. and Conn, P.M. (1986) Endocrinology 118, 1148–1158.
116. Morgan, R.O., Chang, J.P. and Catt, K.J. (1987) J. Biol. Chem. 262, 1166–1171.
117. Hirota, K., Hirota, T., Aguilera, G. and Catt, K.J. (1983) J. Biol. Chem. 260, 3243–3246.
118. Abou-Samra, A.B., Catt, K.J. and Aguilera, G. (1986) Endocrinology 118, 212–217.
119. Negro-Vilar, A. and Lapetina, E.G. (1985) Endocrinology 117, 1559–1564.
120. Hedge, G.A. (1977) Life Sci. 20, 17–34.
121. Canonico, P.L., Cronin, M.J. and MacLeod, R.M. (1985) Life Sci. 36, 997–1002.
122. Camoratto, A.M. and Grandison, L. (1985) Endocrinology 116, 1506–1513.
123. Canonico, P.L., Schettini, G., Valdenegro, C.A. and MacLeod, R.M. (1983) Neuroendocrinology 37, 212–217.
124. Canonico, P.L., Judd, A.M., Koike, K., Valdenegro, C.A. and MacLeod, R.M. (1985) Endocrinology 116, 218–225.
125. Koike, K., Judd, A.M. and MacLeod, R.M. (1985) Endocrinology 116, 1813–1817.
126. Koike, K., Judd, A.M., Login, I.S., Yasumoto, T. and MacLeod, R.M. (1986) Neuroendocrinology 43, 283–290.
127. Kolesnick, R.N., Musacchio, I., Thaw, C. and Gershengorn, C. (1984) Am. J. Physiol. 246, E458–E462.
128. Kolesnick, R.N., Masacchio, I., Thaw, C. and Gershengorn, M.C. (1984) Endocrinology 114, 671–676.
129. Wright, K.C. and Hedge, G.A. (1981) Endocrinology 109, 637–643.
130. Naor, Z., Koch, Y., Bauminger, S. and Zor, V. (1975) Prostaglandins 9, 211–219.
131. Naor, Z. and Catt, K.J. (1981) J. Biol. Chem. 256, 2226–2229.
132. Naor, Z., Vanderhoek, J.Y., Lindner, H.R. and Catt, K.J. (1983) In: Adv. Prostaglandin, Thromboxane and Leukotriene Res., Vol. 12 (Samuelsson, B., Paoletti, R. and Rawwell, P., eds.), pp. 259–263. Raven Press, New York.
133. Hulting, A.L., Lindgren, J.A., Hökfelt, T., Eneroth, P., Werner, S., Patrono, C. and Samuelsson, B. (1985) Proc. Natl. Acad. Sci. USA 82, 3834–3838.
134. Vanderhoek, J.Y., Kiesel, L., Naor, Z., Bailey, J.P. and Catt, K.J. (1984) Prostaglandins Leukotrienes Med. 15, 375–385.

135. Snyder, G.D., Capdevila, J., Chacos, N., Manna, S. and Falck, J.R. (1983) Proc. Natl. Acad. Sci. USA 80, 3504–3507.
136. Volpi, M., Maccache, P.H. and Shaati, R.I. (1980) Biochem. Biophys. Res. Commun. 94, 1231–1237.
137. Abou-Samra, A.B., Catt, K.J. and Aguilera, G. (1986) Endocrinology 119, 1427–1431.
138. Knepel, W. and Meyen, G. (1986) Neuroendocrinology 43, 44–48.
139. Vlaskovska, M., Hertting, G. and Knepel, W. (1984) Endocrinology 115, 895–903.
140. Luini, A.G. and Axelrod, J. (1985) Proc. Natl. Acad. Sci. USA 82, 1012–1014.
141. Vlaskovska, M. and Knepel, W. (1984) Neuroendocrinology 39, 334–342.

B.A. Cooke, R.J.B. King and H.J. van der Molen (eds.)
Hormones and their Actions, Part II

CHAPTER 6

Mechanism of gonadotropin releasing hormone action

LOTHAR JENNES[a,b] and P. MICHAEL CONN[a]

[a]*Department of Pharmacology, University of Iowa, College of Medicine, Iowa City, IA 52242-1109 and* [b]*Department of Anatomy, Wright State University, School of Medicine, Dayton, OH 45435, U.S.A.*

1. *Introduction*

Gonadotropin releasing hormone (GnRH) is a decapeptide which is produced in specific neurons in the brain of all vertebrates. The majority of these GnRH neurons transport the hormone intraaxonally to the median eminence of the ventral hypothalamus where GnRH is released into the perivascular space of fenestrated capillaries. The hormone is then carried via a blood portal system to the anterior pituitary where it stimulates the release of luteinizing hormone (LH) and follicle stimulating hormone (FSH). In addition, GnRH is contained in nerve fibers which project to many intra- and extrahypothalamic areas [1,2] indicating a potential role of the peptide as a neurotransmitter. This is consistent with studies suggesting that GnRH may facilitate or provoke certain reproductive behaviors [3,4].

Because of its importance in endocrine regulation of reproduction, development, and steroid dependent tumorigenesis, GnRH has received increasing attention from various branches of science including pharmacology, physiology, biochemistry, as well as clinical and veterinary medicine [5]. The present article reviews the recent findings on the mechanism of action of GnRH and focuses on the anterior pituitary as the main target of the hormone.

2. *Structure of GnRH*

GnRH is produced as a small portion (10 amino acids) of a large (92 amino acids) precursor molecule. The precursor includes a signal peptide (23 amino acids) fol-

Address correspondence to P. Michael Conn.

lowed by the decapeptide GnRH and an associated peptide consisting of 56 amino acids [6]. During post-translational processing, the amino terminal glutamine residue undergoes cyclization to pyro-glutamic acid, and the carboxyl terminus glycine is amidated. The biologically active structure of GnRH is:

pyro-Glu^1-His^2-Trp^3-Ser^4-Tyr^5-Gly^6-Leu^7-Arg^8-Pro^9-Gly^{10}-NH$_2$ (Fig. 1).

In an aqueous solution, GnRH is believed to assume a configuration resembling the letter C in which the carboxyl-terminus is located close to the amino-terminus [7]. Additional stability in this least energy configuration is generated by the formation of a hydrogen bond between the pyrrolidone carbonyl of pyro-Glu^1 and the glycinamide^{10} [8].

One factor involved in the termination of GnRH action at the pituitary receptor is the degradation of GnRH by proteolytic enzymes [9,10]. The peptide bond between Tyr^5 and Gly^6 appears to be the most susceptible site of cleavage. A proteolytic enzyme which opens this 5–6 bond has been isolated from bovine pituitary. The same enzyme acts secondarily on the Trp^3-Ser^4 peptide bond [11,12]. Another cytoplasmic enzyme which is activated by thiols also attacks the Tyr^5-Gly^6 peptide bond with a secondary action on His^2-Trp^3. An enzymatic activity that cleaves on the amino side of Pro^9 has also been identified. In addition, pyroglutaminases and desamidases have hydrolytic action at both ends of the molecule, although at a slow rate (for review see Ref. 10). Replacement of Gly^6 by hydrophobic D-amino acids results in GnRH analogs which are metabolically stable (and therefore longer acting) [13].

Native sequence GnRH:
pyro-Glu-His-Trp-Ser-Tyr-Gly-Leu-Arg-Pro-Gly-amide

D-Lys^6-GnRH:
pyroGlu-His-Trp-Ser-Tyr-(D-Lys)-Leu-Arg-Pro-Gly-amide

Buserelin (superagonist):
pyroGlu-His-Trp-Ser-Tyr-[D-Ser(*tert*Butyl)]-Leu-Arg-Pro-ethylamide

Leuprolide (superagonist):
pyro-Glu-His-Trp-Ser-Tyr-(D-Leu)-Leu-Arg-Pro-ethylamide

Nafarelin (superagonist):
pyroGlu-His-Tro-Ser-Tyr-[D-Nal(2)]-Leu-Arg-Pro-Gly-amide

Fig. 1. The sequence of naturally occurring GnRH and several analogs described in the text. Buserelin, Leuprolide and Nafarelin are products of Hoescht, Abbott and Syntex, respectively.

Structure-activity studies have shown that both pyro-Glu^1 and Gly^{10} amide are essential for binding of GnRH to its pituitary receptor [8]. Substitution of Gly^{10} amide by ethylamide enhances the affinity of the GnRH analog to the receptor [14]. If both modifications, i.e., a hydrophobic D-amino acid in position 6 and substitution of ethylamide for Gly^{10} amide are combined, a superactive agonist is created. Thus, Buserelin (Hoechst, [D-Ser(tBu)^6-Pro^9-NHEt]-GnRH) is about 20 times as potent as GnRH, Leuprolide (Abbott, [D-Leu^6-Pro^9-NHEt]-GnRH) 50-times and, by increasing the net hydrophobicity, Nafarelin (Syntex, [3-(2-naphthyl)Ala^6]-GnRH) reaches a potency which is about 200-times higher than the naturally occurring releasing hormone (Fig. 1) [15]. However, if the hydrophobicity is further increased as in 3-(dicyclohexyl methyl)-Ala^6-GnRH, the potency is reduced. This may indicate that a hydrophobicity optimum exists for biological activity of GnRH agonists [15].

GnRH antagonists are generally characterized by D-amino acid substitutions in positions 1–3 and 6, based upon some of the same principles of hydrophobicity and degradation resistance employed for agonists. A highly potent modification of the N-terminus is D-p-halo-Phe^{1,2}-D-Trp^3 which is used in combination with the 3-(2-naphthyl)-D-Ala^6 substitution [15,16].

The results of the structure-activity studies indicate that different portions of the GnRH molecule exert different functions. Since pyro-Gly^1 and Gly^{10} are essential for the GnRH molecule to bind to the membrane receptor these 2 amino acids are probably included in the recognition site of the hormone. Amino acids 2 and 3 can be replaced by hydrophobic D-amino acids without total loss of binding, however, such analogs are usually antagonists. Therefore, His^2 and Trp^3 are likely included in an effector activating portion of the GnRH molecule.

3. The biochemical properties of the GnRH receptor

Early attempts to characterize the binding of GnRH to its receptor were problematic since native GnRH is rapidly degraded by proteolytic enzymes. With the availability of enzyme resistant analogs it became clear that GnRH binds to a single class of specific high affinity membrane receptors on pituitary gonadotropes [17]. With the GnRH agonist Buserelin as ligand, a K_a of 2.5×10^{-10} M was measured while the affinity for native GnRH is in the range of $1–2 \times 10^{-9}$ M.

A detailed physico-chemical characterization of the GnRH receptor was not achieved until recently because most conventional detergents destroyed the GnRH binding site of the solubilized proteins. The use of the zwitterionic detergent CHAPS, 3-(3-cholamidopropyl)-dimethylammonio)-1-propane sulfonate, led to the successful isolation of a membrane protein which binds [^{125}I]GnRH with a similar affinity (1.8×10^{-9} M) as a particulate membrane preparation. The binding characteristics of the solubilized protein for various GnRH agonists and antagonists are similar when compared with a membrane preparation indicating that the isolated protein likely is either the functional GnRH receptor or the GnRH binding component of the receptor [18,19]. Photoactivated covalent binding of a ^{125}I-labeled azidobenzoyl-D-Lys^6 GnRH to a pituitary membrane preparation followed by polyacrylamide gel electrophoresis in sodium dodecylsulfate (SDS) showed a single labeled band with an approximate molecular mass of 60 000 Daltons [20]. Similar results were obtained with SDS-PAGE and subsequent immunoblotting [21]. However, after CHAPS solubilization, two labeled bands were identified, one major band with a molecular weight of 59 000 and a minor band at 57 000 [20].

An alternative approach (target size analysis) was used by Conn and Venter [22] in order to estimate the molecular weight of the GnRH receptor. This method relies on the observation that proteins can be destroyed by ionizing radiation. For this approach, the amount of GnRH binding activity remaining after exposure to ionizing radiation is presumed to be inversely related to its molecular weight. The direct destructive effect of ionizing radiation is restricted to a small, 20 Å radius zone but, by transferring vibrational energy, denatures the entire molecule. By comparison with appropriate standard proteins, the molecular weight of the GnRH receptor can be calculated. The relationship between inactivation of GnRH binding and the dose of exposure to ionizing radiation revealed a functional molecular weight of 136 000. Thus, the molecular weight assessed by photoaffinity labeling followed by electrophoresis under reducing conditions (about 60 000) may be identifying only a hormone binding subunit of the receptor.

The chemical composition of the GnRH receptor is not known at present but indirect evidence suggests that the binding protein is glycosylated and contains sialic acid residues. Incubation of pituitary membranes with neuramidase causes a marked reduction in binding of GnRH agonists while the binding of GnRH antagonists is only slightly impaired [23]. Incubation with tunicamycin which inhibits protein glycosylation, causes a time- and dose-dependent reduction in the number of GnRH binding sites without alteration of the affinity of the remaining receptors [24]. It is unclear whether this drug prevents the synthesis of functional receptors or, alternatively, the absence of glycosylation prevents the processing of mature receptor precursors and their appearance at the cell surface as mature, functional receptors.

4. Localization of the GnRH receptor

The binding of GnRH or a GnRH agonist to the pituitary membrane receptor initiates a cascade of events which follow the general pathway of receptor mediated endocytosis. Morphological studies have shown that binding of an agonist leads to aggregation of GnRH receptors at the cell surface followed by the internalization of the ligand.

Initial evidence for such patching and uptake was provided by Hopkins and Gregory [25] who used a D-Lys^6-GnRH coupled via a hemisuccinate bridge to ferritin. The derivative bound to the cell surface of lightly fixed gonadotropes in a random distribution pattern. After 15 min at 37°C the GnRH-marker was concentrated in patches at the cell surface. At the light microscopic level the patches of GnRH receptors could be observed after incubation of living pituitary gonadotropes with a D-Lys^6-rhodamine GnRH conjugate [26]. After 2 h at 4°C the fluorescence was distributed evenly over the entire cell surface. When the cells were warmed for 5 min to 37°C, patching occurred and, after 10 min at 37°C, most of the fluorescent conjugate was localized in intracellular granules. Later electron microscopic studies with [^{125}I]GnRH [27,28], biotinylated GnRH [29] or colloidal gold-GnRH [30,31] confirmed and extended the above findings and increased the resolution of the model. After an initial uniform distribution of GnRH on the cell surface GnRH-colloidal gold markers are taken up by clathrin coated as well as uncoated membrane invaginations. After uptake, GnRH is transported to the lysosomal compartment either directly or after passage through the Golgi apparatus. It appears that receptor-mediated uptake of GnRH does not require the activation of the receptor-effector system since also GnRH antagonists-colloidal gold conjugates [31] and ^{125}I-labeled antagonists [32] are internalized. Uptake of GnRH antagonists is specific, saturable and receptor mediated. However, the rate of uptake of GnRH antagonists is slower when compared to the agonists; antagonists could not be seen in the Golgi apparatus. The differences in the rate of uptake and in the intracellular routing suggest that uptake of antagonists reflects normal membrane protein turnover and that passage through the Golgi apparatus requires the activation of the GnRH receptor by an agonist [31].

The significance of receptor-mediated endocytosis of GnRH by the pituitary gonadotropes is unknown but uptake is not required for the stimulation or maintenance of LH release. If GnRH-containing culture medium is replaced by GnRH-free medium, the LH release stops instantaneously although considerable GnRH has already been internalized [33]. Also, if receptor bound GnRH is prevented from entering the cell, GnRH release continues. This was shown in experiments in which immobilized GnRH (coupled covalently to agarose) was able to stimulate LH release [34]. Incubation of pituitary cells with GnRH in the presence of vinblastin which prevents microtubule formation and thereby receptor patching and capping, was able to stimulate LH release in a proper, dose-dependent manner [33].

5. *Role of receptor microaggregation*

While the above studies show that gross accumulation or patching and endocytosis of the GnRH-receptor complex is not a prerequisite for GnRH action, they do not exclude the possibility that dimerization or 'microaggregation' of the receptors are required to stimulate LH release. For other peptide hormone systems, such as insulin and thyroid stimulating hormone, crosslinking of the receptors appears to be an early step in the hormone action. Incubation of cells with bivalent antibodies to the hormone binding site is able to mimic the hormone effects. The univalent F_{ab} fraction of the same antibodies binds to the receptor in the same manner but it is ineffective in inducing a cellular response. The binding characteristics of the F_{ab} fragment and of the IgG are identical and the only difference is that in the intact IgG, two binding sites are spaced by about 150 Å. From these studies it appears that a key step to activate receptors may involve their microaggregation [35].

The problem to overcome in order to conduct similar studies in the gonadotrope was that no specific antibodies for the GnRH receptor were available. Therefore, two GnRH antagonist molecules were coupled via an ethylene glycol bis-succinimidyl succinate bridge to form a dimer. This antagonist dimer was able to block GnRH action, but was ineffective in stimulating LH release (i.e., a 'pure' antagonist [36]). When the antagonist dimer was bound by specific bivalent antibodies, the antagonist-dimer-antibody complex was able to stimulate LH release. Release was both calcium- and calmodulin-dependent and showed a similar time course to that in response to GnRH. Monovalent F_{ab} preparations of the complex acted as pure antagonists [36]. Incubation of gonadotropes with a GnRH *agonist* dimer-antibody complex resulted in an increase in potency when compared to the agonist alone [37]. Several conclusions can be drawn from these experiments. Receptor aggregation provokes gonadotropin release. The critical distance between two binding sites lies between 15 Å (length of the hemisuccinate spacer between two GnRH antagonist molecules) and 150 Å (distance between two hypervariable regions of an IgG molecule). The chemical nature of the crosslinker is only of secondary importance so long as it permits the proper receptor-receptor interaction.

The importance of crosslinking of GnRH receptors, even by nonspecific means, in the stimulation of LH release was further indicated in experiments in which pituitary cells were incubated with lysine or arginine polymers of varying molecular weights. Polymers of D- or L-lysine and arginine which are positively charged amino acids were able to stimulate LH release [38]. The stimulatory effect required both charge (uncharged amino polymers had no action) and charge separation of at least 120 Å (polymers of molecular weight less than 4000 have no efficacy). Heteropolymers of positively charged and uncharged amino acids in which the distance of 147 Å between two charges is exceeded, are ineffective. Also monomers of D- or L-lysines were unable to stimulate LH release. It appears that a non-specific stimulatory reagent has to fulfill certain criteria: It needs to **a)** have a positive charge, **b)** fulfill a charge separation constraint of ~ 120–150 Å, **c)** have at least two charges within a maximal efficient distance. At present it is, however, not clear where exactly these non-specific LH-releasing polymers bind at the plasma membrane. While the GnRH-receptor is a likely candidate because of its high sialic acid (negative charge) content, other mechanisms are possible. Unspecific crosslinking of membrane proteins could lead to the formation of domains in the plasma membrane in which the chances for random collision of several GnRH receptors is favored [38].

6. Relationships between GnRH receptor number and cellular response

In the intact female rat, GnRH is released from the brain in a pulsatile pattern with low amplitude spikes appearing every 60 min. This pattern is altered during the afternoon of proestrus when an increased, usually longer lasting GnRH surge provokes LH release which is sufficient to induce ovulation [39]. In the pituitary, concomitant changes in the number of GnRH receptors occur throughout the estrus cycle. While the affinity of the GnRH receptors does not change, the binding capacity rises constantly from a low level during estrus to about a 2.5-times high during proestrus [40,41]. A similar parallelism between GnRH receptor number and LH output can be observed during lactation and after castration [42]. The serum LH level in lactating rats is reduced by 50% when compared to regularly cycling rats; a similar reduction is measured in the number of GnRH receptors in the anterior pituitary. On the other hand, serum gonadotropin levels are 5- to 10-fold increased after castration with a concomitant rise in the number of GnRH by 200% [42]. Since the postcastration increase in serum gonadotropin and in the number of GnRH receptors can be prevented by infusion of GnRH antiserum it appears that GnRH itself has an important influence on the number of GnRH receptors in the target cell [43]. Homologous regulation of GnRH receptor could be also shown in vitro in cultured pituitary cells. Incubation with nanomolar concentrations of GnRH or an agonist results in a 150% increase in the number of specific GnRH receptors after 6–9 h [44]. If the GnRH concentration is raised to levels above 10 nM a dramatic loss in the number of specific binding sites is noted. The mechanisms underlying homologous receptor regulation are not fully understood but it appears that up-regulation by low doses of GnRH requires protein synthesis since the increase in the number of GnRH receptors can be prevented by the addition of cyclohexamide [44,45]. Up-regulation is provoked by drugs that allow Ca^{2+} entry and blocked by Ca^{2+} antagonists. The reduction in the number of GnRH receptors after high doses of GnRH may be caused in part by internalization and degradation of the receptors, events which have been described as down-regulation in other hormonal systems.

Simultaneous measurement of the receptor number and cellular response (LH release), showed clearly that under particular circumstances pituitary cell GnRH receptor number did not accurately predict cellular responsiveness. Decreased response to GnRH following prior exposure to the releasing hormone is termed desensitization [46]. The event has important clinical applications since it provides a means to suppress LH release (and consequently steroid synthesis). Situations which result in a dissociation of the number of GnRH receptors present and the response to GnRH are commonly following continuous exposure of gonadotropes to GnRH.

The molecular mechanisms which cause desensitization are not clear and conflicting data exist about the relationship between numbers of GnRH receptors and cellular responsiveness after a desensitizing treatment [47]. Some studies indicate that the reduced cellular response is caused by a loss of receptors (down regulation) [48,49] while other studies show that the total number of membrane GnRH receptors is increased [45,50]. Recent studies favor, however, the view that desensitization can be uncoupled from receptor down-regulation; an uncoupling of the stimulus-effector cascade may be responsible for the reduced cellular response. Thus, under normal physiological conditions occupancy of 20% of the total GnRH receptors by an agonist is sufficient to evoke a full response. If, under down-regulation conditions the total number of available receptors is reduced by 50%, the remaining receptors may be numerous enough to trigger a cellular response [51]. Also, when the internalization of GnRH receptors is prevented by either coupling of GnRH to agarose or by preincubation of the gonadotropes with vinblastine, desensitization persists [52]. Desensitization begins to develop about 20 min after constant GnRH administration and can last for up to 12 h. It takes several days for the gonadotropes to fully recover. This time course does not parallel the changes in the number of GnRH receptors indicating that factors other than variations in the receptor number are responsible [52].

Desensitization requires exposure of the cells to agonists while antagonists do not provoke this event [53]. It appears therefore that occupancy of the GnRH receptor alone is not the step responsible for development of the desensitized state. Agents believed to cause microaggregation results in development of desensitization even when the receptor is occupied by an antagonist [53].

If the culture medium is depleted of calcium (by the addition of EGTA), desensitization to GnRH still occurs even though LH release (calcium-dependent) is virtually completely blocked [54]. Conversely, challenge of pituitary cells with the calcium ionophore A23187 does not cause desensitization, although LH release is evoked [54]. Similarly, incubation of gonadotropes with veratridine or with sn 1,2 dioctanoylglycerol (an activator of protein kinase C) stimulates LH release but does not cause desensitization [55]. Taken together, these experiments show that desensitization is not totally explained by loss of receptor or gonadotropin depletion. The steps governing this event apparently precede mobilization of extracellular Ca^{2+}. A recent study [55a] using the Ca^{2+} ion channel activating toxin maitotoxin to measure the functional state of the ion channel indicates that development of homologous desensitization has two identifiable stages. Early desensitization appears to result from a loss of receptor, but the desensitized state is maintained by loss of functional ion channels even at a time when receptor numbers return to normal.

7. Second messenger systems

Initial studies on second messenger systems which mediate GnRH action focused on cyclic AMP. It was found that addition of GnRH to hemipituitaries causes a 4-

fold increase of intracellular cyclic AMP levels [56]. Similar data could not be obtained in subsequent studies with pituitary cultures. Instead, addition of GnRH to dispersed gonadotropes showed that neither intra- nor extracellular cyclic AMP levels were elevated [57]. Also, cholera toxin (non-specific stimulator of cyclic AMP production), dibutyryl cyclic AMP or cyclic AMP itself were unable to stimulate LH release [58–60].

Cyclic GMP was also suggested to play a role in the signal transduction of GnRH since addition of this peptide to dispersed pituitaries caused a rapid increase in cyclic GMP prior to the release of LH [61]. However, this rise in cyclic GMP was shown to be dependent upon the presence of extracellular Ca^{2+} and it was not required for GnRH action. Neither suppression of cellular cyclic GMP by mycophenolic acid nor elevation of cyclic GMP by sodium nitroprusside stimulated changes in basal or GnRH stimulated LH release [62]. These data indicate that while cyclic GMP levels may rise after GnRH challenge, it is not mandatory for signal transduction and may be involved in secondary, intracellular effects.

8. Calcium as a second messenger

Ca^{2+} plays a major role in the stimulus-secretion coupling of a variety of cells, including neurons, muscle cells or endocrine glands [63]. In order to be accepted as a second messenger, Ca^{2+} must fulfill certain requirements [64]: **a)** removal of extracellular Ca^{2+} must result in an arrest of LH secretion; **b)** the gonadotropes should respond to Ca^{2+} influx with LH release even in the absence of GnRH; **c)** GnRH should provoke the translocation of Ca^{2+} into specific intracellular sites of action that can be related to LH release.

Even before the isolation and synthesis of GnRH were accomplished it was shown that hypothalamic extracts stimulate LH release from hemipituitaries only if extracellular Ca^{2+} was present [65]. Similarly, cultured gonadotropes respond to GnRH only as long as Ca^{2+} is present in the medium. If Ca^{2+} chelators (EDTA, EGTA) are added during the exposure of pituitary cells to GnRH, the release of LH is abolished within 2 min [66]. This inhibition is reversible by addition of extracellular Ca^{2+} [67]. Similar effects were obtained with the Ca^{2+} channel blockers ruthenium red and D-600 (methoxyverapamil), although basal LH release was not impaired [66]. The reduction in LH release in response to Ca^{2+} channel blockers is not caused by an alteration of the GnRH receptors since both the affinity and the number are unchanged [68]. These data indicate that Ca^{2+} influx is mandatory for GnRH action and that the site of action of Ca^{2+} is not the GnRH receptor itself.

The second criterion to accept Ca^{2+} as a second messenger is its ability to stimulate LH release when intracellular concentrations are elevated by any means, even in the absence of GnRH. Incubation of gonadotropes with Ca^{2+} ionophores (A23187 or ionomycin) stimulates LH secretion in a dose- and time-dependent manner

[69,70]. This stimulatory effect of the ionophores is indeed Ca^{2+}-dependent since addition of EGTA to the ionophores leads to an inhibition of the LH release. Also, veratridine which raises intracellular Ca^{2+} levels as well as Ca^{2+} loaded liposomes stimulate LH secretion. These effects are specific for Ca^{2+} since raising the intracellular levels of Mg^{2+} and of monovalent cations shows no effect of LH release [70]. These studies provide evidence that Ca^{2+} influx is required for GnRH action. However, they do not prove that the actual intracellular Ca^{2+} levels are elevated during the GnRH stimulus.

Direct measurement of intracellular Ca^{2+} during GnRH stimulation became recently feasible with the availability of specific fluorescent Ca^{2+} indicators Quin 2 and Fura 2. These dyes can freely enter the cells in the ester form and are then trapped in the cytosol after hydrolysis by nonspecific esterases [71]. Upon GnRH stimulation, the Ca^{2+} binds to the indicators which causes a shift in the emission spectrum of Quin 2 [72]. This effect is rapid (10 s after GnRH stimulation) and it can be elicited only by GnRH agonists but not antagonists [72]. These experiments clearly show that the available intracellular Ca^{2+} levels are elevated during GnRH stimulation. They can not, however, differentiate between incoming Ca^{2+} and Ca^{2+} which has been redistributed from various intracellular storage sites. Exposure of pituitary cells to reagents which stabilize intracellular Ca^{2+} (Dantrolene or 'TMB-8', 8-(N,N-diethylamino)acetyl-3,4,5-trimethyloxy benzoate) had no effect on GnRH stimulated LH release [73]. It can therefore be concluded that GnRH stimulates the influx of extracellular Ca^{2+} and it is this incoming Ca^{2+} and not entirely intracellularly mobilized Ca^{2+} which primarily transmits the signal. A recent study [73a] has identified a temporal portion of LH release dependent on intracellular calcium.

Electrophysiological studies suggest that activation of specific Ca^{2+} channels (or Ca^{2+}-dependent channels) by GnRH binding to the receptor causes the ion influx rather than non-specific diffusion [74]. In gonadotropes, activation of the Ca^{2+} channels does not provoke an action potential but increases the level of membrane voltage noise which is a result of fluctuations in channel conductance. Addition of GnRH to the culture medium outside the high resistance seal of a microelectrode was able to induce the opening of Ca^{2+} channels inside the seal [74]. This finding suggests that GnRH does not bind to the Ca^{2+} channel directly but that instead the stimulatory action of GnRH is transmitted to the Ca^{2+} channel by another intracellular mechanism.

Various Ca^{2+} antagonists have different potencies in the anterior pituitary when compared to their effects in muscle tissue. In gonadotropes, methoxyverapamil is more potent than verapamil in inhibiting veratridine induced depolarization while diltiazem and nifedepine have little effect [75]. These differences in the order of potencies indicate that the pituitary Ca^{2+} channels are distinct from corresponding channels in, for example, muscle cells. In summary, the results of the above studies on the role of Ca^{2+} strongly suggest that this cation enters the gonadotrope during GnRH stimulation through specific membrane channels which are distinct from the GnRH receptor.

One major intracellular target for incoming Ca^{2+} is calmodulin, a protein which regulates a variety of enzyme systems [76]. Calmodulin binds Ca^{2+} with high affinity and, upon binding, undergoes a conformational change allowing it to bind and activate specific intracellular enzymes. The role of calmodulin in the transmission of GnRH stimulation of LH release was evaluated first with a radioimmunoassay measuring cytosol and plasma membrane content of this calcium binding protein. Upon addition of GnRH, calmodulin redistributes in a time- and dose-dependent manner from the cytosolic to the plasma membrane fraction. The total amount of calmodulin was not changed suggesting that de novo synthesis of the protein is not responsible for the increase of calmodulin in the membrane fraction [77]. The translocation of calmodulin was also demonstrated morphologically with immunofluorescence which showed a close spatial association of GnRH receptors with calmodulin containing membrane patches [78].

The accumulation of calmodulin at the plasma membrane upon GnRH stimulation does not prove that calmodulin is a required integral part of the stimulus secretion cascade. However, addition of calmodulin inhibitors, such as penfluoridol or pimozide, to cultured gonadotropes results in a dose-dependent reduction of GnRH induced LH release [79]. Also, pimozide inhibits Ca^{2+}-ionophore induced LH secretion probably by blocking the Ca^{2+} binding sites of calmodulin. The inhibitory effects are not caused by an interaction of the drugs with the GnRH receptor directly since the affinity and number of the GnRH binding sites are unaltered. These results suggest that Ca^{2+} binding to calmodulin is a mandatory step for GnRH action and provide evidence that the third criterion for Ca^{2+} as second messenger in the GnRH stimulus transduction cascade is fulfilled [80].

9. Phospholipids

In many Ca^{2+} activated hormonal or neurotransmitter systems, binding of an agonist stimulates rapid changes in the metabolism of phospholipids [81]. In general, polyphosphoinositides are likely hydrolyzed by phospholipase C to sn-1,2-diacylglycerols (DAGs) and inositol triphosphate [82,83]. DAGs and inositol triphosphate are formed in a 1:1 ratio and follow independent regenerative pathways. Both breakdown products have been suggested to play an informational role in cells. Inositol-1,4,5-triphosphate has been reported to release Ca^{2+} from intracellular pools and DAGs can stimulate protein kinase C. After further metabolism by lipases, DAGs can be metabolized to free arachidonic acid typically at the C-2 position of the diacyl-compound [84]. Phosphorylation of DAG by diglyceride kinase results in the formation of phosphatidic acid, which is suggested to act as an endogenous Ca^{2+} ionophore [85].

In the anterior pituitary, stimulation with a GnRH agonist causes an increase in phosphatidylinositol turnover [86–88]. After stimulation with GnRH, prelabeled

($^{32}PO_4$) cells rapidly incorporate the label into phosphatidylinositols while the levels of phosphatidylcholine, lysophosphatidylcholine and phosphatidic acid are not affected [86]. The time course of the incorporation of the tracer into phosphatidylinositol showed a 10 min lag period when compared to the onset of LH release. This delay may be caused by a masking effect by other pituitary cells which account for the vast majority of cells in this gland [86]. This possibility was supported by subsequent experiments with gonadotrope enriched cultures which showed a rapid (< 1 min) incorporation of ^{32}P into phosphatidic acid and phosphatidylinositol [87]. Later, the additional appearance of ^{32}P tracer in phosphatidylinositol-4-phosphate and phosphatidylinositol-4,5-biphosphate was described [89]. The incorporation of the ^{32}P tracer into the additional metabolites reported by Andrews and Conn [89] was probably caused by the different equilibrium conditions these authors had chosen for the experiments. Since in the studies by Snyder et al. [86] and Raymond et al. [87], a short, non-equilibrium reaching incubation with ^{32}P orthophosphate was used, the pool for de novo synthesis was not reached and the results can only be interpreted as an indication of an increase in turnover of phosphatidylinositol. When equilibrium reaching preincubation conditions (48 h) with ^{33}P orthophosphate are chosen, followed by a 1 h pulse of ^{32}P orthophosphate, a detailed description of turnover as well as mass changes is possible. With this experimental design, a rapid reduction after the GnRH stimulus was noted in ^{33}P-labeled phosphatidylinositol, phosphatidylinositol-4-phosphate and phosphatidylinositol-4,5-bisphosphate [89]. This decline was followed by a subsequent rise in ^{32}P content of inositol phospholipids and in [^{33}P]phosphatidic acid indicating that GnRH had caused a relative mass increase in phosphatidylinositols. Also, the increase in phosphatidic acid suggests an involvement of phospholipase C in the increased turnover of the inositol phospholipids [89].

10. Diacylglycerols

The second branch of the hydrolysis reaction of inositol phospholipids leads to the formation of diacylglycerols (DAGs). Since GnRH was shown to increase inositol phosphate turnover and net production it could be expected that, simultaneously, the DAG turnover is increased. Prelabeling to equilibrium conditions with [^3H]arachidonic acid followed by stimulation with GnRH results in a rapid (1 min) increase in the rate of [^3H]DAG production. This effect appears to be independent of Ca^{2+} influx and of LH release since the addition of the Ca^{2+} channel blocker D-600 did not interfere with the GnRH induced DAG production. On the other hand, the Ca^{2+} ionophore A23187 or phorbol myristate acetate, both of which stimulate LH release in the absence of GnRH, did not stimulate DAG synthesis [89].

The effects of GnRH on the metabolism of inositol phospholipids were studied by preloading the gonadotropes with [^3H]myoinositol. In the presence of Li^+ which prevents the enzymatic breakdown of inositol-1-phosphate, the accumulation of the tritium in inositol phosphates can be used as an index of their production. Addition of GnRH to pituitary cultures stimulated the incorporation of the tracer into inositol phosphate in a time- and dose-dependent manner. Only agonists or dimerized antagonist conjugates which induce GnRH receptor microaggregation were effective. This stimulation was neither Ca^{2+}- [90,91] nor calmodulin-dependent (at doses which reduce LH release by 20 to 60%) and was inhibited by activators of protein kinase C (phorbol myristate acetate and sn-1,2-dioctanoglycerol) [90]. However, since LH release, but not [^{3}H]inositol phosphate production, is maintained after protein kinase C activation it appears that inositol phosphate production may not be mandatory for an appropriate stimulus/secretion coupling [90].

11. GTP binding proteins

Recent studies in other releasing hormone systems showed that the TRH receptor which, upon stimulation by an agonist, induced hydrolysis of inositol phosphates, may be coupled to a GTP binding protein or 'G-protein' [92,93]. In permeabilized gonadotropes, GTP and stable GTP analogs cause a dose-dependent release of LH as well as inositol phosphate hydrolysis [94]. These effects were inhibited in the presence of GnRH-antagonists indicating that a G-protein is closely associated with the GnRH recognition site of the membrane receptor [94]. Reagents which stimulate cyclic AMP production by ADP-ribosylation of G-protein subunits and thereby inhibit inositol phosphate turnover in some systems, did not affect GnRH- or GTP-induced LH release. The treatment was, however, effective in elevating cyclic AMP levels which indicates that the G-protein coupled to the GnRH receptor is probably different from the G-proteins which are involved in adenylate cyclase regulation [94].

12. Protein kinase C

DAGs have been shown to stimulate a Ca^{2+}-dependent protein kinase (PKC) by increasing the affinity of this enzyme for phospholipids and Ca^{2+} [95,96]. Initial evidence for a similar mechanism in the GnRH stimulus transmission was obtained from experiments in which phorbol dibutyrate and phorbol 12-myristate 13-acetate were able to stimulate LH release [53,97,98]. This stimulatory effect of phorbol esters is probably caused by binding and activation of PKC, thereby mimicking endogenous DAGs [99].

In order to obtain more direct evidence for an activational role of DAGs on PKC, water soluble DAGs were synthesized and added to pituitary cells [100]. Synthetic DAGs which had an acyl chain length between 4 and 10 carbons in the sn-1 and 2 position were able to stimulate PKC activity as well as LH release. Structure/activity studies indicate a close correlation between the ability to stimulate LH release and PKC activation [100]. The most potent compound was sn-1,2-dioctanylglycerol (diC$_8$). Replacement of the hydroxyl group by hydrogen, chloride or sulfydryl groups rendered compounds which were inactive in both induction of LH release and stimulation of PKC activity [100].

Phorbol esters and diC$_8$ do not require Ca^{2+} influx in order to activate PKC activity and LH release and are fully active in the presence of EGTA in the culture medium [53,100]. However, when Ca^{2+} efflux from the gonadotropes is stimulated by the addition of EGTA and the Ca^{2+} ionophore A23187, phorbol esters are ineffective. Conversely, minimally effective doses (ED$_{10}$) of phorbol 12-myristate 13-acetate (PMA) and diC$_8$ promote synergistically A23187 stimulated LH release [98]. Similar results were obtained with 1-oleoyl-2-acetylglycerol and 12-O-tetradecanoyl-phorbol-13-acetate which also stimulate Ca^{2+} activated, phospholipid-dependent protein kinase [101].

Upon stimulation by an agonist but not an antagonist, PKC redistributes from the cytosolic pool to the particulate fraction. This redistribution could be measured in vivo [102] and in vitro [103–105] and was shown to be dependent upon Ca^{2+} influx [102]. These data suggest that a minimal intracellular concentration of Ca^{2+} is required for adequate PKC stimulation by DAGs. Since DAG production and translocation of PKC are parallel events it appears that PKC mediates endogenous DAG stimulation of LH secretion.

In order to test this hypothesis the effects of an inhibition of PKC on LH release were studied. Since selective drugs which inhibit PKC activity are not available, exhaustion of the cytosolic PKC pool (probably via proteolysis) by the continuous presence of PMA was used as a model. Preincubation with PMA causes reduced LH release upon subsequent PKC stimulation and loss of all measurable PKC. However the LH response to GnRH or Ca^{2+} ionophores was not impaired as a result of pretreatment with PMA [99]. These data indicate that PKC activation or phosphoinositide hydrolysis is *not* a mandatory event required for GnRH action. It appears therefore that the GnRH agonist induced signal is transmitted through two pathways, one leading to Ca^{2+} influx and binding of Ca^{2+} to calmodulin, the other to the stimulation of inositol phospholipid hydrolysis by a G-protein activated phospholipase C which in turn activates protein kinase C. Both pathways are probably coupled at some point because of a synergism between Ca^{2+} and DAG in the stimulation of LH release. Similar studies using PKC-depleted cells indicate that homologous receptor regulation and desensitization are not mediated by protein kinase C [106].

13. Conclusion

Binding of GnRH to its plasma membrane receptor [108] results in regulation of receptors and of cellular responsiveness, as well as release of the gonadotropins,

LH and FSH. Because the ability to regulate gonadotropin release is important for human and veterinary medicine, there has been a great deal of interest in understanding the molecular basis of GnRH action. Such information is useful for identification of new sites for intervention and anticipation of side-effects. The availability of a large number of agonists and antagonists of GnRH, some of which are metabolically stable and can be labeled to high specific activity, has been useful to understand the subcellular locus, physiological regulation and chemical nature of the receptor. A current model for GnRH action is presented in Fig. 2 [107]. The

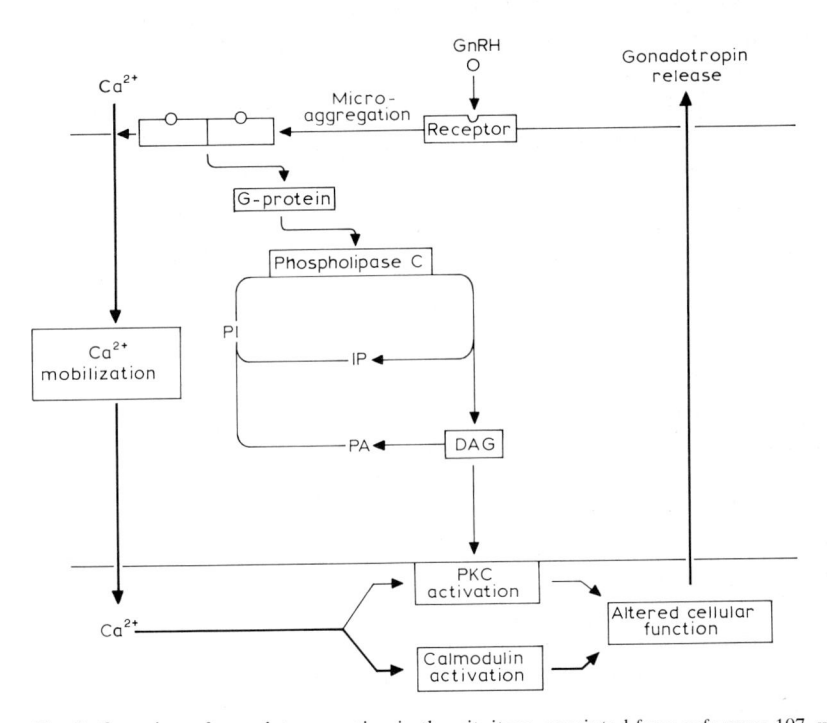

Fig. 2. Overview of gonadotrope action in the pituitary, reprinted from reference 107, with permission of Academic Press. In this model the plasma membrane GnRH receptor [108] is functionally coupled to a calcium ion channel; structure-activity studies with ion channel antagonists indicate that this channel is similar, but not identical, to that observed in nerve and muscle tissue. Calcium mobilized from the extracellular space fulfills the requirement of a second messenger or mediator of GnRH action on gonadotropin release since it is required for release, is mobilized in response to GnRH, and can activate LH release. Calmodulin appears to behave as an intracellular messenger for calcium mobilized in response to GnRH. The receptor is also intimately coupled to a GTP binding protein ("G-protein") which can activate phospholipase C leading to the production of diacylglycerols and, presumably, activate protein kinase C. The production of diacylglycerols, redistribution of protein kinase C, and observation that its activation can synergize with calcium ionophores to provoke gonadotropin release make a role for this enzyme in gonadotropin release attractive. However, the observation that PKC levels and IP production do not appear closely coupled to LH release indicates that the role of this enzyme may be secondary to release or it may subserve other functions. Mathematic models support this theoretical scheme [109–110].

receptor is coupled to a calcium ion channel; structure-activity studies with ion channel antagonists indicate that this channel is similar, but not identical, to that observed in nerve and muscle tissue. Calcium mobilized from the extracellular space fulfills the requirement of a second messenger or mediator of GnRH action on gonadotropin release since it is required for release, is mobilized in response to GnRH, and can activate LH release. Calmodulin appears to behave as an intracellular messenger for calcium mobilized in response to GnRH. The receptor is also intimately coupled to a GTP binding protein ('G-protein') which can activate phospholipase C leading to the production of diacylglycerols and, presumably, activate protein kinase C. The production of diacylglycerols, redistribution of protein kinase C, and observation that its activation can synergize with calcium ionophores to provoke gonadotropin release make a role for this enzyme in gonadotropin release attractive. However, the observation that PKC levels and IP production do not appear closely coupled to LH release indicates that the role of this enzyme may be secondary or it may subserve other functions. Mathematical models support this theoretical scheme [109, 110].

Acknowledgement

This work was supported by HD19899.

References

1. Barry, J. (1979) Int. Rev. Cytol. 60, 179–221.
2. Jennes, L. and Stumpf, W.E. (1980) Cell Tissue Res. 209, 239–256.
3. Moss, R.L. and McCann, S.M. (1973) Science 181, 177–179.
4. Pfaff, D.W. (1973) Science 183, 1148–1149.
5. Conn, P.M. (1986) Endocrine Reviews: GnRH, Vol. 5, pp. 3–10.
6. Seeburg, P.H. and Adelman, J.P. (1984) Nature 311, 666–668.
7. Momany, F.A. (1976) J. Am. Chem. Soc. 98, 2990–2996.
8. Coy, D.H., Seprodi, J., Vilchez-Martinez, J.A., Pedroza, E., Gardner, J. and Schally, A.V. (1979) In: Central Nervous Systems Effects of Hypothalamic Hormones and Other Peptides. (Collu, R., Barbeau, A., Ducharme, J.R. and Rochefort, J.G., eds.) pp. 317–323. Raven Press, New York.
9. Griffiths, E.C. and Kelly, A.J. (1979) Mol. Cell. Endocrinol. 14, 3–17.
10. Lapp, C.A. and O'Connor, J.L. (1986) Neuroendocrinology 43, 230–238.
11. Wilk, S. and Orlowski, M. (1980) J. Neurochem. 35, 1172–1182.
12. Horsthemke, B. and Bauer, K. (1981) Biochem. Biophys. Res. Commun. 103, 1322–1328.
13. Coy, D.H., Vilchez-Martinez, E.J., Coy, E.J. and Schally, A.V. (1976) J. Med. Chem. Soc. 19, 423–425.
14. Fujino, M., Fukada, S., Shinagawa, S., Kobayashi, S., Yamazaki, I. and Nakayama, R. (1974) Biochem. Biophys. Res. Commun. 60, 406–413.
15. Nestor, J.J. Jr., Ho, T.L., Tahilramani, R., McRae, G.I. and Vickery, B.H. (1984) In: LHRH and Its Analogues. (Labrie, F., Belanger, A. and Dupont, A., eds.) pp. 24–35. Elsevier, New York.

16. Erchegyi, J., Coy, D.H., Nekola, M.V., Coy, E.J., Schally, A.V., Mezo, I. and Teplan, I. (1981) Biochem. Biophys. Res. Commun. 100, 915–920.
17. Clayton, R.N. and Catt, K.J. (1981) Endocr. Rev. 2, 186–209.
18. Capponi, A.M., Aubert, M.L. and Clayton, R.N. (1984) Life Sci. 34, 2139–2144.
19. Hazum, E., Schvartz, I., Waksman, Y. and Keinan, D. (1986) J. Biol. Chem. 261, 13043–13048.
20. Hazum, E. (1981) Endocrinology 109, 1281–1283.
21. Eidne, K.A., Hendricks, D.T. and Millar, R.P. (1985) Endocrinology 116, 1792–1795.
22. Conn, P.M. and Venter, J.C. (1985) Endocrinology 116, 1324–1328.
23. Hazum, E. (1982) Mol. Cell. Endocrinol. 26, 217–222.
24. Schvartz, I. and Hazum, E. (1985) Endocrinology 116, 2341–2346.
25. Hopkins, C.R. and Gregory, H. (1977) J. Cell Biol. 75, 528–540.
26. Hazum, E., Cuatrecasas, P., Marian, J. and Conn, P.M. (1980) Proc. Natl. Acad. Sci. U.S.A. 77, 6692–6695.
27. Pelletier, G., Dube, D., Guy, J., Seguin, G. and Lefebvre, F.A. (1982) Endocrinology 111, 1068–1076.
28. Duello, T.M., Nett, T.M. and Farquhar, M.G. (1983) Endocrinology 112, 1–9.
29. Childs (Moriarty), G.V., Naor, Z., Hazum, E., Tibolt, R., Westlund, K.N. and Hancock, M.B. (1983) Peptides 4, 549–555.
30. Jennes, L., Stumpf, W.E. and Conn, P.M. (1983) Endocrinology 113, 1683–1689.
31. Jennes, L., Coy, D.H. and Conn, P.M. (1986) Peptides 7, 459–463.
32. Wynn, P.C., Suarez-Quian, C.A., Childs, G.V. and Catt, K.J. (1986) Endocrinology 119, 1852–1863.
33. Conn, P.M. and Hazum, E. (1981) Endocrinology 109, 2040–2045.
34. Conn, P.M., Smith, R.G. and Rogers, D.C. (1981) J. Biol. Chem. 256, 1098–1100.
35. Jacobs, S., Chang, K.J. and Cuatrecasas, P. (1978) Science 200, 1283–1284.
36. Conn, P.M., Rogers, D.C., Stewart, J.M., Niedel, J. and Sheffield, T. (1982) Nature 296, 653–655.
37. Conn, P.M., Rogers, D.C. and McNeil, R. (1982) Endocrinology 111, 335–337.
38. Conn, P.M., Rogers, D.C., Seay, S.G. and Stanley, D. (1984) Endocrinology 115, 1913–1917.
39. Levine, J.E. and Ramirez, V.D. (1982) Endocrinology 111, 1439–1448.
40. Clayton, R.N., Solano, A.R., Garcia-Vela, A., Dufau, M.L. and Catt, K.J. (1980) Endocrinology 107, 699–706.
41. Marian, J., Cooper, R.L. and Conn, P.M. (1981) Mol. Pharmacol. 19, 399–405.
42. Clayton, R.N., Detta, A., Naik, S.I., Young, L.S. and Charlton, H.M. (1985) J. Steroid Biochem. 23, 691–702.
43. Clayton, R.N., Popkin, R.M. and Fraser, H.M. (1982) Endocrinology 110, 1116–1123.
44. Conn, P.M., Rogers, D.C. and Seay, S.G. (1982) Mol. Pharmacol. 25, 51–55.
45. Smith, M.A., Perrin, M.H. and Vale, W.W. (1983) Mol. Cell. Endocrinol. 30, 85–96.
46. Badger, T.M., Loughlin, J.S. and Naddaff, P.G. (1983) Endocrinology 112, 793–799.
47. Schuiling, G.A., Moes, H. and Koiter, T.R. (1984) Acta Endocrinol. 106, 454–458.
48. Nett, T.M., Crowder, M.E., Moss, G.E. and Duello, T.M. (1981) Biol. Reprod. 24, 1145–1155.
49. Heber, D., Dodson, R., Stoskopf, C., Peterson, M. and Swerdloff, R.S. (1982) Life Sci. 30, 2301–2308.
50. Keri, G., Nikolics, K., Teplan, I. and Molnar, J. (1983) Mol. Cell. Endocrinol. 30, 109–120.
51. Naor, Z., Clayton, R.N. and Catt, K.J. (1980) Endocrinology 107, 1144–1152.
52. Gorospe, W.C. and Conn, P.M. (1987) Endocrinology 120, 222–229.
53. Smith, W.A. and Conn, P.M. (1984) Endocrinology 114, 553–559.
54. Smith, W.A. and Conn, P.M. (1983) Endocrinology 112, 408–410.
55. Jinnah, H.A. and Conn, P.M. (1986) Endocrinology 118, 2599–2604.
55a. Conn, P.M., Staley, D.D., Yasumoto, T., Huckle, W.R. and Janovick, J.A. (1987) Mol. Endocrinol. 1, 154–159.
56. Borgeat, P., Chavancy, G., Dupont, A., Labrie, F., Arimura, A. and Schally, A.V. (1972) Proc. Natl. Acad. Sci. U.S.A. 69, 2677–2681.
57. Conn, P.M., Morrell, D.V., Dufau, M.L. and Catt, K.J. (1979) Endocrinology 104, 448–453.

58. Wakabayashi, K., Date, Y. and Tamaoki, B. (1973) Endocrinology 92, 698–704.
59. Sundberg, D.K., Fawcett, C.P. and McCann, S.M. (1974) J. Int. Res. Commun. 2, 1178–1188.
60. Tang, L.K.L. and Spies, H.G. (1974) Endocrinology 94, 1016–1021.
61. Naor, Z., Fawcett, C.P. and McCann, S.M. (1978) Am. J. Physiol. 235, 586–590.
62. Naor, Z. and Catt, K.J. (1980) J. Biol. Chem. 255, 342–344.
63. Douglas, W.W. and Poissner, A.M. (1964) J. Physiol. 172, 1–18.
64. Robison, G.A., Butcher, R.W. and Sutherland, E.W. (1971) Cyclic AMP. Academic Press, New York.
65. Samli, M.H. and Geshwind, I.I. (1968) Endocrinology 82, 225–231.
66. Marian, J. and Conn, P.M. (1979) Mol. Pharmacol. 16, 196–201.
67. Conn, P.M. and Rogers, D.C. (1979) Life Sci. 24, 2461–2466.
68. Marian, J. and Conn, P.M. (1980) Life Sci. 27, 87–92.
69. Hopkins, C.R. and Walker, A.M. (1978) Mol. Cell. Endocrinol. 12, 189–195.
70. Conn, P.M., Rogers, D.C. and Sandhu, F.S. (1979) Endocrinology 105, 1122–1127.
71. Tsien, R.Y., Pozzan, T. and Rink, T.J. (1982) J. Cell Biol. 94, 325–334.
72. Clapper, D.A. and Conn, P.M. (1985) Biol. Reprod. 32, 269–278.
73. Bates, M.D. and Conn, P.M. (1984) Endocrinology 115, 1380–1385.
73a. Hansen, J., McArdle, C. and Conn, P.M. (1987) Mol. Endocrinol. 1, 808–815.
74. Mason, W.T. and Waring, D.W. (1986) Neuroendocrinology 43, 205–212.
75. Conn, P.M., Rogers, D.C. and Seay, S.G. (1983) Endocrinology 113, 1592–1595.
76. Cheung, W.Y. (1980) Science 207, 19–27.
77. Conn, P.M., Chafouleas, J.G., Rogers, D.C. and Means, A.R. (1981) Nature 292, 264–265.
78. Jennes, L., Bronson, D., Stumpf, W.E. and Conn, P.M. (1985) Cell Tissue Res. 239, 311–315.
79. Conn, P.M., Rogers, D.C. and Sheffield, T. (1981) Endocrinology 109, 1122–1126.
80. Conn, P.M., Bates, M.D., Rogers, D.C., Seay, S.G. and Smith, W.A. (1984) In: The Role of Drugs and Electrolytes in Hormonogenesis (Fotherby, K. and Pal, S.B., eds.) pp. 85–103. Walter de Gruyter, Berlin.
81. Sekar, M.C. and Hokin, L.E. (1986) J. Membrane Biol. 89, 193–210.
82. Michell, R.H. (1975) Biochim. Biophys. Acta 415, 81–147.
83. Berridge, M.J. and Irvine, R.F. (1984) Nature 312, 315–321.
84. Farese, R.V. (1983) Endocr. Rev. 4, 78–95.
85. Putney, J.W. Jr., Weiss, S.J., Van De Walle, C. and Haddas, R.A. (1980) Nature 284, 345–347.
86. Snyder, G.D. and Bleasdale, J.E. (1982) Mol. Cell. Endocrinol. 28, 55–63.
87. Raymand, V., Leung, P.C.K., Veilleux, R., Lefevre, G. and Labrie, F. (1984) Mol. Cell. Endocrinol. 36, 157–164.
88. Kiesel, L. and Catt, K.J. (1984) Arch. Biochem. Biophys. 231, 202–210.
89. Andrews, W.V. and Conn, P.M. (1986) Endocrinology 118, 1148–1158.
90. Huckle, W.R. and Conn, P.M. (1987) Endocrinology 120, 160–169.
91. Naor, Z., Azrad, A., Limor, R., Zakut, H. and Lotan, M. (1986) J. Biol. Chem. 261, 12506–12512.
92. Gilman, A.G. (1986) Trends Neurosci., 460–463.
93. Straub, R.E. and Gershengorn, M. (1986) J. Biol. Chem. 261, 2712–2717.
94. Andrews, W.V., Staley, D.D., Huckle, W.R. and Conn, P.M. (1986) Endocrinology 119, 2537–2546.
95. Nishizuki, Y. (1984) Science 225, 1365–1370.
96. Takai, Y., Kishimoto, A., Iwasa, Y., Kawahara, Y., Mori, T. and Nishizuki, Y. (1979) J. Biol. Chem. 254, 3692–3695.
97. Smith, M.A. and Vale, W.W. (1980) Endocrinology 107, 1425–1431.
98. Harris, C.E., Staley, D. and Conn, P.M. (1985) Mol. Pharmacol. 27, 532–536.
99. McArdle, C.A., Huckle, W.R. and Conn, P.M. (1987) J. Biol. Chem. 262, 5028–5035.
100. Conn, P.M., Ganong, B.R., Ebeling, J., Staley, D., Neidel, J.E. and Bell, R.M. (1985) Biochem. Biophys. Res. Commun. 126, 532–539.

101. Naor, Z. and Eli, Y. (1985) Biochem. Biophys. Res. Commun. 130, 848–853.
102. McArdle, C.A. and Conn, P.M. (1986) Mol. Pharmacol. 29, 570–576.
103. Hirota, K., Hirota, T., Aguilera, G. and Catt, K.J. (1985) J. Biol. Chem. 260, 3243–3246.
104. Hirota, K., Hirota, T., Aguilera, G. and Catt, K.J. (1986) Arch. Biochem. Biophys. 249, 557–562.
105. Naor, Z., Zer, J., Zakut, H. and Hermon, J. (1985) Proc. Natl. Acad. Sci. U.S.A. 82, 8203–8207.
106. McArdle, C.A., Gorospe, W.C., Huckle, W.R. and Conn, P.M. (1987) Mol. Endocrinol. 53, 131–140.
107. Conn. P.M., Huckle, W.R., Andrews, W.V. and McArdle, C.A. (1987) Rec. Progr. Hormone Res. 43, 29–68.
108. Marian, J. and Conn, P.M. (1983) Endocrinology 112, 104–112.
109. Blum, J.J. and Conn, P.M. (1982) Proc. Natl. Acad. Sci. U.S.A. 79, 7307–7311.
110. Leiser, J., Conn, P.M. and Blum, J.J. (1986) Proc. Natl. Acad. Sci. U.S.A. 83, 5963–5967.

CHAPTER 7

The mechanisms of action of luteinizing hormone. I. Luteinizing hormone-receptor interactions

B.A. COOKE[a] and F.F.G. ROMMERTS[b]

[a]*Department of Biochemistry and Chemistry, Royal Free Hospital School of Medicine, University of London, Rowland Hill Street, London NW3 2PF, England and* [b]*Department of Biochemistry, The Medical Faculty, Erasmus University, Rotterdam, The Netherlands*

1. Introduction

Luteinizing hormone (LH, lutropin) is a glycoprotein and is secreted by the anterior pituitary in the male and female mammal in response to LH releasing hormone (LHRH, gonadotropin releasing hormone (GnRH)) from the hypothalamus. Its site of action in the male is the Leydig cells in the testis, where it stimulates the production of the male androgen testosterone. The latter controls the male secondary sex characteristics and spermatogenesis. In the female, LH interacts with the theca and the granulosa cells and, together with follicle stimulating hormone (FSH), controls the differentiation of the follicles, ovulation and steroidogenesis. In the male, FSH specifically controls the differentiation and activity of the Sertoli cells in the seminiferous tubules.

There are many similarities in the action of LH on Leydig and ovarian cells; LH couples to a similar specific receptor and to similar transducing systems in the plasma membrane of these cells. Progress has been made in the elucidation of these regulatory pathways and it is now recognized that, in addition to cyclic AMP, other factors such as calcium, calmodulin, inositol phosphates and arachidonic acid metabolites are involved. Recently, progress has also been made in the search for the elusive protein regulators of steroid biosynthesis. In addition to the stimulatory effects of LH on steroid production, other actions of this protein hormone, e.g., its trophic effects and desensitizing effects have been studied in more detail.

It is the purpose of this review to discuss some of the advances made in our understanding of the above actions of LH. Although both the ovary and the testis will be dealt with, more emphasis will be placed on the latter. Detailed aspects in relation to the action of FSH on the ovary are discussed in Chapter 9. For other recent reviews on the testis and the ovary the reader is referred to Refs. 1–6.

2. The structure of LH

LH is a glycoprotein hormone which like thyroid stimulating hormone (TSH), FSH and chorionic gonadotropins (CGs) contains two non-identical polypeptide chains (α and β). The carbohydrate content is approximately 15–30% of the molecular mass for the pituitary hormones, whereas the placental hormone contains up to 45% carbohydrate. The carbohydrate residues are linked to the polypeptide chains through N-glycosidic linkages to either the amide group of asparagine or the hydroxyl group of serine or threonine residues. Qualitative and quantitative variations occur in the carbohydrate contents of the glycoproteins from different species. Sequence analysis of the amino acid residues has shown that the α chains of the glycoproteins are remarkably similar and the β chains are quite different. Dissociation of LH into the α and β subunits can be achieved under nondenaturing conditions indicating that they are linked by noncovalent bonds. The dissociated units are biologically inactive. Reassociation at 4°C can be achieved with restoration of most of the original activity [7]. It has been suggested that the β subunit of LH determines the biological specificity of the molecule, but this can only be achieved in the presence of the α subunit.

The total molecular weight of LH is approximately 30 000, with the α subunit being 15 750 and the β subunit 15 350. The number of amino acid residues for each subunit is different; 89 and 129, respectively. The molecular weight of hCG is 36 000. The biological properties of hCG are similar to those of LH, but hCG is not so rapidly metabolized in vivo as LH and it also binds to more high affinity LH binding sites than LH (see Section 3.2). Because relatively large amounts of purified material are available and because of the stability of hCG and its similar action to LH it is often used in place of LH in in vivo and in vitro studies on the testes and ovaries.

The functional significance of the carbohydrate moieties in LH and the CGs has been investigated by deglycosylating these hormones chemically – the most effective way being with anhydrous hydrogen fluoride [8]. This results in removal of approximately 80% of the carbohydrates with a recovery of the deglycosylated hormone of up to 70%. The polypeptide chains and the quaternary structure apparently remain intact. The affinity of deglycosylated LH for rat luteal, porcine granulosa and testis preparations remains unchanged. However, deglycosylated hormones have a low stimulatory or no effect on adenylate cyclase and have a low steroidogenic potency. Their immunological activity is unaffected [8]. They are thus extremely useful tools for investigations into the mechanisms of action of the native hormones.

3. The LH receptor

3.1. Purification and characterization

In the male, the plasma membrane receptor for LH is located in the testis Leydig cells and in the female in the ovarian theca, granulosa and luteal cells. Progress in the purification and characterization of the LH receptor has not been so rapid as with other peptide hormone receptors. This is because there are relatively few receptors per cell and they are very labile. Consequently the yields of purified receptors in the early studies were very low. In addition the purified receptors had a lower affinity for LH compared with receptors in intact membranes. However, the use of techniques developed for other receptors and the application of the stabilizing effect of glycerol on the LH receptor [9] has now led to the isolation and purification of the LH receptor to homogeneity from ovarian and Leydig cell preparations in good yields and retention of full binding activity. However there is disagreement about the existence of subunits and their molecular size.

The studies of the testis LH receptor indicated that it had a molecular weight of 194 000 and that it exists as a dimer with both subunits having a molecular weight of 100 000 [10–13]. In contrast, photoaffinity labelling [14,15] and chemical cross-linking [16] of LH with ovarian membranes suggests that the rat ovarian receptor consists of three or four subunits. It is possible that in addition to tissue differences there are species differences in the numbers and size of the subunits, because different results have been obtained from bovine [17] and porcine ovaries [18].

Bruch et al. [19] have purified the LH receptor in high yield from superovulated rat ovaries. Using two different methods it was found that the receptor consisted of four non-identical subunits with molecular weights of 79 300, 66 300, 55 300 and 46 700 as indicated by PAGE electrophoresis under reducing conditions. Under non-reducing conditions, a molecular weight of 268 000 was obtained. The isolated receptor retained its LH/hCG binding properties. In disagreement with this conclusion are the results of Rajaniemi and co-workers [20] and Kusuda and Dufau [21]; they report that the rat ovarian receptor is a monomer of molecular weight 98 000 and 73 000, respectively. A recent report from Kim et al. [22] also concludes that the LH receptor in Leydig tumour cells is a monomer of molecular weight 92 000. They point out that the collagenase preparation commonly used to diperse gonadal cells contains contaminating enzyme(s) that cause partial proteolysis of the LH/hCG receptor. Also, the plasma membranes of granulosa/luteal cells cause similar proteolysis of the LH/hCG receptor particularly where molecular weights of less then 90 000 have been found.

The data available therefore indicate that the LH receptor has a molecular weight of approximately 73–100 000 in the testes and the ovaries. Oligomers may exist but further confirmatory studies are required.

3.2. Interaction of LH with its receptor

It has been shown by Dighe and Moudgal [23], using antibodies to the two polypeptide subunits, that the initial site of interaction of hCG with the LH receptor is through the β subunit. Further studies [24], again using antibodies to the subunits, found that after binding the α subunit of the hormone became masked. This suggested a two step model for the binding of the hormone to its receptor. Initially the binding of hCG is a specific low affinity interaction of the β unit, which brings about a high affinity binding of the α subunit to the LH receptor. Hwang and Menon [16] came to similar conclusions in their studies.

Many studies have been carried out to determine the number of LH binding sites per cell and binding affinity of the LH receptor. Conflicting results have been obtained especially with regard to the number of receptors per cell. Using mature rat testis Leydig cells, approximately 13–30 000 LH receptors per cell have been found [25,26] and 23 000 per cell were reported using rat Leydig cell homogenates [27]. In porcine Leydig cells 60 000 receptors/cell have been found [28]. Various values have also been reported for the binding affinity of LH for its receptor although this is consistently high and in the range of 10^9 to 10^{11} M^{-1} [29–31].

The reasons for the reported variability in the numbers of binding sites per Leydig cell is possibly due to species differences and different procedures especially when intact cells are used. At temperatures greater than 4°C an over-estimation of the numbers of receptors is likely to occur. This is because of the temperature sensitive endocytic pathway in rat Leydig cells (see Section 3.3) [32]. At 4°C the cell associated [^{125}I]hCG remains bound only to cell surface receptors and thus gives a more accurate measure of total receptor numbers/cell; i.e., approximately 4000 [33]. This is much lower than the values obtained in the studies referred to above.

The binding of several polypeptide hormones including LH to their plasma membrane receptors has been shown not to be completely reversible (see Ref. 34). Katikineni et al. [35] have investigated this phenomenon in whole testis homogenates and shown that there is a progressive change in the degree of reversibility of binding of the LH/hCG-receptor complex. There are differences in the binding and effects of CG and LH from different species. Huhtaniemi and Catt [30] have reported differential binding affinities in rat testis homogenates for hCG, human LH (hLH) and ovine LH (oLH). Scatchard analysis of the binding data revealed a single set of binding constants for hCG and hLH with a K_a of 2.5–7.6 × 10^{10} M^{-1}. In contrast two sets of binding sites were found for oLH; about 20% with a K_a of 7 × 10^{10} M^{-1} and the remainder with a K_a of 7×10^8 M^{-1}. The total numbers of binding sites were however found to be the same for all three hormones. The higher affinity of the hCG contributes to the higher bioactivity of the hCG compared with oLH. Solubilization of the homogenates was found to decrease the differences between hCG and LH binding. This suggested that the differences were not due to heterogeneity of the receptor molecule itself but to membrane-mediated interactions between the receptors.

3.3. LH receptor recycling and synthesis

During short term exposure (up to 24 h) of Leydig cells in vitro or in vivo to LH/hCG there is no net loss of LH receptors [36]. It has now been demonstrated that although there is no detectable change in receptor numbers, the LH receptor is apparently in a highly dynamic state being continually internalized and recycled back to the cell surface [32,37] (Fig. 1). The endocytic pathway was found to have two temperature sensitive steps. At 4°C movement of the hormone receptor-complex inside the cell did not occur, and at 21°C hormone accumulated within the cytoplasm but was not degraded or released from the cell. At 34°C, internalization, degradation and loss of the degraded hormone fragments from the cell occurred [32]. Whether recycling of the LH receptor plays a part in the continued response of the Leydig cell to LH over relatively short periods of time has yet to be determined. The importance of receptor recycling in the ovarian cells has not been investigated. Also the fact that the actual number of receptors may be far fewer than previously thought (i.e., there may be not so many 'spare receptors') may have important im-

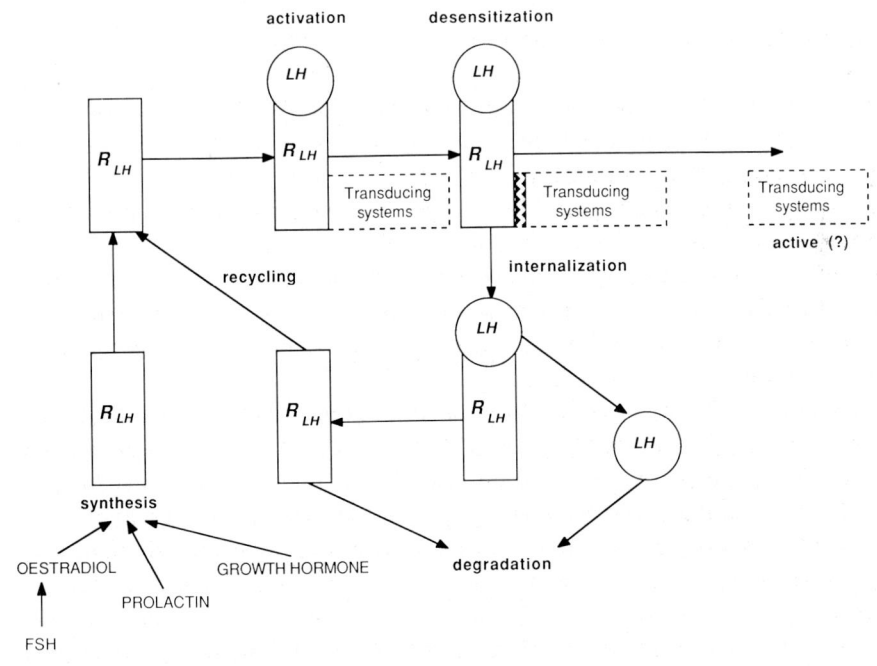

Fig. 1. A model for the control of the luteinizing hormone receptor. LH interacts with a specific receptor (R_{LH}) in Leydig cells which results in the activation of several transducing systems. This is followed by an uncoupling (desensitization) from one or more of the transducing systems. The latter may transiently remain active. The LH receptor complex is internalized and then dissociated in endocytic vesicles. LH is degraded in the lysosomes. The LH receptor synthesis can be controlled by several hormones including oestradiol, prolactin and growth hormone.

plications for the relationship between receptor occupancy, cellular control of surface receptor number and the response of the cell to hormonal stimulation.

3.4. Regulation of LH receptors

In the ovarian follicular cells there is a continuous change in LH receptor numbers depending on the stage of the ovarian cycle. FSH seems to be the initial signal for the synthesis of receptors involved in the differentiation of the granulosa cells. This trophic hormone is capable of inducing receptors for itself, LH, prolactin, EGF, lipoproteins and possibly β-adrenergic ligands (for review see Ref. 5). Once induced other hormone regulated receptors are capable of heteroregulatory effects, e.g., prolactin, insulin and platelet-derived growth factor (PDGF) have positive regulatory effect on LH receptors. The intracellular steroid receptors also play a role in membrane receptor regulation, especially oestradiol receptors, which are present in the granulosa cells [5]. Knecht et al. [38] have shown that inhibition of oestrogen synthesis by an aromatase inhibitor in immature rats prevented FSH stimulated increases in LH receptors thus indicating an essential role for oestrogens.

Cyclic AMP is one of the mediators of FSH action on LH receptor induction because dibutyryl cyclic AMP will increase granulosa receptors in vitro. Other factors may also be involved (e.g., oestrogens) because the cyclic AMP analogues were less potent than FSH. Once induced, the continual presence of FSH is required to maintain normal levels of LH receptors [5].

The role of oestrogens in controlling LH receptor synthesis in the testis is not known but oestradiol receptors are present in rat Leydig cells [39]. Glucocorticoids inhibit the induction of LH receptors; corticosteroid receptors have been demonstrated in rat testis [40] and treatment of prepubertal rats [41] and immature hypophysectomized rats with corticosteroids decreased the numbers of LH receptors and steroid production.

In the immature rat the Leydig cells gradually acquire LH receptors during development; the binding of hCG increases between 2 and 11 days, although the numbers of Leydig cells remain constant during that time. Later in development there is a good correlation between the number of LH receptors per gram of tissue and the volume density of the interstitial tissue indicating that the number of receptors per Leydig cell is relatively constant [42]. The neonatal testis responds to a low dose of hCG treatment with a large increase in LH receptors, whereas no such effect, or only a minor increase was seen in the adult. Other studies have shown however that the initial effects of LH/hCG administration to mature rats is accompanied by changes in receptor number which occurs in two distinct phases. First, within a few hours there is a transient rise in the LH receptor content. After several hours the elevated receptor number returns to normal and subsequently decreases below control levels [44]. This initial up-regulation has also been shown in vitro providing the cells are attached to plastic culture wells [45]. There were no corresponding changes in cyclic AMP or testosterone production indicating that these additional receptors may not be functionally coupled.

After hypophysectomy there is a rapid loss of LH receptors and androgen secretion by the testis. Treatment with LH alone will restore Leydig cell steroidogenesis but also causes even further decreases in testicular LH receptors. Continued treatment with LH alone leads to Leydig cell hyperplasia and an increase in LH receptors. However, even with marked Leydig cell hyperplasia in response to sustained elevations of endogenous LH, as in the tfm rat (see Ref. 6 for other references), the LH receptor content of the testis remains subnormal.

Prolactin has been shown to increase LH receptor numbers in dwarf mice, seasonally repressed hamsters and hypophysectomized rats (see Ref. 6 for other references). In the hypophysectomized rats the combined effects of prolactin, growth hormone and LH were necessary to maintain the LH receptors [46]. The induction of hyperprolactinemia leads to increased LH receptors. Decreases in serum prolactin levels caused by treatment with compounds that inhibit the release of prolactin (dopamine analogues) decrease LH receptors (see Ref. 6 for other references).

FSH also increases LH receptor numbers in rat testes but this has only been observed in immature hypophysectomized rats. This effect of FSH can be blocked by administration of oestrogens which is the opposite of their effects in the ovary. The mechanism of action of this effect of FSH on LH receptors is not known. Presumably FSH is acting via stimulation of the Sertoli cells because there are no receptors for FSH on Leydig cells. Paracrine factors secreted by the seminiferous tubules and having a positive effect on Leydig cell function have been demonstrated [4]. The described effects of FSH may have been due to the small amounts of LH contaminating the FSH preparations [47]. Treatment with the estimated quantities of LH alone had no effect on the in vitro LH responses [48]. The availability of very pure FSH should now enable workers to determine the exact effect of FSH alone.

References

1. Birnbaumer, L. and Cooke, B.A. (1983) Research in Reproduction Wall Map. (Edwards, R.G., ed.) IPPF, 18-20 Lower Regent St., London.
2. Hsueh, A.J.W., Adashi, E.Y., Jones, P.B.C. and Welsh, T.H. (1984) Endocr. Rev. 5, 76–127.
3. Cooke, B.A. (1983) Curr. Topics Membrane Trans. 18, 143–177.
4. Sharpe, R.M. (1986) Clin. Endocrinol. Metab. 15, 185–207.
5. Adashi, E.Y. and Hsueh, A.J.W. (1984) The Receptors, Vol. I. (Conn, P.J., ed.) pp. 587–636. Academic Press, London.
6. Catt, K.J., Harwood, J.P., Clayton, R.N., Davies, T.F., Chan, V., Katikineni, M., Nozu, K. and Dufau, M.L. (1980) Rec. Progr. Horm. Res. 36, 557–622.
7. Parsons, T.F., Strickland, T.W. and Pierce, J.G. (1985) Methods in Enzymology, Hormone Action, Part I. (Birnbaumer, L. and O'Mally, B.W., eds.) pp. 736–749. Academic Press, London.

8. Manjunath, P. and Sairam, M.R. (1985) Methods in Enzymology, Part I, Peptide Hormones, (Birnbaumer, L. and O'Mally, B.W., eds.) pp. 725–735. Academic Press, London.
9. Ascoli, M. (1983) Endocrinology 113, 2129–2134.
10. Dufau, M.L., Charreau, E.H. and Catt, K.J. (1973) J. Biol. Chem. 248, 6973–6982.
11. Dufau, M.L., Ryan, D.W., Baukal, A.J. and Catt, K.J. (1975) J. Biol. Chem. 250, 4822–4824.
12. Pahnke, V.G. and Leidenberger, F.A. (1978) Endocrinology 102, 814–823.
13. Rebois, R.V., Omedeosale, F., Brady, R.O. and Fishman, P. (1981) Proc. Natl. Acad. Sci. U.S.A. 78, 2086–2091.
14. Ji, T.H. and Ji, H. (1981) Proc. Natl. Acad. Sci. U.S.A. 78, 5465–5469.
15. Rapoport, B., Hazum, E. and Zor, U. (1984) J. Biol. Chem. 259, 4267–4271.
16. Hwang, J. and Menon, K.M.J. (1984) Proc. Natl. Acad. Sci. U.S.A. 81, 4667–4671.
17. Dattatreyamurty, B., Rathnam, P. and Saxena, B.B. (1983) J. Biol. Chem. 258, 3140–3158.
18. Wimalasena, J., Abel, J.A., Wiebe, J.P. and Chen, T.T. (1986) J. Biol. Chem. 261, 9416–9420.
19. Bruch, R.C., Thotakura, R. and Bahl, Om.P. (1986) J. Biol. Chem. 261, 9450–9460.
20. Keinanen, K.P. and Rajaniemi, H.J. (1986) Biochem. J. 239, 83–87.
21. Kusuda, S. and Dufau, M.L. (1986) J. Biol. Chem. 261, 16161–16168.
22. Kim, I.C., Ascoli, M. and Segaloff, D.L. (1987) J. Biol. Chem. 262, 470–477.
23. Dighe, R.R. and Moudgal, N.R. (1983) Arch. Biochem. Biophys. 225, 490–499.
24. Milius, R.P., Midgley, A.R. and Birken, S. (1983) Proc. Natl. Acad. Sci. U.S.A. 80, 7375–7379.
25. Tsitouras, P.D., Kowatch, M.A. and Harman, S.M. (1979) Endocrinology 105, 1400–1405.
26. Dehejia, A., Nozu, K., Catt, K.J. and Dufau, M.L. (1982) J. Biol. Chem. 257, 13781–13786.
27. Huhtaniemi, I.T., Nozu, K., Warren, D.W., Dufau, M.L. and Catt, K.J. (1982) Endocrinology 111, 1711–1720.
28. Mather, J.P., Saez, J.M. and Haour, F. (1982) Endocrinology 110, 933–940.
29. Huhtaniemi, I.T. (1983) Clin. Endocrinol. Metab. 12, 117–132.
30. Huhtaniemi, I.T. and Catt, K.J. (1981) Endocrinology 108, 1931–1938.
31. Mendelson, C., Dufau, M. and Catt, K.J. (1975) J. Biol. Chem. 250, 8818–8823.
32. Habberfield, A.D., Dix, C.J. and Cooke, B.A. (1986) Biochem. J. 233, 369–376.
33. Habberfield, A.D., Dix, C.J. and Cooke, B.A. (1987) Mol. Cell. Endocrinol. 51, 153–161.
34. Abramowitz, J., Iyengar, R. and Birnbaumer, L. (1979) Mol. Cell. Endocrinol. 16, 129–146.
35. Katikineni, M., Davies, T.F., Huhtaniemi, I.T. and Catt, K.J. (1980) Endocrinology 107, 1980–1987.
36. Catt, K.J., Harwood, J.P., Aguilera, G. and Dufau, M.L. (1979) Nature (London) 280, 109–116.
37. Ascoli, M. (1982) J. Biol. Chem. 257, 13306–13311.
38. Knecht, M., Brodie, A.M.H. and Catt, K. (1985) Endocrinology 117, 1156–1161.
39. Brinkman, A.O., Mulder, E., Lamers-Stahlhoten, G.T.M., Mechielsen, M.J. and van der Molen, H.J. (1972) FEBS Lett. 26, 301–305.
40. Evain, D., Morera, A.M. and Saez, J.M. (1976) J. Steroid Biochem. 7, 1135–1151.
41. Saez, D., Morera, A.M., Haour, F. and Evain, D. (1977) Endocrinology 101, 1256–1263.
42. Bambino, T.H. and Hsueh, A.J.W. (1981) Endocrinology 108, 2142–2148.
43. Huhtaniemi, I.T., Nozu, K., Warren, D.W., Dufau, M.L. and Catt, K.J. (1982) Endocrinology 111, 1711–1720.
44. Huhtaniemi, I.T., Katikineni, M., Chan, V. and Catt, H.J. (1981) Endocrinology 108, 58–64.
45. Baranao, J.L.S. and Dufau, M.L. (1983) J. Biol. Chem. 258, 7322–7330.
46. Zipf, W.B., Payne, A.H. and Kelch, R.P. (1978) Endocrinology 103, 595–600.
47. Purvis, K., Clamsson, O.P.F. and Hansson, V. (1979) J. Reprod. Fertil. 56, 657–665.
48. van Beurden, W.M.O., Roodnat, B.S. and van der Molen, H.J. (1978) Int. J. Androl. (Suppl.) 2, 374–382.

CHAPTER 8

The mechanisms of action of luteinizing hormone. II. Transducing systems and biological effects

F.F.G. ROMMERTS[a] and B.A. COOKE[b]

[a]*Department of Biochemistry, The Medical Faculty, Erasmus University, Rotterdam, The Netherlands and* [b]*Department of Biochemistry and Chemistry, Royal Free Hospital School of Medicine, University of London, Rowland Hill Street, London NW3 2PF, England*

1. LH receptor transducing systems

Receptors for protein hormones can be coupled to different transducing systems. Most of the data from studies approximately 10 years ago showed that many receptors are coupled to the adenylate cyclase system. However in the early 1980s it became clear that calcium fluxes and phospholipid turnover could also be important in conveying hormonal signals. During the last few years the number of papers describing the effects of hormones on phospholipid metabolism and calcium fluxes has increased tremendously and has outnumbered studies on cyclic AMP.

Recently various investigators have also reported that LH can raise levels of intracellular calcium and inositol 1,4,5-trisphosphate in addition to cyclic AMP production. It is difficult to conclude anything from the number of papers about the relative importance of the various messenger systems. It merely shows that the concepts of regulatory systems depend largely on the available techniques and are significantly influenced by trends and fashion [1]. Moreover, since it is now recognized that hormones exert pleiotropic effects inside the cell, which finally result in different physiological responses (protein synthesis, enzyme activation, secretion, growth, differentiation, cellular movements, etc.), it is difficult to link specific transducing systems to particular responses. This is emphasized by the fact that the various transducing systems may interact at the cell membrane level resulting in inhibition or amplification of the initial signals.

These introductory remarks are important when the available experimental data on the mechanisms of action of LH are interpreted.

1.1. Formation of cyclic AMP

Stimulation of Leydig cells or LH responsive ovarian cells with LH results in the stimulation of adenylate cyclase activity [2–4]. Stimulation occurs within several minutes after addition of the hormone, is dose dependent and can also be demonstrated in isolated membranes. The stimulatory effects of LH on the adenylate cyclase system are most probably mediated by the stimulatory GTP binding protein (G_s). Cholera toxin, which specifically activates the G_s protein, increases cyclic AMP production and steroidogenesis in intact Leydig cells [5,6]. hCG and LH stimulation of adenylate cyclase in plasma membranes from Leydig cells is increased in the presence of GTP and non-hydrolysable GTP analogues, again indicating the involvement of a G_s protein [7]. The coupling between the receptor and the adenylate cyclase system can also be negatively controlled by protein kinase C as was shown with tumour-promoting phorbol esters [8].

The inhibitory GTP binding protein (G_i) which has been demonstrated in many different cell types is also present in Leydig cells. Evidence for this was obtained from studies with cultured Leydig cells in which it was shown that the inhibitory effects of arginine vasotocin on steroidogenesis could be abolished by pertussis toxin, which inactivates G_i via ADP ribosylation [9]. However, effects of LH with or without vasotocin were not reported. Similar studies with forskolin [10] demonstrated that low concentrations of this compound inhibited hCG-stimulated cyclic AMP production and this inhibition was prevented by pertussis toxin. Recently, direct evidence for the presence of G_i in rat Leydig cells has been obtained by the demonstration of a 41 000 protein (α_i) after [^{32}P]ADP ribosylation [11].

The level of cyclic AMP inside the cell is the balance between synthesis, breakdown and release. Increased intracellular levels of cyclic AMP can therefore also be obtained by inhibition of cyclic AMP metabolism with phosphodiesterase inhibitors. There are many reports of the use of these inhibitors to demonstrate increased steroidogenesis at submaximal (but not maximal) concentrations of LH, thus confirming the involvement of cyclic AMP.

Effects of LH on other cyclic nucleotides have not been reported. Cyclic GMP production in mouse Leydig cells has recently been demonstrated but the stimulation of its synthesis could only be obtained with atriopeptins [12].

1.2. The phosphoinositide cycle

The formation of inositol phosphates, including inositol 1,4,5-trisphosphate (IP_3) has been demonstrated in ovarian cells [13,14]. The levels of these compounds are raised for periods of seconds to minutes after addition of LH to isolated cells. However, these LH-induced changes are an order of magnitude smaller than the changes in cyclic AMP. The transient and small effects of LH on polyphosphoinositide turnover has made it difficult to establish accurate kinetics and dose response characteristics. Similarly, correlations of these primary signals with steroid production are difficult. In view of the LH-induced increases in intracellular calcium that do occur it is possible that IP_3 is acting in this capacity as shown in other cell types. However, cyclic AMP may be a more important regulator of intracellular calcium especially in Leydig cells [15] (see Section 1.4).

Another consequence of the activated breakdown of phospholipids by phospholipase C is the liberation of diacylglycerol. This compound can directly stimulate protein kinase C and can be further metabolised to release arachidonic acid and eicosanoids. There are no reports on LH-induced changes in intracellular diacylglycerol levels in gonadal cells, although effects on production of eicosanoids have been described [16,17].This does not, however, necessarily mean that phospholipase C is implicated since liberation of arachidonic acid and further metabolism to the prostaglandins can also be triggered by phospholipase A_2. Effects of added diacylglycerols and phorbol esters on gonadal function have been reported [18–20] which, together with the found effects of LH on inositol phosphates, indicate that this hormone activates the polyphosphoinositide pathway. However its quantitative importance in regulating steroidogenesis remains to be established.

1.3. Arachidonic acid: release and metabolism to prostaglandins and leukotrienes

As indicated above arachidonic acid is released from phospholipids by phospholipase C and/or phospholipase A_2. The latter enzyme may well be a key site of action of hormone-induced release of this important fatty acid. Phospholipase A_2 is activated by calcium and protein kinase C (in neutrophils) and inhibited by corticosteroid-controlled lipomodulin (in connective tissues). There is little information on control mechanisms in the gonads.

Arachidonic acid is metabolised to the prostaglandins, prostacyclins and thromboxanes via the cyclooxygenase pathway and to the leukotrienes via the lipoxygenase pathway. The prostaglandins are formed in the ovaries and testes. However, apart from a possible role in follicle rupture [21], their physiological status in these glands remains uncertain [16,22]. Recent evidence suggests that the leukotrienes on the other hand may play an important role in the synthesis and release of hormones in endocrine glands [16,23,24]. The relative roles of the prostaglandins and leukotrienes in steroidogenesis in Leydig cells and ovarian cells has been investigated using the differential effects of inhibitors of the cyclooxygenase and lipoxygenase pathways; the former are fairly specific but the latter inhibit both pathways. In testis and tumour Leydig cells it was found that inhibitors of the lipoxygenase but not the cyclooxygenase pathway inhibited LH, LHRH and dibutyryl cyclic AMP stimulated steroidogenesis [16,25]. Chromatographic evidence for the formation of lipoxygenase products has been found in the rat ovary [26] and rat Leydig cells [16]. Further work using a radioimmunoassay for leukotriene B_4 showed that this compound is synthesised and secreted by Leydig cells [27].

1.4. Control and action of intracellular calcium

Extracellular calcium is required to give maximum LH stimulated steroidogenesis in isolated Leydig cells [15,18]. Using the intracellular fluorescent calcium indicator, Quin 2, it has been shown that lowering the extracellular calcium concentration from 2.5 mM to 1.1 μM lowered the LH stimulated intracellular calcium concentrations from 300–500 nM to only 52 nM. These studies showed that LH increases intracellular calcium and that this increase is dependent mainly on extracellular levels of calcium. Cyclic AMP analogues were also shown to increase intracellular calcium to the same concentrations as those obtained with LH, indicating that cyclic AMP is one of the mediators of LH action on calcium mobilization [15].

Studies with ovarian cells have also shown that extracellular calcium is required to give maximum steroidogenesis [29]. The calcium channel blockers, verapamil or diltiazem also suppressed LH-stimulated progesterone production. The calcium ionophore A23187 significantly enhanced the stimulatory effects of LH. Recently it was shown that FSH, prostaglandin E_2 and $F_{2\alpha}$ have distinct effects on steady state calcium exchange in ovarian cells [30].

In chicken granulosa cells, TMB-8, a putative antagonist of intracellular calcium mobilization, inhibited progesterone synthesis in calcium free medium and also inhibited LH stimulated ^{45}Ca release from digitonin permeabilized cells [31]. The latter occurred in two phases: after 1–2 min followed by a slower phase after 5 min. Forskolin and 8-Br-cyclic AMP stimulated only the slow phase of calcium mobilization. These studies indicate that LH may initially stimulate calcium mobilization in granulosa cells via a cyclic AMP-independent pathway followed by a slower cyclic AMP-dependent pathway.

In testis Leydig cells the LH stimulated cyclic AMP production was unaffected by lowering the extracellular calcium concentration [15]. However, three inhibitors of calmodulin all inhibited LH stimulated cyclic AMP production, indicating that there may well be a requirement for calcium. In swine granulosa cells, calcium deprivation suppressed the LH dose-dependent accumulation of cyclic AMP by 50% [29].

2. Steroidogenesis

The transport of cholesterol to, and its side-chain cleavage in, the mitochondria are the most important steps in the hormonal control of steroidogenesis. The enzymatic activities in the endoplasmic reticulum are also under hormonal control. However, expression of hormonal effects in this compartment requires several hours and is therefore part of the chronic, trophic effects of LH [3,4,32].

It is clear from the previously discussed transducing systems for LH action that several second messenger transducing systems may play a role in regulation of steroidogenesis. In connection with this, different classes of protein kinases (protein kinase A, C and calcium and calmodulin-dependent protein kinases) have been implicated in the intracellular signal transduction. Many proteins that can be phosphorylated under the influence of LH have been identified but it has been difficult so far to identify specific LH-dependent phosphoproteins that play an essential role in the regulation of cholesterol side-chain cleavage enzyme activity. The investigations on the latter are complicated further by the fact that the regulatory system probably involves a concerted action of different pathways in which rapidly turning over proteins are essential.

2.1. Second messengers

The many LH-dose and -time correlations that have been shown between the levels of cyclic AMP and steroid production and the positive effects of phosphodiesterase inhibitors have been interpreted as strong evidence that cyclic AMP is directly involved in the regulation of steroid production. However, there are several observations that do not fit this general concept.

1. At low concentrations of cholera toxin, cyclic AMP production is stimulated in purified Leydig cells without any detectable change in steroidogenesis [5,6].

2. Isoproterenol can stimulate cyclic AMP production more than 10-fold in freshly isolated Leydig cells without any change in steroid production, whereas LH can stimulate steroid production more than 10-fold without any detectable change in cyclic AMP [33].

These and other discrepancies between cyclic AMP and steroidogenesis have often been rationalized by assuming the existence of intracellular pools. However, it is known that cyclic AMP can pass through the cell membrane and can accumulate in the extracellular spaces. It is difficult therefore to envisage such pools of cyclic AMP.

Experimental data obtained with dibutyryl cyclic AMP have also been considered to support the obligatory role of cyclic AMP. Since it has been shown that cyclic AMP analogues can stimulate IP_3 formation and calcium release it is possible that cyclic AMP exerts some of its actions by modulating calcium levels. As discussed in Section 1.4, calcium plays an important part in the regulation of steroidogenesis and is regulated by LH. Increases in calcium levels alone are, however, not sufficient to give maximum steroidogenesis; the calcium ionophore A23187 and the LHRH analogues also increase intracellular calcium levels but not cyclic AMP in rat Leydig cells [34] – the maximum testosterone production obtained is approximately 30–40% of that with LH.

In addition to calcium, the products of phospholipid metabolism, especially the leukotrienes, could also be regulators of steroidogenesis. The roles of calcium and the leukotrienes is supported by the observed correlations of the inhibitions of calmodulin action [35], leukotriene formation [16] and steroidogenesis. However, it remains difficult to prove the specificity of the inhibitors. Positive correlations between inhibited activities are therefore less convincing than correlations between stimulated activities.

The evidence obtained recently by Pereira et al. [36] using cyclic AMP agonists and antagonists in Leydig tumour cells is consistent with cyclic AMP being an obligatory mediator of hCG action. Furthermore, they concluded that if other signalling systems are involved they would act together with cyclic AMP rather than independently.

To summarize, the present evidence is consistent with cyclic AMP and calcium and, possibly, phospholipid metabolites, e.g., leukotrienes being second messengers for the stimulation of steroidogenesis by LH in the ovaries and testes. However, critical experiments to investigate their specific roles in controlling steroidogenesis can only be carried out when it is known how extramitochondrial factors regulate cholesterol side-chain cleavage enzyme activity.

2.2. Formation and possible roles of specific (phospho)proteins

LH activation of the various transducing systems in gonadal cells results in the rapid activation of protein kinases and subsequent phosphorylation or sometimes dephosphorylation of specific proteins [37–41]. The following proteins have been detected in subcellular fractions after incubation of intact cells with LH and [^{32}P]phosphate: 17 (nucleus); 20,22,43 and 76 (cytosol); and 24, 33 and 57 (endoplasmic reticulum) (units are kDa). No LH-dependent phosphoproteins were found in the mitochondria. This indicates that protein kinases do not directly regulate cholesterol side-chain cleavage. However, phosphorylation of the cholesterol side-chain cleavage enzyme was demonstrated directly in isolated mitochondria [42]. This does not give any guarantee that this occurs in intact cells; it is well known that isolated proteins have many more phosphorylation sites than proteins in situ.

With regard to the possible roles of the detected phosphoproteins in steroidogenesis, the phosphorylation of the nuclear 17 kDa protein occurs too slowly for it to be directly involved. It may be involved in regulation of gene expression and consequently the trophic effects of LH. It has been shown that RNA synthesis is not required in the acute stimulation of steroidogenesis [43,41].

Rapid phosphorylation of the other detected phosphoproteins does occur but no definite roles have yet been ascribed to them. The 33 kDa protein may be the S_6 ribosomal protein involved in the control of protein synthesis. The 57 kDa protein has been identified as the regulatory subunit of the cyclic AMP-dependent protein kinase [44]. Of the other proteins the 76, 43 and 20 kDa may be connected with the microfilaments (43 kDa actin, 76 kDa myosin light chain kinase and 20 kDa myosin light chain) but this must be further investigated. These proteins may only play a permissive role in steroidogenesis. The fact that the pattern of protein phosphorylation is very similar after stimulation of protein kinase C with phorbol esters supports this because the latter only marginally increase steroidogenesis [18].

The stimulation of steroid synthesis in gonadal cells is dependent on ongoing protein synthesis. Experiments with protein synthesis inhibitors have shown the presence of one or more proteins with half lives of 5–15 min which are essential for the expression of the stimulating hormone effect on steroidogenesis [42,45–47]. Many attempts to identify the protein(s) have been made but none have been detected that satisfy all the functional and kinetic criteria required. Using two-dimensional gel electrophoresis a 28 kDa protein has been identified recently [48], however its functional properties are unknown. It is also not known if the active protein is synthesised de novo or is derived from a precursor protein by covalent modification [49].

It may be that the short half life protein has been missed because of the electrophoresis methods used. For example, a 3 kDa protein has recently been detected in adrenal cells that can stimulate steroid production in isolated mitochondria [50].

The LH effect on steroid synthesis is not mediated by a general increase in protein synthesis. The increase in the synthesis of several specific proteins in Leydig cells that have been detected occur only after a few hours of stimulation and they are therefore probably connected with the trophic effects of LH [18,51–53].

2.3. Control mechanisms in mitochondria

In both the adrenal and the gonads, a labile protein and phosphoproteins have been postulated to play an important role, and a good supply of cholesterol is essential. Adrenal cells and luteinized ovarian cells are most active in the amounts of steroids produced and they are therefore most dependent on an external supply of cholesterol from the lipoproteins. In ovarian cells it has been shown that LH can regulate the supply of lipoproteins probably by regulating their numbers of receptors [54]. Leydig cells are much less dependent on external cholesterol, probably because they can synthesize enough cholesterol de novo [55]. Also, unlike the adrenal gland, there is no evidence to suggest that the liberation of cholesterol from cholesterol esters in Leydig cells contributes significantly to the cholesterol pool [56]. Irrespective of the systems that operate to ensure the supply of cholesterol, sufficient amounts of this steroid must be present at the site of the cholesterol side-chain cleavage enzyme in the inner mitochondrial membrane of the mitochondria. The mechanisms that control the pool at this level are not understood but are generally considered to be one of the major sites of action of LH [57]. Measurements of changes in the 'active pool of cholesterol' in cells or in mitochondria are extremely difficult because 'bound' cholesterol is a natural constituent of cell membranes. It is possible to measure hormone effects on the association between cholesterol and the cytochrome *P*450-side-chain cleavage enzyme by spectral analysis, but this does not give information about the individual reactions that give rise to the formation of this complex.

A lot of attention has been given to the steroid carrier protein (SCP_2) which is located in all cells in which cholesterol is synthesised or converted into steroids or bile acids [58]. It has been shown that this protein can be regulated by LH and that it can stimulate steroid production in isolated mitochondria. Its mechanism of action is not clear, however – it probably does not act as a classical binding protein since no affinity constant for cholesterol has yet been reported.

Another possibility for the stimulation of the conversion of cholesterol at a fixed substrate concentration is to stimulate the cholesterol side-chain cleavage enzyme activity. Calcium has been shown to do this but very high levels are required [59]. Other factors have been shown to have similar effects but most of the claims made in the past have not been confirmed. This applies to the rate limiting production of NADPH, the effects of specific cytosolic fractions on isolated mitochondria [60] and the stimulatory actions of specific phospholipids [61,62].

Recently the structure of a 3 kDa protein isolated from Leydig tumour cells was described [51]. This peptide, which has a short half life, stimulated steroid production in isolated mitochondria. Although these results look promising, its specificity must be tested because it has been reported that positively charged polymers containing lysine can also stimulate steroid production in isolated mitochondria [63,64].

Although there are uncertainties about the exact mechanisms for stimulation of cholesterol side-chain cleavage enzyme activity, the rapid hormonal effects cannot be explained by the stimulation of enzyme synthesis, since this takes more than several hours and represents therefore a trophic rather than an acute stimulatory LH effect [65]. Experiments using hydroxycholesterols as substrates, especially those with 25-hydroxycholesterol, have shown that the capacity of the cholesterol converting enzyme and the availability of cofactors and oxygen in intact cells is higher than necessary for converting endogenous cholesterol [65,66]. It was therefore surprising to find that the capacity of this system in rat Leydig cells was further increased in the presence of interstitial fluid and that protein synthesis was not required [67]. LHRH analogues and phospholipase C also had similar effects [18].

It is difficult to incorporate these observations into a general model for the control of mitochondrial side-chain cleavage enzyme. However, the data may serve to illustrate that this is a multiple regulated system controlled by different endocrine hormones and paracrine factors each using totally or partly different regulatory pathways. The stimulation of these pathways although transient in intact cells appears to be stable once relayed into the mitochondria. Experimental evidence for this is the observation that mitochondria isolated from stimulated cells can retain their stimulated steroid production during fractionation at 4°C [62] and even during a preincubation period of 1 h at 37°C without substrate [68].

3. Desensitization and down regulation

In addition to the stimulatory actions of LH and hCG on steroidogenesis in Leydig cells and ovarian cells, these hormones also cause a refractoriness or desensitization of that same steroidogenic response. This may involve a loss of LH receptors (down regulation), an uncoupling of the LH receptor from the adenylate cyclase, an increase in the metabolism of cyclic AMP due to an increased phosphodiesterase activity and a decrease in the activities in some of the enzymes in the pathways of steroidogenesis (see Ref. 69 for other references).

3.1. Uncoupling of the LH receptor from the adenylate cyclase system

It has been established that there is a time- and dose-dependent loss of LH receptors after exposure of gonadal cells to hCG in vivo [69]. However, high doses are required and it takes about 24 h before the decreases occur. During a short term (3–4 h) exposure to LH or hCG in vitro there is no net loss of LH receptors although there is considerable receptor internalization and recycling in rat Leydig cells.

Evidence from in vivo experiments after administration of hCG suggested that a lesion develops between the LH receptor and the adenylate cyclase in rat testis cells and ovarian cells (see Ref. 69 for other references). Further work carried out in vitro with tumour Leydig cells demonstrated that LH produced a decrease in LH stimulation of adenylate cyclase in intact cells. This loss of response persisted in plasma membranes prepared from the desensitized cells [70]. The response of the intact cells to cholera toxin and of the plasma membranes to fluoride and p(NH)ppG were not decreased. These studies suggested that the lesion occurred between the LH receptor and G_s, whereas the coupling between G_s and the adenylate cyclase catalytic unit was intact. Such a conclusion is in agreement with that of other workers investigating the catecholamine-induced desensitization of S49 lymphoma cells [71,72]. This lesion may be necessary before internalization of the receptor can occur.

Hunzicker-Dunn et al. [73] and Ezra and Salomon [74] suggested that the desensitization of the ovarian adenylate cyclase system by LH involves a phosphorylation reaction. The latter may alter the conformation of the receptor and/or G_s. If phosphorylation is the mechanism it may well be mediated by protein kinase C because phorbol esters, which are presumed to activate this enzyme, mimic LH induced desensitization in Leydig tumour cells [75,76]. There are, however, differences between LH- and phorbol ester induced desensitization [76]. Cyclic AMP-dependent protein kinase is probably not involved because dibutyryl cyclic AMP does not mimic LH induced desensitization in Leydig cells [70].

The physiological relevance of most of the studies on desensitization in ovarian and testis cells is not clear because they were carried out with high unphysiological levels of hormones and over long periods of time. Experiments with superfused

Leydig cells have shown however, that exposure to low levels of LH delivered as a 10 min pulse every 2 h does desensitize the steroidogenic response [77]. Desensitization of the LH-induced cyclic AMP production and internalization of the LH receptor does occur with testis Leydig cells in vitro at physiological concentrations of LH and these processes paradoxically may be necessary to maintain the normal physiological response of the Leydig cell to LH [78].

3.2. Reversal of desensitization

Several reports indicate that LH induced desensitization can be reversed. Reversal of LH-induced desensitization of adenylate cyclase in testis Leydig cells has been found to be dependent on the time and concentration of LH used [78]; at low physiological concentrations of LH the desensitization was completely reversed during incubation of cells for 2 h in the absence of further hormone. With higher concentrations of LH however, the cells remained unresponsive to further challenges with the hormone over the same time period.

Addition of phosphoprotein phosphatase can reverse LH induced desensitization of follicular adenylate cyclase [73]. Addition of human erythrocyte membranes to plasma membranes from LH desensitized Leydig tumour cells restores their response to LH [79].

3.3 Inhibition of steroidogenesis in LH desensitized cells

In addition to a decrease in cyclic AMP production and down regulation of receptors, a high dose of hCG to mature rats leads to a decrease in testicular 17,20-desmolase and 17α-hydroxylase activity which results in a lower conversion of progesterone and 17α-hydroxyprogesterone to testosterone (see Ref. 80 for other references). The steroid biosynthetic changes in rats induced by hCG resemble those caused by oestrogens which inhibit 17α-hydroxylase and 17,20-desmolase activities [81]. It is established that oestradiol-17β is formed in rat testes. Several workers demonstrated that the steroidogenic lesions caused by injection of hCG could be induced by oestradiol-17β. These and other data support the concept that oestradiol-17β and hCG can both induce steroidogenic lesions but this does not prove that hCG acts via oestradiol-17β [82,83].

In contrast to the rat testis where there is a 'late lesion' in the steroidogenic pathway after a high dose of hCG in vivo, desensitization of the steroidogenic response in the mouse testis is associated with decreased pregnenolone synthesis ('early lesion'). Metabolism of pregnenolone to androgens is unchanged. In the mouse tumour Leydig cells it has been demonstrated that the rate of hCG-induced progesterone production is dependent on the level of intracellular cholesterol and desensitization is associated with the depletion of cellular cholesterol [84]. Desensitization was reversed by the addition of cholesterol in the form of human low-density lipoproteins. Similar results have been obtained with mouse testis Leydig cells [85].

In rat Leydig cells, in addition to the 'late lesion', the 'early lesion' has also been reported. However the latter was not overcome by the addition of mevalonate or lipoprotein cholesterol [86]. Hattori et al. [87] reported that a heat-labile inhibiting factor is present in mitochondria from normal rat Leydig cells and is markedly activated or increased by hCG treatment. This substance competitively modulates cholesterol side-chain cleavage activity and could contribute to the early steroidogenic lesion in the rat cells. Further studies by the same group confirmed that the early steroidogenic lesion was not due to an inappropriate supply of cholesterol since the level of the latter in the inner mitochondrial membrane was increased from rats treated 24 h earlier in vivo with hCG or the cholesterol side-chain cleavage enzyme inhibitor, aminoglutethimide [88].

4. Other effects of LH

From the endocrine point of view, the Leydig cells are only important for the production of androgens, with LH being the most important hormone for short- and long-term regulation of the steroidogenic activities. However, it is now recognised that Leydig cells also have other functions which may be important for the local regulation of testis function. LH may play a role in the regulation of these paracrine activities although locally produced factors may exert equal or more important actions. In this respect it is worth noting the small amount of oestrogen production by Leydig cells is regulated in a different way from androgens [89].

The development of ovarian cells and Leydig cells is also under the influence of LH [3,90]. In the testis the development of the Leydig cells is completed during puberty and no further development occurs. In the ovary, the development of LH responsive cells is cyclic during the active period of reproduction. In both gonads there are still many uncertainties about the exact origin of the steroid producing cells and about the final steps that ultimately lead to the mature cells.

The development of the LH responsive cells is not only controlled by LH but also depends on FSH action. This effect is mediated by paracrine factors since Leydig cells and ovarian interstitial cells do not contain FSH receptors. Many efforts are currently being made to elucidate the structure and mechanism of action of these intragonadal regulators [91]. It seems that, in developing organs, FSH is continuously required, whereas for the fully differentiated cells FSH is no longer required. This was clearly shown by the fact that Leydig cells from hypophysectomized mature rats, which were completely regressed, could be restored functionally and morphologically after treatment with LH [92]. Similar observations have been made with isolated Leydig cells and granulosa cells in culture. With these cells it was found that LH could induce synthesis of the mitochondria and microsomal cytochrome *P*450 enzymes via stimulation of transcription [65].

In the testis, the interstitial fluid bathes the Leydig cells, blood vessels and the outside of the seminiferous tubules. It is via this fluid that all hormones and nutrients are transported from the blood stream to testicular cells. All processes in the interstitial tissue and the gametogenic compartment depend on the supply of oxygen, nutrients and hormones from the blood. The testicular capillaries are highly permeable and the protein composition of the testicular interstitial fluid is almost the same as that in blood plasma [91]. Similarly, in the ovary the follicular fluid bathes the granulosa cells. Changes in the rate of blood, interstitial or follicular fluid flow will alter the concentrations of hormones or nutrients to which testicular or ovarian cells are exposed. It is not clear at the present time which gonadal factors regulate the development and the properties of the gonadal blood vessels [91]. Prostaglandins can influence the properties of the blood vessels in the ovary [92] and the testis [93] and, since they are produced by the steroid producing cells under the influence of LH [17], they may be considered as paracrine regulators.

LH and hCG increase blood flow in the testis [91], although it takes 16 h to produce an effect and the changes are small in magnitude. In contrast, large changes take place in interstitial fluid volumes 4–8 h after injection of LH or hCG (with amounts in the upper physiological range) reaching a peak by 16–24 h and returning to normal by 40 h [95,96]. This increase in interstitial fluid volume is a result of an increase in vascular permeability [95,96].

Hypophysectomy leads to an 80% decrease in interstitial fluid volume which is returned to normal by injection of hCG. The mediator of this effect is either produced directly by the Leydig cells in response to LH or originates in other cell types, e.g., macrophages or Sertoli cells in response to Leydig cell factors. The increases in interstitial fluid volume may be caused by mechanisms similar to those inducing oedema in acute inflammation. After hCG treatment, polymorphonuclear leukocytes (PMNs) accumulate in testicular blood vessels prior to the increases in the interstitial fluid volume and then migrate into the interstitial spaces 4 and 8 h after injection; this invasion by PMNs may be caused by the release of a chemotactic agent [97].

In the ovary the ovulatory surge of LH increases blood flow [98]. This is accompanied by an increase in vascular permeability in the follicles resulting in oedema which persists through the time of follicle rupture and ovulation. The effect of LH can be mimicked by histamine but not by FSH or serotonin. Leukocytes accumulate in Graafian follicles near the time of ovulation and fibroblasts migrate from the theca layer to the stratum granulosum.

It is possible that leukotrienes (in addition to the prostaglandins) play an important role in the above effects. Of particular relevance is LTB_4 which is a powerful chemotactic signal for PMNs [99]. Thus the actions of hCG and LH, on testis and ovarian fluid volumes and migration of PMNs etc may be mediated by leukotrienes especially LTB_4. Recently it has been demonstrated that rat testis Leydig cells can produce LTB_4 [27]. There is also evidence that LH, without involving the Leydig cells can directly activate yet unknown factors in isolated testicular fluid which can modify the permeability of the blood vessels [100].

Apart from secreting small molecules like steroids and arachidonic acid metabolites, Leydig cells can also produce peptides like oxytocin [101], angiotensin [102] and opiates [103]. Although there is a limited knowledge about the functional properties of these proteins, oxytocin could play a role in the regulation of tubular contractions, angiotensin might be involved in the regulation of the vasculature and products from the pro-opiomelanocortin precursor might influence Sertoli cell function.

5. *LH action on gonadal cells in perspective*

The effects of LH on the functional properties of gonadal cells have been described mainly in terms of effects of receptors, cyclic AMP, Ca^{2+} and steroid production within a time of several minutes to several days. Many attempts have been made to understand LH regulation of cell function by correlation studies, e.g., changes in receptor numbers or cyclic AMP production with steroid production. In this context terms such as stimulation, desensitization, down regulation, inhibition, etc., have been used to describe changes in cellular activities. However, since it is now recognized that LH exerts a pleiotropic response which comprises initially many transducing systems (Ca^{2+}, cyclic AMP, IP_3, leukotrienes, prostaglandins) and many final responses (growth, steroid production, formation of angiotensin, oxytocin, opioids, etc.) (see Fig. 1), the terminology for describing effects of LH may be confusing. Furthermore the specific functional properties of the Leydig cells do not always respond to LH in the same way. This will be illustrated with a few examples.

1. After one injection of hCG to rats, testosterone production is maximally stimulated within 2 h, whereas steroidogenic lesions develop between 12 and 24 h. The effect of hCG on prostaglandin production is different; within 2 h no effect can be demonstrated, whereas between 12 and 24 h, $PGF_{2\alpha}$ and PGE_2 levels rise more than 5-fold [17].

2. In Leydig cells from mature rats isolated 10 days after two injections of hCG (administered at day 0 and day 7) LH receptors are 'down regulated' and adenylate cyclase is 'desensitized', however, the LH-dependent steroidogenic capacity is increased [104].

3. The down regulation of LH receptors which occurs after administration of hCG in vivo to mature rats does not occur in 5-day-old rats. In the latter the LH receptors can be up-regulated [105].

4. HCG can also exert tropic effects in addition to short term regulation of Leydig cell activity; during the first few days after one large dose of hCG, when adenylate cyclase activity and steroid production are 'desensitized' the differentiation of the Leydig cells is strongly activated – the numbers of Leydig cells increase by 30%

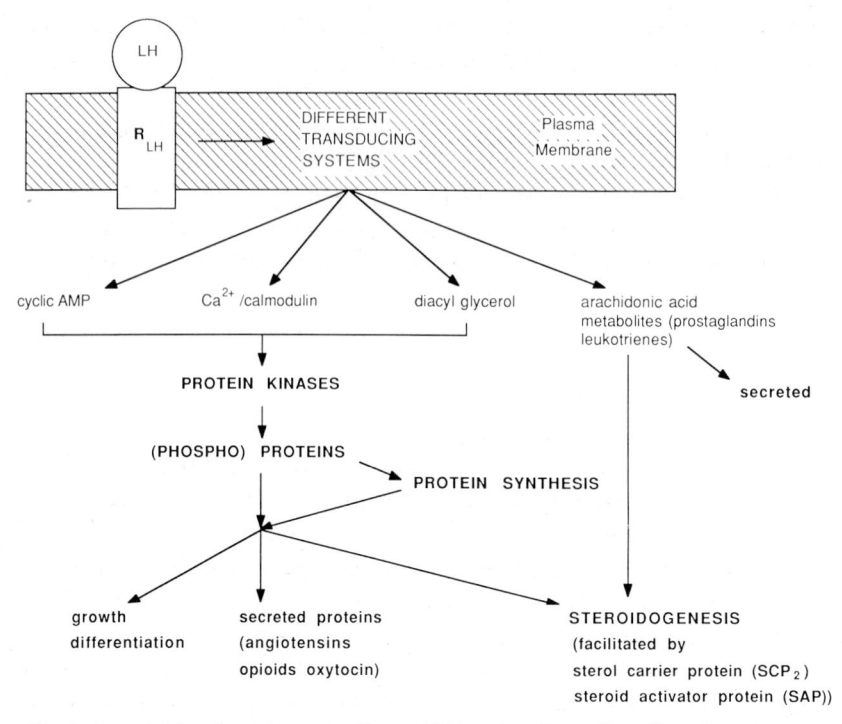

Fig. 1. A model for the pleiotropic effects of LH on functions of Leydig cells. LH interacts with its specific receptor in the plasma membrane of the Leydig cell which results in the activation of several transducing systems and the formation of several second messengers (cyclic AMP, Ca^{2+}, diacylglycerol and arachidonic acid metabolites). Protein kinases (A, C and calmodulin dependent) are activated resulting in the phosphorylation of specific proteins and the synthesis of specific proteins. The (phospho)proteins are involved in the transport of cholesterol to, and the control of, cholesterol metabolism in the inner mitochondrial membrane. Arachidonic acid metabolites (prostaglandins, leukotrienes) may also control steroidogenesis. LH can also regulate the secretion of proteins. The trophic effects of LH are manifested in the growth and differentiation of the Leydig cells.

in 2 days. During the same time period there is also a 12-fold increase in thymidine incorporation in interstitial cells [106] and the Leydig cells are hypertrophied [90,107].

The hCG-induced changes in the composition of Leydig cell populations may be a complicating factor when investigating long-term effects of LH/hCG. In granulosa cells it has been shown that growing cells have different properties from fully differentiated cells [108]. Similarly LH effects on differentiating cells or fully differentiated cells may be completely different and there may be no relation between some observed biochemical activities in such a heterogeneous cell population.

It is normal to classify gonadal cells as desensitized/down regulated when adenylate cyclase activity and receptor numbers are decreased following exposure to LH/hCG. However in the light of the above this terminology is inappropriate. A more general and neutral term is 'adaptation'. This term includes stimulatory and inhibitory changes in intracellular activities that may take place at the same time. Major questions to be resolved are:

1. Does the type of response depend on the concentration of LH?

2. Is there one or are there several forms of the LH receptor that interact with the different transducing systems involved in adaptation?

3. Are the qualitative and quantitative aspects of adaptation controlled by the same mechanisms?

4. How do paracrine factors influence adaptation?

If uncoupling and internalization of receptors are interpreted not only as part of a desensitization event but also as an adaptive process, other processes, possibly stimulatory ones may concurrently occur. Uncoupling of the receptor from the adenylate cyclase may enable the receptor to couple to other proteins within the membranes or, after internalization, within the cell. At the same time the uncoupled hormone-activated adenylate cyclase may transiently retain the activation signal. Although this model is speculative supporting data are available. In the studies of Habberfield et al. [78] with LH-desensitized testis Leydig cells it was shown that although there was a complete loss of response to LH, there was a residual 'hormone-insensitive' cyclic AMP production; this accounted for up to 50% of the maximal LH-stimulated production. These results and those of LH receptor internalization and recycling led to the conclusion that these processes may be part of an amplification system for cyclic AMP production in the presence of physiological amounts of LH. This may be necessary because of the low numbers of LH receptors in rat Leydig cells. By this means the receptor may be rapidly recycled and activate many G_s-adenylate cyclase units. The latter uncouple from the receptor and continue to produce cyclic AMP as indicated by the 'LH-independent' cyclic AMP production that occurs.

Stimulation, inhibition, desensitization, internalization and adaptation have been discussed so far with the assumption that all cells of one type (e.g., Leydig and granulosa cells) show the same increased response (i.e., analogous) when hormone levels are increased. However, all cells may not have the same functional properties [109,110] and it may also be possible that individual cells show a quantum response (i.e., all or none) [112]. Recent data obtained with liver cells regulated with vasopressin have even shown that this quantum response causes oscillations of intracellular calcium. Moreover the oscillatory system was mainly regulated by frequency modulation rather than by regulation of the amplitude [113]. Similar studies have not been carried out with gonadal cells although it is known that LH is secreted in a pulsatile manner. It is therefore important to investigate LH effects at the level of individual cells for a proper understanding of the basic mechanisms of LH action.

References

1. Tata, J.R. (1984) Mol. Cell. Endocrinol. 36, 17–27.
2. Rommerts, F.F.G., Cooke, B.A. and van der Molen, H.J. (1974) J. Steroid Biochem. 5, 279–285.
3. Erickson, G.F., Magoffin, D.A., Dyer, C.A. and Hofeditz, C. (1985) Endocr. Rev. 6, 371–399.
4. Hsueh, A.J.W., Adashi, E.Y., Jones, P.B.C. and Welsh, T.H. (1984) Endocr. Rev. 5, 76–127.
5. Dufau, M.L., Horner, K.A., Hayashi, K., Tsuruhara, T., Conn, P.M. and Catt, K.J. (1978) J. Biol. Chem. 253, 3721–3729.
6. Cooke, B.A., Lindh, L.M. and Janszen, F.H.A. (1977) FEBS Lett. 73, 67–71.
7. Dufau, M.L., Baukal, A.J. and Catt, K.J. (1980) Proc. Natl. Acad. Sci. U.S.A. 77, 5837–5841.
8. Mukhopadhyay, A.K. and Schumacher, M. (1985) FEBS Lett. 187, 56–60.
9. Adashi, E.Y., Resnick, C.E., Cronin, M.J. and Hewlett, E.L. (1984) Endocrinology 115, 839–841.
10. Khanum, A. and Dufau, M.L. (1986) J. Biol. Chem. 261, 11456–11459.
11. Platts, E.A., Schulster, D. and Cooke, B.A. (1988) Biochem. J. 253, in press.
12. Mukhopadhyay, A.K., Schumacher, M. and Leidenberger, F.A. (1986) Biochem. J. 239, 463–467.
13. Lowitt, S., Farese, R.V., Sabir, M.A. and Root, A.W. (1982) Endocrinology 111, 1415–1417.
14. Davis, J.S., Weakland, L.L., West, L.A. and Farese, R.V. (1986) Biochem. J. 238, 597–604.
15. Sullivan, M.H.F. and Cooke, B.A. (1986) Biochem. J. 236, 45–51.
16. Dix, C.J., Habberfield, A.D., Sullivan, M.H.F. and Cooke, B.A. (1984) Biochem. J. 219, 529–537.
17. Haour, F., Kovenetzova, B., Dray, F. and Saez, J.M. (1979) Life Sci. 24, 2151–2158.
18. Themmen, A.P.N., Hoogerbrugge, J.W., Rommerts, F.F.G. and van der Molen, H.J. (1986) FEBS Lett. 203, 116–120.
19. Kawai, Y. and Clark, M.R. (1985) Endocrinology 116, 2320–2326.
20. Welsh, T.H., Jones, P.B.C. and Hsueh, A.J.W. (1984) Cancer Res. 44, 885–892.
21. Reich, R., Haberman, S., Abisogun, A.O., Sofer, Y., Grossman, S., Adelmann-Grill, B.C. and Tsafriri, A. (1986) In: Proc. Serono Symp. The control of follicle development, ovary and luteal function. (Naftolin, F. and De Cherney, A.H., eds.) (in press) Raven Press, New York.
22. Reich, R., Kohen, F., Naor, Z. and Tsafriri, A. (1983) Prostaglandins 26, 1011–1020.
23. Turk, J., Colca, J.R. and McDaniel, M.L. (1985) Biochim. Biophys. Acta 834, 23–36.
24. Kiesel, L., Przylipiak, A., Emig, E., Rabe, T. and Runnebaum, B. (1987) Life Sci. 40, 847–851.
25. Sullivan, M.H.F. and Cooke, B.A. (1985) Biochem. J. 232, 55–59.
26. Reich, R., Kohen, F., Slager, R. and Tsafriri, A. (1986) Prostaglandins 30, 581–590.
27. Sullivan, M.H.F. and Cooke, B.A. (1985) Biochem. J. 230, 821–824.
28. Janszen, F.H.A., Cooke, B.A., van Driel, M.J.A. and van der Molen, H.J. (1976) Biochem. J. 160, 433–437.
29. Veldhuis, J.D. and Klase, P.A. (1982) Endocrinology 111, 1–6.
30. Veldhuis, J.D. (1987) Endocrinology 120, 445–449.
31. Asem, E.K., Molnar, M. and Hertelendy, F. (1987) Endocrinology 120, 853–859.
32. van der Molen, H.J. and Rommerts, F.F.G. (1981) In: The Testis (Burger, H. and de Kretser, D., eds.) pp. 213–238. Raven Press, New York.
33. Cooke, B.A., Golding, M., Dix, C.J. and Hunter, M.G. (1982) Mol. Cell. Endocrinol. 27, 221–232.
34. Sullivan, M.H.F. and Cooke, B.A. (1984) Biochem. J. 218, 621–624.
35. Hall, P.F., Osawa, S. and Mrotek, J. (1981) Endocrinology 109, 1677–1682.
36. Pereira, M.E., Segaloff, D.L., Ascoli, M. and Eckstein, F. (1987) J. Biol. Chem. 262, 6093–6100.

37. Cooke, B.A., Lindh, L.M. and Janszen, F.H.A. (1977) Biochem. J. 168, 43–48.
38. Richards, J.S., Sehgal, N. and Tash, J.S. (1983) J. Biol. Chem. 258, 5227–5232.
39. Bakker, G.H., Hoogerbrugge, J.W., Rommerts, F.F.G. and van der Molen, H.J. (1981) Biochem. J. 198, 339–346.
40. Bakker, G.H., Hoogerbrugge, J.W., Rommerts, F.F.G. and van der Molen, H.J. (1982) Biochem. J. 204, 809–815.
41. Bakker, G.H., Hoogerbrugge, J.W., Rommerts, F.F.G. and van der Molen, H.J. (1983) FEBS Lett. 161, 33–36.
42. Vilgrain, I., Defaye, G. and Chambaz, E.M. (1984) Biochem. Biophys. Res. Commun. 125, 554–561.
43. Cooke, B.A., Janszen, F.H.A., van Driel, M.J.A. and van der Molen, H.J. (1979) Mol. Cell. Endocrinol. 14, 181–189.
44. Cooke, B.A., Lindh, L.M. and van der Molen, H.J. (1979) J. Endocrinol. 83, 32P.
45. Cooke, B.A., Janszen, F.H.A., Clotscher, W.F. and van der Molen, H.J. (1975) Biochem. J. 150, 413–418.
46. Cosier, A.J. and Younglai, E.V. (1981) J. Steroid Biochem. 14, 285–293.
47. Bakker, G.H., Hoogerbrugge, J.W., Rommerts, F.F.G. and van der Molen, H.J. (1985) J. Steroid Biochem. 22, 311–314.
48. Pon, L.A. and Orme-Johnson, N.R. (1986) J. Biol. Chem. 6594–6599.
49. Cooke, B.A., Lindh, L.M. and van der Molen, H.J. (1979) Biochem. J. 184, 33–38.
50. Pedersen, R.C. and Brownie, A.C. (1987) Science 236, 188–190.
51. Janszen, F.H.A., Cooke, B.A. and van der Molen, H.J. (1977) Biochem. J. 162, 341–346.
52. Janszen, F.H.A., Cooke, B.A., van Driel, M.J.A. and van der Molen, H.J. (1978) Biochem. J. 170, 9–15.
53. Janszen, F.H.A., Cooke, B.A., van Driel, M.J.A. and van der Molen, H.J. (1978) Biochem. J. 172, 147–153.
54. Toaff, M.E., Schleyer, H. and Strauss, J.F. (1982) Endocrinol. 111, 1785–1790.
55. Pignataro, O.P., Radicella, J.P., Calvo, J.C. and Charreau, E.H. (1983) Mol. Cell. Endocrinol. 33, 53–67.
56. Freeman, D.A. (1987) Eur. J. Biochem. 164, 351–356.
57. Simpson, E.R. (1979) Mol. Cell. Endocrinol. 13, 213–227.
58. Van Noort, M., Rommerts, F.F.G., Van Amerongen, A. and Wirtz, K.W.A. (1986) J. Endocrinol. 109, R13–R16.
59. van der Vusse, G.J., Kalkman, M.L., van Winsen, M.P.I. and van der Molen, H.J. (1975) Biochim. Biophys. Acta 398, 28–38.
60. Farese, R.V. (1967) Biochemistry 6, 2052–2065.
61. Farese, R.V., Sabir, A.M., Vandor, S.L. and Larson, R.E. (1980) J. Biol. Chem. 255, 5728–5734.
62. Terpstra, P., Rommerts, F.F.G. and van der Molen, H.J. (1985) J. Steroid Biochem. 22, 773–780.
63. Kido, T. and Kimura, K. (1981) J. Biol. Chem. 256, 8561–8568.
64. Mason, J.I., Arthur, J.R. and Boyd, G.S. (1978) Mol. Cell. Endocrinol. 10, 209–223.
65. Waterman, M.R. and Simpson, E.V. (1985) Mol. Cell. Endocrinol. 39, 81–89.
66. Bakker, G.H., Hoogerbrugge, J.W., Rommerts, F.F.G. and van der Molen, H.J. (1983) Mol. Cell. Endocrinol. 33, 243–253.
67. Rommerts, F.F.G., Hoogerbrugge, J.W. and van der Molen, H.J. (1986) J. Endocrinol. 109, 111–117.
68. Schumacher, M., Ludolph, J., Schwarz, M. and Heidenberger (1987) Acta Endocrinol. (Suppl.) 283, 2–3.
69. Catt, K.J., Harwood, J.P., Aguilera, G. and Dufau, M.L. (1979) Nature (London) 280, 109–116.
70. Dix, C.J., Schumacher, M. and Cooke, B.A. (1982) Biochem. J. 202, 739–745.
71. Iyengar, R., Vhat, M.K., Riser, M.E. and Birnbaumer, L. (1981) J. Biol. Chem. 256, 4810–4815.
72. Green, D.A. and Clark, R.B. (1981) J. Biol. Chem. 256, 2105–2108.
73. Hunzicker-Dunn, M., Derda, D., Jungmann, R.A. and Birnbaumer, L. (1979) Endocrinology 104, 1785–1793.

74. Ezra, E. and Salomon, Y. (1980) J. Biol. Chem. 255, 653–658.
75. Rebois, R.V. and Patel, J. (1985) J. Biol. Chem. 260, 8026–8031.
76. Dix, C.J., Habberfield, A.D. and Cooke, B.A. (1987) Biochem. J. 243, 373–377.
77. Wu, F.C.W., Zhang, G.Y., Williams, B.C. and de Kretser, D.M. (1985) Mol. Cell. Endocrinol. 40, 45–56.
78. Habberfield, A.D., Dix, C.J. and Cooke, B.A. (1987) J. Endocrinol. 114, 415–422.
79. Dix, C.J. and Cooke, B.A. (1982) Biochem. J. 204, 613–616.
80. Catt, K.J., Harwood, J.P., Clayton, R.N., Davies, T.F., Chan, V., Katikineni, M., Nozu, K. and Dufau, M.L. (1980) Rec. Progr. Hormone Res. 36, 557–622.
81. Samuels, L.T., Vchikawa, T., Zain-Abedin, M. and Huseby, R.A. (1969) Endocrinology 85, 96–100.
82. Brinkman, A.O., Leemborg, F.G. and van der Molen, H.J. (1981) Mol. Cell. Endocrinol. 24, 65–72.
83. Brinkman, A.O., Leemborg, F.G., Rommerts, F.F.G. and van der Molen, H.J. (1982) Endocrinology 110, 1834–1836.
84. Freeman, D.A. and Ascoli, M. (1982) Proc. Natl. Acad. Sci. U.S.A. 79, 7796–7800.
85. Schumacher, M., Schwarz, M. and Leidenberger, F. (1985) Biol. Reprod. 33, 335–345.
86. Charreau, E.E., Calvo, J.C., Nozu, K., Pignataro, C., Catt, K.J. and Dufau, M.L. (1981) J. Biol. Chem. 256, 12719–12724.
87. Hattori, M., Aquilano, D.R. and Dufau, M.L. (1986) J. Steroid Biochem. 21, 265–277.
88. Aquilano, D.R., Tsai-Morris, C.H., Hattori, M.A. and Dufau, M.L. (1985) Endocrinology 116, 1745–1754.
89. Rommerts, F.F.G., de Jong, F.H., Brinkman, A.O. and van der Molen, H.J. (1982) J. Reprod. Fertil. 65, 281–288.
90. Christensen, A.K. and Peacock, K.C. (1980) Biol. Reprod. 22, 383–391.
91. Sharpe, R.M. (1983) Quart. J. Exp. Physiol. 68, 265–287.
92. Ewing, L.L. and Zirkin, B. (1982) Rec. Progr. Hormone Res. 39, 599–632.
93. Aami, T.I. and O'Shea, J.D. (1984) Lab. Invest. 51, 206–217.
94. Free, M.J. and Jaffe, R.A. (1972) Prostaglandins 1, 377–382.
95. Setchell, B.P. and Sharpe, R.M. (1981) J. Endocrinol. 91, 245–254.
96. Sharpe, R.M. (1984) Biol. Reprod. 30, 29–49.
97. Damber, J., Bergh, A. and Daehlin, L. (1985) Endocrinology 117, 1906–1913.
98. Espey, L.L. (1980) Biol. Reprod. 22, 73–106.
99. Samuelsson, B. (1982) Ang. Chemie. 21(12), 902.
100. Veijola, M. and Rajaniemi, H. (1986) Mol. Cell. Endocrinol. 45, 113–118.
101. Nicholson, H.D., Worley, R.T.S., Guldenaar, S.E.F. and Pickering, B.T. (1987) J. Endocrinol. 112, 311–316.
102. Pandey, K.N. and Inagami, T. (1986) J. Biol. Chem. 261, 3934–3938.
103. Pintar, J.E., Schachter, B.S., Herman, A.B., Durgerian, S. and Krieger, D.T. (1984) Science 225, 632–634.
104. Calvo, J.C., Radicella, J.P., Pignataro, O.P. and Charreau, E.H. (1984) Mol. Cell. Endocrinol. 34, 31–38.
105. Huhtaniemi, I.T., Dwarren, D.W. and Catt, K.J. (1984) Ann. N.Y. Acad. Sci. 438, 283–303.
106. Teerds, K.J., de Rooij, D.G., Rommerts, F.F.G. and Wensing, C.J.G. (1988) J. Androl. in press.
107. Hodgson, Y.M. and De Kretser, D.M. (1984) Mol. Cell. Endocrinol. 35, 75–82.
108. Orly, J., Sato, G. and Erickson, G.F. (1980) Cell 20, 817–827.
109. Cooke, B.A., Magee-Brown, R., Golding, M. and Dix, C.J. (1981) Int. J. Androl. 4, 355–366.
110. Hsueh, A.J.W., Dufau, M.L., Katz, S.I. and Catt, K.J. (1976) Nature 261, 710–713.
111. Rommerts, F.F.G. and Brinkman, A.O. (1981) Mol. Cell. Endocrinol. 21, 15–28.
112. Moyle, W.R., Kuczek, T. and Bailey, C.A. (1985) Biol. Reprod. 32, 43–69.
113. Woods, N.M., Cuthbertson, K.S.R. and Cobbold, P.H. (1986) Nature 319, 600–602.

CHAPTER 9

Mechanism of action of FSH in the ovary

KRISTINE D. DAHL and AARON J.W. HSUEH

Department of Reproductive Medicine, School of Medicine, M-025, University of California, San Diego, La Jolla, CA 92093, U.S.A.

1. Introduction

It is well known that the pituitary gonadotropins, luteinizing hormone (LH) and follicle-stimulating hormone (FSH), are indispensable for ovulation. FSH, in particular, is necessary for the stimulation of preovulatory ovarian estrogen production and the initiation of follicle maturation. This chapter will discuss the mechanisms underlying the FSH stimulation of ovarian cell differentiation and maturation.

2. Biochemistry of FSH

2.1. α, β subunits

FSH and LH are glycoprotein hormones which contain two noncovalently linked dissimilar subunits [1–4]. The gonadotropic hormones, including those from the placenta, consist of a common subunit designated α and a hormone-specific β subunit. In addition, pituitary thyroid-stimulating hormone (TSH) also belongs in the same protein family and shares the α subunit. Beginning in the early 1970s, the primary amino acid sequence of the gonadotropins was determined [5–8], while only recently the genes for the subunits were isolated and sequenced [9–11]. Although all subunits are believed to be derived from a common ancestral gene, each subunit is encoded for by separate genes. Postulated evolutionary origins of FSH, LH, TSH, human chorionic gonadotropin (hCG) and pregnant mare serum gonadotropin (PMSG) are shown in Fig. 1. The α-subunit gene is conserved, while the ancestral β-chain genes evolve into specific β subunits for each hormone.

2.2. Carbohydrate content

The primary structure of FSH consists of the polypeptide backbone, and oligosaccharide moieties. FSH molecules from all species including those from lower ver-

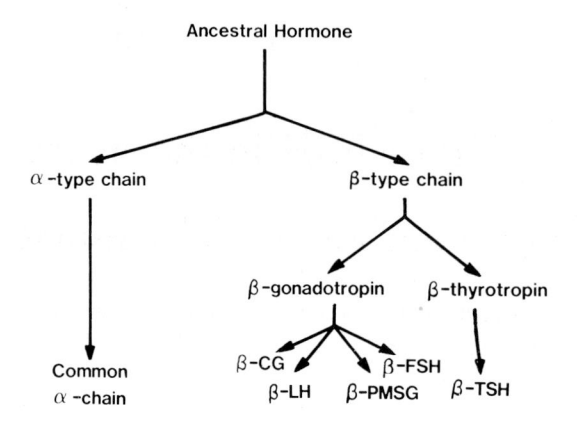

Fig. 1. The postulated evolutionary origins of FSH, LH, TSH, hCG and PMSG.

tebrates contain carbohydrate units covalently linked to the polypeptide [2,4,12]. The α subunit of FSH in different species generally has more carbohydrate side chains than the β subunit, and ovine FSH has four carbohydrate moieties which are N-glycosylitic linked. Several studies have dealt with the structure-function relationship of FSH [13–17] and the essential role of sialic acid in maintaining the in vivo biological activity of the hormone. Removal of this sugar residue by neuraminidase treatment decreases the in vivo biological activity of FSH due to its rapid elimination from the circulation [18,19], but its in vitro biological response is retained [20–23]. In contrast, removal of the majority of carbohydrate moieties by specific enzymes or chemical treatment [14,24] decreases the bioactivity of the molecules [14].

Figure 2 depicts the current concept of the mode of action of FSH on granulosa or Sertoli cells. The protein domain of the FSH molecule binds to specific FSH receptors in plasma membrane, while the carbohydrate moieties of FSH are believed to interact with membrane components either within or adjacent to the receptor molecule. Isolation and characterization of the FSH receptors will contribute to elucidating the binding site and the role of the carbohydrate moieties in the transduction of the hormonal signal. The binding of FSH to the receptors is followed by the activation of adenyl cyclase resulting in the accumulation of cAMP.

3. *FSH receptors in target cells*

FSH molecules secreted by the anterior pituitary interact with specific receptors in the gonadal cells [25–27]. The identification of target cells for FSH in the ovary and testis was made possible by the development of techniques for radiolabeling protein hormones while retaining their biological activities and by their detection using autoradiography or direct radioligand binding assays.

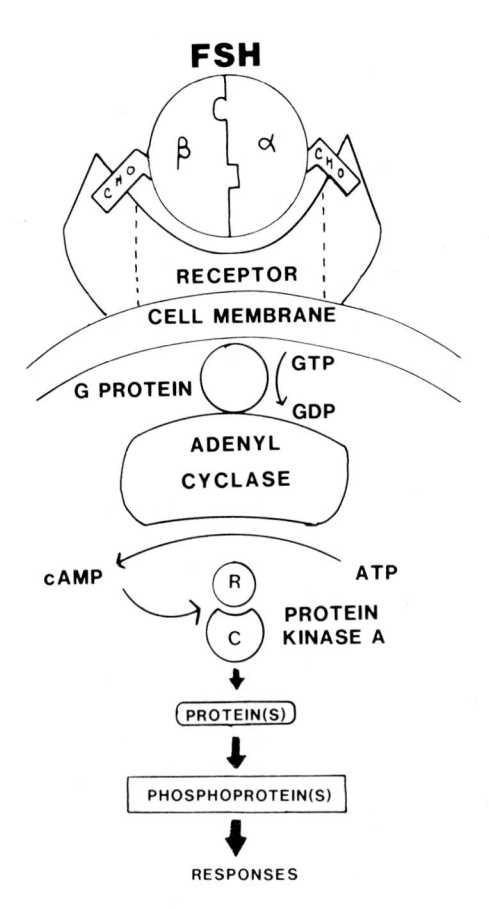

Fig. 2. The mode of action of FSH on the granulosa cell.

3.1. Radioligand receptor assay

Specific receptors for FSH have been found exclusively in the granulosa cells of the ovary and in the Sertoli cells of the testes [28,29]. Similar to other protein hormones, FSH binding to target cell receptors is hormone-specific and of high affinity ($K_d = 10^{-9}$ to 10^{-10} M). With this discovery, several investigators developed sensitive and specific radioligand receptor assays based on the ability of [^{125}I]FSH to bind specific receptors in rat testis homogenates [30,31].

3.2. Agonistic and antagonistic effects of FSH analogs

Under in vitro conditions, granulosa or Sertoli cells respond to the addition of FSH by prompt accumulation of cyclic AMP (cAMP) [29,32]. Sairam and associates [14] have demonstrated that removal of 80% of the carbohydrate moieties in FSH increases the receptor binding ability and immunological activities of the hormone in radioreceptor and radioimmunoassays, respectively. Several other groups also reported that deglycosylated FSH binds better than its intact counterpart [33]. In contrast, the deglycosylated FSH molecules have a greatly diminished capacity for stimulation of cAMP production by gonadal cells [14]. In fact, the deglycosylated FSH, when added together with the intact hormone, interfere with the ability of intact FSH to elicit steroidogenesis and are potent and specific antagonists in in vitro bioassays [14,15,17,34]. It is believed that the carbohydrate side-chains of FSH interact with putative cell membrane lectins to allow the coupling of the hormone-receptor complex to the adenyl cyclase, and thus the transduction of the biological signal (Fig. 2).

4. Activation of the protein kinase A pathway

It is generally accepted that cAMP is the second messenger for FSH action. Treatment with membrane soluble cAMP analogs or cAMP-inducing agents induces multiple physiological responses in granulosa cells [35]. Furthermore, the FSH action in granulosa cells is enhanced by cotreatment with phosphodiesterase inhibitors that minimize cAMP breakdown [36]. Following FSH stimulation, the second messenger formed binds to the regulatory subunit of protein kinase A. This results in the activation of the catalytic subunit of the kinase and the phosphorylation of uncharacterized proteins, leading to genomic activation.

4.1. Coupling between the FSH receptor and adenylate cyclase

It is now clear that at least three distinct plasma membrane components are involved in the FSH stimulation of cAMP production: the hormone receptor, the nucleotide regulatory or G protein (guanine nucleotide binding protein) [37,38] and the catalytic moiety of adenyl cyclase (Fig. 2). The receptor and cyclase are responsible for hormone recognition and enzymatic activity, respectively, and the G protein, acting in conjunction with GTP, modulates the activities of the other two components. The G proteins are heterotrimeric complexes containing three distinct chains. The FSH activated G protein enhances cyclase activity and is termed G_s to distinguish it from other analogous proteins found in diverse signal transducing systems [37,39]. On stimulation by the occupied and activated receptor molecule, the G protein releases GDP and binds GTP. In its GTP-bound conformation, the activated G protein is capable of stimulating adenyl cyclase. This active complex takes on the activity of a GTPase. Hydrolysis of the bound GTP to GDP terminates the action of the G protein.

4.2. Stimulation of protein kinase A

Following FSH stimulation of cAMP production, cAMP activates protein kinases, which result in subsequent phosphorylation of key proteins involved in granulosa cell differentiation [40]. It has been demonstrated that the action of FSH is mediated by R_{II}, the regulatory subunit of type II protein kinase [41]. Estrogens also synergize with FSH in the induction of granulosa cell differentiation. Estrogen treatment enhances the FSH-dependent cAMP accumulation without concomitant changes in the number of FSH binding sites [42]. Furthermore, estrogens augment FSH stimulation of the formation of cAMP binding proteins which may be identical to the regulatory subunit of protein kinase [42].

5. *FSH induction of granulosa cell differentiation*

Ovarian follicles, the basic unit of the ovary, consist of an outer layer of theca interna cells which encircle inner layers of granulosa cells. Granulosa cells, in turn, surround the innermost oocyte-cumulus complex. This complex array of cell layers is essential for the maturation and ovulation of the ovum. Significantly, the granulosa cell, by virtue of its close interaction with the outer theca and inner oocyte-cumulus compartment, serves to integrate information necessary for the achievement of successful follicular development.

Although the granulosa cell is subject to different varieties of signaling systems, FSH is the most important hormone for its differentiation. Specifically, FSH stimulates the granulosa cells to secrete estrogens and progestins, as well as various nonsteroidal substances [43]. The induction of various hormone receptors in the granulosa cell is also FSH dependent.

5.1. LH and PRL receptors, and β-adrenergic responsiveness

Negligible binding of LH is observed in granulosa cells not previously primed with FSH [44]. Several in vivo [45] and in vitro [44] studies have demonstrated the ability of FSH to induce LH receptors in cultured granulosa cells. As in the case of LH receptors, immature granulosa cells display prolactin (PRL) binding prior to priming with FSH [46]. Both in vivo and in vitro [47,48] studies have demonstrated the ability of FSH to increase the number of PRL receptors in granulosa cells. Following the acquisition of LH and PRL receptors, these hormones are capable of inducing multiple granulosa cell functions [44,47].

Several investigators have demonstrated the presence of stereospecific β-adrenergic binding sites in corpus luteum [49] and granulosa cells [50]. Although the regulation of the acquisition of β-adrenergic receptors during granulosa cell differentiation remains unknown, several studies suggest a role for FSH in the induction of β-adrenergic responsiveness [50,51]. The coupling of granulosa cell β-adrenergic receptors to progesterone production [50,51] suggests that their acquisition during granulosa cell differentiation sets the stage for the luteotropic role of adrenergic transmitters.

5.2. Lipoprotein receptors

More than 90% of serum cholesterol is carried by lipoproteins. Two major lipoproteins are involved in ovarian steroidogenesis, namely, high density lipoprotein (HDL) and low density lipoprotein (LDL). The action of these two lipoproteins are mediated through distinct pathways via different cell membrane receptors [52,53]. In the granulosa cell, FSH enhances the ability of granulosa cells to utilize serum lipoproteins by increasing lipoprotein receptor content in the cell membrane [54]. Lipoprotein dependency varies with the stage of granulosa cell development. Since the granulosa cells inside the follicle do not have ready access to the general circulation, these cells are not bathed by the plasma lipoproteins. Due to differences in size, high concentrations of HDL, but not LDL, are found in follicular fluid [55]. After ovulation, vascularization of the luteinized granulosa cells take place and the luteal cell progestin biosynthesis is highly dependent on serum lipoproteins.

5.3. Gap junction and microvilli formation

Profound changes in granulosa cell morphology takes place during follicular development [56]. The avascular nature of granulosa cells necessitates intercellular contacts between neighboring cells. During follicular development, extensive gap junctions are found among granulosa cells [57]. FSH treatment increases the number of gap junctions as well as the amount of junctional membranes in granulosa cells [58].

FSH treatment in vitro also causes the formation of microvilli in granulosa cells and it has been suggested that LH receptors are localized mainly on microvilli. Accompanied by the FSH stimulation of steroidogenesis in cultured granulosa cells, morphological studies indicate the presence of smooth endoplasmic reticulum, mitochondria with tubular cristae, lipid droplets and Golgi apparatus in these cells [59].

6. FSH stimulation of steroidogenic enzymes

Granulosa cells are capable of de novo synthesis of cholesterol. Side chain cleavage enzyme in the mitochondria are responsible for the conversion of cholesterol to pregnenolone, then enzymes in the smooth endoplasmic reticulum convert pregnenolone to progesterone. Granulosa cells lack the necessary enzyme 17α-hydroxylase/17,20 lyase to convert progesterone to androgens but has aromatases for the

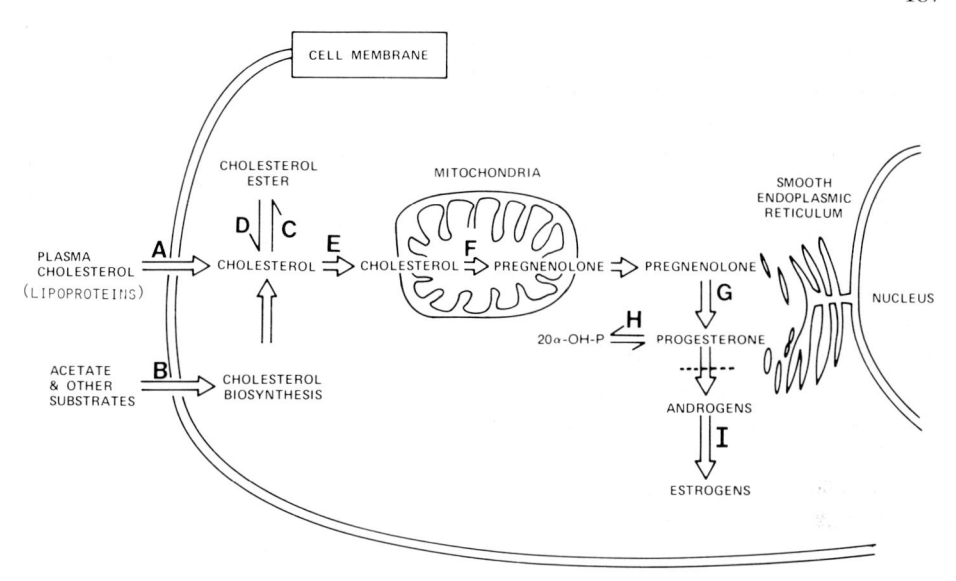

Fig. 3. Steroidogenic pathway in granulosa cells. A. Lipoprotein in receptors. B. 3-Hydroxy-3-methylglutaryl coenzyme A reductase (HMG-CoA reductase). C. Acyl-coenzyme A (cholesterol acyl transferase). D. Cholesterol esterase. E. Cholesterol transport to the mitochondria. F. Cholesterol side-chain cleavage enzymes (phospholipid membrane environment and enzyme levels). G. 3β-Hydroxysteroid dehydrogenase (3β-HSD). H. 20α-Hydroxysteroid dehydrogenase (20α-HSD). I. Aromatases.

metabolism of androgens to estrogens. The steroidogenic pathway in the granulosa cells and the enzyme complexes which are stimulated by FSH are depicted in Fig. 3.

6.1. Aromatase induction

6.1.1. Enzyme induction

Dorrington and colleagues [60] first demonstrate the ability of FSH to induce aromatases in granulosa cells. Treatment with FSH from various mammalian species increases aromatase activity in granulosa cells from rats [61]. It is known that the 'aromatase' system is comprised of a specific form of cytochrome P-450 and the flavoprotein NADPH-cytochrome P-450 reductase. Recently, the cytochrome P-450 aromatase enzyme has been cloned from human placenta, thus allowing further investigation of the molecular mechanisms that mediate the regulation of aromatase activity [62].

6.1.2. Two-cell, two-gonadotropin theory

Mainly based on rat studies, a two-cell, two-gonadotropin hypothesis for ovarian estrogen biosynthesis has been proposed [63]. According to this model, LH stimulates the biosynthesis of androgens from cholesterol in the theca interna compartment [64,65]. Androgens diffuse across the basal lamina and are converted to estrogens by the aromatase enzymes in the granulosa cells. While the two-cell, two-gonadotropin hypothesis provides a useful model for understanding follicular estrogen biosynthesis in rat, porcine, bovine, and rabbit ovaries [66–68], the thecal compartment may represent the primary site of estrogen formation during certain stages of follicle development in the horse and human.

6.1.3. Granulosa cell aromatase bioassay for FSH

The study of potential paracrine modulators (estrogens, androgens and growth factors) that could enhance FSH action led to the development of an in vitro granulosa cell aromatase bioassay (GAB). The combined action of enhancing hormones and factors resulted in an assay highly sensitive to the aromatase-inducing function of FSH [69]. This assay is hormone-specific and sensitive, and has been applied to the measurement of FSH bioactivity in serum [70] and urine samples [71] from humans and diverse animal species.

6.2. Induction of cholesterol side-chain cleavage enzymes

The rate-limiting step in progestin biosynthesis is the side-chain cleavage of cholesterol to pregnenolone taking place in the mitochondria. At least three steps in the cholesterol side-chain cleavage (SSC) reaction may be under hormonal regulation [72,73]. FSH stimulates the synthesis of the SSC cytochrome P-450 which results in the stimulation of SSC activity [74].

6.3. Induction of 3β-hydroxysteroid dehydrogenase enzyme

3β-Hydroxysteroid dehydrogenase regulates the conversion of pregnenolone to progesterone. This enzyme has been identified in the granulosa cells by histochemical staining and direct enzyme assay. The activity of this enzyme is stimulated by FSH in cultured granulosa cells [75]. Since the conversion of pregnenolone to progesterone is irreversible, this is also one of the regulatory steps in progesterone biosynthesis.

7. FSH stimulation of inhibin biosynthesis

Inhibin, an ovarian protein hormone, has been shown to selectively decrease FSH but not LH secretion by the anterior pituitary cell. The gonadal source of inhibin has been localized to the testis Sertoli cells [76] and ovarian granulosa [77]. In cultured granulosa cells, FSH increases inhibin production via stimulation of the protein kinase A pathway. In vivo treatment with injections of gonadotropin also increases the circulating levels of inhibin [78], whereas treatment with inhibin

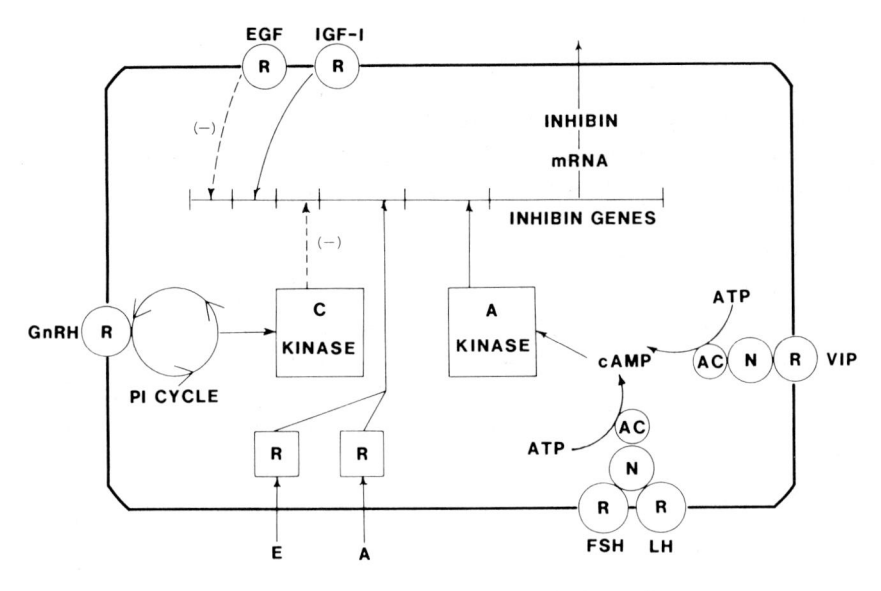

Fig. 4. Hormonal regulation of inhibin biosynthesis by granulosa cells.

antibodies causes an increase in circulating FSH levels [79]. These studies validate the existence of a feedback loop between ovarian inhibin-producing granulosa cells and pituitary FSH secreting gonadotropes. FSH stimulation of inhibin is modulated by multiple endocrine and paracrine hormones. The stimulatory effect of FSH on inhibin biosynthesis is augmented by estrogens, androgens and IGF-I, whereas EGF and GnRH inhibit FSH action. An overview of granulosa cell inhibin regulation is presented in Fig. 4.

8. FSH stimulation of tissue-type plasminogen activator

The release of oocytes from Graafian follicles during ovulation is essential for the propagation of the species. This phenomenon is associated with substantial increases in the activity of follicular proteases, plasminogen activators, and can be blocked by serine protease inhibitors [80]. The activity of plasminogen activators, specifically tissue-type plasminogen activator, is produced by gonadotropin-stimulated granulosa cells [81,82] and increases in the follicular fluid prior to ovulation. In addition, oocytes also produce tissue-type plasminogen activator and the hormonal regulation is mediated through the cumulus cells which possess FSH receptor [83,84]. It is believed that granulosa cells elaborate increasing amounts of plasminogen activators in response to FSH and LH. These enzymes, in turn, act on the available plasminogen in follicular fluid, generating plasmin within the follicle. Plasmin may activate procollagenase into the active collagenase, which initiates destruction of the follicular wall leading to ovum release [85].

9. Conclusion

Granulosa cells of the ovary are specific target cells for FSH. The action of FSH on these cells results from the binding of FSH to its specific receptors and subsequent activation of the protein kinase A pathway and the accumulation of cAMP. This activation induces multiple diverse physiological responses such as steroidogenesis, LH and PRL receptor induction, inhibin production and tPA secretion. The FSH-dependent physiological responses, which can be enhanced by steroids and growth factors may provide a model for studying glycoprotein hormone action. Purification and cloning of the FSH receptor, in conjunction with available molecular probes to FSH responsive proteins such as aromatase, side-chain cleavage enzyme and tissue-type plasminogen activator, will aid in the future understanding of the molecular mechanisms of FSH action at the gene level.

References

1. Pierce, J.G., Liao, T.H., Carlsen, R.B. and Reino, T. (1971) J. Biol. Chem. 246, 866–872.
2. Liu, W.K. and Ward, D.N. (1975) Pharmacol. Ther. 1, 545–567.
3. Vaitukaitus, J.L., Ross, G.T., Braunstein, G.D. and Rayford, P.L. (1976) Rec. Prog. Horm. Res. 32, 289–331.
4. Sairam, M.R. (1983) In: Hormonal Proteins and Peptides, Vol. XI, Chapter 1 (Li, C.H., ed.) pp. 1–79. Academic Press, New York.
5. Saxena, B.B. and Rathnam, P. (1976) J. Biol. Chem. 251, 993–1002.
6. Papkoff, H., Sairman, M.R., Farmer, S.W. and Li, C.H. (1973) Rec. Prog. Horm. Res. 29, 563–592.
7. Sairam, M.R. (1981) Biochem. J. 197, 535–552.
8. Sairam, M.R., Seidah, N.G. and Chretiem, M. (1981) Biochem. J. 197, 541–552.
9. Godine, J.E., Chin, W.W. and Habener, J.R. (1980) J. Biol. Chem. 255, 8780–8783.
10. Godine, J.E., Chin, W.W. and Habener, J.F. (1982) J. Biol. Chem. 257, 8368–8371.
11. Chin, W.W., Godine, J.E., Klein, D.R., Chang, A.S., Tank, L.K. and Habener, J.R. (1983) Proc. Natl. Acad. Sci. U.S.A. 80, 4649–4653.
12. Sairam, M.R. and Papkoff, H. (1974) In: Handbook of Physiology, Sect. 7, Vol. IV, Part 2 (Knobil, E. and Saywer, W.H., eds.) pp. 111–131. Am. Physiol. Soc., Washington, DC.
13. Giudice, L.C. and Pierce, J.G. (1978) In: Structure and Function of the Gonadotropins (McKerns, K.W., ed.) pp. 81–110. Plenum Press, New York.
14. Manjunath, P., Sairam, M.R. and Sairam, J. (1982) Mol. Cell. Endocrinol. 28, 125–138.
15. Sairam, J.R. and Manjunath, P. (1982) Mol. Cell. Endocrinol. 28, 139–150.
16. Chappel, S.C., Ulloa-Aguirre, A. and Coutifaris, C. (1983) Endocr. Rev. 4, 179–211.
17. Keutmann, H.T., Johnson, L. and Ryan, R.J. (1985) FEBS Lett. 185, 333–338.
18. Blum, W.F.P. and Gupta, D. (1985) J. Endocrinol. 105, 29–37.
19. Peckham, W.D. and Knobil, E. (1976) Endocrinology 98, 1054–1060.
20. Ryle, M., Chaplan, M.F., Gray, C.J. and Kennedy, J.F. (1970) In: Gonadotropins and Ovarian Development (Butt, W.R., Crooke, A.C. and Ryle, M., eds.) pp. 98–106. Livingstone, London.
21. Chappel, S.C., Ulloa-Aguirre, A. and Ramaley, J.A. (1983) Biol. Reprod. 28, 195–205.
22. Ulloa-Aguirre, A., Mejia, J.J., Dominguez, R., Guevara-Aguirre, J., Diaz-Sanchez, V. and Larra, F. (1986) J. Endocrinol. 110, 539–549.
23. Ulloa-Aguirre, A., Miuer, C., Hyland, L. and Chappel, S. (1984) Biol. Reprod. 30, 382–387.

24. Manjunath, P. and Sairam, M.R. (1985) Methods Enzymol. 109, 725–735.
25. Salomon, Y., Ezra, E., Nimrod, A., Amir-Zaltsman, Y. and Lindner, H.R. (1980) In: Chorionic Gonadotropins (Segal, S.J., ed.) pp. 345–369. Plenum Press, New York.
26. Moyle, W.R. (1980) Oxford Rev. Reprod. Biol. 2, 123–204.
27. McIlroy, P.J. and Ryan, R.J. (1983) In: Hormonal Proteins and Peptides, Vol. IX (Li, C.H., ed.) pp. 93–133. Academic Press, New York.
28. Catt, K.J., Dufau, M.L. and Tsuruhara, J. (1972) J. Clin. Endocrinol. Metab. 34, 1972–1976.
29. Means, A.R., Dedman, J.R., Tash, J.S., Tindall, D.J., Van Sickle, M. and Welsh, M.J. (1980) Ann. Rev. Physiol. 42, 59–70.
30. Reichert, L.E. Jr. and Bhalla, V.K. (1974) Endocrinology 94, 483–491.
31. Cheng, K.W. (1975) J. Clin. Endocrinol. Metab. 41, 581–589.
32. Fritz, L.B. (1979) In: Biochemical Actions of Hormones, Vol. 5 (Litwack, E., ed.) pp. 249–281. Academic Press, New York.
33. Calvo, F.O., Keutmann, H.T., Bergert, E.R. and Ryan, R.S. (1986) Biochemistry 25, 3938–3943.
34. Keutmann, H.T., Johnson, L. and Ryan, R.J. (1985) FEBS Lett. 185, 333–338.
34. Liu, W.K., Young, J.D. and Ward, D.N. (1984) Mol. Cell. Endocrinol. 37, 29–39.
35. Marsh, J.M. (1975) Adv. Cyclic Nucleotide Res. 6, 137–199.
36. Welsh, T.H. Jr., Jia, X.-C. and Hsueh, A.J.W. (1984) Mol. Cell. Endocrinol. 37, 51–60.
37. Ross, E.M. and Gilman, A.G. (1980) Ann. Rev. Biochem. 49, 533–564.
38. Limbird, L.E. (1981) Biochem. J. 195, 1–13.
39. Cassel, D., Leukovitz, H. and Selinger, Z. (1977) J. Cyclic Nucleotide Res. 3, 393–406.
40. Richards, J.S., Jonassen, J.A., Rolfes, A.I., Kersey, K. and Reichert, L.E. Jr. (1979) Endocrinology 104, 765–773.
41. Richards, J.S., Sehgal, N.A. and Tash, J.S. (1983) J. Biol. Chem. 258, 5227–5232.
42. Richards, J.S. and Rolfes, A.I. (1980) J. Biol. Chem. 255, 5481–5489.
43. Hsueh, A.J.W., Adashi, E.Y., Jones, P.B.C. and Welsh, T.H. Jr. (1984) Endocr. Rev. 4, 76–127.
44. Erickson, G.E., Wang, C. and Hsueh, A.J.W. (1979) Nature 279, 336–337.
45. Hillier, S.G., Zeleznik, A.J. and Ross, G.T. (1978) Endocrinology 102, 937–946.
46. Dunaif, A.E., Zimmerman, E.A., Friessen, H.G. and Frantz, A.G. (1982) Endocrinology 110, 1465–1471.
47. Wang, C., Hsueh, A.J.W. and Erickson, G.F. (1979) J. Biol. Chem. 254, 11330–11336.
48. Navickis, R.J., Jones, P.B.C. and Hsueh, A.J.W. (1982) Mol. Cell. Endocrinol. 27, 77–88.
49. Harwood, J.P., Reichert, N.D., Dufau, M.L. and Catt, K.J. (1980) Endocrinology 107, 280–288.
50. Aguado, L.I., Pentrovic, S.L. and Ojeda, S.R. (1982) Endocrinology 110, 1124–1132.
51. Adashi, E.Y. and Hsueh, A.J.W. (1981) Endocrinology 108, 2170–2178.
52. Gwynne, J.T. and Strauss, J.F. III (1982) Endocr. Rev. 3, 299–329.
53. Carr, B.R., MacDonald, P.C. and Simpson, E.R. (1982) Fert. Steril. 38, 303–311.
54. Strauss, J.F. III, MacGregor, L.C. and Gwynne, J.T. (1982) J. Steroid Biochem. 16, 525–531.
55. Chang, S.C.S., Jones, J.D., Ellefson, R.D. and Ryan, R.J. (1976) Biol. Reprod. 15, 321–325.
56. Bjersing, L. (1978) In: The Vertebrate Ovary (Jones, R.E., ed.) pp. 181–214. Plenum Press, New York.
57. Albertini, D.F. and Anderson, E. (1974) J. Cell Biol. 63, 234–250.
58. Burghardt, R.C. and Matheson, R.L. (1982) Dev. Biol. 94, 206–215.
59. Erickson, G.F. (1983) Mol. Cell. Endocrinol. 29, 21–49.
60. Dorrington, J.H., Moon, Y.S. and Armstrong, D.T. (1975) Endocrinology 97, 1328–1331.
61. Hsueh, A.J.W., Erickson, G.F. and Papkoff, H. (1983) Arch. Biochem. Biophys. 225, 505–511.
62. Evans, G.T., Ledesma, D.B., Schultz, T.Z., Simpson, E.R. and Mendelson, C.R. (1986) Proc. Natl. Acad. Sci. U.S.A. 83, 6387–6391.
63. Dorrington, J.H. and Armstrong, D.T. (1979) Rec. Prog. Horm. Res. 35, 301–342.
64. Tsang, B.K., Armstrong, D.T. and Whitfield, J.F. (1980) J. Clin. Endocrinol. Metab. 51, 1407–1411.

65. Makris, A. and Ryan, K.J. (1975) Endocrinology 96, 694–701.
66. Channing, C.P., Schaerf, F.W., Anderson, L.D. and Tsafriri, A. (1980) Int. Rev. Physiol. 22, 117–139.
67. Hillier, S.C. (1981) J. Endocrinol. 89, 36–44.
68. Liu, Y.X. and Hsueh, A.J.W. (1986) Biol. Reprod. 35, 27–36.
69. Jia, X.-C. and Hsueh, A.J.W. (1986) Endocrinology 119, 1570–1577.
70. Jia, X.-C., Kessel, B., Yen, S.S.C., Tucker, E.M. and Hsueh, A.J.W. (1986) J. Clin. Endocrinol. Metab. 62, 1243–1249.
71. Dahl, K.D., Czekala, N.M., Lim, P. and Hsueh, A.J.W. (1987) J. Clin. Endocrinol. Metab. 64, 486–493.
72. Sulimovici, S. and Boyd, G.S. (1968) Eur. J. Biochem. 3, 332–345.
73. Toaff, M.E., Strauss, J.R. III and Hammond, J.M. (1983) Endocrinology 112, 1156–1157.
74. Funkenstein, B., Waterman, M.R., Masters, B.S.S. and Simpson, E.R. (1983) J. Biol. Chem. 258, 10187–10191.
75. Jones, P.B.C. and Hsueh, A.J.W. (1982) Endocrinology 110, 1663–1671.
76. Steinberger, A. and Steinberger, C. (1976) Endocrinology 99, 918–921.
77. Erickson, G.F. and Hsueh, A.J.W. (1978) Endocrinology 103, 1960–1963.
78. Bicsak, T.A., Tucker, E.M., Cappel, S., Vaughn, J., Rivier, J., Vale, W. and Hsueh, A.J.W. (1986) Endocrinology 119, 2711–2719.
79. Rivier, C., Rivier, J. and Vale, W. (1986) Science 234, 205–208.
80. Strickland, S. and Beers, W.H. (1986) J. Biol. Chem. 251, 5694–5702.
81. Wang, C. and Leung, A. (1983) Endocrinology 112, 1201–1208.
82. Ny, T., Bjersing, L. and Hsueh, A.J.W. (1985) Endocrinology 116, 1666–1668.
83. Liu, Y.-X. and Hsueh, A.J.W. (1986) Endocrinology 119, 1570–1577.
84. Liu, Y.-X. and Hsueh, A.J.W. (1987) Biol. Reprod. 36, 1055–1063.
85. Werbs, Z., Mainardi, C.L., Vater, C.A. and Harris, E.D. (1977) New Engl. J. Med. 296, 1017–1023.

CHAPTER 10

The mechanism of action of ACTH in the adrenal cortex

PETER J. HORNSBY

Department of Cell and Molecular Biology, Medical College of Georgia, Augusta, GA 30912, U.S.A.

1. ACTH and the cyclic AMP intracellular messenger system

1.1. The intracellular messenger for ACTH

ACTH (adrenocorticotropic hormone, corticotropin) is a 39-amino-acid peptide synthesized and secreted by the corticotrope cells of the anterior lobe of the pituitary gland. ACTH acts on several target tissues, including the adrenal cortex, adipose tissue and brain. It is synthesized as part of proopiomelanocortin (POMC) as amino acids 132–170 of this molecule, which is proteolytically cleaved to produce ACTH [1].

ACTH stimulates the synthesis of steroids by the cells of the adrenal cortex. The steroidogenic action of ACTH is mediated primarily by the intracellular messenger cyclic AMP acting via cyclic AMP-dependent protein kinase. The evidence for this is that (i) ACTH stimulates cyclic AMP production in intact adrenocortical cells and in plasma membrane preparations; (ii) cyclic AMP analogues added to adrenocortical cells stimulate steroidogenesis to the same extent as ACTH; and (iii) mutant adrenocortical cells with defective cyclic AMP-dependent protein kinase lack stimulation of steroidogenesis by ACTH [2].

Although the primary mode of action of ACTH is via cyclic AMP and cyclic AMP-dependent protein kinase (A-kinase), calcium probably plays a secondary role in enhancing and modulating the action of ACTH via the A-kinase pathway. This is discussed later in the consideration of the interaction of the A-kinase pathway of activation by ACTH and other second messenger systems involving calcium and protein kinase C.

Unlike some other hormones which act on the adrenal cortex, such as angiotensin II and acetylcholine, ACTH probably does not have a major portion of its action via the hydrolysis of phosphatidylinositol 4,5-bisphosphate (PIP_2) and the generation of the intracellular messengers inositol 1,3,5-trisphosphate (IP$_3$) and diacylglycerol (DG) [3]. Thus, protein kinase C probably does not mediate any of the effects of ACTH on the adrenal cortex. However, there are important interactions between the protein kinase C system and the cyclic AMP-dependent protein kinase system activated by ACTH, which are discussed later in this chapter.

1.2. Spare cyclic AMP generating capacity and its function

Much lower concentrations of ACTH are required for full stimulation of steroidogenesis than for full stimulation of cyclic AMP accumulation in adrenocortical cells. This spare capacity for second messenger production has led some investigators to consider that ACTH does not necessarily exert all of its action on the adrenal cortex via generation of intracellular cyclic AMP. However, spare second messenger production confers on cells enhanced sensitivity to circulating hormone. Moreover, ACTH-stimulated steroidogenesis generally parallels the extent of activation of cyclic AMP-dependent protein kinase. The large difference in sensitivity to ACTH for steroidogenesis and for cyclic AMP production is consistent with the requirement for sensitive and fast response to changes in circulating ACTH levels [2,4].

1.3. The interaction of the ACTH receptor with adenylate cyclase

ACTH interacts with a 225 kDA protein in the adrenocortical plasma membrane which is thought to be the ACTH receptor on the basis of (i) binding to antibodies raised against a peptide encoded by a nucleic acid sequence complementary to that encoding ACTH, and (ii) crosslinking of ACTH with a photoactivatable bifunctional reagent [5]. On treatment with urea and 2-mercaptoethanol, the protein dissociates into subunits of 83 kDa, 64 kDa, 52 kDa and 22 kDa; ACTH binding is associated with the 83 kDa subunit. Although the receptor is a cell-surface protein, it does not appear to be extensively glycosylated. Binding studies using ^{125}I-labeled ACTH and purified receptors indicate two binding sites with affinities of 3.4×10^{-10} M^{-1} and 1.0×10^{-9} M^{-1}, with ten times more low-affinity sites than high-affinity sites. However, studies using intact cells show only a single class of ACTH-binding sites of the lower affinity [4]. The low-affinity binding sites may be physiologically relevant for ACTH activation of adenylate cyclase because their affinity for ACTH binding is consistent with the EC$_{50}$ for cyclic AMP production. Extracellular calcium is required for binding of ACTH to the receptor [4].

In plasma membrane preparations, binding of ACTH to its receptor activates adenylate cyclase. In common with other receptors which activate adenylate cyclase, the interaction of ACTH with adenylate cyclase appears to require the action of a G-protein, presumably G$_s$, since ACTH stimulation of cyclic AMP synthesis in plasma membrane preparations is enhanced by GTP, by non-hydrolysable GTP

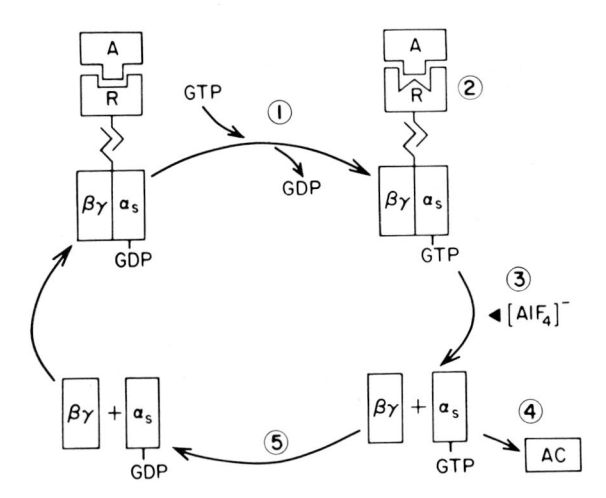

Fig. 1. Scheme for activation of adenylate cyclase by adrenocortical plasma membrane ACTH receptors, based on that elucidated in other tissues. ACTH (A) binds to receptor molecules (R) which then form a complex with G_s proteins, which have bound GTP (steps 1 and 2). G_s comprises α_s, β and γ subunits. Fluoride ions, via AlF_4^-, mimic receptor activation by dissociation of the G-protein (step 3). α_s with bound GTP binds adenylate cyclase, forming an active cyclic AMP-producing complex (step 4) until the GTPase activity of the G-protein hydrolyses GTP to GDP (step 5). The G-protein is now inactive in adenylate cyclase activation until it binds again to a hormone-receptor complex and GDP is exchanged for GTP (step 1). From Ref. 7.

analogues, and by fluoride ions [6]. The scheme shown in Fig. 1 is typical for receptors which interact via G-proteins with adenylate cyclase, and is probably valid for the interaction of the ACTH receptor with adenylate cyclase.

1.4. Cyclic AMP-dependent protein kinase in the adrenal cortex

Both type I and type II cyclic AMP-dependent protein kinase forms are found in the adrenal cortex [8]. The predominant form in the Y1 adrenocortical tumor cell line is type I [2]. Y1(Kin) mutants with RI subunits which have a much lower affinity for cyclic AMP lack ACTH-stimulated cyclic AMP production, thus demonstrating the involvement of the type I form of cyclic AMP-dependent protein kinase in the action of ACTH.

1.5. The pathway of biosynthesis of steroids in the adrenal cortex

1.5.1. The enzymes of steroidogenesis
The short-term action of ACTH on the adrenocortical cell is the stimulation of the conversion of cholesterol to glucocorticoid, mineralocorticoid or androgen-precursor steroids (Fig. 2). The conversion of cholesterol to the end-product steroids involves two mitochondrial cytochrome *P*-450 enzymes, cytochrome *P*-450$_{SCC}$ (cho-

```
                                Fatty acids
    Lipid droplet              ─────────→        Mitochondrion
    ┌─────────────┐                              ┌──────────────┐
    │ Cholesteryl │─────────────────────────────→│ Cholesterol  │
    │   esters    │                              │      │       │
    └─────────────┘                              │      │ P-450_scc
                                                 │      ↓       │
    ┌─────────────┐                              │ Pregnenolone │
    │ Pregnenolone│←─────────────────────────────│              │
    │      │ P-450_17α                           │              │
    │      ↓                                     │              │
    │ 17α-OH-Pregnenolone                        │              │
    │      │ dehydrogenase        ──────────────→│ 11-Deoxycortisol
    │      │ and isomerase                       │      │       │
    │      ↓                                     │      │ P-450_11β
    │ 17α-OH-Progesterone                        │      ↓       │
    │      │ P-450_C21                           │   Cortisol   │
    │      ↓                                     │              │
    │ 11-Deoxycortisol ──────────────────────────│              │
    └─────────────┘                              └──────┬───────┘
    Endoplasmic reticulum                               │
                                                        ↓
                                                     Cortisol
```

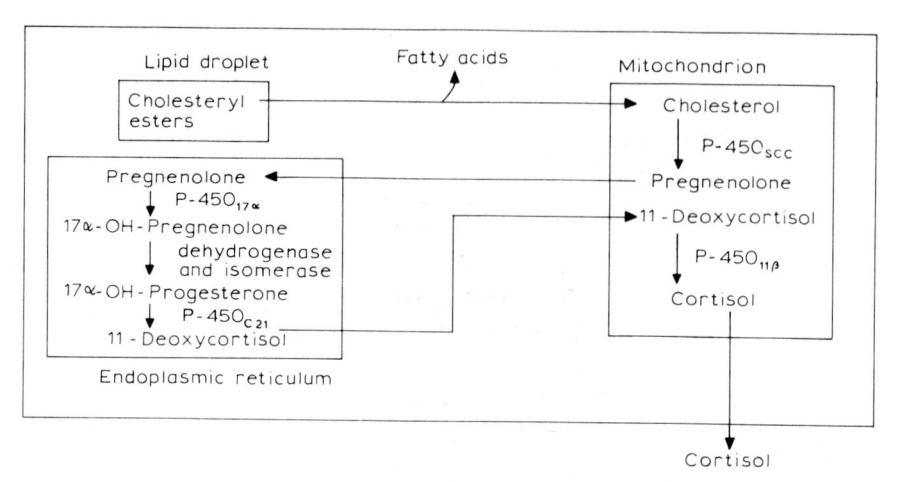

Fig. 2. Pathway of biosynthesis of the glucocorticoid, cortisol, in the adrenal cortex. Cholesterol, from stores in cholesteryl esters or from other sources (see text) is converted via mitochondrial cytochrome $P\text{-}450_{SCC}$ (cholesterol side-chain cleavage enzyme) to pregnenolone, which then is successively converted by the microsomal enzymes cytochrome $P\text{-}450_{17\alpha}$ (17α-hydroxylase), 3β-hydroxysteroid dehydrogenase/isomerase and cytochrome $P\text{-}450_{C21}$ (21-hydroxylase) to 11-deoxycortisol, followed by conversion by the mitochondrial cytochrome $P\text{-}450_{11\beta}$ (11β-hydroxylase) to cortisol. The short-term action of ACTH in stimulation of steroidogenesis is to increase the availability of cholesterol for conversion by cytochrome $P\text{-}450_{SCC}$. From Ref. 9.

lesterol side-chain cleavage enzyme) and cytochrome $P\text{-}450_{11\beta}$ (11β-hydroxylase); two microsomal cytochrome P-450 enzymes, cytochrome $P\text{-}450_{17\alpha}$ (17α-hydroxylase) and cytochrome $P\text{-}450_{C21}$ (21-hydroxylase); a microsomal enzyme with 3β-hydroxysteroid dehydrogenase and 3-oxosteroid Δ^5-isomerase activity; and a cytoplasmic enzyme with dehydroepiandrosterone (DHEA) sulfotransferase activity. The pathway of conversion of cholesterol to the major glucocorticoid product of the human adrenal cortex, cortisol, is shown in Fig 2. The synthesis of the major mineralocorticoid, aldosterone, is similar, but does not involve a 17α-hydroxylation step; corticosterone is converted to aldosterone in the mitochondria by the corticosterone methyl oxidase (CMO) activity of cytochrome $P\text{-}450_{11\beta}$ [10]. In the synthesis of the major human adrenal androgen, DHEA sulfate (DHEAS), pregnenolone is converted by cytochrome $P\text{-}450_{17\alpha}$, via 17α-hydroxypregnenolone, to DHEA and then by sulfotransferase to DHEAS [11].

1.5.2. Zonation of steroidogenesis
In the adrenal cortex, the outermost zona glomerulosa secretes the mineralocorticoid aldosterone, and corticosterone. The middle zona fasciculata and the inner zona reticularis secrete predominately the glucocorticoids cortisol and corticosterone. In the human, the inner zona reticularis secretes large amounts of the adrenal androgens, DHEA and DHEAS, as well as cortisol [12]. Whereas steroidogenesis in the

zona fasciculata and zona reticularis is regulated predominantly by ACTH, steroidogenesis in the zona glomerulosa is coordinately regulated by ACTH and several other hormones. This is discussed in more detail later in this chapter.

The relative levels of the enzymes expressed by the individual adrenocortical cell determine the series of hydroxylation steps that occurs and the ultimate steroid that is produced. Thus, the rate of production of a given adrenocortical steroid is determined by the product of
   (i) the total rate of synthesis of all steroids in the part of the cortex involved in its synthesis and
   (ii) the proportion of total steroid output that the steroid represents within this part.

The rate of total steroidogenesis is determined by the rate of supply of cholesterol to cytochrome P-450$_{SCC}$. The rate of flux through this step determines the rate of synthesis of the sum of the steroid products, but does not determine the rate of synthesis of any individual steroid. The pattern of steroidogenesis, i.e. which steroids are produced and in what ratio, is determined by the relative activities of the enzymes of the steroidogenic pathway beyond the formation of pregnenolone.

1.6. The regulation by ACTH of the rate-limiting step of steroidogenesis, the conversion of cholesterol to pregnenolone

1.6.1. Nature of the rate-limiting step: limitation on cellular movement of cholesterol
The rate-limiting step of the synthesis of the adrenocortical steroids is the conversion of cholesterol to pregnenolone. The rate-limiting nature of this step does not result from a limitation of the activity of cholesterol side-chain cleavage enzyme (cytochrome P-450$_{SCC}$), but from limitation on the access of cholesterol to the substrate site of cytochrome P-450$_{SCC}$. This may readily be demonstrated on addition of mono-hydroxylated cholesterol to isolated adrenocortical cells or to cholesterol side-chain cleavage enzyme preparations; the substrate is converted at a high rate to pregnenolone [13]. In the pathway of steroidogenesis, sterols and steroids beyond the initial hydroxylation performed by cytochrome P-450$_{SCC}$ are sufficiently soluble in aqueous media to diffuse freely within the cell (Fig. 3). The advantageous control point for steroidogenesis in the adrenocortical cell is therefore clearly the conversion of cholesterol to pregnenolone, the transformation of the non-diffusible cholesterol to the subsequent diffusible steroids. Once past this step, steroids move freely from one enzyme to the next in the pathway, and finally diffuse out of the cell into the blood stream. Prior to this step, the poor solubility of cholesterol in aqueous media requires that cholesterol be bound to carrier proteins for transport within the cell, from lipid droplets after hydrolysis or from endocytosed lipoproteins (Fig. 3). Sterol carrier protein SCP$_2$, which is present in adrenocortical cell cytoplasm, is likely involved in this transport [14]. Cytoskeletal elements may also be involved in intracellular movement of cholesterol to mitochondria [15].

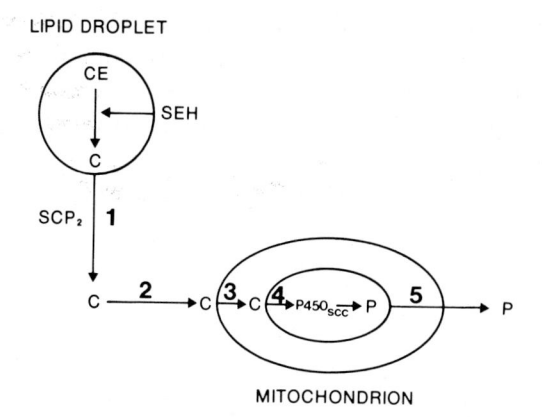

Fig. 3. The rate-limiting step of steroidogenesis under ACTH regulation. The transfer of cholesterol (C) from the outer to the inner mitochondrial membrane under ACTH regulation (step 3) makes cholesterol available to cytochrome P-450_{SCC} for conversion to pregnenolone (step 4), which diffuses out of the mitochondrion (step 5). Because of its insolubility in aqueous media, cholesterol must be transported to mitochondria, probably by SCP_2, from a precursor pool (step 2). Here, cholesterol in the precursor pool is shown as being formed from cholesterol esters (CE) by cholesterol ester hydrolase (CEH) (step 1); other possible pathways are shown in Figs. 4 and 6. From Ref. 14.

1.6.2. Supply of cholesterol to the precursor pool available for steroidogenesis

Cholesterol for ACTH-stimulated steroidogenesis may be provided to the immediate precursor pool for steroidogenesis both by endogenous synthesis from acetate and by uptake from plasma lipoproteins (Fig. 4). The extent to which cholesterol is supplied from these two sources depends on the species and on the availability of lipoproteins. When human adrenocortical cells are supplied with high levels of low-density lipoprotein (LDL), de novo cholesterol synthesis via the rate-limiting enzyme 3-hydroxy-3-methylglutaryl coenzyme A reductase (HMG CoA reductase) is suppressed and most steroidogenesis is from exogenous cholesterol [16]. When cells are deprived of lipoproteins, HMGCoA reductase is induced and most steroidogenesis is from endogenous synthesis. Exogenous lipoproteins may be supplied both as LDL and as high-density lipoprotein (HDL), the latter being the preferred substrate in some species, such as the rat [17]. Intracellularly, cholesterol may be stored in esterified form in lipid droplets, serving as a reserve pool of cholesterol which may be drawn on in periods of high rates of utilization. Such intracellular storage is particularly important when most cholesterol for steroidogenesis is from de novo synthesis. Cholesterol ester hydrolase then releases cholesterol to a cellular pool from which it is available for steroidogenesis (Fig. 4).

1.6.3. ACTH regulation of the cholesterol pool

Several actions of ACTH, which are unlikely to be rate-limiting for the immediate conversion of cholesterol to steroids, maintain the precursor pool of cholesterol and

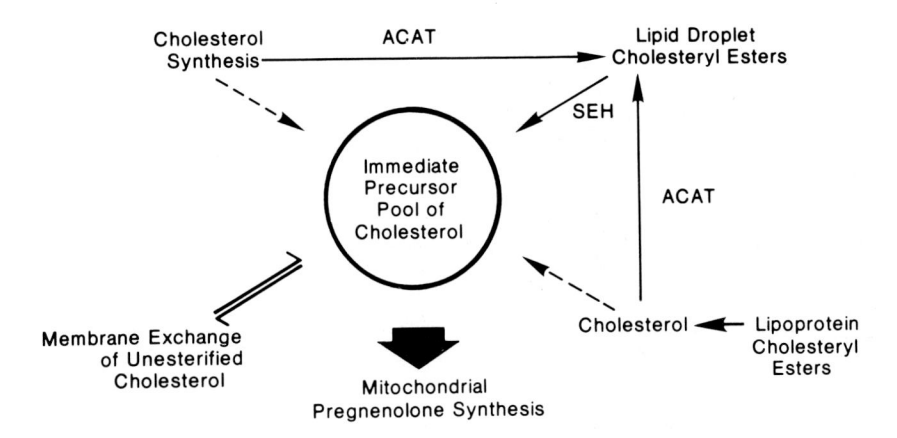

Fig. 4. Potential sources of the immediate precursor pool of unesterified cholesterol available for mitochondrial steroidogenesis. ACAT, microsomal acyl coenzyme A cholesterol acyltransferase; SEH, sterol ester hydrolase. From Ref. 14.

thereby maintain a constant rate of supply of cholesterol to mitochondria for steroidogenesis. ACTH increases the synthesis of sterol carrier protein SCP_2 [18]; increases HMGCoA reductase activity [16]; increases the synthesis of LDL receptors [19]; and increases the activity of cholesterol ester hydrolase [14]. Apart from the last, which results from an increase in phosphorylation by cyclic AMP-dependent protein kinase, these actions may result from compensatory changes caused by the increased rate of utilization of cholesterol stimulated by ACTH and the consequent depletion of the cellular cholesterol pool.

1.6.4. Regulation of the rate-limiting step by cyclic AMP-dependent protein kinase
Despite intensive study, the mode by which cyclic AMP-dependent protein kinase activates the movement of cholesterol from the immediate supply pool to cytochrome P-450_{SCC}, i.e. the mode by which ACTH regulates the rate-limiting step of steroidogenesis, is still uncertain. The involvement of a labile, protein factor has been implicated since the discovery of the sensitivity of ACTH-stimulated steroidogenesis to inhibitors of protein synthesis [8]. The labile factor may be a cholesterol-binding protein, which acts to increase the availability of cholesterol to cytochrome P-450_{SCC}, or it may be a regulatory factor which activates a cholesterol-binding protein such as SCP_2, which has been shown to increase pregnenolone formation when added to adrenocortical mitochondria [14]. A strong candidate for the labile factor is the steroidogenesis-activator polypeptide isolated from steroidogenic tissues which is increased in concentration after increases in intracellular cyclic AMP and which enhances pregnenolone formation in mitochondria [20]. The role of cyclic AMP-dependent protein kinase in the activation of the rate-limiting step has not been positively identified. One hypothesis for the role of ACTH-stimulated cyclic AMP in the activation of cholesterol conversion to pregnenolone is presented in Fig. 5 [21].

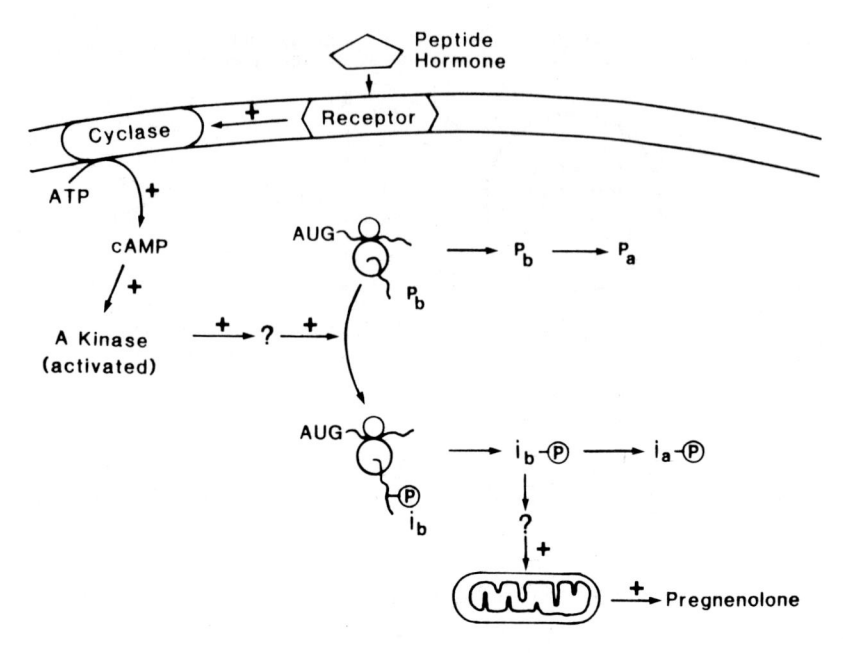

Fig. 5. Hypothesis for the role of phosphorylation in the stimulation of steroidogenesis by ACTH. ACTH activation of cyclic AMP-dependent protein kinase results in the co-translational phosphorylation of protein p_b to protein i_b, postulated to be the effector of the increased availability of cholesterol to mitochondrial cytochrome $P\text{-}450_{SCC}$. p_a and i_a are hypothesized to be derived from p_b and i_b. From Ref. 21.

1.7. The regulation of the synthesis of the steroidogenic enzymes by ACTH

1.7.1. The integration of the short- and long-term actions of ACTH to provide increased steroidogenesis

Apart from its short-term actions on cholesterol availability to cytochrome $P\text{-}450_{SCC}$, ACTH maintains the steroidogenic capacity of the adrenocortical cell by long-term effects on the synthesis of the enzymes involved in steroidogenesis. To varying extents, all of the steroidogenic cytochrome P-450 enzymes (Fig. 2) are inducible by ACTH, acting via cyclic AMP [9]. For example, the rate of synthesis of cytochrome $P\text{-}450_{17\alpha}$ is extremely low in the absence of intracellular cyclic AMP, and very high after cells have been exposed to ACTH or other factors which raise intracellular cyclic AMP [22]; on the other hand, cytochrome $P\text{-}450_{SCC}$ synthesis proceeds at a measurable rate in the absence of cyclic AMP stimulation and is increased by cyclic AMP only approximately 2-fold, implying the involvement of other regulatory factors, perhaps cholesterol itself in this instance (Fig. 6). As discussed later in this chapter, other intracellular messenger systems may coordinately regulate the synthesis of some of the steroidogenic enzymes in conjunction with ACTH and cyclic AMP. The synthesis of the accessory proteins involved in electron transport, adrenodoxin, adrenodoxin reductase and cytochrome P-450 reductase, is also regulated by ACTH acting via cyclic AMP [9].

1.7.2. Mechanism of enzyme induction by cyclic AMP
The mode by which ACTH, via cyclic AMP, causes induction of the steroidogenic enzymes has not been fully elucidated. Nuclear run-off experiments show that increased intracellular cyclic AMP increases the rate of transcription of the steroidogenic enzyme genes and that the increase in transcription accounts for the majority of the elevation in levels of the mRNA and of the enzyme protein [23]. However, changes in mRNA stability may also be a secondary means of regulation in the case of cytochrome P-450$_{SCC}$ [24]. The mode of increase of transcription is not clear. Presumably, an as yet unidentified regulatory protein is a substrate for cyclic AMP-dependent protein kinase, is phosphorylated and then acts as a signal for increased transcription. Y1(Kin) cell mutants with impairment of cyclic AMP-dependent protein kinase activity show poor induction of transfected 21-hydroxylase genes [25]. Upstream sequences in steroidogenic cytochrome P-450 genes have been identified which are required for ACTH-dependent transcription [25]. These

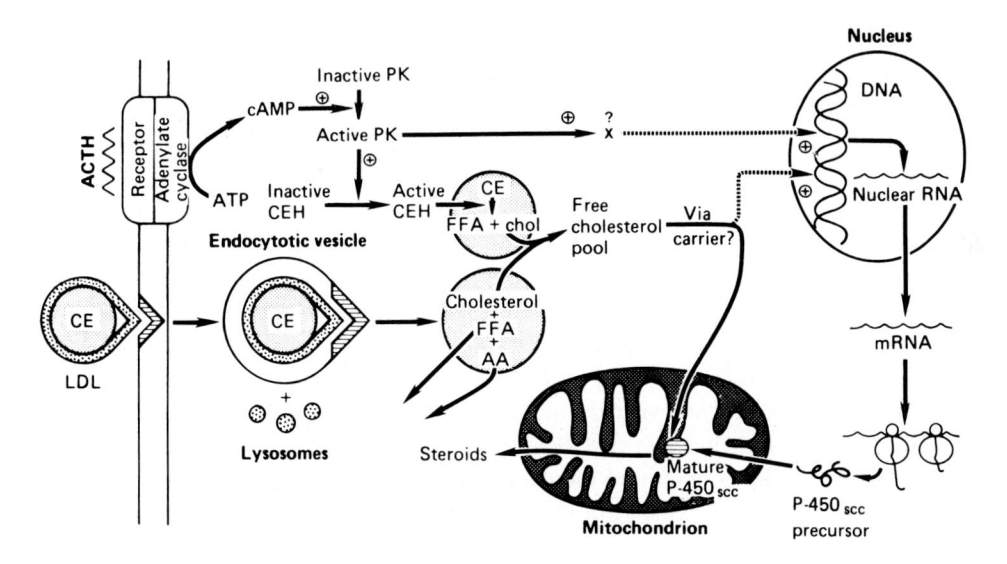

Fig. 6. The integration of the short- and long-term actions of ACTH in the supply of cholesterol and the level of cytochrome P-450$_{SCC}$ protein. ACTH stimulation of adenylate cyclase and activation of cyclic AMP-dependent protein kinase result in increased supply of cholesterol to the rate-limiting step of conversion to pregnenolone by cytochrome P-450$_{SCC}$. Cholesterol may also be supplied to the precursor pool from cholesterol esters (CE) endocytosed in lipoproteins (LDL). Cholesterol esters after hydrolysis by cholesterol ester hydrolase (CEH) yield free cholesterol and free fatty acids (FFA), including arachidonic acid (AA). In the long term, an unknown intermediate (x) serves as a nuclear signal for increased synthesis of mRNA for cytochrome P-450$_{SCC}$; cholesterol itself may also independently form a signal for increased cytochrome P-450$_{SCC}$ synthesis. From Ref. 9.

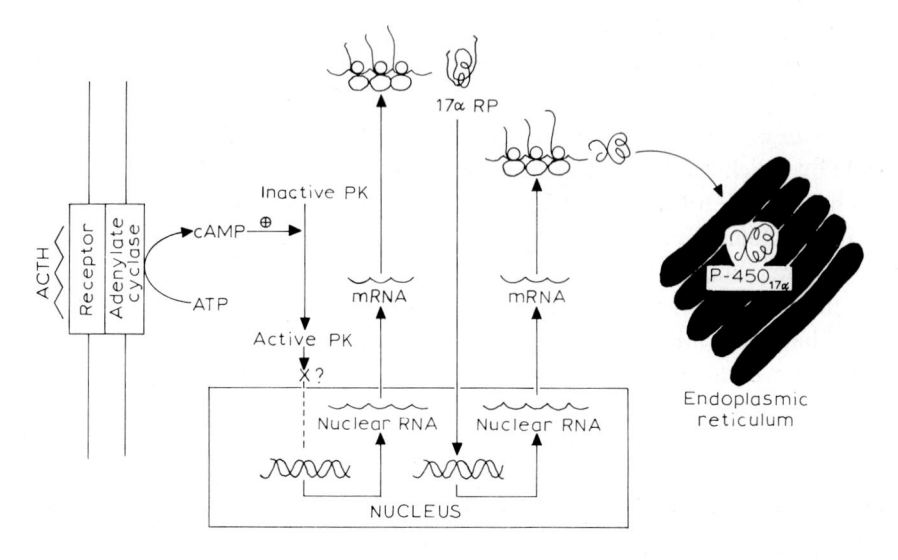

Fig. 7. Hypothesis for the action of ACTH in increasing the level of steroidogenic enzymes, in this case 17α-hydroxylase. In this model, cyclic AMP activates cyclic AMP-dependent protein kinase which by a series of unknown steps results in the accumulation of mRNA coding for a regulatory protein (17α-RP). After translation of this mRNA, the 17α-RP is hypothesized to translocate to the nucleus, where it activates the transcription of the cytochrome $P\text{-}450_{17\alpha}$ gene. This is one of several hypotheses which account for the sensitivity of cytochrome P-450 gene expression to inhibition of protein synthesis. From Ref. 27.

sequences may serve as binding sites for cyclic AMP-regulated DNA-binding regulatory proteins which enhance transcription; such sequences have been identified for other cyclic AMP-regulated genes [26]. Transcription of the steroidogenic cytochrome P-450 genes is greatly reduced when protein synthesis is inhibited. This may indicate that the initial effect of cyclic AMP is to increase the synthesis of such a regulatory protein (Fig. 7) [24]. Alternatively, the sensitivity of transcription to inhibitors of protein synthesis may indicate that a labile protein factor is required for transcription but that this factor is not itself regulated by cyclic AMP. Transfection experiments with cloned mouse 21-hydroxylase also indicate the existence of tissue-specific factors which are required for expression of the gene in response to ACTH [28].

1.8. Indirect actions of ACTH on growth and metabolism

ACTH is directly anti-mitogenic for adrenocortical cells, but indirectly stimulates the hypertrophy and hyperplasia of the adrenal cortex (reviewed in Ref. 29). Although the mechanism of the indirect stimulation of proliferation of adrenocortical cells by ACTH is not known it appears to be mediated by cyclic AMP [29]. A plausible model for this indirect growth stimulation is that ACTH stimulates the growth of the adrenal vascular system, and that the increased vascularization and production of basement membrane are permissive for the growth of adrenocortical cells. ACTH-stimulated vascularization could result from activation of neural reflexes, resulting in local release of growth factors from nerves, or from ACTH-induced release of angiogenic factors, such as bFGF and IGF-II [29]. bFGF is synthesized by adrenocortical cells [30], and ACTH stimulates the synthesis of IGF-II by adrenocortical cells [31].

ACTH has several other effects on adrenocortical cell metabolism. It increases the synthesis of adrenocortical cell phospholipids; this may also have a long-term supportive role for increased steroidogenic capacity. It increases the rate of glycolysis in the adrenocortical cell; this may have a supportive role in the supply of pyruvate for the mitochondrial steroid hydroxylases [32].

2. Interactions of the ACTH/cyclic AMP system with other hormones and intracellular messengers

2.1. Zonal differences

The interactions between the ACTH-stimulated cyclic AMP-dependent protein kinase system and intracellular messengers stimulated by other steroidogenic agents depend on the zonal origin of the cell. Like the calcium intracellular messenger system, the level of activation of the protein kinase C system appears to be higher in the zona glomerulosa than in the zona fasciculata, and higher in the zona reticularis than in the zona fasciculata. The potential for multiple interactions of other systems with ACTH/cyclic AMP-dependent protein kinase, at various levels from receptors/G-proteins to the final point of regulation of steroidogenesis, is therefore greater for the zona glomerulosa cell. Regulation by multiple hormones activating A-kinase versus C-kinase is shown in Fig. 8. The probable interactions of the intracellular messengers, cyclic AMP, calcium, DG and IP_3, is shown in Fig. 9 for the three primary stimuli of the adrenocortical zona glomerulosa cell, ACTH, angiotensin II and potassium. Zonation of intracellular messenger systems in the cortex may result from zonation of the receptors which activate these intracellular messengers or zonation of the components of the second messenger systems themselves. This zonation may play a critical role in the interplay of regulation of mineralocorticoid, glucocorticoid and androgen synthesis by ACTH and by the other steroidogenic agents in the three zones of the cortex.

The cyclic AMP-dependent protein kinase system is involved in the effects of ACTH in both the short term (stimulation of conversion of cholesterol to pregnenolone) and in the long term (increased synthesis of steroidogenic enzymes). In the interaction of the ACTH/cyclic AMP system with other intracellular messengers,

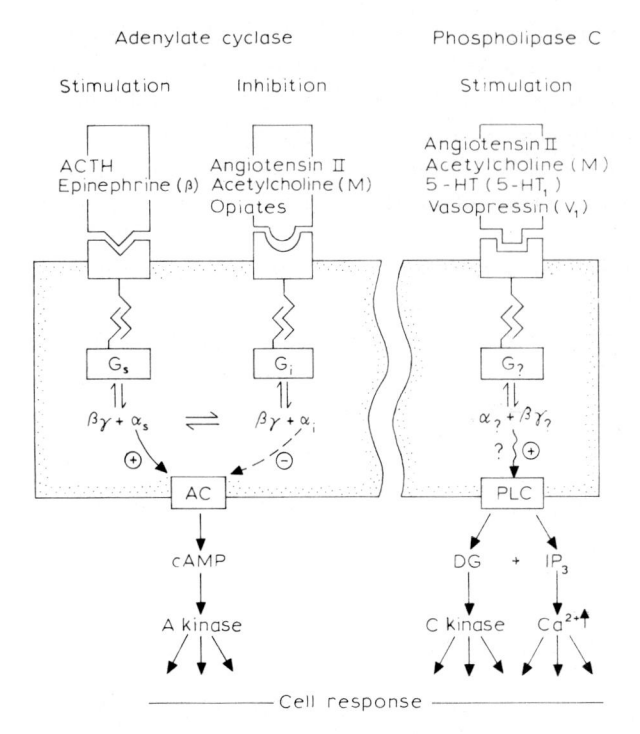

Fig. 8. Known and potential interactions of the ACTH/adenylate cyclase/cyclic AMP-dependent protein kinase system in the adrenocortical cell with other hormones and intracellular messengers. Epinephrine activates adenylate cyclase in the adrenocortical cell [33]. Adrenocortical cells have receptors for several hormones which may activate G_i, including angiotensin II [34], acetylcholine [35], and endogenous opioid peptides [36]. Angiotensin II, acetylcholine and vasopressin [37–39] have all been demonstrated to activate the breakdown of PIP_2 in adrenocortical cells and to stimulate steroidogenesis; 5-hydroxytryptamine is also a known steroidogenic agent [40]. Probable receptor subtypes involved are indicated (β; M (muscarinic); 5-HT_1; and V_1). This is not a comprehensive diagramming of all stimuli or all possible interactions. Modified from Ref. 7.

effects may differ in the short term and in the long term. For example, angiotensin II, via protein kinase C, stimulates cortisol synthesis in zona fasciculata-reticularis cells, probably acting like ACTH on the conversion of cholesterol to pregnenolone, but inhibits the induction of 17α-hydroxylase by cyclic AMP [42,43], as discussed below.

2.2. Interactions at adenylate cyclase

Activators of plasma membrane G_i protein may inhibit adenylate cyclase. The zona glomerulosa has higher levels of receptors which activate G_i, such as those for angiotensin and somatostatin [44,45], which may inhibit ACTH-stimulated adenylate cyclase [34]. Alternatively, activation of protein kinase C and phosphoryla-

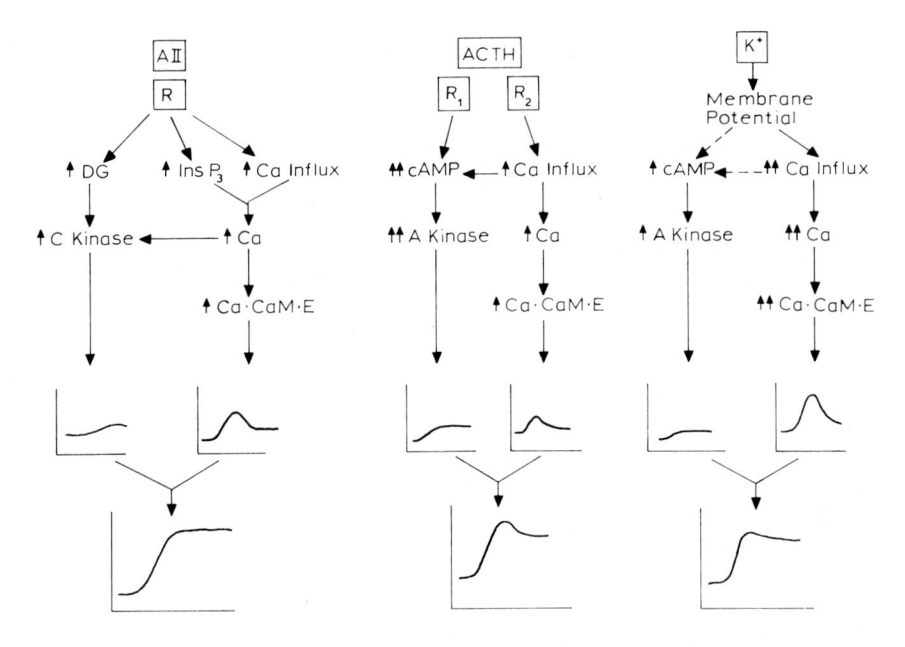

Fig. 9. The interaction of ACTH with the cyclic AMP and calcium intracellular messenger systems in the regulation of steroidogenesis in the adrenocortical zona glomerulosa cell: comparison with angiotensin II and potassium. ACTH activates both adenylate cyclase and calcium influx, here shown as involving two receptor subtypes (R_1 and R_2) although such receptor subtypes have not been identified. The A-kinase and calmodulin systems produce individual responses of characteristic amplitudes and timecourses, which combine to give the observed response of the intact cell. The sequence of events for ACTH is compared to those for the other two major stimuli of steroidogenesis in the zona glomerulosa cell, angiotensin II and potassium. From Ref. 41.

tion of an unknown substrate may increase ACTH stimulation of adenylate cyclase; angiotensin II has this action in the bovine adrenocortical cell [46].

2.3. ACTH and cyclic GMP

The role of cyclic GMP in the adrenal cortex is unknown. The only physiological regulators of cyclic GMP in the adrenal cortex appear to be atrial natriuretic factor (ANF) and related peptides. ANF stimulates guanylate cyclase in the adrenocortical cell and inhibits steroidogenesis, with a greater effect on the zona glomerulosa than on the zona fasciculata-reticularis, corresponding to a higher level of ANF receptors and guanylate cyclase in the outer cortex [47]. Cyclic GMP-dependent protein kinase has been purified from the adrenal cortex [48]. However, added cyclic GMP derivatives and increases in intracellular cyclic GMP by chemical agents do not alter steroidogenesis, casting doubt on the role of cyclic GMP as an intracellular mediator of the action of ANF [49]. ANF inhibits both ACTH-stimulated and angiotensin II-stimulated steroidogenesis, suggesting that the site of inhibition lies beyond the point of intracellular messenger generation [50].

2.4. ACTH and the calcium intracellular messenger system

2.4.1. Zonal differences

Although the primary mode of action of ACTH is via cyclic AMP and cyclic AMP-dependent protein kinase, calcium probably plays a secondary role in enhancing and modulating the action of ACTH via the A-kinase pathway. The extent to which calcium acts as an intracellular messenger for ACTH may vary according to the zonal origin of the cell and on the species. Although both zona glomerulosa and zona fasciculata cells have a voltage-dependent calcium conductance [51] zona glomerulosa cells may have a higher concentration of plasma membrane calcium channels [52]. Other steroidogenic stimuli to the zona glomerulosa cell, angiotensin II and potassium, activate calcium influx as part of their steroidogenic action [41]. In the zona glomerulosa cell, ACTH appears to act similarly, but perhaps only in some species [41,53]; moreover, the extent to which calcium may play a part in the action of ACTH on zona fasciculata-reticularis cells is uncertain.

2.4.2. The calcium second messenger system in the adrenal cortex

However, it is clear that the calcium/calmodulin pathway can be activated by ACTH [54]. ACTH, like other steroidogenic hormones, increases the influx of calcium into the adrenocortical cell, probably by an action on voltage-sensitive calcium channels [41]. Because stimuli which act only to increase cyclic AMP, such as forskolin, do not enhance calcium influx, it is probable that to some extent the ACTH receptor is coupled to a calcium channel, with intermediacy of a G-protein, or may act by some other mechanism (for example, inactivation of a potassium channel [55]).

Calcium, via calmodulin-dependent protein kinase, may activate and synergize with elements of the steroidogenic response also activated by cyclic AMP. Increased intracellular calcium in response to ACTH activates phospholipase A_2 with release of arachidonic acid and the production of prostaglandins [56]. There is no evidence to support an obligatory role of prostaglandins in the steroidogenic action of ACTH, but they may play a synergistic role, since several prostaglandins activate adrenocortical adenylate cyclase [57]. The role of lipoxygenase products of arachidonic acid is currently uncertain. Some experiments suggest that they may play a role in the action of ACTH [58,59].

2.4.3. Cyclic AMP phosphodiesterase

Calcium, via calmodulin, also activates cyclic AMP phosphodiesterases which inactivate cyclic AMP by metabolism to AMP. Inhibition of phosphodiesterase yields a small increase in ACTH-stimulated steroidogenesis [2]. This is relatively slight in the adrenocortical zona fasciculata-reticularis cell, which has a low phosphodiesterase level; the zona glomerulosa cell has a higher activity [60,61], which may result from higher level of activation of the calcium/calmodulin intracellular messenger system in the zona glomerulosa.

2.5. ACTH and protein kinase C

2.5.1. C-kinase in the adrenal cortex: presence and steroidogenic effects
Several steroidogenic hormones activate phospholipase C, causing breakdown of PIP_2 to two intracellular messengers, DG and IP_3 (Fig. 8). Calcium is mobilized from intracellular stores by IP_3 and DG activates the calcium- and phospholipid-dependent protein kinase C (C-kinase) [3].

Protein kinase C is present in the adrenocortical cell, and phorbol esters, which are potent activators of protein kinase C, stimulate adrenocortical steroidogenesis [62] and cell growth [63].

2.5.2. Coordinate regulation of adrenal enzyme synthesis by A- and C-kinases
Studies on cultured human adrenocortical cells suggest that protein kinase C plays an important role in modulating the long-term ACTH-stimulated maintenance of the steroidogenic enzyme of the adrenal cortex [43,63]. Long-term elevation of intracellular cyclic AMP increases the synthesis of both androgens and glucocorticoids, whereas activation of protein kinase C changes the balance of products, with increased non-17α-hydroxylated steroid synthesis and decreased androgen and cortisol biosynthesis. ACTH, via activation of cyclic AMP-dependent protein kinase, is the only stimulus required for adrenal androgen synthesis [11]. In the human adrenocortical cell, physiological activators of protein kinase C may act as 'androgen-inhibiting hormones'.

2.5.3. Mechanisms for regulation of steroidogenic enzymes that differ in activity between the different zones
Although ACTH-stimulated cAMP production in membrane particles from the glomerulosa and fasciculata is similar [64,65], there may normally be lower cyclic AMP levels in the glomerulosa. This may be due to the differences between the zones in the activity of phosphodiesterase and higher levels in the glomerulosa of receptors which activate G_i, as discussed above. 17α-Hydroxylase is strongly inducible by cyclic AMP, as discussed above. The glomerulosa maintains a lower intracellular cyclic AMP level than the rest of the cortex, thus lowering A-kinase activity, which, together with higher C-kinase activity, prevents induction of 17α-hydroxylase [42,43,63]. C-kinase activation causes inhibition of induction of enzymes in some other steroidogenic cell types [66,67]. On the other hand, the combination of second messenger activity levels in the outer cortex enhances 3β-hydroxysteroid dehydrogenase and the glomerulosa-specific CMO. 3β-Hydroxysteroid dehydrogenase is induced by both A-kinase and C-kinase activation [43,63]. CMO is induced by potassium in a non-cyclic AMP-dependent process [68].

It may be hypothesized that low activation of A-kinase and high activation of the other intracellular messengers in the outermost cortex results in mineralocorticoid synthesis, whereas high activation of A-kinase and low C-kinase in the innermost

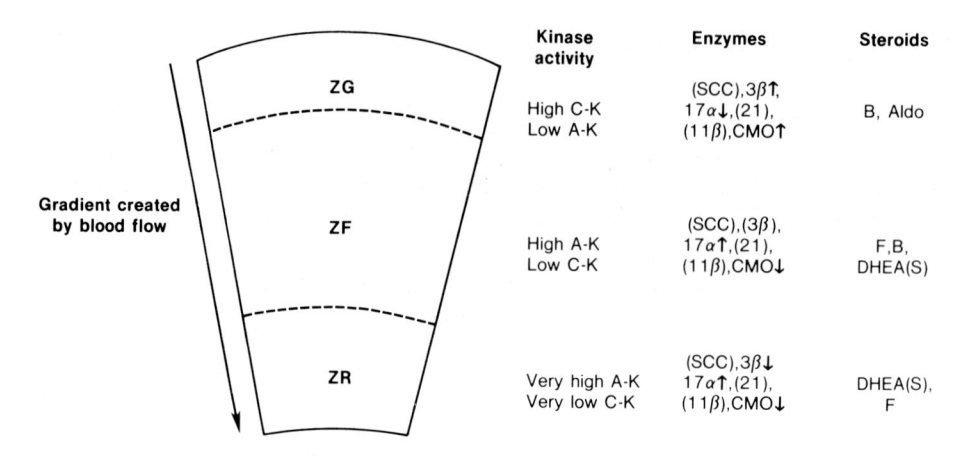

	Kinase activity	Enzymes	Steroids
ZG	High C-K Low A-K	(SCC),3β↑, 17α↓,(21), (11β),CMO↑	B, Aldo
ZF	High A-K Low C-K	(SCC),(3β), 17α↑,(21), (11β),CMO↓	F,B, DHEA(S)
ZR	Very high A-K Very low C-K	(SCC),3β↓ 17α↑,(21), (11β),CMO↓	DHEA(S), F

Gradient created by blood flow

Fig. 10. Hypothesis for the interaction of the A-kinase (A-K) system activated by ACTH with the C-kinase system (C-K) in the long-term regulation of the enzymes of steroidogenesis throughout the adrenal cortex. The primary determinant of zonation of A-kinase and C-kinase activities, via zonation of cell surface receptors or other mechanisms, is hypothesized to be a gradient (e.g., of steroids) created by the pattern of blood flow in the adrenal cortex. The resultant levels of induction of steroidogenic enzymes are indicated by ↑ to show particular elevation and by ↓ to show particular lack of induction or suppression of induction. Other enzymes involved in steroidogenesis are shown in parentheses. SCC=cholesterol side-chain cleavage enzyme; 3β=3β-hydroxysteroid dehydrogenase; 17α=17α-hydroxylase; 21=21-hydroxylase; 11β=11β-hydroxylase; CMO= corticosterone methyl oxidase activity of 11β-hydroxylase. Secreted steroids are indicated as: B=corticosterone; Aldo=aldosterone; F=cortisol; DHEA(S)= dehydroepiandrosterone (sulfate).

cortex results in adrenal androgen synthesis, and that cortisol synthesis in the middle region of the cortex results from a balance of A-kinase and C-kinase activation which is compatible with induction of 17α-hydroxylase without complete suppression of 3β-hydroxysteroid dehydrogenase (Fig. 10).

2.5.4. The origin of zonation in the adrenal cortex
Adrenocortical cells may create a gradient of a substance or substances in the blood stream which alter adrenocortical cell function and morphology to create the zonation of the adrenal cortex [69] (Fig. 10). Adrenal arterial plasma steroid concentrations are, of course, those of the general peripheral circulation, whereas very high concentrations of steroids are found in the adrenal venous effluent. Thus, the gradient substance may be steroids secreted by the adrenocortical cell [29]. Several mechanisms by which adrenocortical cells may sense their position in the steroid gradient, and thus their position within the cortex, have been proposed [29,69,70]. Irrespective of the nature of the gradient and of the sensing mechanism, the end result may be the zonation of the intracellular messenger systems, thus providing a system for the control by ACTH of mineralocorticoid, glucocorticoid, and androgen synthesis, in conjunction with the other stimuli that regulate adrenocortical steroidogenesis.

References

1. Lowry, P.J. (1984) Biosci. Rep. 4, 467–82.
2. Schimmer, B.P. (1980) Adv. Cyclic Nucleotide Res. 13, 181–214.
3. Abdel-Latif, A.A. (1986) Pharmacol. Rev. 38, 227–272.
4. Ramachandran, J. (1984) Endocr. Res. 10, 347–63.
5. Bost, K.L. and Blalock, J.E. (1986) Mol. Cell. Endocrinol. 44, 1–9.
6. Saez, J.M., Durand, P. and Cathiard, A.M. (1984) Mol. Cell. Endocrinol. 38, 93–102.
7. Taylor, C.W. and Merritt, J.E. (1986) Trends Pharmacol. Sci. 7, 238–242.
8. Garren, L.D., Gill, G.N., Masui, H. and Walton, G.M. (1970) Rec. Progr. Hormone Res. 27, 433–455.
9. Waterman, M.R. and Simpson, E.R. (1985) Mol. Cell. Endocrinol. 39, 81–89.
10. Wada, A., Okamoto, M., Nonaka, Y. and Yamano, T. (1984) Biochem. Biophys. Res. Commun. 119, 365–371.
11. Hornsby, P.J. and Aldern, K.A. (1984) J. Clin. Endocr. Metab. 58, 121–130.
12. Neville, A.M. and O'Hare, M.J. (1982) The Human Adrenal Cortex. Pathology and Biology – An Integrated Approach, Springer-Verlag, Berlin.
13. Mason, J.I., Arthur, J.R. and Boyd, G.S. (1978) Biochem. J. 174, 1045–1051.
14. Vahouny, G.V., Chabderbhan, R., Noland, B.J. and Scallen, T.J. (1984) Endocr. Res. 10, 473–505.
15. Hall, P.F. (1984) Can. J. Biochem. Cell Biol. 62, 653–665.
16. Mason, J.I. and Rainey, W.E. (1987) J. Clin. Endocr. Metab. 64, 140–146.
17. Gwynne, J.T. and Strauss, J.F. (1982) Endocrine Rev. 3, 299–330.
18. Trzeciak, W.H., Simpson, E.R., Scallen, T.J., Vahouny, G.V. and Waterman, M.R. (1987) J. Biol. Chem. 262, 3713–3717.
19. Kovanen, P.T., Faust, J.R., Brown, M.S. and Goldstein, J.L. (1979) Endocrinology 104, 599–609.
20. Pedersen, R.C. and Brownie, A.C. (1987) Science 236, 180–190.
21. Pon, L.A., Hartigan, J.A. and Orme-Johnson, N.R. (1986) J. Biol. Chem. 261, 13309–16.
22. Zuber, M.X., John, M.E., Okamura, T., Simpson, E.R. and Waterman, M.R. (1986) J. Biol. Chem. 261, 2475–2482.
23. John, M.E., John, M.C., Boggaram, V., Simpson, E.R. and Waterman, M.R. (1986) Proc. Natl. Acad. Sci. USA 83, 4715–4719.
24. Waterman, M.R., Mason, J.I., Zuber, M.X., John, M.E., Rodgers, R.J. and Simpson, E.R. (1986) Endocr. Res. 12, 393–408.
25. Parker, K.L., Chaplin, D.D., Wong, M., Seidman, J.G. and Schimmer, B.P. (1986) Endocr. Res. 12, 409–428.
26. Montminy, M.R. and Bilezikjian, L.M. (1987) Nature 328, 175–178.
27. Zuber, M.X., Simpson, E.R. and Waterman, M.R. (1985) Ann. N.Y. Acad. Sci. 458, 252–61.
28. Parker, K.L., Chaplin, D.D., Wong, M., Seidman, J.G., Smith, J.A. and Schimmer, B.P. (1985) Proc. Natl. Acad. Sci. USA 82, 7860–7864.
29. Hornsby, P.J. (1985) In: Adrenal Cortex (Anderson, D.C. and Winter, J.S.D., eds.), pp. 1–31, Butterworth, London.
30. Schweigerer, L., Neufeld, G., Friedman, J., Abraham, J.A., Fiddes, J.C. and Gospodarowicz, D. (1987) Endocrinology 120, 796–800.
31. Voutilainen, R. and Miller, W.L. (1987) Proc. Natl. Acad. Sci. USA 84, 1590–1594.
32. Kowal, J. (1970) Rec. Progr. Hormone Res. 26, 623–630.
33. Shima, S., Komoriyama, K., Hirai, M. and Kouyama, H. (1984) Endocrinology 114, 325–325.
34. Marie, J. and Jard, S. (1983) FEBS Lett. 159, 97–101.
35. Hadjian, A.J., Ventre, R. and Chambaz, E.M. (1981) Biochem. Biophys. Res. Commun. 98, 892–900.
36. Lymangrover, J.R., Keku, E. and Eldridge, J.C. (1983) Life Sci. 33, 1605–1612.
37. Farese, R.V. (1984) Mol. Cell. Endocrinol. 35, 1–14.

38. Hadjian, A.J., Culty, M. and Chambaz, E.M. (1984) Biochim. Biophys. Acta 804, 427–433.
39. Balla, T., Enyedi, P., Spat, A. and Antoni, F.A. (1985) Endocrinology 117, 421–423.
40. Muller, J. and Ziegler, W.H. (1968) Acta Endocrinol. 59, 23–35.
41. Kojima, I., Kojima, K. and Rasmussen, H. (1985) J. Biol. Chem. 260, 4248–4256.
42. Hornsby, P.J., Hancock, J.P., Vo, T.P., Nason, L.M., Ryan, R.F. and McAllister, J.M. (1987) Proc. Natl. Acad. Sci. USA 84, 1580–1584.
43. McAllister, J.M. and Hornsby, P.J. (1988) Endocrinology 122, 2012–2018.
44. Brecher, P., Tabacchi, M., Pyun, H.Y. and Chobanian, A.V. (1973) Biochem. Biophys. Res. Commun. 54, 1511–1520.
45. Aguilera, G., Parker, D.S. and Catt, K.J. (1982) Endocrinology 111, 1376–1384.
46. Morera, A.M., Andoka, G. and Chauvin, M.A. (1984) C.R. Acad. Sci. (III) 298, 507–512.
47. Tremblay, J., Gerzer, R., Pang, S.C., Cantin, M., Genest, J. and Hamet, P. (1986) FEBS Lett. 194, 210–214.
48. Gill, G.N. and Kanstein, C.B. (1975) Biochem. Biophys. Res. Commun. 63, 1113–22.
49. Aguilera, G. (1987) Endocrinology 120, 299–304.
50. Chartier, L., Schiffrin, E., Thibult, G. and Garcia, R. (1984) Endocrinology 115, 2026–2028.
51. Quinn, S.J., Cornwall, M.C. and Williams, G.H. (1987) Endocrinology 120, 903–914.
52. Aguilera, G. and Catt, K.J. (1986) Endocrinology 118, 112–118.
53. Braley, L.M., Menachery, A.I., Brown, E.M. and Williams, G.H. (1986) Endocrinology 119, 1010–1019.
54. Brown, B.L. (1982) In: Cyclic Nucleotides. Part II: Physiology and Pharmacology (Kebabian, J.W. and Nathanson, J.A., eds.), pp. 623–650, Springer-Verlag, Berlin.
55. Payet, M.D., Benabderrazike, M. and Gallo-Payet, N. (1987) Endocrinology 121, 875–880.
56. Shaw, J.E. and Ramwell, P.W. (1966) In: Prostaglandins (Bergstrom, S. and Samuelsson, B., eds.), pp. 293–299, Interscience, New York.
57. Saez, J.M. (1975) J. Clin. Invest. 56, 536–545.
58. Hirai, A., Tahara, K., Tamura, Y., Saito, H., Terano, T. and Yoshida, S. (1985) Prostaglandins 30, 749–768.
59. Kojima, I., Kojima, K. and Rasmussen, H. (1985) Endocrinology 117, 1057–1066.
60. Gallant, S., Kauffman, F.C. and Brownie, A.C. (1974) Life Sci. 14, 937–944.
61. Koletsky, R.J., Brown, E.M. and Williams, G.H. (1983) Endocrinology 113, 485–490.
62. Vilgrain, I., Cochet, C. and Chambaz, E.M. (1984) J. Biol. Chem. 259, 3403–3406.
63. McAllister, J.M. and Hornsby, P.J. (1987) Endocrinology 121, 1908–1910.
64. Shima, S., Kawashima, Y. and Hirai, M. (1979) Endocrinol. Japan. 26, 219–25.
65. Shima, S., Kawashima, Y. and Hirai, M. (1979) Acta Endocrinol. 90, 139–146.
66. Welsh, T.H., Jr., Jones, P.B. and Hsueh, A.J. (1984) Cancer Res. 44, 885–92.
67. Shinohara, O., Knecht, M. and Catt, K.J. (1985) Proc. Natl. Acad. Sci. USA 82, 8518–8522.
68. Hornsby, P.J. and O'Hare, M.J. (1977) Endocrinology 101, 997–1005.
69. Hornsby, P.J. (1987) J. Steroid Biochem. 27, 1161–1171.
70. Anderson, D.C. (1980) Lancet 2, 454–456.

CHAPTER 11

Mechanism of action of angiotensin II

PAULA Q. BARRETT, WENDY B. BOLLAG and
HOWARD RASMUSSEN

Departments of Internal Medicine, Physiology and Cell Biology, Yale University School of Medicine, New Haven, CT 06510, U.S.A.

1. Introduction

The octapeptide angiotensin II (AII) is the primary effector hormone of the renin-angiotensin-aldosterone system, regulating not only the release of aldosterone from the adrenal but also that of renin from the kidney. By inhibiting the secretion of renin, the enzyme which catalyses the proteolysis of angiotensinogen to the AII precursor angiotensin I (AI), AII effectively regulates its own formation as well via a negative feedback loop. Formed as a result of the hydrolysis of AI by the converting enzyme, AII has a wide variety of physiological actions, the combined effects of which are to coordinately regulate whole body sodium and potassium balance and blood pressure-volume homeostasis. Particularly well described are the actions of AII on adrenal glomerulosa, vascular smooth muscle and hepatic cells, in which increases in steroidogenesis, contraction and glucose production, respectively, are observed with hormonal stimulation. AII receptors have also been identified in the kidney, brain, uterus, placenta, pituitary and reticuloendothelium (reviewed in Ref. 1). However, in these latter tissues the mechanism of AII action has not been as extensively investigated as in the liver, adrenal glomerulosa and vascular smooth muscle, and therefore the focus of this review will be on the action of AII on these three major target tissues.

The response of these tissues to AII is initiated by the interaction of AII with its receptor and the coupling of this receptor to plasma membrane enzymes which transduce the hormonal signal. These membrane events result in the generation of second messengers whose role is the amplification of the hormonal signal and its transmission to the cell interior. Upon their intracellular diffusion, these messengers are then received by proteins whose activation mediates the cellular response. Information about this sequence of events initiated by AII will be reviewed and integrated into a model which illustrates the importance of temporal and spatial patterns of messenger-mediated hormonal stimulation.

2. AII receptors

Saturable binding of radiolabeled AII is demonstrable in adrenal glomerulosa [2], vascular smooth muscle [1,3] and hepatic [4] plasma membranes. In adrenal and hepatic tissue, two classes of AII receptors have been described; a high-affinity site with a K_d in the range 0.2–0.35 nM (possibly linked to phosphoinositide-specific phospholipase C), and a low-affinity site with a K_d between 2.9 and 6.3 nM (linked to adenylate cyclase) [2,4,5]. Conflicting reports [6,7] of only a single binding site may be explained by the recent finding of Fishman et al. [8]. These investigators described a cytosolic protein which in the presence of calcium specifically reduces the binding of AII to its receptor. Therefore, preparation of membranes in buffers lacking EDTA may permit the calcium-dependent association of this inhibitor protein with one class of receptors and prevent its detection.

Unlike adrenal and hepatic tissue, vascular smooth muscle apparently possesses only a single receptor type. The observed affinity of this receptor varies among different muscle preparations from a K_d of 15–50 nM in aorta [1] to a K_d of 2–5.5 nM in mesenteric artery and cultured vascular cells [9]. Nonetheless, it appears that, in at least some smooth muscle cell types, AII alters the activity of both phospholipase C and adenylate cyclase, presumably via a single receptor type.

2.1. Regulation of receptor affinity

The biological effect of AII on its target tissues can be altered by changing either receptor affinity or receptor number. In the adrenal, liver and vasculature the affinity of the receptor for AII is modulated by cations [2,4,9] such as sodium, calcium or magnesium, and by reducing agents [3,5,10]. Extreme examples of the importance of each of these modifiers of receptor affinity have been documented. For instance, in adrenal plasma membranes there is an absolute dependence of high-affinity AII binding on sodium [2], so that in the absence of sodium no high-affinity binding is observed. In hepatic membranes high-affinity binding which is abolished by EDTA exposure can be recovered by treatment with either calcium, magnesium or manganese [4]. And in mesenteric arterial membranes, the enhancement of binding observed with monovalent sodium and divalent manganese is additive [9].

Unlike mono- and divalent cations, which enhance high-affinity binding, agents such as dithiothreitol (DTT) reduce high-affinity binding. As suggested by the work of Sen et al. [10], this DTT-induced inhibition likely reflects a direct effect on the conformation of the AII receptor. These investigators have isolated from hepatic tissue an AII-binding protein of molecular mass 66 kDa which migrates as a larger species when subjected to electrophoresis under reducing conditions. Moreover, when exposed to DTT this protein does not bind AII, suggesting that the proper conformation of this receptor is maintained by disulfide bonds. In addition, since DTT inhibits glycogenolysis in liver [5] and contraction in vascular smooth muscle [3], this receptor appears to be coupled to the physiological response.

2.2. Regulation of receptor number

The biological effect of AII on its target tissues can also be altered via changes in receptor number. Such a phenomenon has been observed with AII receptors, whose number is regulated in response to changes in sodium intake. In these studies, receptor number was determined without regard to receptor subtype so it is not clear whether the high- and low-affinity receptors are equally affected by dietary status. Nonetheless, in vascular smooth muscle elevations in circulating AII levels, produced as a consequence of dietary sodium deprivation, result in a sustained reduction in receptor number [1], an example of classic receptor down-regulation. In contrast, in hepatocytes a change in dietary sodium results in a biphasic response. Down-regulation of the hepatic AII receptor is only a transient phenomenon, reaching a nadir within 36 hours of sodium restriction. This decrease in receptor number is followed by a sustained increase which is fully maximal at approximately 8.5 days [6]. In adrenal tissue down-regulation is not observed even as a short-term response to reduced dietary sodium. Instead, a monophasic increase in receptor concentration is observed after 36 hours of sodium restriction and is further augmented with longer periods of sodium deprivation. This increase in the number of AII binding sites in the adrenal is accompanied as well by an increase in the sensitivity and magnitude of the AII-elicited aldosterone secretory response [11]. Furthermore, in all three tissues both the down- and the up-regulation are a result of a direct action of AII on the target cells, rather than a direct effect of sodium per se, as they are prevented by the administration of inhibitors of the converting enzyme and are reproduced by physiological infusions of AII [1,6,11].

The number of receptors on the cell surface can also be regulated by altering their rate of recycling following receptor-mediated endocytosis. Many hormones upon binding to their receptors trigger internalization of the hormone-receptor complex [12]. Upon internalization, the ligand-receptor complex dissociates and the freed receptor either undergoes proteolysis or is recycled to the plasma membrane. Recently in adrenocortical cells, Crozat et al. [13] have described the internalization of the AII receptor. Based upon the high rate of internalization and degradation of radiolabeled AII, these investigators calculated that at hormone concentrations near the K_d (10^9) one half of the surface receptors should disappear in 20 minutes. However, they observed no reduction in either surface binding sites or internalized radioligand under steady-state conditions. They concluded, therefore, that in order to maintain this constant level of both surface-bound and internalized radioactivity, receptor recycling must be occurring at an extremely rapid rate. Interestingly, such a high rate of receptor-ligand internalization could explain the very short half-life of AII in plasma and provide the mechanism via which AII could interact with its previously described nuclear receptors [1].

3. Receptor–guanine nucleotide interactions

The prototype for the mechanism by which hormones regulate the activity of transducing enzymes is provided by the model of receptor coupling to the adenylate cyclase enzyme system. In this model receptor interaction with adenylate cyclase is mediated by a GTP-binding protein [14]. Initially, a higher-affinity receptor state is preserved by the association of the receptor with a G-protein containing bound GDP. Ligand binding to the receptor promotes the exchange of GDP for GTP. The GTP thus bound facilitates the dissociation of the receptor G-protein complex and the induction of a lower-affinity state of the receptor. This state is maintained until hydrolysis of the bound GTP, catalysed by the GTPase of the G-protein, permits the reassociation of the G-protein and the receptor, and the return of the higher-affinity state. Thus, GTP effects on receptor affinity are an indication of G-protein involvement in receptor coupling [14].

Guanine nucleotides do indeed affect the binding of AII to its receptor in all three target tissues. GTP is found to reduce AII binding with an inhibitory K_i of 0.2 μM in the adrenal [15], 10.0 μM in the hepatocyte [16] and 3.0 μM in the mesenteric artery [9]. This range of values may reflect a real difference in the sensitivity to GTP inhibition or merely a difference in the conditions of the binding assays, i.e. the omission or inclusion of a GTP-regenerating system in the assay to replace hydrolysed GTP. Traditionally such guanylnucleotide effects on binding would be interpreted as an indication of coupling to the adenylate cyclase enzyme complex. And indeed, AII inhibits the activity of adenylate cyclase in most AII-responsive tissues, including the adrenal, liver and vasculature [5,17,18]. However, recently receptors not coupled to adenylate cyclase have also been shown to be regulated by GTP (see, for example, Refs. 19 and 20). Therefore, the effects of GTP on AII receptor binding may not solely reflect coupling to adenylate cyclase, but also likely indicate coupling via a G-protein to another transducing enzyme, such as phospholipase C. In keeping with this conclusion are studies showing that pertussis-toxin pretreatment of adrenal glomerulosa cells blocks the AII-induced inhibition of adenylate cyclase activity but does not block the effect of AII on either GTPase activity or aldosterone secretion [22].

4. Transducing enzyme activation

AII receptors are coupled to at least two transducing enzyme systems, phospholipase C and adenylate cyclase. The traditional model of transducing enzyme activation is based upon receptor activation of adenylate cyclase (see Birnbaumer, Chapter 1). In this model the dissociation of the G-protein from its receptor induces a lower-affinity receptor state and facilitates the interaction of the freed G-protein with the catalytic unit of adenylate cyclase, producing either a stimulation or an inhibition of activity. Inhibitory information generated by ligand binding to adenylate cyclase-coupled receptors is conveyed to the catalytic unit of adenylate cyclase through GTP-binding proteins termed G_i, whereas stimulatory information is conveyed through a separate G-protein, G_s. The functions of these G-proteins can be affected by bacterial toxins (cholera toxin, pertussis toxin) which catalyse the covalent modification of the α subunit of the G-proteins. The ADP-ribosylation of G_s induced by cholera toxin permanently activates G_s, whereas pertussis toxin-catalysed ADP-ribosylation of G_i permanently blocks receptor activation of G_i but does not prevent the activation of G_i induced by nonhydrolysable GTP analogues [15].

While the coupling functions of G_s and G_i are well characterized, additional pertussis toxin-sensitive G-proteins (for example G_o) have been identified whose functions are still unknown [21]. Interestingly, in certain systems pertussis-toxin treatment prevents the activation of phospholipase C both by hormones and by nonhydrolysable guanine nucleotides [21]. Thus, a pertussis-toxin-sensitive G-protein other than G_i (such as G_o) may couple receptors to this transducing enzyme. In other systems activation of phospholipase C is not sensitive to pertussis toxin, indicating that if a G-protein mediates the activation of PLC in these systems, it must be via yet another member of the G-protein family.

4.1. Adenylate cyclase

In adrenal, hepatic and vascular membranes, AII reduces both basal and stimulated adenylate cyclase activity via the action of G_i [5,17,18]. To date, however, this inhibition is of unknown physiological significance. In the adrenal pertussis-toxin pretreatment prevents the inhibition of adenylate cyclase induced by AII, but does not block AII-induced steroidogenesis. This result suggests that adenylate cyclase inhibition is not a critical component of the cellular response [22]. Moreover, in the hepatocyte the high-affinity receptor, which is not coupled to adenylate cyclase, mediates the cellular response [5]. Adenylate cyclase inhibition instead is mediated by the low-affinity AII receptor [5]. Interestingly, in the hepatocyte the high-affinity receptor is more GTP-sensitive than the low-affinity receptor, emphasizing again that G-proteins can be involved in receptor coupling to transducing enzymes other than adenylate cyclase.

In contrast to adrenal and hepatic tissue, adenylate cyclase inhibition may be physiologically relevant in renal vasculature, since pertussis toxin attenuates AII-induced vasoconstriction [23]. The physiological role for adenylate cyclase inhibition may therefore depend upon the differential response of the three tissues to the second messenger, cAMP. In vascular smooth muscle, adenylate cyclase inhibition should enhance AII-induced contraction by reducing cellular cAMP, the messenger of relaxation, while in the adrenal and the hepatocyte cAMP mediates a stimulatory response and adenylate cyclase inhibition would be expected to oppose AII action.

4.2. Phospholipase C

An early signalling event associated with the cellular response to calcium-linked hormones is a rapid hydrolysis of a class of plasma membrane lipids known collectively as the phosphoinositides. This class of lipids consists of interconvertible compounds which differ only in the phosphorylation state of the inositol head group. By far the largest proportion is the parent compound, phosphatidylinositol (PI). Phosphorylation of its inositol ring in the 4-position by a kinase yields phosphatidylinositol 4-phosphate (PIP), which upon further phosphorylation at the 5-position generates phosphatidylinositol 4,5-bisphosphate (PIP$_2$). These phosphorylation events can be reversed by the action of phosphomonoesterases which convert the polyphosphoinositides back to PI. The concentration of these phospholipids can additionally be regulated by the hydrolytic activity of phosphodiesterase enzymes, such as phospholipase C (PLC), which cleaves the ester bond linking the inositol head group to the diglyceride backbone. A water-soluble sugar and a lipid-soluble diacylglycerol are therefore generated as a consequence of PLC activation (see Guy and Kirk, Chapter 2).

4.2.1. Activation via G-proteins
In a variety of tissues, including the adrenal, liver and vascular smooth muscle, the activity of phosphoinositide-specific phospholipase C is regulated by AII via a receptor mechanism. Although a consensus has not yet been reached concerning G-protein involvement in receptor-coupling to PLC, strong evidence of such an involvement is provided by studies using permeabilized adrenal cells and hepatic membranes. In these systems nonhydrolysable analogues of GTP enhance PLC activity, resulting in the accumulation of inositol phosphates and a reduction in polyphosphoinositides [24,25]. The identity of the G-protein involved in this coupling of PLC to the AII receptor is still unknown. However, in the adrenal complete ADP-ribosylation of endogenous pertussis-toxin-sensitive substrates fails to affect the AII-induced generation of inositol phosphates [22], indicating that in this system a G-protein other than G_i mediates receptor activation of PLC.

4.2.2. Substrate(s)
The major substrates of AII-activated PLC remain a matter of controversy. Purified PLC activated by calcium has been shown to directly hydrolyse PI, PIP and PIP$_2$ in vitro to yield diacylglycerol and, respectively, inositol monophosphate (IP$_1$), inositol bisphosphate (IP$_2$) and inositol trisphosphate (IP$_3$). In intact cells the substrate specificity of receptor-activated PLC has been investigated by monitoring the loss of radiolabeled inositol incorporated into phosphoinositides. In the target tissues under discussion, activation of PLC by AII results in an immediate (approx. 15 s) reduction in the levels of PIP$_2$ and PIP and a subsequent loss (approx. 5 min) of PI [26–28]. These losses could result either from the direct PLC-catalysed hy-

drolysis of PI, PIP and PIP_2 or from their interconversion by phosphatases and kinases. Thus, for example, a reduction in PIP could be a consequence of (1) a direct action of PLC, (2) the phosphorylation of PIP by a kinase to form PIP_2, and/or (3) the dephosphorylation of PIP by a monoesterase to yield PI. However, in addition to the AII-induced reductions in phosphoinositides, an accumulation of inositol phosphates is also observed, with a dramatic increase (2–6-fold) in the cytosolic levels of radiolabeled IP_3 and IP_2 and a more delayed rise in the concentration of IP_1 [26–28]. The rapidity of the rise in the cytosolic levels of IP_3 and IP_2 and their immediate fall upon agonist removal [29] indicate that PIP_2 and/or PIP is hydrolysed by receptor-activated PLC. Because no known function exists for IP_2, it is generally believed that PIP_2 is the physiologically important substrate of receptor-activated PLC.

Despite a continued hydrolysis of PIP_2, a significant reduction in PIP_2 is not maintained with prolonged hormonal stimulation of the target tissues. The transient nature of the fall in PIP_2 content contrasts sharply with the sustained reduction in PI content seen with continuous exposure to agonist [30]. This sustained loss of PI has been assumed to result from the kinase-catalysed phosphorylation of PI to PIP and then to PIP_2 as a replacement for the phosphoinositides hydrolysed by receptor-activated PLC [31]. However, recent evidence suggests that during prolonged AII stimulation, PI may not be consumed solely through the action of kinases but may also be hydrolysed directly by PLC. It has therefore been proposed that maintained AII stimulation is accompanied by a time-dependent relaxation of the substrate specificity of PLC [30], so that initially PIP_2 is the preferred substrate but at later times PI as well as PIP_2 is hydrolysed. Alternatively, there may be two PLC enzymes with different substrate specificities and time courses of activation [30]. In either case, during a sustained cellular response to AII, PI may be a quantitatively more important substrate than PIP_2, particularly in terms of the generation of diacylglycerol.

4.2.3. Products (second messengers)
Although the reduction in PIP_2 is transient, the elevation in diacylglycerol (DG) in response to AII is sustained in all three target tissues [26,27,32]. The time course of DG formation in cultured vascular smooth muscle cells has been described in detail by Griendling et al. [30], who demonstrated that AII-stimulated DG production is biphasic, peaking initially at 15 seconds and again at 5 minutes. This time course can be correlated with the observed changes in phosphoinositides and inositol phosphates to suggest a time-dependent change in the phosphoinositide hydrolysed: it appears that the first peak of DG production is the result of PIP_2 hydrolysis while the second may originate from PI hydrolysis [30]. However, regardless of its source, the DG produced from PLC-catalysed phosphoinositide turnover functions as a lipid-soluble messenger which increases the activity of membrane-associated protein kinase C (PKC). PKC is a calcium-activated and phospholipid-de-

pendent protein kinase which catalyses the phosphorylation of a set of cellular proteins thought to mediate the sustained AII response. Thus, continuous production of DG allows maintained activation and participation of PKC in the cellular response.

The hydrolysis of PIP_2 catalysed by receptor-activated PLC yields an additional messenger, inositol 1,4,5-trisphosphate (1,4,5-IP_3). This messenger couples receptor activation to the mobilization of calcium from a dantrolene-sensitive intracellular pool, thought to be a component of the endoplasmic reticulum (ER) [31]. Thus, in permeabilized cells preloaded with radiocalcium, the addition of 1,4,5-IP_3 results in efflux of the radiolabel. This IP_3-induced release appears to be mediated by a mechanism involving specific intracellular receptors for 1,4,5-IP_3 such as those identified in various AII target tissues [33]. Noteworthy is the fact that these receptors for 1,4,5-IP_3 are recovered largely in a crude plasma membrane fraction [33], suggesting that the component of the ER from which 1,4,5-IP_3 mobilizes calcium is located close to the plasma membrane.

Recently, an additional IP_3 isomer, inositol 1,3,4-trisphosphate (1,3,4-IP_3), which is ineffective in releasing calcium from the ER, has been identified. Unlike 1,4,5-IP_3, 1,3,4-IP_3 is thought not to be a product of the direct hydrolysis of an isomer of PIP_2, but rather a result of the action of a 5-phosphatase on a more polar inositol phosphate, inositol 1,3,4,5-tetrakisphosphate (IP_4) [34]. At this time neither 1,3,4-IP_3 nor its precursor IP_4, which is formed as a result of a 3-kinase-catalysed phosphorylation of 1,4,5-IP_3, has a clear physiological role, although IP_4 has been implicated in the regulation of plasma membrane calcium influx (see Rasmussen and Barrett, Chapter 4).

In adrenal glomerulosa cells, AII induces the production of both 1,4,5-IP_3 and 1,3,4-IP_3 with different time courses of formation [35], as has been demonstrated in many cell types in response to various hormones (for example, see Ref. 36). The concentration of 1,4,5-IP_3 rises immediately, but transiently, peaking with a 2-fold increase in radiolabel at approximately 5 seconds and returning to a value not significantly different from control by 10 minutes. On the other hand, the concentration of 1,3,4-IP_3 rises progressively during the first 60 seconds following AII stimulation and remains elevated for up to 20 minutes (the entire observation period). This rate of 1,3,4-IP_3 production is evidence that its precursor IP_4, although not detectable, is formed and rapidly metabolized. Lithium, a known inhibitor of inositol phosphatase, affects only the accumulation of the 1,3,4-IP_3 isomer, causing a further elevation in the levels of this isomer with little or no alteration in 1,4,5-IP_3 radiolabel. This progressive accumulation of 1,3,4-IP_3 with lithium treatment indicates that, despite the decline in 1,4,5-IP_3 levels, 1,4,5-IP_3 is being continuously generated and is rapidly metabolized first to IP_4 and then 1,3,4-IP_3. Such effective metabolism of the active isomer may ensure the rapid termination of the calcium-mobilizing signal and thereby minimize the cells' exposure to the potentially toxic effects of calcium (see Rasmussen and Barrett, Chapter 4).

Although in the hepatocyte IP_4 and the two IP_3 isomers are known to be formed in response to certain hormones, the effect of AII on these isomers has yet to be investigated. There is also little information available to date concerning changes in the levels of these inositol phosphates following AII stimulation of vascular smooth muscle cells.

5. AII-induced changes in calcium metabolism

The addition of AII to the target tissues under discussion leads to three changes in cellular calcium metabolism: (1) a transient increase in cytosolic calcium; (2) a reduction in total cell calcium; and (3) a sustained increase in the rate of calcium influx across the plasma membrane. Each of these changes in cellular calcium metabolism is discussed below.

5.1. Intracellular calcium concentration

A change in cytosolic calcium concentration has been observed in response to AII in adrenal [37,38], hepatic [27] and vascular smooth muscle [32] tissue, regardless of whether this change is measured with Fura 2, which increases its fluorescence upon binding calcium, or aequorin, a jellyfish photoprotein which emits light in a calcium-dependent manner. AII induces an immediate and marked elevation in cytosolic calcium from approximately 80–120 nM to 600–1600 nM, followed by a sharp decline to a plateau level which is maintained throughout the sustained cellular response. The amount of calcium which has entered the cytosol can be estimated by determining the peak area of the dye signal, and its source assessed by monitoring the response in the absence of extracellular calcium. When extracellular calcium is acutely lowered, the peak height of the AII-induced calcium transient is nearly identical to that seen in the presence of calcium; however, the peak width is reduced in tissues or cells incubated in the absence of extracellular calcium. These results indicate that calcium entering the cytosol from an intracellular compartment determines the magnitude of the early phase of the calcium transient response, but that the later phase of the measured transient is also affected by the influx of extracellular calcium into the cell.

5.2. Calcium mobilization

Although slightly attenuated, the rise in cytosolic calcium proceeds in the absence of extracellular calcium (as mentioned above), indicating that upon AII stimulation calcium is released into the cytosol from an internal pool. The identity of this mobilized internal pool was initially inferred from cellular studies in which treatment with dantrolene inhibited the redistribution of intracellular calcium [39,40]. Because this drug is known to prevent calcium release from the sarcoplasmic reticulum of skeletal muscle [41], the analogous endoplasmic reticulum (ER) is its likely target of action, and therefore a component of the ER is the likely source of the released calcium. Further confirmation of the involvement of the ER in this calcium release is provided by extensive studies in permeabilized adrenal glomerulosa, hepatic and vascular smooth muscle cells. In these studies the application of 1,4,5-IP_3 to permeabilized cells whose ER pools have been preferentially labeled with radiocalcium induces the release of the previously sequestered label [26,42,43]. Taken together, the above observations argue that a component of the ER is the principal source of the mobilized calcium.

Once mobilized, a large proportion of the cytosolic calcium load is extruded from the cell across the plasma membrane. This extrusion can be demonstrated by measuring the efflux of radiocalcium from previously labeled cells. AII stimulation of adrenal, hepatic and vascular smooth muscle cells preloaded with radioactive calcium induces a marked increase in efflux of the radiolabel. This increased efflux is transient, peaking between 4 and 5 minutes after AII addition and is observed in the absence of extracellular calcium [39]. Furthermore, under conditions of zero calcium, treatment of cells with dantrolene prior to hormonal stimulation abolishes the AII-induced calcium efflux [40], confirming that the radiocalcium lost from the cell is mobilized from a component of the ER.

5.3. Total cell calcium

The mobilization of calcium results not only in the observed transient rise in intracellular free calcium and enhanced cellular efflux, but also in a net loss of calcium from the cell (Fig. 1). Thus, total cell calcium declines with AII stimulation of adrenal and vascular smooth muscle cells [44]. Furthermore, total cell calcium remains low throughout the duration of exposure to AII, suggesting that the continued formation of small amounts of 1,4,5-IP_3 prevents refilling of the ER pool. Upon the removal of AII and the immediate reduction in IP_3 concentration, total cell calcium rapidly recovers to prestimulation levels without a detectable change in cytosolic free calcium, as measured by calcium-sensitive dyes. This observation has been taken as evidence that the IP_3-releasable ER pool is in direct communication with the plasma membrane and that extracellular calcium refills the pool without entering the bulk cytosol (see Ref. 45). The location of this pool within the cell (cytosolic vs. adjacent to the plasma membrane) remains a matter of controversy (see Rasmussen and Barrett, Chapter 4).

5.4. Calcium entry

AII induces a sustained increase in the rate of calcium influx across the plasma membrane of adrenal and hepatic cells. In hepatocytes AII enhances the rate of

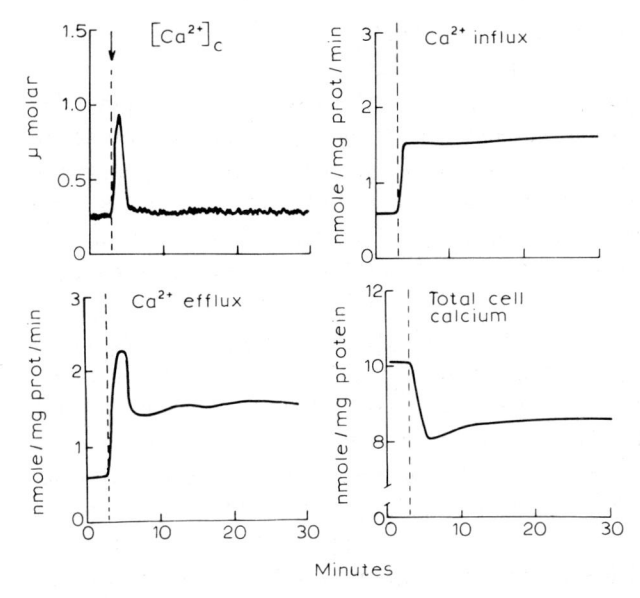

Fig. 1. AII-induced alterations in cellular calcium metabolism. Indicated are the increase in cytosolic calcium (upper left panel); enhanced calcium influx (upper right panel); the loss of total cell calcium (lower right panel); and the calculated compensatory increase in calcium efflux as a function of time after AII addition (at arrow).

calcium influx in a dose-dependent manner with a K_m of approximately 0.8 nM. The magnitude of this enhancement is dependent on alterations in the influx mechanism, measured kinetically as a seven-fold increase in the V_{max} of calcium uptake which is partially opposed by a ten-fold elevation (0.1 to 1.5 mM extracellular calcium) in K_m [46]. In contrast, in AII-stimulated adrenal glomerulosa cells only the increase in V_{max} is observed [47]. Despite this difference, in both systems the AII effect on calcium influx is sustained in the continued presence of AII. Because during a sustained cellular response, total cell calcium does not increase [44], this enhanced influx must be matched by a compensatory increase in calcium efflux, thus establishing a high rate of calcium cycling across the plasma membrane. Although the mechanism by which the sustained increase in calcium influx is balanced by a compensatory increase in calcium efflux is not known, it has been demonstrated that the calcium efflux pump of the plasma membrane can be activated by a direct interaction with calcium-calmodulin [48] and by an interaction with protein kinase C (J. Smallwood, unpublished results). The latter activation event is presumed to involve protein phosphorylation.

In the adrenal the AII-regulated calcium uptake pathway has been extensively characterized. AII-induced calcium influx into glomerulosa cells is dependent on the concentration of extracellular potassium within the range 2 to 8 mM and with a threshold value of 2 mM [47]. Because glomerulosa cells are exquisitely sensitive to depolarization by potassium, potassium modulation of their uptake pathway reflects a voltage-dependence of the calcium influx mechanism. However, at concentrations of 10 mM potassium or greater AII does not further augment calcium influx above an already elevated basal level. These results argue that AII and potassium act on a similar calcium channel which is both voltage-dependent and receptor-regulated. In addition, calcium uptake via this channel is inhibited by the calcium-channel antagonist, nitrendipine, at a concentration of 1 μM [47]. Nitrendipine blocks both the voltage-dependent influx of calcium (i.e., the effect of potassium) and the receptor-dependent influx (i.e. the effect of AII).

To date, two populations of voltage-dependent calcium channels have been identified in the adrenal glomerulosa plasma membrane [49]. These two channel types exhibit properties similar to those of the T- and L-type channels found in excitable cells [50]. In the adrenal AII increases calcium current through the T-type channels by 50% and slows their rate of deactivation [49]. Because these transient channels have also been shown to be blocked by nitrendipine at the high dose which inhibits calcium influx, they likely mediate the AII-elicited enhancement of calcium entry. Although the molecular mechanism by which AII regulates T-channel activity is not known, there is precedence for at least three possible mechanisms. AII may increase calcium influx by: (1) a direct messenger-mediated modulation of channel activity; the two candidates presently under serious consideration are 1,4,5-IP_3 and IP_4, as observed in lymphocytes and oocytes, respectively [51,52]; (2) a messenger-mediated phosphorylation event catalysed by an AII-activated kinase such as PKC; and/or (3) a direct receptor-coupled activation of the channel possibly involving a specific G-protein. Regardless of the mechanism, these channels appear to be critical to the sustained cellular response, as discussed below.

6. Integration of signals and cellular response

In the second messenger model originally proposed by Sutherland et al. [53], a hormone or extracellular messenger upon binding to its receptor activates (transducing event) a membrane-associated enzyme (adenylate cyclase) to generate a soluble intracellular messenger (cAMP), which diffuses into the cell interior to be received by a specific class of cytosolic receptor proteins (R_2C_2) possessing enzymatic activity (cAMP-dependent protein kinase). Implicit in this classical model are several assumptions: (1) there is a single transducing event that initiates a single pathway of information transfer; (2) the second messenger is diffusible throughout the cytosol; (3) the concentration of the messenger remains elevated to sustain cell activation; and (4) the initiation of the cellular response is mediated by the same second messenger as is the maintenance of the response.

Clearly AII violates these assumptions in several respects. First, AII functions through two types of receptors, one of which inhibits adenylate cyclase and the other stimulates PLC. Second, this activated PLC by hydrolysing PIP_2 produces not one but two messengers. The first, $1,4,5-IP_3$, acts as a typical diffusible cytosolic signal; however, the second, DG, is confined to the plasma membrane by its lipid solubility. Not only do these messengers have a different spatial distribution within the cell, but they also differ in their accumulation with time. Thus, the cytosolic concentration of $1,4,5-IP_3$ is increased only transiently whereas elevated membrane levels of DG are sustained. This difference in time course suggests that the role of these messengers in the cellular response changes with time. Therefore, the classical second messenger model of hormone action patently does not describe the sequence of activation events occurring in response to AII.

An alternative model recognizes the importance of temporal and spatial domains [54]. Although a continuum of information transfer is appreciated, in this model the cellular response is divided into two temporal phases to illustrate the profound differences in the intracellular events which are responsible for the initial or sustained effects of hormonal stimulation. These temporal phases in turn dictate a spatial compartmentalization to the activation events which mediate the cellular response. As proposed by this alternative model, the segregation of the cellular response into two phases serves as a useful construct within which to approach the complexity underlying AII stimulation of its target tissues.

6.1. Initiation of response

During the initiation of the response the $1,4,5-IP_3$ generated as a result of AII binding to its receptor induces a mobilization of calcium from a component of the ER. This mobilized calcium is predominantly responsible for raising the cytosolic free calcium from the observed basal value of approximately 100 nM to the peak stimulation level of 1000 nM. The rise in cytosolic calcium acts predominantly via calmodulin-dependent response elements in the cell cytosol to initiate the cellular response. Of particular importance is the activation of calmodulin-dependent protein kinases whose activity produces the altered state of protein phosphorylation which is observed in adrenal, hepatic and vascular smooth muscle cells stimulated with AII. The activity of these protein kinases can also be increased artificially by raising the cytosolic calcium concentration using calcium ionophores. These agents induce an altered state of cellular protein phosphorylation essentially identical to that initially elicited by AII. However, in order to maintain the activation of these calcium-dependent protein kinases, the concentration of calcium must remain elevated. Yet because of the continued operation of the calcium efflux pump and the finite pool of $1,4,5-IP_3$-releasable calcium (estimated to be between 25 and 50 μmoles), the marked rise in cytosolic calcium is only transient, returning within minutes to a low steady-state value. As a consequence, the degree of phosphorylation of the substrates of calmodulin-dependent protein kinases also declines to original basal values, albeit with a slower relaxation time than the decrease in cytosolic calcium [55].

The importance of the rise in cytosolic calcium to the initiation of the response is illustrated in studies in which the rise is either prevented or reproduced through the use of pharmacological agents. Prevention of the calcium transient with dantrolene alters the cellular response so that the onset of response is delayed and the magnitude of the sustained phase is reduced [39]. On the other hand, calcium ionophores such as A23187, which reproduce the calcium transient, mimic the initiation of the response induced by AII [55]. Moreover, manipulations (calcium-channel blockers or a calcium-free medium) which block the AII-induced increase in calcium influx rate, but do not counter the AII elicited rise in cytosolic calcium, do not prevent the initial AII-induced response [39]. These results strongly suggest that the rise in cytosolic calcium is the major early-phase regulator.

6.2. Maintenance of response

In contrast to the initiation, the maintenance of the cellular response by AII is dependent upon extracellular calcium. Manipulations which alter calcium channel activity modulate the sustained AII response. Thus, although nitrendipine has no effect on the initial response, this calcium-channel antagonist inhibits the sustained phase of the response converting the sustained response into a transient response. On the other hand, the calcium-channel agonist, BAY K 8644, enhances prolonged cellular activation. In fact, the AII-dependent modulation of the adrenal calcium channels described in Section 5.4 appears to be critical for the maintenance of the response, since the voltage-dependence of these channels parallels the voltage-dependence of aldosterone secretion [49]. Moreover, countering calcium channel activity by removing extracellular calcium abolishes the sustained phase response in adrenal and vascular smooth muscle [39,56].

This obvious dependence on extracellular calcium is somewhat unexpected because: (1) the sustained enhancement of calcium influx rate is adequately balanced by an increase in calcium efflux rate so that (2) the calcium concentration in the bulk cytosol is maintained near the basal value. This apparent paradox may be resolved by a model [54] which postulates that during the sustained phase of cellular response the high rate of calcium cycling across the plasma membrane raises the calcium concentration in a region just below the plasma membrane, often called the submembrane domain (see Rasmussen and Barrett, Chapter 4). Because the elevated calcium level in this domain is not conducted into the bulk cytosol, it cannot activate calcium-dependent response elements in the cytosol. Rather it regulates the activity of calcium-sensitive, plasma membrane-associated enzymes such as the calcium pump and PKC, the previously described phospholipid-dependent, calcium-activated protein kinase.

There are many reasons to believe that plasma membrane-associated PKC responds to the increase in the calcium concentration beneath the plasma membrane and in so doing regulates the sustained phase of the cellular response. First, during the sustained response the elevated level of plasma membrane DG would function to increase the sensitivity of PKC to activation by elevated calcium levels in the submembrane domain (see Rasmussen and Barrett, Chapter 4). Second, phorbol esters or synthetic diacylglycerols, which substitute for natural DG and activate PKC, elicit a sustained response, but only when used in combination with a calcium ionophore or calcium-channel agonist, either of which reproduces the calcium cycling phenomenon [55]. Third, in certain systems where AII fails to sustain an increase in plasma membrane DG content, no sustained response occurs [58]. Fourth, neither phorbol esters alone nor calcium ionophores alone can reproduce an AII-induced response (in the adrenal) [54]. Taken together these observations imply that

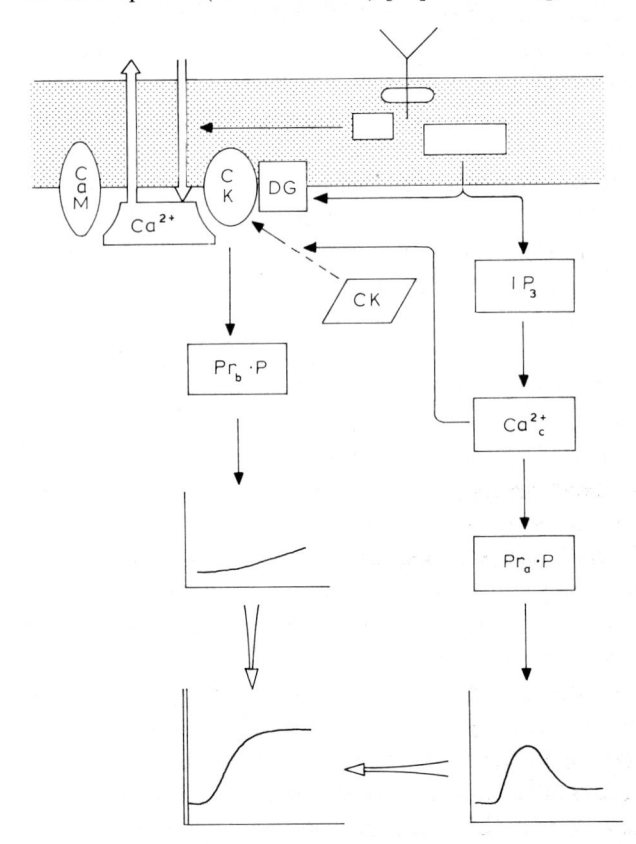

Fig. 2. Messengers mediating the initial and sustained phases of the AII-induced cellular response. Initial phase: AII-elicited hydrolysis of PIP_2 induces a transient rise in cytosolic calcium (via IP_3), a transient activation of calcium-, calmodulin-dependent protein kinases, a transient increase in the phosphorylation of early-phase phosphoproteins (P_{ra}·P), and a transient cellular response. Sustained response: AII-elicited hydrolysis of phosphoinositides generates a sustained increase in the diacylglycerol (DG) content of the plasma membrane. In conjunction with a sustained increase in plasma membrane calcium cycling, DG induces the sustained activation of protein kinase C (CK), the sustained increase in the phosphorylation of late-phase phosphoproteins (P_{rb}·P) and the sustained cellular response.

during the sustained phase of cellular response calcium cycling and DG function as dual second messengers of AII action.

Confirmation of the central role of PKC in mediating the sustained response to AII is provided by studies in the adrenal in which the state of cellular protein phosphorylation was evaluated [55]. Addition of AII leads to a changing temporal pattern of protein phosphorylation, with increased phosphate incorporation into six early- and four late-phase phosphoproteins, only one of which is common to both. This increase in the phosphorylation of the early-phase protein subset is immediate but transient, paralleling the transient rise in cytosolic calcium; while the increase in the phosphorylation of the late-phase protein subset is delayed but sustained (Fig. 2). These changes in protein phosphorylation elicited by AII are reproduced by the pharmacological agents TPA and A23187. Because the only known receptor for TPA is protein kinase C, the reproduction of the AII-induced patterns of protein phosphorylation by the combination of TPA and A23187 is strong evidence that PKC is activated during the sustained response. Moreover, the ability of TPA and A23187 to mimic the AII-induced secretory response implies not only that PKC is activated but also that it mediates this phase of the cellular response. It would appear then that in the two-phase model, the AII response is initiated by the activation of calcium-calmodulin-dependent protein kinases, and the response is maintained via the activation of PKC.

6.3. Temporal relationship of the two phases

In terms of the aldosterone secretory response, there is no evidence for distinct phases. Nevertheless, as has been previously discussed, time-dependent alterations in phosphoinositide turnover, calcium metabolism and protein phosphorylation are observed in response to AII, indicating a changing temporal pattern of messengers and phosphoproteins. These changes are integrated in such a way that a monotonic rather than a biphasic aldosterone secretory response is observed (Fig. 3). To fully appreciate the interrelationships between the two phases and their integration, three additional points must be considered: (1) the importance of the initial intracellular calcium transient in determining the magnitude of the sustained response; (2) the probable role of protein kinase C in the feedback regulation of phospholipase C activity; and (3) the possible role of PKC in the phenomenon of cellular memory.

One of the ways in which integration to yield a monotonic response is achieved is through the initial calcium transient. As already discussed, the transient rise in intracellular calcium is responsible for stimulating calcium-calmodulin dependent kinases to bring about the initiation of the response. However, the magnitude of the sustained phase of the response also correlates with the magnitude of the initial calcium transient. Several lines of evidence indicate that this correlation reflects an effect of calcium on PKC activation. It has been demonstrated in studies with isolated red blood cell membranes that the amount of PKC which becomes associated

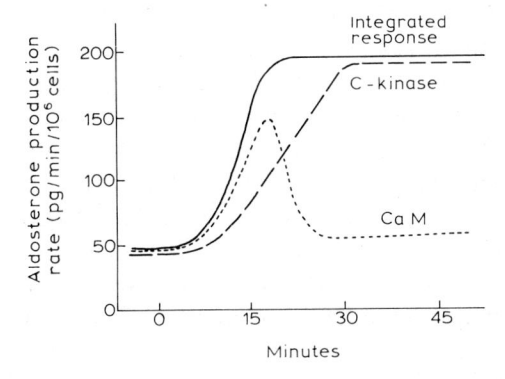

Fig. 3. Temporal summation of response. Although the initial (calmodulin-dependent, CaM) and the sustained (C-kinase) phases of AII-induced aldosterone secretion are mediated by different protein kinases, their activities are integrated to produce a monotonic rather than a biphasic secretory response.

with the plasma membrane is a function of both the DG content and the calcium concentration [58]. Therefore, it is likely that in the intact cell the initial calcium transient in part determines the extent of the association of PKC with the plasma membrane. Because only the membrane-associated PKC, by sensing increased calcium cycling, participates in the sustained phase of the cellular response, it is not surprising that the magnitude of the initial rise in intracellular calcium helps determine the amplitude of the sustained response.

In addition, a reciprocal interaction between the two phases occurs, in that PKC exerts important feedback effects during the sustained phase of the response. Since activation of PKC leads to an enhancement of calcium efflux and a reduction in the calcium concentration within the submembrane domain, PKC essentially limits its own activity. Furthermore, in adrenal glomerulosa and cultured vascular smooth muscle cells, PKC activation during the sustained phase acts as a feedback modulator of the initial transducing events. The artificial activation of PKC with phorbol esters prevents these initial events (the calcium transient and PIP_2 hydrolysis) [59,60]; yet at least in cultured smooth muscle cells PI hydrolysis is not blocked [30]. Thus, PKC activation may result in a shift in the substrate specificity of phospholipase C.

This shift in the substrate specificity of phospholipase C may persist for some time after the removal of agonist and may contribute to the phenomenon of cellular memory in the adrenal. When glomerulosa cells are sequentially exposed to AII (20 minutes), no agonist (10 minutes), and again AII, the character of the response elicited by the second addition of AII differs dramatically from that elicited by the first: the cell seems to 'remember' its prior exposure to AII [29]. Although this second addition of AII induces a smaller calcium transient, the rate of aldosterone secretion increases more rapidly and reaches a higher plateau value than is seen in response to the first exposure to the hormone. This result suggests either that the second addition of AII elicits a smaller increase in 1,4,5-IP_3 as a result of an altered coupling between the receptor and phospholipase C, or that 1,4,5-IP$_3$ is generated but fails to mobilize as much calcium from the ER pool. Furthermore, a low dose of the calcium-channel agonist, BAY K 8644, which has no effect on control cells, increases the aldosterone secretory rate in AII-pretreated cells (20 min) with memory. This result suggests that during the sustained phase of the response the nature of the association of PKC with the membrane is altered so that a prompt dissociation does not occur upon the removal of agonist and the resultant fall in calcium cycling. However, despite an apparent alteration in the nature of this association, PKC retains its sensitivity to activation by calcium so that BAY K 8644, by increasing calcium cycling, increases aldosterone secretion [29]. The molecular basis for this altered interaction between PKC and the plasma membrane is not known.

On the basis of results obtained from studies in adrenal glomerulosa, vascular smooth muscle and hepatic cells, it is evident that the major effects of AII on cell function are mediated via an activation of the calcium messenger system. Furthermore, the data, although far from complete, provide convincing evidence that a temporal and spatial pattern of events underlies hormonal action.

References

1. Mendelsohn, F.A.O. (1985) J. Hypertens. 3, 306–316.
2. Glossmann, H., Baukal, A. and Catt, K.J. (1974) Science 185, 281–283.
3. Devynck, M.-A. and Meyer, P. (1976) Am. J. Med. 61, 758–767.
4. Campanile, C.P., Crane, J.K., Peach, M.J. and Garrison, J.C. (1982) J. Biol. Chem. 257, 4951–4958.
5. Gunther, S. (1984) J. Biol. Chem. 259, 7622–7629.
6. Sernia, C., Sinton, L., Thomas, W.G. and Pascoe, W. (1985) J. Endocrinol. 106, 103–111.
7. Marie, J. and Jard, S. (1983) FEBS Lett. 159, 97–101.
8. Fishman, J.B., Dickey, B.F., McCrory, M.F. and Fine, R.E. (1986) J. Biol. Chem. 261, 5810–5816.
9. Wright, G.B., Alexander, R.W., Ekstein, L.S. and Gimbrone, M.A., Jr. (1982) Circ. Res. 50, 462–469.
10. Sen, I., Bull, H.G. and Soffer, R.L. (1984) Proc. Natl. Acad. Sci. USA 81, 1679–1683.
11. Aguilera, G., Schirar, A., Baukal, A. and Catt, K.J. (1980) Circ. Res. Suppl. 1 46, I118–I127.
12. Pastan, I.H. and Willingham, M.C. (1981) Science 214, 504–509.
13. Crozat, A., Penhoat, A. and Saez, J.M. (1986) Endocrinology 118, 2312–2318.
14. Ross, E.M. and Gilman, A.G. (1980) Ann. Rev. Biochem. 49, 533–564.
15. Glossmann, H., Baukal, A. and Catt, K.J. (1974) J. Biol. Chem. 249, 664–666.
16. Crane, J.K., Campanile, C.P. and Garrison, J.C. (1982) J. Biol. Chem. 257, 4959–4965.
17. Woodcock, E.A. and Johnston, C.I. (1984) Endocrinology 115, 337–341.
18. Anand-Srivastava, M.B. (1983) Biochem. Biophys. Res. Commun. 117, 420–428.
19. Hepler, J.R. and Harden, T.K. (1986) Biochem. J. 239, 141–146.
20. Smith, C.D., Cox, C.C. and Snyderman, R. (1986) Science 232, 97–100.
21. Putney, J.W., Jr. (1987) Am. J. Physiol. 252, G149–G157.
22. Woodcock, E.A. and McLeod, T.K. (1986) Endocrinology 119, 1697–1702.
23. Hackenthal, E., Aktories, K. and Jakobs, K.H. (1985) Mol. Cell. Endocrinol. 42, 113–117.
24. Enyedi, P., Musci, I., Hunyady, L., Catt, K.J. and Spät, A. (1986) Biochem. Biophys. Res. Commun. 140, 941–947.

25. Wallace, M.A. and Fain, J.N. (1985) J. Biol. Chem. 260, 9527–9530.
26. Kojima, I., Kojima, K., Kreutter, D. and Rasmussen, H. (1984) J. Biol. Chem. 259, 14448–14457.
27. Williamson, J.R., Cooper, R.H., Joseph, S.K. and Thomas, A.P. (1985) Am. J. Physiol. 248, C203–C216.
28. Alexander, R.W., Brock, T.A., Gimbrone, M.A., Jr. and Rittenhouse, S.E. (1985) Hypertension 7, 447–451.
29. Barrett, P.Q., Kojima, I., Kojima, K., Zawalich, K., Isales, C.M. and Rasmussen, H. (1986) Biochem. J. 238, 905–912.
30. Griendling, K.K., Rittenhouse, S.E., Brock, T.A., Ekstein, L.S., Gimbrone, M.A., Jr. and Alexander, R.W. (1986) J. Biol. Chem. 261, 5901–5906.
31. Berridge, M.J. (1983) Biochem. J. 212, 849–858.
32. Bouscarel, B. and Exton, J.H. (1986) Biochim. Biophys. Acta 888, 126–134.
33. Guillemette, B., Balla, T., Baukal, A.J., Spät, A. and Catt, K.J. (1987) J. Biol. Chem. 262, 1010–1015.
34. Hawkins, P.T., Stephens, L. and Downes, C.P. (1986) Biochem. J. 238, 507–516.
35. Balla, T., Baukal, A.J., Guillemette, G., Morgan, R.O. and Catt, K.J. (1986) Proc. Natl. Acad. Sci. USA 83, 9323–9327.
36. Batty, I.R., Nahorski, S.R. and Irvine, R.F. (1985) Biochem. J. 232, 211–215.
37. Apfeldorf, W.J. and Rasmussen, H. (1988) Endocrinology. 122, 1460–1465.
38. Capponi, A.M., Lew, P.D., Jornot, L. and Vallotton, M.B. (1984) J. Biol. Chem. 259, 8863–8869.
39. Foster, R. and Rasmussen, H. (1983) Am. J. Physiol. 245, E281–E287.
40. Kojima, I., Kojima, K. and Rasmussen, H. (1985) Am. J. Physiol. 248, E36–E43.
41. Putney, J.W., Jr. and Branchi, C.P. (1974) J. Pharmacol. Exp. Ther. 189, 202–212.
42. Joseph, S.K., Thomas, A.P., Williams, R.J., Irvine, R.F. and Williamson, J.R. (1984) J. Biol. Chem. 259, 3077–3081.
43. Somlyo, A.V., Bond, M., Somlyo, A.P. and Scarpa, A. (1985) Proc. Natl. Acad. Sci. USA 82, 5231–5235.
44. Kojima, I., Kojima, K. and Rasmussen, H. (1985) J. Biol. Chem. 260, 9177–9184.
45. Putney, J.W., Jr. (1986) Cell Calcium 7, 1–12.
46. Mauger, J.-P., Poggioli, J. and Claret, M. (1985) J. Biol. Chem. 260, 11635–11642.
47. Kojima, I., Kojima, K. and Rasmussen, H. (1985) J. Biol. Chem. 260, 9171–9176.
48. Waisman, D.M., Gimble, J.M., Goodman, D.B.P. and Rasmussen, H. (1981) J. Biol. Chem. 256, 409–414.
49. Cohen, C.J., McCarthy, R.T., Barrett, P.Q. and Rasmussen, H. (1987) Biochem. J. 51, 224a.
50. Nowycky, M.C., Fox, A.P. and Tsien, R.W. (1985) Nature 316, 440–446.
51. Kuno, M. and Gardner, P. (1987) Nature 326, 301–304.
52. Irvine, R.F. and Moor, R.M. (1986) Biochem. J. 240, 917–920.
53. Sutherland, E.W., Robison, G.A. and Butcher, R.W. (1968) Circulation 37, 279–306.
54. Rasmussen, H. (1986) N. Engl. J. Med. 314, 1094–1101, 1164–1170.
55. Barrett, P.Q., Kojima, I., Kojima, K., Zawalich, K., Isales, C.M. and Rasmussen, H. (1986) Biochem. J. 238, 893–903.
56. Deth, R. and Van Breemen, C. (1974) Pflügers Arch. 348, 13–22.
57. Danthuluri, N.R. and Deth, R.C. (1986) J. Pharmacol. 126, 135–139.
58. Wolf, M., LeVine, H., III, Mays, W.S., Jr., Cuatrecases, P. and Sahyoun, N. (1985) Nature 317, 546–549.
59. Kojima, I., Shibata, H. and Ogata, E. (1986) Biochem. J. 237, 253–258.
60. Brock, T.A., Rittenhouse, S.E., Powers, C.W., Ekstein, L.S., Gimbrone, M.A., Jr. and Alexander, R.W. (1985) J. Biol. Chem. 260, 14158–14162.

CHAPTER 12

Mechanisms of action of glucagon

J.H. EXTON

The Howard Hughes Medical Institute and the Department of Molecular Physiology and Biophysics, Vanderbilt University School of Medicine, Nashville, TN 37232, U.S.A.

1. Introduction

Glucagon is a 29 amino acid single chain polypeptide secreted by the A (α) cells of the islets of Langerhans. It is also found in the gastrointestinal mucosa together with glicentin a larger peptide containing glucagon within its C-terminal portion. The physiological role of gastrointestinal glucagon or glicentin is presently unknown. Pancreatic glucagon secretion is increased by a rise in plasma amino acids, e.g., following a protein meal, and by enhanced sympathetic nervous activity due to activation of β-adrenergic receptors, whereas an increase in plasma glucose inhibits its release. Several other agents have been reported to be stimulatory or inhibitory to secretion, but it is unclear whether or not their effects are physiological.

Glucagon's major target in the body is the liver where it promotes glycogenolysis, gluconeogenesis and ketogenesis at physiological concentrations. At higher concentrations, it has also been shown to affect the heart, endocrine pancreas and adipose tissue. The primary mechanism of glucagon action has not been investigated in detail in these tissues, but it appears to be the same as that in liver. This chapter is concerned exclusively with the mechanisms involved in the hepatic actions of glucagon.

The mechanism of action of glucagon is generally believed to invariably involve an increase in the intracellular second messenger adenosine 3′,5′-cyclic monophosphate (cAMP). Although one recent report points to the possibility of another mechanism [1], the supporting evidence has been questioned [2] (Section 7). There is now much evidence that glucagon also increases the cytosolic Ca^{2+} concentration [2–5] by mobilizing intracellular calcium [2,5–10] and stimulating Ca^{2+} entry (see Section 7), but these changes are probably secondary to the increase in cAMP [2,5,9].

Glucagon increases intracellular cAMP by binding to a surface receptor which is coupled to adenylate cyclase by a stimulatory guanine nucleotide binding protein (G_s). The sole target of cAMP in mammalian cells is the cAMP-dependent protein kinase which exists in isozymic forms. In non-stimulated cells, the kinase exists mainly as an inactive holoenzyme comprising two catalytic subunits bound to a regulatory subunit dimer. cAMP binds to two sites on each regulatory subunit causing a conformational change releasing the catalytic subunits from the inhibitory action of regulatory dimer. cAMP-dependent protein kinase acts on many enzymes and other proteins in the cell. Through phosphorylation of specific serine or threonine residues in these proteins, their activities are markedly altered. Since many of the proteins are key enzymes involved in the regulation of carbohydrate and lipid metabolism, this explains the large effects exerted by glucagon on these processes.

In the sections that follow, the key components involved in the mechanism(s) by which glucagon exerts its regulatory effects in the liver are described in sequence.

2. The glucagon receptor

Binding sites for glucagon have been identified in plasma membranes from liver [11–14], adipose tissue [13], heart [15] and insulin-secreting tumors [16]. The binding sites in liver and adipose tissue are distinct from those for secretin and vasoactive intestinal polypeptide, which are homologous peptides [13]. Glucagon receptors have been solubilized from rat liver plasma membranes using Lubrol PX [14,17]. The glucagon receptor has also been covalently labeled with [^{125}I]monoiodoglucagon using the cross-linker hydroxysuccinimidyl p-azidobenzoate [18,19] and solubilized using Lubrol-PX or CHAPS [18]. The covalently labeled receptor has an M_r of 63 000 as determined by SDS-polyacrylamide gel electrophoresis [18,19], whereas the non-denatured form has an M_r of 119 000 as determined from its hydrodynamic characteristics [18]. The labeled receptor can also be adsorbed to wheat germ lectin-Sepharose [8] and contains at least four N-linked glycans [19]. Proteolytic digestion indicates that the glucagon binding activity of the receptor and its capacity to interact with G_s are contained within a non-glycosylated fragment of 21 000 Da [19]. These data indicate that the receptor is a glycoprotein and exists in the membrane as a dimer with a subunit of M_r 63 000. As will be discussed below, the liver receptor exists in two affinity forms. The high affinity form has a K_d for glucagon of approximately 1 nM and the total binding sites correspond to 2.6 pmol per mg of membrane protein.

Synthetic glucagon peptides and chemically modified forms of glucagon have been used to probe the binding site of the liver receptor [20–24]. These studies have shown that both the amino-terminal and carboxy-terminal regions are required for full binding and biological activity [20–23]. Des-His[1] glucagon has a 15-fold lower affinity for the receptor [21], and removal of two amino acid residues from the carboxy-terminus reduces the affinity 50-fold [21]. However, a glucagon fragment encompassing the first six amino acids can bind to the receptor and activate adenylate cyclase, but with 1000-fold less potency than the native hormone [22].

Glucagon binding to isolated rat river parenchymal cells has also been studied [25–27]. Two classes of binding sites were found in one study [25]: 20 000 sites per cell with K_d values of 0.7 nM and 200 000 sites with K_d values of 13 nM. In another study [26] of the kinetics of glucagon binding, the receptors were found to exist in two interconvertible affinity states. Interconversion of the two states was time- and temperature-dependent and noncooperative [26]. The transition of the low-affinity glucagon-receptor complex to the high-affinity form was deduced to be rate-limiting for glucagon binding [27]. When chemically modified forms of glucagon were tested, it was concluded that reduced binding potency (with full agonism) was due to reduced affinity of the unoccupied receptor for the ligand and not due to an alteration in interconversion [27]. On the other hand, decreased intrinsic activity was due to decreased conversion of the receptor from the low to high affinity state [27].

Glucagon receptors are also involved in internalization and degradation of the hormone [28–30]. Internalized glucagon becomes associated with lysosomes where it is degraded [29,30]. This is demonstrated by quantitative autoradiography and subcellular fractionation of [^{125}I]glucagon incubated with hepatocytes [29] and examination of the effects of lysosomotropic agents [30].

3. Guanine nucleotide binding regulatory protein

Rodbell and associates [31–32] first recognized the importance of GTP and its analogues in regulating the binding of glucagon to its liver plasma membrane receptor and in the activation of adenylate cyclase. These seminal observations have now been extended to all agonists that stimulate or inhibit adenylate cyclase via receptor mechanisms. It is now clear that the effects of the guanine nucleotides are mediated by regulatory proteins to which they bind, and that there are regulatory proteins (G_s or N_s) which stimulate adenylate cyclase and regulatory proteins (G_i or N_i) which inhibit the enzyme (Fig. 1) [33]. The mechanisms by which glucagon activates G_s and adenylate cyclase are described below.

GTP and its poorly hydrolyzed analogues GMP-PNP, GMP-PCP and GTPγS inhibit the binding of glucagon to its hepatic receptor [19,31]. The effect is observed with micromolar concentrations of the nucleotides and is due to a decrease in the affinity of the receptors for glucagon which results from a stimulation of the rate of dissociation of the hormone [31,34,35]. Guanine nucleotides are also obligatorily required for glucagon activation of adenylate cyclase [32,35] and, by themselves, the poorly hydrolyzable GTP analogues activate the enzyme [32]. Glucagon has also been found to stimulate a low K_m GTPase activity in rat liver plasma membranes [36].

Based on these findings and a large number of parallel observations of β-adrenergic receptor-mediated activation of adenylate cyclase in other cells [33,37], a model has been developed to explain the interactions between receptors that are

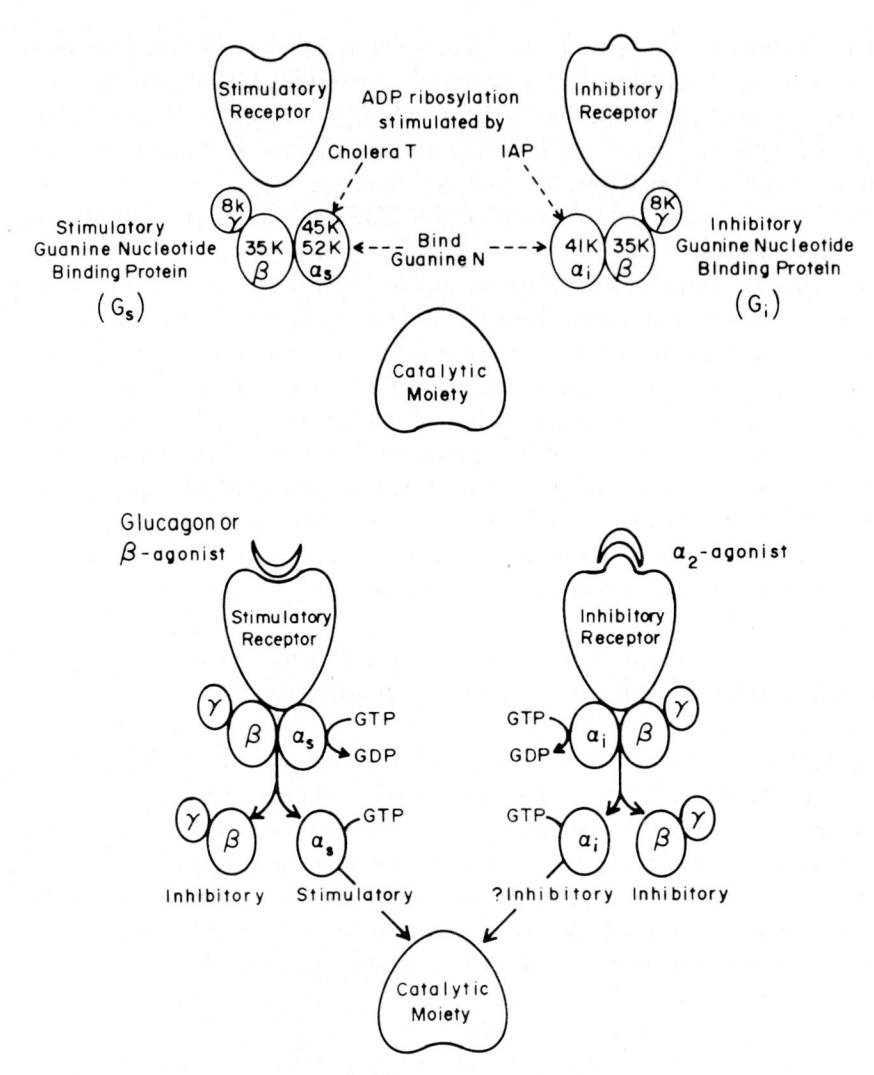

Fig. 1. Mechanisms of activation and inactivation of adenylate cyclase. Abbreviations not given in the text or figure are: cholera T, *Vibrio cholerae* toxin; IAP, islet activating protein, a *Bordetella pertussis* toxin; guanine N, guanine nucleotide; β- and $α_2$-agonists, β- and $α_2$-adrenergic agonists; $β_s$ and $α_i$, α subunits of G_s and G_i, respectively.

stimulatory to adenylate cyclase, G_s and the catalytic subunit of the cyclase (Fig. 1). G_s belongs to a family of signal-transducing proteins which have a heterotrimeric structure (αβγ) [33]. The α-subunits are structurally distinguishable in the different guanine nucleotide-binding proteins, and these subunits exist in several forms in G_s with M_r values ranging between 52 000 and 45 000 [33]. Slight differences also exist in the β-subunits which have M_r values of 36 000 and 35 000 [38,39], and differences have been found in the γ-subunits (M_r of 8 000). The α-subunits contain the guanine nucleotide binding site(s) [40,41] and can be ADP-ribosylated under the influence of specific bacterial toxins [33,41–44]. For example, the *Vibrio cholerae* toxin catalyzes the ADP-ribosylation of the α-subunit of G_s in the presence of NAD^+ [42,44]. This leads to activation of G_s and stimulation of adenylate cyclase [42].

The cycle of events involved in hormonal activation of adenylate cyclase is depicted in Fig. 1. Occupancy of a stimulatory receptor by its agonist is believed to lead to interaction of the receptor with G_s to cause an alteration in guanine nucleotide binding to the $α_s$-subunit. This entails the release of GDP and the binding of GTP [45]. Binding of GTP to the $α_s$-subunit is associated with dissociation of this subunit from the βγ-complex [40]. Since the βγ-complex is inhibitory to the $α_s$-subunit [46], the dissociation leads to activation of this subunit, i.e., it becomes stimulatory to the catalytic subunit of adenylate cyclase [33,40]. Non-hydrolyzable GTP analogues and AlF_4^- (which interacts with GDP bound to $α_s$) also cause dissociation of G_s with resulting activation of the cyclase [33,30].

Interaction of the receptor with G_s also alters the receptor to a state with high affinity for agonist [37,47–49], whereas addition of nonhydrolyzable GTP analogues disrupts the association and converts the receptor to low affinity state [47–49]. Thus the increased binding of GTP to G_s induced by occupancy of the receptor by hormone [45] leads secondarily to dissociation of the receptor from G_s and a return of the receptors to a low affinity state [37,49]. Due to the GTPase activity of G_s, $α_s$·GTP is also converted to $α_s$·GDP which reassociates with the βγ complex to form inactive G_s·GDP. As a result of the conversion of $α_s$·GTP to $α_s$·GDP, activation of adenylate cyclase ceases unless hormone continues to be present to initiate another cycle of activation.

It is obvious from the above description that many mechanistic and molecular details of the hormonal activation of adenylate cyclase are still unknown, but it seems unlikely that the scheme presented will require major modification.

4. Adenylate cyclase catalytic subunit

The catalytic component of the adenylate cyclase complex, i.e., the subunit which catalyzes the conversion of ATP to cAMP, has proved more difficult to purify than the other components. It has recently been purified from myocardium and brain [50–52] using affinity chromatography on forskolin linked to agarose and other chromatographic procedures. Forskolin is a diterpene which can activate the catalytic moiety directly [53]. The heart enzyme was purified approximately 60000-fold and exhibited two major peptides of M_r 150000 and 42000 on $NaDodSO_4$ polyacrylamide gel electrophoresis [50]. The lower M_r peptide was probably $α_s$, whereas the higher M_r peptide was probably the catalytic subunit. Crosslinking of the partially purified catalytic subunit with a G_s preparation, after [^{32}P]ADP-ribosylation of the latter by cholera toxin, gave a single M_r 190000 labeled peptide corresponding to the α_s·catalytic subunit complex [50].

The brain enzyme has been purified about 13000-fold by one group [51]. This enzyme has an M_r of 120000 or 150000 depending on the method of gel electrophoresis [51]. Its catalytic activity is markedly stimulated by G_s and forskolin, and to a lesser extent by Ca^{2+}-calmodulin, whereas the α-subunits of other guanine nucleotide-binding proteins (G_i, G_o) are inert [51]. Addition of the $\beta\cdot\gamma$ subunit complex inhibits the Mg^{2+}-supported activity due probably to contamination of the enzyme with α_s-subunits. Another group has purified the calmodulin-sensitive form of the enzyme 3000-fold from cerebral cortex [52]. This enzyme has a native M_r of 328000 and is stimulated by Ca^{2+}-calmodulin, forskolin, GMPPNP and NaF. It contains major polypeptides of M_r 150000, 47000 and 35000 on $NaDodSO_4$ polyacrylamide gel electrophoresis [52]. Photoaffinity labeling with azido [^{125}I]calmodulin gives one product at M_r 170000. It therefore appears that the catalytic subunit of this enzyme has an M_r of 150000 and can bind calmodulin directly. The smaller M_r peptides are probably the α_s- and β_s-subunits since the purified enzyme responds to GMPPNP and the M_r 47000 peptide is a substrate for cholera toxin [52].

Reconstitution of one of the highly purified catalytic subunit preparations [51] with preparations of G_s and β-adrenergic receptors in phospholipid vesicles yielded a system in which β-adrenergic agonists stimulated adenylate cyclase activity more than 2-fold [54]. Stimulation was dependent upon GTP and showed appropriate selectivity and stereospecificity for β-adrenergic agonists and antagonists. There have been other reports of the successful reconstitution of purified β-adrenergic receptors and G_s with a preparation of the catalytic unit resolved from the other components [55], and of the reconstitution of the partially purified or pure receptors with pure G_s to give high affinity agonist binding, hormone-sensitive GTPase activity or enhanced activation of G_s [47,48,56,57].

5. cAMP and cAMP-dependent protein kinase

There is abundant evidence that glucagon elevates cAMP levels in isolated liver parenchymal cells, in perfused liver and in the liver in vivo [58,59]. As illustrated in Fig. 2, this occurs rapidly and with concentrations of the hormone [59] within the range found in portal venous blood in vivo i.e., $0.2 - 2 \times 10^{-10}$ M. When sufficiently sensitive and accurate methods are employed to measure cAMP, an increase in the nucleotide is consistently observed in situations where the hormone induces metabolic responses [58,59]. However, an increase of only 2- to 3-fold is capable of inducing full stimulation of some major hepatic responses, e.g., phosphorylase activation (Fig. 2) and gluconeogenesis [58,59]. Since higher concentrations of the hormone can elevate cAMP 10-fold or more [59] it appears that there is considerable receptor reserve for these responses.

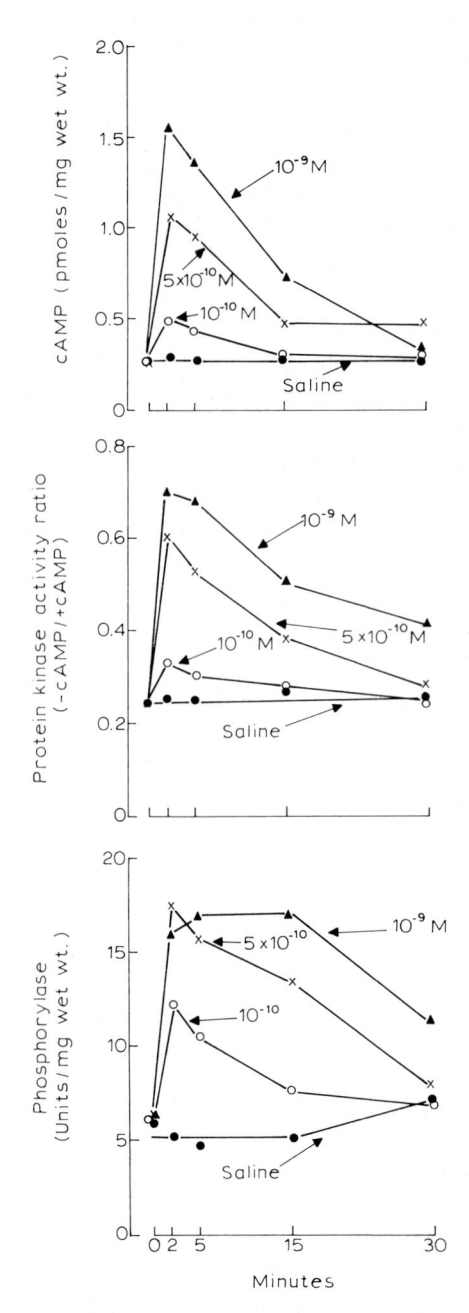

Fig. 2. Effects of glucagon (10^{-10}, 5×10^{-10} and 10^{-9} M) on cAMP levels, cAMP-dependent protein kinase activity ratio and phosphorylase a activity in isolated rat hepatocytes. Reproduced from Ref. 58 by permission of the author and publisher.

The cAMP-dependent protein kinase appears to be the sole mediator of cAMP actions in mammalian cells. It exists in tissues in two major isozymic forms (Types I and II) which can be separated by ion exchange chromatography [60]. Liver possesses both types in approximately equal amounts. In resting cells, the enzyme exists predominantly in a holo form comprising a regulatory dimer (R) of M_r 49 000 (Type I) or 54 000–56 000 (Type II) subunits and two catalytic subunits (C) of M_r 39 000–42 000 [60]. This is illustrated in Fig. 3. When the intracellular cAMP concentration rises due to hormonal stimulation, the nucleotide binds to two sites on each subunit of the regulatory dimer [60]. This binding is believed to result in a conformational change which frees the catalytic subunits from the inhibitory action of the regulatory dimer [60]. Consequently the activity of the enzyme increases.

The changes in cAMP induced by glucagon in isolated hepatocytes are well correlated with the changes in the activation state of the protein kinase [58,59]. This is illustrated in Fig. 2. Careful examination of the correlations between the increases in cAMP and cAMP-dependent protein kinase activity induced by very low concentrations of glucagon illustrates some cooperativity in the effect of the nucleotide on the kinase [59] consistent with the synergistic interaction between the

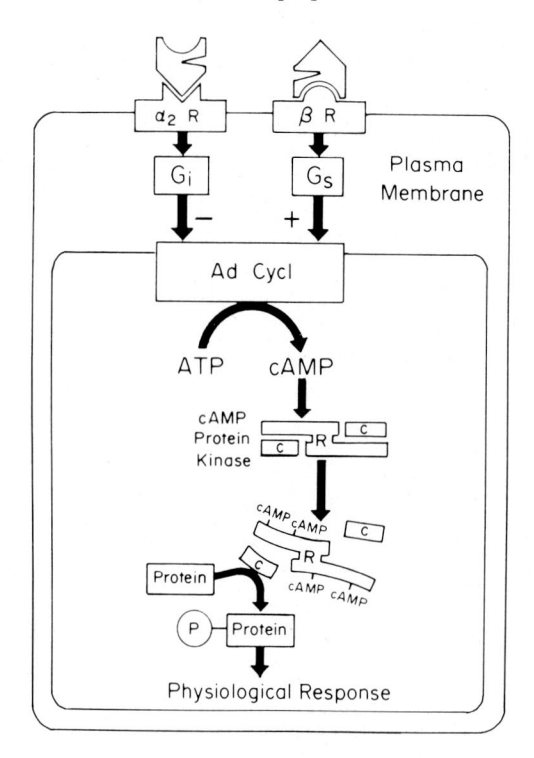

Fig. 3. Mechanism of hormonal activation of cAMP-dependent protein kinase. Abbreviations not defined in the text are: α_2R and βR, the receptors for α_2- and β-adrenergic agonists, respectively; Ad Cycl, adenylate cyclase.

two cAMP binding sites observed in vitro and in vivo (vide infra).

The two binding sites for cAMP on each regulatory monomer differ in their nucleotide specificities and cAMP dissociation rates [61–63] and are located in different domains, as shown by proteolytic fragmentation [59]. One site (Site 1) has a preference for cAMP analogues modified at C-8, while the other (Site 2) prefers a modification at C-6 [62,63]. Binding of cAMP at Site 1 stimulates binding at Site 2 and vice versa [62,64], and synergism between cAMP analogues which interact at the two sites is also observed in activation of kinase activity [64,65,67,68]. Types I and II isozymes display differences in the relative effectiveness of Site 1- or Site 2- directed cAMP analogues to activate kinase activity [64,66,68], implying differences in these binding sites in the two isozymes. The synergism between Site 1 and Site 2 selective cAMP analogues in activating the protein kinase in vivo [64,65,67,68] is also seen when these analogues are used to stimulate cAMP-dependent processes in intact cells, including hepatocytes [67,68].

The catalytic subunit of the cAMP-dependent protein kinase has been purified to homogeneity from several tissues including liver [69]. It has an M_r of 39 000–42 000 (depending on the method used) and appears to be similar in all tissues in terms of its chemical, physical, catalytic and immunological properties [60]. It also has a similar substrate specificity and can interact with either the Type I or Type II regulatory subunit.

It has a broad substrate specificity [60], but the preferred amino acid sequence around the phosphorylated serine or threonine includes a pair of basic amino acids on the amino terminus side [60]. Two arginine residues removed by one residue from the phosphorylated amino acid or a Lys-Arg sequence removed by two residues are preferred.

6. Substrates of cAMP-dependent protein kinase in liver

6.1. Phosphorylase b kinase

Two key regulatory enzymes involved in the control of glycogen metabolism were first recognized as targets of cAMP and cAMP-dependent protein kinase in liver and skeletal muscle. These are phosphorylase b kinase and glycogen synthase. The molecular details of the phosphorylation and regulation of these enzymes are better understood in muscle than in liver since the liver enzymes have only recently been purified to homogeneity in the native form. However, it appears that they share many key features in common.

Both liver and muscle phosphorylase b kinase are large M_r (1.3 million) proteins with a tetrameric structure $(\alpha\beta\gamma\delta)_4$ [70,71]. The molecular weights of the subunits are: α, 140 000–145 000; β, 116 000–130 000; γ, 41 000–45 000; δ, 17 000. White and red muscles contain isozymic forms differing in the M_r of the α-subunit. The δ subunit is virtually identical to calmodulin [72] and the γ-subunit is believed to contain the catalytic site, since its primary structure is homologous to that of the catalytic subunit of cAMP-dependent protein kinase [73]. Studies of the separated subunits also indicate that the γ-subunit is catalytically active [74–76]. The α- and β-subunits appear to have a regulatory role since activation of the enzyme by cAMP-dependent protein kinase is correlated with phosphorylation of these subunits [71,77,78]. Furthermore, in the skeletal muscle enzyme, both the α- and β-subunits can bind exogenous calmodulin, which stimulates the kinase activity [79]. There is evidence that the β-subunit contains a binding site for ATP [80], but it is uncertain that this is a catalytic site.

The activation of skeletal muscle phosphorylase b kinase by cAMP-dependent protein kinase is concomitant with an initial phosphorylation of the β-subunit [77,78]. After 2 mol phosphate per $(\alpha\beta\gamma\delta)_4$ have been incorporated into this subunit, phos-

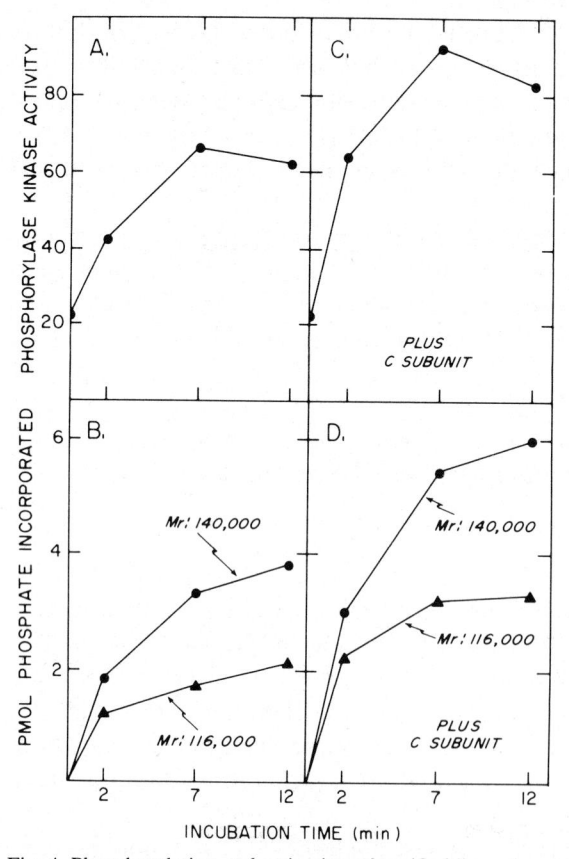

Fig. 4. Phosphorylation and activation of purified liver phosphorylase kinase by cAMP-dependent protein kinase. Panels A and B are without, and Panels C and D are with the catalytic subunit of cAMP-dependent protein kinase. M_r:140 000 and M_r:116 000 refer to the α- and β-subunits of liver phosphorylase kinase. Reproduced from Ref. 71 by permission of the authors and publisher.

phorylation of the α-subunit begins. In the liver and heart enzyme, phosphorylation of the two subunits occurs in parallel (Refs. 71,81 and Fig. 4).

The major substrate of phosphorylase b kinase is phosphorylase b which is phosphorylated on a single serine residue at position 14, resulting in conversion to the more catalytically active form phosphorylase a [70]. Phosphorylation of skeletal muscle phosphorylase also results in conversion of the M_r 200000 dimeric b form to the M_r 400000 tetrameric a form, whereas phosphorylation of the liver enzyme does not alter its dimeric structure [82]. Phosphorylase a is much less dependent than phosphorylase b upon the allosteric activator AMP [82]. Since the activity of phosphorylase is rate-limiting for glycogen breakdown, its activation by phosphorylase b kinase results in enhanced glycogenolysis and glucose release from the liver.

As illustrated in Fig. 2, the changes in phosphorylase activity induced by low concentrations of glucagon are relatively greater than the changes in cAMP-dependent protein kinase, i.e., there is signal amplification. Phosphorylase b kinase can also phosphorylate and inactivate glycogen synthase a, but the rate of phosphorylation is much slower than that of phosphorylase b [71,82–84]. In addition, it is unclear what role this phosphorylation plays in the action of glucagon and other cAMP-elevating hormones on glycogen synthesis since cAMP-dependent protein kinase can phosphorylate and inactivate glycogen synthase directly (vide infra). No other physiological substrates for phosphorylase b kinase have been identified in liver.

6.2. Glycogen synthase

As mentioned above, cAMP-dependent protein kinase phosphorylates and inactivates liver and muscle glycogen synthase [85,86]. Phosphorylation occurs at several serine residues in the muscle enzyme designated 1a, 1b, 2, 3 and 4 [87–90] and it is probable that the phosphorylation pattern is similar in the liver enzyme [91,92]. Phosphorylation of site 1 of the muscle enzyme occurs more rapidly than the other sites [87], and sites 3 and 4 have only recently been recognized as phosphorylation sites [89]. All the sites are surrounded by amino acid sequences characteristic for cAMP-dependent protein kinase [90,93,94].

Phosphorylation of liver and muscle glycogen synthase by cAMP-dependent protein kinase results in reduced catalytic activity measured in the presence of low concentrations of glucose-6-P [85,86] (Fig. 5). This is an allosteric activator of the enzyme which markedly decreases the K_m of the enzyme for its substrate uridine diphosphate glucose [95]. In the case of both the muscle and liver enzymes, phosphorylation causes a large increase in the concentration of glucose-6-P required to half-maximally activate the enzyme [95–97]. In the case of the muscle enzyme, the initial effect of phosphorylation is an increase in the K_m for uridine diphosphate glucose measured in the absence of glucose-6-P [96,97]. At high phosphorylation states, there is also a decrease in V_{max} [97]. The kinetic changes induced in the liver enzyme by cAMP-dependent phosphorylation have not been analyzed, but phos-

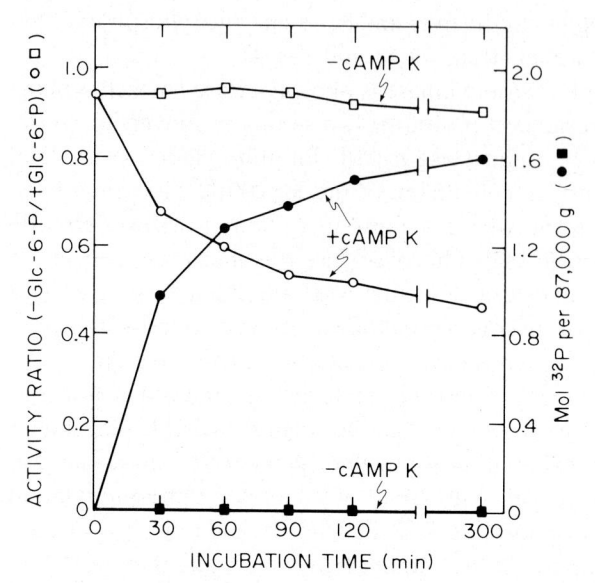

Fig. 5. Time course of phosphorylation and inactivation of purified liver glycogen synthase by cAMP-dependent protein kinase. Reproduced from Ref. 92 by permission of the authors and publisher.

phorylation by other kinases induces a decrease in V_{max} with little change in the K_m for uridine diphosphate glucose [98]. The liver enzyme has a similar M_r to the muscle enzyme [91,98–100], but is not recognized by antisera to the muscle enzyme and exhibits slightly different CNBr and tryptic peptide patterns [91,92] implying differences in amino acid sequence.

Many studies have shown that glucagon inactivates glycogen synthase in intact liver preparations [58] (Fig. 6). There is an increase in the concentration of glucose-6-P for half-maximal activation of the enzyme and an increase in the K_m for uridine diphosphate glucose, with no detectable change in the V_{max} [101,102]. There is also phosphorylation of the enzyme on multiple sites [101,103]. Some of these sites correspond to those phosphorylated by cAMP-dependent protein kinase [91,92], but other sites are also phosphorylated [101,103]. The action of cAMP-dependent protein kinase on glycogen synthase is undoubtedly responsible for the inhibition of liver glycogen synthesis by glucagon [58] since this enzyme is rate-controlling for this process.

6.3. Pyruvate kinase

Early studies of the stimulatory effects of glucagon on hepatic gluconeogenesis using different gluconeogenic substrates and measuring the changes in the concentrations of intermediary metabolites identified the substrate cycles between pyruvate and P-enolpyruvate and between fructose-1,6-P_2 and fructose-6-P as major sites of

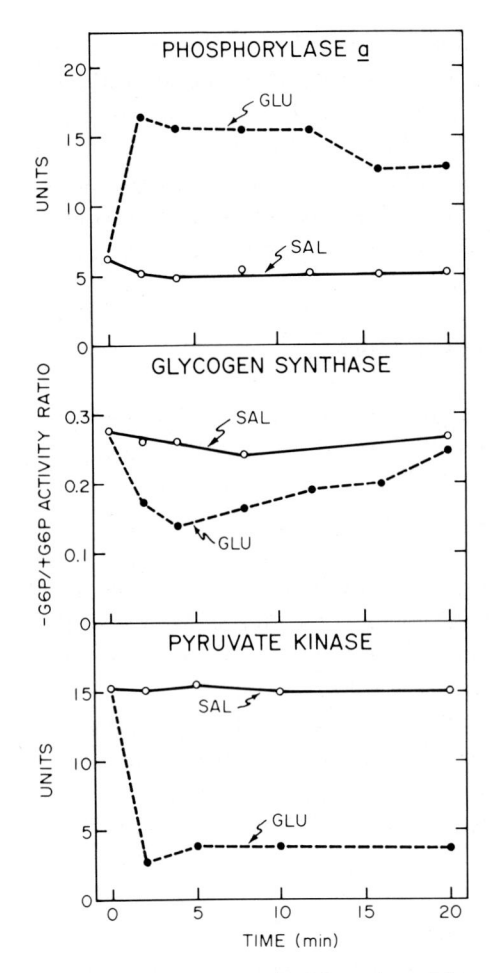

Fig. 6. Effects of glucagon (1 nM) on the activities of phosphorylase a, glycogen synthase and pyruvate kinase in isolated rat hepatocytes. Reproduced from Ref. 58 by permission of the author and publisher.

stimulation of the pathway [58]. The work of Engstrom and co-workers [104] first identified the L-type isozyme of pyruvate kinase as a substrate of cAMP-dependent protein kinase. Phosphorylation causes an increase in the K_m for P-enolpyruvate and in the Hill coefficient for this substrate. However, in the presence of the allosteric activator fructose-1,6-P_2, the K_m for P-enolpyruvate is greatly decreased and the effect of phosphorylation largely disappears [104,105]. The phosphate content of the enzyme can be increased by cAMP-dependent protein kinase by 3 mol/mol [106] and the amino acid sequence around the major phosphorylated serine corresponds to that preferred by the kinase [107].

There is now abundant evidence that the activity of pyruvate kinase is inhibited by glucagon in intact liver preparations [58,108] (Fig. 6), leading to reduced flux of

P-enolpyruvate to pyruvate, i.e., inhibition of glycolysis [109,110] and stimulation of gluconeogenesis [108,109]. The inhibition of pyruvate kinase occurs with physiological concentrations of glucagon and is well correlated with the increase in gluconeogenesis 108]. The kinetic changes induced in the enzyme by glucagon in hepatocytes [108,110,111] correspond to those induced by cAMP-dependent protein kinase in vitro [104,105]. There have also been several demonstrations that glucagon increases the phosphorylation of the enzyme in intact liver preparations and in vivo [58].

6.4. 6-Phosphofructo 2-kinase/fructose 2,6-bisphosphatase

As alluded to above, early studies provided evidence of additional hormonal control of hepatic glycolysis and gluconeogenesis at the fructose-1,6-P_2/fructose-6-P substrate cycle [58]. Since most of the physiologically important gluconeogenic substrates enter the gluconeogenic pathway at or prior to the P-enolpyruvate/pyruvate cycle, e.g., lactate, pyruvate, alanine and certain other amino acids, it is probable that hormonal regulation of this cycle is of greater physiological significance. However, glycerol, fructose and certain other sugars enter the gluconeogenic pathway above the P-enolpyruvate/pyruvate cycle and therefore their conversion to glucose could be regulated at the fructose-1,6-P_2/fructose-6-P cycle [58].

The regulation of the fructose-1,6-P_2/fructose-6-P cycle by glucagon occurs indirectly through a decrease in fructose-2,6-P_2 [112] (Fig. 7), which activates 6-phos-

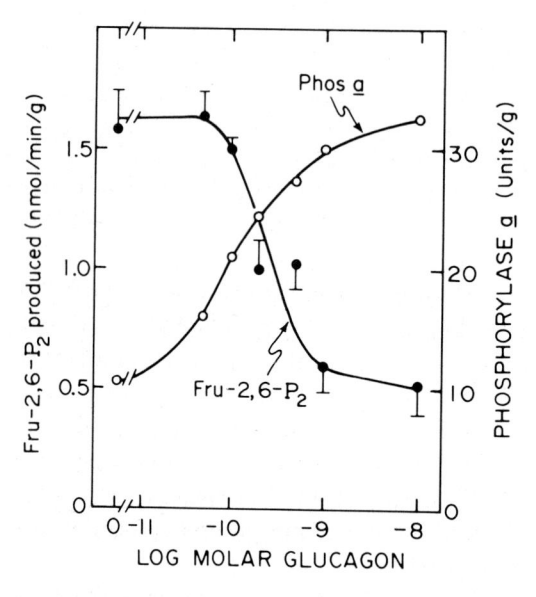

Fig. 7. Effects of different concentrations of glucagon on phosphorylase a and fructose 2,6-P_2 levels in isolated rat hepatocytes. Reproduced from Ref. 112 by permission of the authors and publisher.

phofructo 1-kinase and inhibits fructose 1,6-bisphosphatase [113–116]. The decrease in fructose 2,6-P_2 occurs because cAMP-dependent protein kinase regulates the activities of both 6-phosphofructo 2-kinase, which synthesizes fructose 2,6-P_2 from fructose 6-P [117–119], and fructose 2,6-bisphosphatase, which hydrolyzes fructose 2,6-P_2 to fructose 6-P [120,121]. These enzyme activities are now known to reside on the same protein which therefore exerts bidirectional control of fructose 2,6-P_2 levels [122].

6.5. Acetyl-CoA carboxylase, ATP citrate lyase

Glucagon inhibits fatty acid synthesis in liver [123] and the inhibition is correlated with decreased activity of acetyl-CoA carboxylase [124,125]. The inhibition of the enzyme is due to its phosphorylation by cAMP-dependent protein kinase [125,126] which causes a decrease in the V_{max} and an increase in the K_a for citrate [126].

Incubation of hepatocytes with glucagon also increases the phosphorylation of another lipogenic enzyme ATP-citrate lyase [127,128]. The purified enzyme is also phosphorylated by cAMP-dependent protein kinase in vitro [128,129] resulting in a 2-fold increase in the K_m for ATP [130]. However, it is unclear what role this plays in the inhibition of lipogenesis.

Glucagon decreases cholesterol synthesis in isolated hepatocytes [131,132] apparently because it reduces the fraction of hydroxymethylglutaryl-CoA reductase in the active form [131,132]. This is due to an increase in reductase kinase activity [133]. However, there is no evidence that cAMP-dependent protein kinase phosphorylates either the reductase, reductase kinase or reductase kinase kinase [134]. It has been proposed that the phosphorylation state of these enzymes is indirectly controlled through changes in the activity of protein phosphatase I [132,134]. This phosphatase can dephosphorylate and activate the reductase [134,135] and its activity can be controlled by a heat stable inhibitor (inhibitor 1), the activity of which is increased by cAMP-dependent phosphorylation [136,137]. Since the phosphorylated forms of acetyl-CoA carboxylase, ATP-citrate lyase, pyruvate kinase, phosphorylase, phosphorylase kinase and glycogen synthase are also substrates for protein phosphatase I [135], this mechanism could also contribute to their phosphorylation by glucagon.

7. Effects of glucagon on cell calcium

Early experiments utilizing the perfused rat liver [6] demonstrated that glucagon alters cell Ca^{2+} fluxes. These observations have been confirmed in numerous ways, but the physiological significance of the effect remains unclear. Although initial indirect measurements of cytosolic Ca^{2+} in hepatocytes indicated no effects of glucagon [138], more recent studies utilizing the fluorescent indicator quin2 have shown

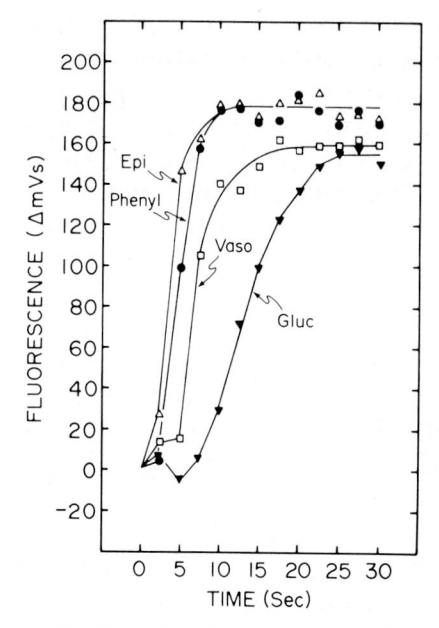

Fig. 8. Effects of agonists on cytosolic Ca^{2+} (measured by fluorescence) in isolated rat hepatocytes loaded with Quin 2. The concentrations of agonists are: Epi, 1 μM epinephrine; Phenyl, 10 μM phenylephrine; Vaso, 10 nM vasopressin; Gluc, 10 nM glucagon. Reproduced from Ref. 3 by permission of the authors and publisher.

that the hormone can elicit a 2- to 3-fold increase in cytosolic Ca^{2+} within 10–20 s [2–5] (Fig. 8). This response is slower than those elicited by vasopressin, epinephrine and angiotensin II [2–5,139]. In general, the changes in total cellular Ca^{2+} induced by glucagon in isolated hepatocytes or the perfused liver are also smaller than those caused by these other Ca^{2+}-mobilizing agonists [5,7,140] (Fig. 9). Furthermore, the half-maximal glucagon concentration for elevation of cytosolic Ca^{2+} or mobilization of cellular Ca^{2+} (0.2–0.5 nM) [4,5,10] is slightly higher than that for phosphorylase activation (approx. 0.1 nM) [5,7]. In addition, the effects of glucagon on glycogenolysis and gluconeogenesis are not impaired if hepatocytes are depleted of Ca^{2+} and incubated in low Ca^{2+} media [7,140–142]. For these reasons, it has been questioned whether the alterations in cell Ca^{2+} play any role in these physiological actions of glucagon [58]. However, it is possible that the Ca^{2+} changes are important in other actions of the hormone.

Glucagon increases cytosolic Ca^{2+} both by mobilizing intracellular Ca^{2+} stores and promoting Ca^{2+} influx across the plasma membrane. The internal Ca^{2+} stores mobilized by glucagon appear to be the same as those acted on by other Ca^{2+}-mobilizing hormones [5,9,143] (Fig 9), and there is evidence that the plasma membrane Ca^{2+} channel(s) controlled by glucagon are the same as those controlled by vasopressin or phenylephrine [144,145].

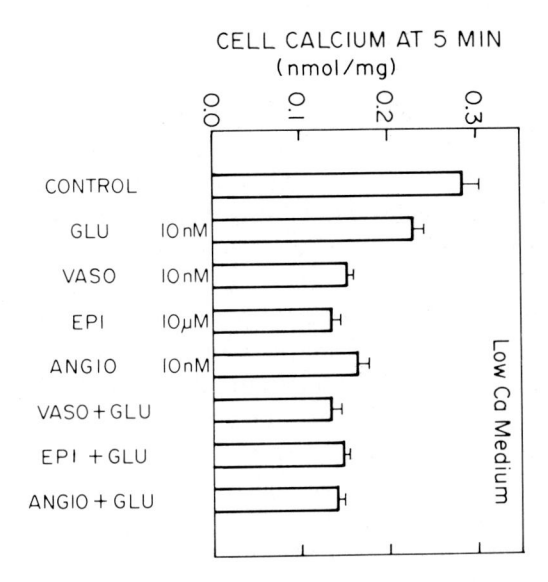

```
                    CELL CALCIUM AT 5 MIN
                           (nmol/mg)
                      o    o    o    o
                      o    ─    ἰυ   ὼ
                      ┬────┬────┬────┬
  CONTROL       ┃                  ┠┤
  GLU    10 nM  ┃              ┠┤
  VASO   10 nM  ┃     ┠┤
  EPI    10 μM  ┃    ┠┤                  L
  ANGIO  10 nM  ┃       ┠┤                o
                                          w
  VASO+GLU      ┃   ┠┤                    C
                                          a
  EPI +GLU      ┃    ┠┤                   M
                                          e
  ANGIO + GLU   ┃    ┠┤                   d
                                          i
                                          u
                      ┴────┴────┴         m
```

Fig. 9. Effects of hormones on the calcium content of isolated rat hepatocytes. Hepatocytes were incubated for 5 min in medium containing 100 μM Ca^{2+} with hormones at the concentrations shown and the calcium content measured by atomic absorption spectroscopy. Glu is glucagon and Angio is angiotensin II. Reproduced from Ref. 5 by permission of the authors and publisher.

The mechanism(s) by which glucagon alters cellular Ca^{2+} are somewhat controversial. Wakelam et al. [1] have studied the actions of glucagon and a glucagon analogue (TH-glucagon or 1-N-α-trinitrophenylhistidine, 12-homoarginine-glucagon) on the production of inositol phosphates in hepatocytes and have concluded that these cells possess two types of glucagon receptor, namely one coupled to adenylate cyclase and another coupled to inositol phospholipid breakdown. They reported that glucagon concentrations between 0.2 and 5 nM increased inositol phosphates, whereas higher concentrations did not [1]. On the other hand, TH-glucagon increased inositol phosphates at all concentrations above 0.1 nM, but did not detectably elevate cAMP [1]. The inhibition of inositol phosphate accumulation observed with high concentrations of glucagon was attributed to an elevation of cAMP since it was mimicked by cholera toxin [1].

Although the hypothesis of Wakelam et al. is attractive, certain of their findings do not agree with reports from other laboratories. For example, glucagon has been observed to elevate inositol trisphosphate [5] and mobilize Ca^{2+} [4,5,10,142] at concentrations (1–100 nM) reported to be inhibitory by Wakelam et al. [1]. The original report of Corvera et al. [146] also showed that TH-glucagon did not significantly increase glycogenolysis in hepatocytes at concentrations between 0.1 and 10 nM, whereas the increase in inositol trisphosphate reported by Wakelam et al. [1] should have caused Ca^{2+} mobilization and phosphorylase activation [147]. In addition, Ca^{2+} depletion of hepatocytes did not impair the effects of TH-glucagon on glycogenolysis and gluconeogenesis [146], which is inconsistent with it acting through Ca^{2+} mobilization.

An alternative explanation for the effects of glucagon on hepatocyte Ca^{2+} fluxes is that they are secondary to the increase in cAMP [2,5,9,144]. Evidence supporting this comes from the potentiation of the effects by an inhibitor of cAMP phosphodiesterase [5] and their mimickry by cAMP analogues [2,5,9,144] and by forskolin [5,9] which activates adenylate cyclase (see Fig. 13). A key point is whether or not glucagon or other agents that elevate intracellular cAMP raise inositol trisphosphate. Blackmore and Exton [5] have provided evidence that glucagon, forskolin and 8-p-chlorophenylthio cAMP produce small increases in inositol trisphosphate in hepatocytes that are of sufficient magnitude to mobilize intracellular Ca^{2+} [147] (Fig. 10). Other groups have obtained evidence that glucagon stimulates phosphatidylinositol 4,5-bisphosphate breakdown and inositol trisphosphate formation [2,148] and a small effect of glucagon to elevate 1,2-diacylglycerol in hepatocytes has been reported [149]. On the basis of these results, it has been proposed that glucagon mobilizes internal Ca^{2+} through cAMP-dependent phosphorylation and activation of the guanine-nucleotide binding protein or other membrane component involved in signal transduction to the phospholipase C which catalyzes the

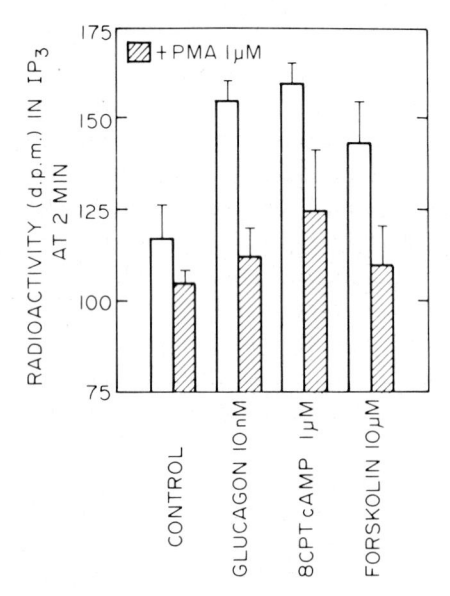

Fig. 10. Effects of glucagon, cAMP analogue and forskolin, in the presence or absence of phorbol ester, on [^{3}H]IP$_3$ levels in isolated rat hepatocytes. Hepatocytes were incubated for 90 min with [^{3}H]myo-inositol to label the inositol phospholipids and then for 2 min with the designated compounds at the concentrations shown in the presence or absence of 1 μM 4β-phorbol 12β-myristate 13α-acetate (PMA). [^{3}H]IP$_3$ was measured as described in Ref. 139. 8CPTcAMP is 8-p-chlorophenylthioadenosine 3',5'-cyclic monophosphate. Reproduced from Ref. 5 by permission of the authors and publisher.

breakdown of phosphatidylinositol 4,5-bisphosphate to inositol 1,4,5-trisphosphate and 1,2-diacylglycerol.

In addition to mobilizing internal Ca^{2+}, glucagon promotes the net entry of extracellular Ca^{2+} into hepatocytes [144,145]. The effect is apparently due to increased influx of Ca^{2+} through a plasma membrane channel(s) [144,145], but there is also inhibition of the plasma membrane Ca^{2+} ATPase-pump [150–153]. The mechanism by which glucagon stimulates Ca^{2+} entry is unknown, but it almost certainly involves cAMP since the effect can be mimicked by forskolin, dibutyryl cAMP and β-adrenergic agonists [144,145]. It may involve cAMP-dependent phosphorylation of a Ca^{2+} channel analogous to the situation in cardiac and skeletal muscle [154–156], but this is strictly speculative.

Glucagon causes a 30–40% inhibition of ATP-dependent Ca^{2+} transport activity and $(Ca^{2+}-Mg^{2+})$-ATPase activity in liver plasma membranes [150–152]. however, much higher glucagon concentrations (0.1–10μM) are required to produce these changes [150–152] than to activate adenylate cyclase (0.1–100 nM), and the inhibition of the ATPase is not mimicked by cAMP or its analogues [157]. The effects of several glucagon derivatives on the ATPase are also very different from their effects on adenylate cyclase [150]. All of these observations indicate that the two effects are not related.

The inhibitory effect of glucagon on the ATPase is completely blocked by pretreatment of the animals with cholera toxin, whereas pertussis toxin or exogenous cAMP is without effect [152]. This suggests the involvement of a guanine nucleotide binding protein, but it is uncertain whether or not this is G_s. A most interesting development in this work is the recognition that it is a proteolytically generated fragment of glucagon that is responsible for the inhibition of the $(Ca^{2+}-Mg^{2+})$-ATPase [153]. This is shown by the fact that bacitracin, an inhibitor of glucagon degradation, abolishes the effect of the hormone on the ATPase. Furthermore, a tryptic peptide of glucagon corresponding to amino acids 19 through 29 potently inhibited the ATPase and also ATP-dependent Ca^{2+} uptake by liver plasma membrane vesicles, but was without effect on adenylate cyclase [153]. The K_i for the peptide on the ATPase was 0.75 nM compared with 700 nM for native glucagon. Other peptides corresponding to residues 22 through 29 and 1 through 21 were even less potent than glucagon.

The physiological significance of the proteolytically generated glucagon fragment is unclear because of its very recent discovery. It could be generated by membrane-associated proteases in other tissues and therefore have additional physiological effects. The accumulated evidence indicates that it acts by a mechanism not involving the glucagon receptor linked to adenylate cyclase, but involving a guanine nucleotide binding protein that is a substrate of cholera toxin. Since the interaction of glucagon with adenylate cyclase appears to secondarily result in Ca^{2+}-mobilization and Ca^{2+} channel opening in liver cells, as described above, the contribution of the inhibition of the Ca^{2+} pump by the glucagon fragment to the elevation of cytosolic

Ca^{2+} by glucagon is presently unclear. However, the peptide could have additional effects in this and other tissues.

8. Synergistic interaction between glucagon and calcium-mobilizing agonists in liver

An important aspect of the action of glucagon in liver is its potentiation of the effects of calcium-mobilizing agonists on cell Ca^{2+} fluxes and on certain physiological processes. Thus low concentrations of glucagon (0.1–1 nM) amplify the actions of low concentrations of epinephrine, vasopressin and presumably other Ca^{2+}-mobilizing agonists [5,143–145,158]. Calcium mobilization is potentiated [5,158] (Fig. 11) as is net Ca^{2+} entry [143–145], and the synergism is mimicked by other means of raising intracellular cAMP [5,144,145,158]. The potentiation of cytosolic Ca^{2+} elevation by glucagon is reflected in enhanced phosphorylase activation and glycogenolysis [5,158].

Blackmore and Exton [5] have provided much evidence that the potentiation by glucagon of the intracellular Ca^{2+} mobilization induced by vasopressin is due to an enhancement of inositol trisphosphate accumulation (Fig. 11). This is probably true for its synergistic interaction with other Ca^{2+}-mobilizing agonists. All the actions of

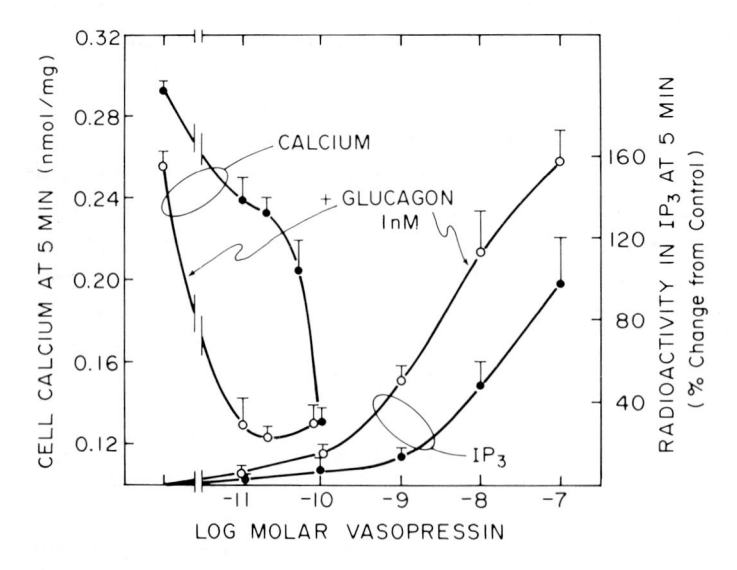

Fig. 11. Potentiation by glucagon of the actions of vasopressin on the calcium and [^{3}H]IP$_3$ content of isolated rat hepatocytes previously labeled with [^{3}H]myo-inositol. Hepatocytes were incubated with [^{3}H]myo-inositol to label the inositol phospholipids and then for 5 min with the concentrations of vasopressin shown in the presence or absence of 1 nM glucagon. Total cell calcium and [^{3}H]IP$_3$ were measured as described in Refs. 139 and 140. Reproduced from Ref. 5 by permission of the authors and publisher.

these agonists can be mimicked by AlF_4^- [5,159], which activates guanine nucleotide binding proteins directly, i.e., by a non-receptor mechanism. Since glucagon also potentiates the actions of AlF_4^- [5], it has been proposed that the hormone acts through cAMP-dependent phosphorylation and sensitization of the guanine nucleotide binding protein that couples the Ca^{2+}-mobilizing receptors to the phospholipase C which breaks down phosphatidylinositol 4,5-P_2 or of another protein involved in their interaction [5].

The synergistic interaction between glucagon or β-adrenergic agonists and Ca^{2+}-mobilizing agonists in the stimulation of Ca^{2+} influx into hepatocytes has been studied in detail by Claret and associates [143–145]. The interaction involves the same Ca^{2+} channels as revealed by kinetic analysis [144,145]. Unlike the situation with Ca^{2+} mobilization [5,9,143], the cAMP-mediated hormones increased the maximal Ca^{2+} uptake induced by the Ca^{2+}-mobilizing hormones without significantly altering their half-maximally effective concentrations and vice versa [144,145]. Since the possible role of inositol polyphosphates in the regulation of Ca^{2+} entry into liver cells is presently unclear, it is uncertain whether the action of glucagon on Ca^{2+} entry is due to the enhanced elevation of inositol trisphosphate [5] or of one of its metabolites. Alternative explanations are that cAMP-dependent phosphorylation of a Ca^{2+}-channel or of a component involved in its regulation could partly activate

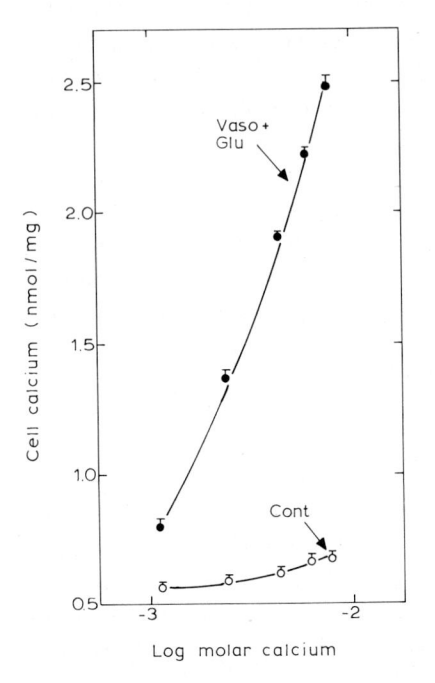

Fig. 12. Effect of extracellular Ca^{2+} concentration on the accumulation of Ca^{2+} by isolated rat hepatocytes incubated for 5 min with 10 nM vasopressin plus 10 mM glucagon. Reproduced from Ref. 160 by permission of the authors and publisher.

the channel and render it more responsive to Ca^{2+}-mobilizing hormones.

The combination of glucagon with a high concentration of a Ca^{2+}-mobilizing hormone can reverse the loss of Ca^{2+} from liver cells [159] or lead to a large accumulation of Ca^{2+} in these cells [160]. This is particularly pronounced when the extracellular Ca^{2+} concentration is increased [160] (Fig. 12) and presumably reflects a predominance of the stimulation of Ca^{2+} influx [143–145] over the mobilization of intracellular Ca^{2+} stores [5] induced by the combined hormones. Analysis of the Ca^{2+} content of subcellular fractions indicates that the extra Ca^{2+} accumulated by the cells under these conditions is found principally in the mitochondria [160]. As expected, the cytosolic Ca^{2+} concentration in the presence of the hormone combination is much greater than seen with either hormone alone (P.F. Blackmore, unpublished observations) and reaches the level (> 600 nM) at which mitochondria begin to accumulate Ca^{2+}.

9. Inhibitory action of phorbol esters on glucagon-induced calcium mobilization

Tumor promoting phorbol esters, e.g., 4β-phorbol 12β-myristate 13α-acetate (PMA) potently inhibit the actions of glucagon on liver cell Ca^{2+} fluxes [2,5,161]. However, they have little effect on glucagon-stimulated cAMP accumulation [5,162]. Their ability to inhibit Ca^{2+} mobilization is also observed when this is induced by forskolin or cAMP analogues [2,5,161] (Fig. 13), other physiological Ca^{2+} mobilizing agonists, especially α_1-adrenergic agonists [5,162,163], or AlF_4^- [5]. Because they are capable of reducing the accumulation of inositol trisphosphate induced by all these agents [5,162], it has been proposed that their action is exerted on the guanine nucleotide binding protein controlling the phospholipase C hydrolyzing phosphatidylinositol 4,5-P_2 or on a related regulatory component [5]. Since a major target of active phorbol esters in most cells is protein kinase C, it is probable that their action involves phosphorylation of these proteins by this kinase [5].

10. Other actions of glucagon

There have been reports of other glucagon actions in the liver which can be related to the elevation of cAMP, but whose molecular mechanisms are not well defined. Examples are the stimulations of ketogenesis, ureogenesis, amino acid transport, respiration and ion fluxes, the rapid changes in pyruvate dehydrogenase and pyruvate carboxylase, and the induction of P-enolpyruvate carboxykinase and other enzymes.

The ketogenic action of glucagon is probably related in part to cAMP-dependent phosphorylation and inactivation of acetyl-CoA carboxylase [125,126]. This would

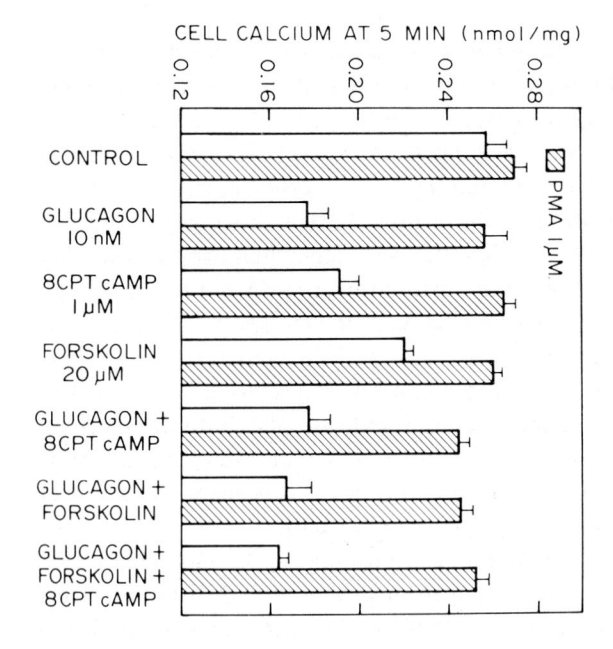

Fig. 13. Inhibition by PMA of calcium mobilization induced by glucagon, cAMP analogue or forskolin in isolated rat hepatocytes. Hepatocytes were incubated for 5 min and total cell calcium measured as described in Ref. 140. Abbreviations are those defined in the legend to Fig. 10. Reproduced from Ref. 5 by permission of the authors and publisher.

explain the decrease in hepatic malonyl-CoA induced by glucagon [123]. Since malonyl-CoA is a reversible inhibitor of carnitine acyltransferase I (the initial step in mitochondrial fatty acid oxidation) [165,166], the reduction in its level by glucagon would be expected to enhance the oxidation of fatty acids. There is also evidence that glucagon can phosphorylate and activate carnitine acyltransferase [167]. Since this enzyme is rate-limiting for mitochondrial fatty acid oxidation, its activation by both mechanisms would lead to increased acetyl-CoA production and, with the concurrent impairment in the utilization of acetyl-CoA for fatty acid synthesis, these metabolites would be diverted into ketone body synthesis. The enhanced oxidation of fatty acids could explain the effect of glucagon to decrease hepatic fatty acid esterification [168] and hence triglyceride output, but other factors are probably involved [169].

In the in vivo situation, the ketogenic action of glucagon is most prominent in states of insulin deficiency. This can be explained because insulin normally suppresses the effect of glucagon on hepatic cAMP levels [170] and inhibits the action of the hormone on lipolysis, i.e., fatty acid release in adipose tissue [171].

The stimulation of ureogenesis by glucagon is poorly understood. It may be the result of the actions of the hormone on hepatic proteolysis, amino acid uptake and amino acid gluconeogenesis. The proteolytic action is thought to be due to activation of lysosomes [172,173] and the stimulation of amino acid transport may be secondary to enhanced Na^+ uptake (vide infra). Amino acid gluconeogenesis would be enhanced because of the action of glucagon on the two substrate cycles of the gluconeogenic pathway (see Sections 6.3 and 6.4) and also because of the enhancement of the conversion of glutamine and glutamate to 4-carbon citric acid cycle intermediates because of the activation of α-oxoglutarate dehydrogenase [174,175].

Glucagon and exogenous cAMP stimulate the Na^+-dependent transport of alanine and certain other amino acids into the perfused liver [176] and isolated hepatocytes [177–179]. There is a rapid initial stimulation of the transport [177, 178] which is probably related to the stimulation of (Na^{2+}-K^+)-ATPase activity and membrane hyperpolarization [177]. This is followed after 30–90 min by a larger increase which is blocked by cycloheximide [178]. Kinetic analysis indicates that both the short and long term actions of glucagon result in an increase in the V_{max} for transport [177,179,180], and it has been proposed that the slower effect is due to the synthesis of a high affinity amino acid transport component [179,180].

The biochemical basis for the stimulation of respiration by glucagon [181,182] is poorly defined and many explanations have been proposed and disputed. Interested readers are referred to recent reports [183–186]. Mitochondria isolated from glucagon-treated livers or hepatocytes show increased pyruvate dehydrogenase [187–189] and pyruvate carboxylase activities [190,191]. The increase in pyruvate dehydrogenase activity in perfused livers is well correlated with an increase in mitochondrial Ca^{2+} content [189], and this is most marked when glucagon is combined with a Ca^{2+}-mobilizing hormone [160,189]. The probable explanation for the correlation is that the enzyme is activated because pyruvate dehydrogenase phosphatase is stimulated by Ca^{2+} [192]. As alluded to above [174,175], glucagon also increases the activity of another liver mitochondrial enzyme, namely α-oxoglutarate dehydrogenase [189]. This enzyme is also Ca^{2+}-sensitive [192], and changes in its activity are also well correlated with the increase in mitochondrial Ca^{2+} [189]. The mechanism of activation of pyruvate carboxylase by glucagon is not known, but the effect is seen with other Ca^{2+}-mobilizing hormones and could be related to the changes in cell Ca^{2+}.

The observations that glucagon and Ca^{2+}-mobilizing hormones activate pyruvate dehydrogenase and increase mitochondrial Ca^{2+} [189,192] are in contradiction to earlier reports that these agents decrease mitochondrial Ca^{2+} [193–197]. However, in the studies reporting that Ca^{2+} was decreased in crude mitochondrial fractions isolated from livers treated with hormones [193–195], it is likely that the fractions were contaminated with the endoplasmic reticulum components that lose Ca^{2+} in response to these agents. In the other studies [195–197] indirect measurements of mitochondrial Ca^{2+} content were employed. The idea that pyruvate dehydrogenase is activated because of an increase in mitochondrial Ca^{2+} which is secondary to the increase in cytosolic Ca^{2+} receives support from experiments with PMA [198]. This phorbol ester inhibits the activation of pyruvate dehydrogenase by glucagon or phenylephrine, but not vasopressin (25 nM) consistent with its effects on the increases in cytosolic Ca^{2+} induced by these agonists.

Glucagon causes hyperpolarization and large fluxes in Na^+ and K^+ across the hepatic plasma membrane [6,193,199–202], and the effect can be partly attributed to the stimulation of $(Na^+\text{-}K^+)$-ATPase by the hormone [203,204]. This effect can be mimicked by cAMP and its analogues, but its biochemical basis remains obscure. It is possible that the ATPase or a regulatory protein is activated by cAMP-dependent phosphorylation. The initial K^+ change caused by glucagon is an uptake, which can be attributed to the stimulation of $(Na^+\text{-}K^+)$-ATPase, but there is a subsequent large K^+ release [6,199,201] (Fig. 14). This could be due to the opening of Ca^{2+}-dependent K^+ channels, which have been shown to exist in the livers of most species [205–208].

Although most investigations of glucagon action have focused on its rapid effects, which occur within seconds or minutes, it is clear that it also has longer term actions involving changes in protein synthesis. Glucagon causes changes in levels of many hepatic enzymes which are related to alterations in enzyme synthesis [209].

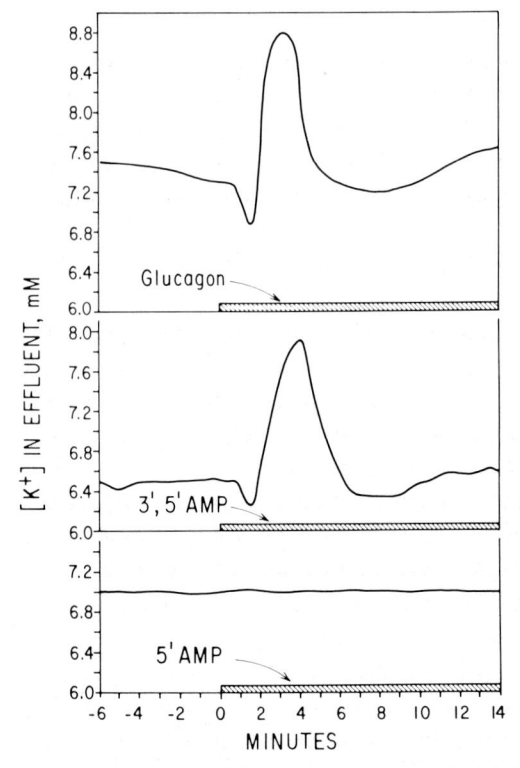

Fig. 14. Effects of glucagon (3 nM), adenosine 3',5'-cyclic monophosphate (3',5' AMP) (1 mM) and 5' AMP (1 mM) on K^+ fluxes in isolated rat livers perfused with non-recirculating medium. Reproduced from Ref. 199 by permission of authors and publisher.

The most extensively studied enzyme is the cytosolic form of P-enolpyruvate carboxykinase, which is induced within 30 to 60 min by glucagon or cAMP in liver or hepatoma cells [210–213]. The increase in synthesis is due to an elevation of functional mRNA [214–218] which is due in turn to increased transcription of the gene [219–221]. The cAMP regulatory element of the P-enolpyruvate carboxykinase gene has been determined by deletion mutagenesis to be located between basepairs -109 and -62 relative to the transcriptional start site [222–224]. The regulatory element of the P-enolpyruvate carboxykinase contains a 12 basepair core sequence that is virtually identical or very similar to sequences in the 5′ flanking regions of many other genes [223,224] suggesting a common regulatory mechanism. Since the ability of cAMP analogues to induce P-enolpyruvate carboxykinase correlates very well with their ability to activate cAMP-dependent protein kinase [225] it appears that cAMP-dependent phosphorylation is involved in the regulation of this gene, but the nuclear and other events involved have yet to be defined.

11. Summary

The liver is the major target of glucagon in vivo and most of the information about its mechanism of action has come from studies with this tissue. The actions of glucagon are entirely explicable in terms of the generation of cAMP. Although there is also a rapid increase in cytosolic Ca^{2+}, the evidence that this is mediated by a second type of glucagon receptor is tenuous, and it appears that it is secondary to the increase in cAMP. The glucagon receptor has not yet been purified to homogeneity, but has a binding subunit of 63 kDa. It exists in two affinity states and appears to be involved in the internalization and degradation of the hormone. Like other receptors mediating cAMP accumulation, the glucagon receptor is linked to adenylate cyclase through a stimulatory guanine nucleotide binding protein (G_s). This has a heterotrimeric ($\alpha\beta\gamma$) structure. There are different forms of G_s containing α-subunits of different M_r values. The activation of adenylate cyclase is thought to occur as follows. Interaction of glucagon with its receptor activates G_s by stimulating the dissociation of GDP and the binding of GTP to its α-subunit (α_s) causing dissociation of α_s-GTP from the inhibitory $\beta\gamma$ complex. The α_s-GTP subunit activates the catalytic subunit of adenylate cyclase. Activation is terminated by the GTPase activity of α_s which converts α_s-GTP to inactive α_s-GDP which reassociates with the $\beta\gamma$ complex.

The two major isozymic forms of cAMP-dependent protein kinase mediate all the known effects of cAMP in eukaryotes. In resting cells the kinase exists in a holoform consisting of a regulatory dimer and two catalytic subunits. cAMP activates the enzyme by binding to two sites on each subunit of the regulatory dimer. This is believed to cause a conformational change which frees the catalytic subunits from the inhibitory action of the regulatory dimer. The active catalytic subunits phosphorylate a variety of cellular proteins on serine or threonine residues contained within a preferred amino acid sequence.

Major targets of cAMP-dependent protein kinase in liver include the key enzymes of carbohydrate metabolism. The kinase phosphorylates and activates phosphorylase b kinase leading to phosphorylation and activation of phosphorylase b with resulting glycogen breakdown and glucose release. Liver phosphorylase b kinase has a tetrameric structure $(\alpha\beta\gamma\delta)_4$ and cAMP-dependent protein kinase phosphorylates both the α- and β-subunits leading to activation. Phosphorylase b kinase can also phosphorylate and inactivate glycogen synthase a, but the importance of this reaction is unclear. cAMP-dependent protein kinase can directly phosphorylate liver glycogen synthase a on several sites leading to inactivation, which explains the inhibition of hepatic glycogen synthesis by glucagon.

Glucagon also inhibits hepatic glycolysis and stimulates gluconeogenesis due to the actions of cAMP-dependent protein kinase on two rate-controlling enzymes in these pathways. The kinase phosphorylates and inactivates the L-type isozyme of pyruvate kinase leading to reduced glycolytic flux between P-enolpyruvate and pyruvate. It also phosphorylates the 6-phosphofructo 2-kinase/fructose 2,6-bisphosphatase bifunctional enzyme resulting in depression of its kinase activity and enhancement of its phosphatase activity. As a result, the level of fructose 2,6-P_2 is decreased. Since this compound is an activator of 6-phosphofructo 1-kinase and an inhibitor of fructose 1,6-bisphosphatase, its reduced concentration leads to decreased glycolytic flux from fructose 6-P to fructose 1,6-P_2 and a corresponding increased flux in the reverse direction.

Glucagon affects hepatic lipid metabolism. A major effect is inhibition of fatty acid synthesis, which is mainly due to the phosphorylation and inhibition of acetyl-CoA carboxylase by cAMP-dependent protein kinase. ATP-citrate lyase is also phosphorylated, but it is unclear that this is involved in the inhibition of lipogenesis. Glucagon also inhibits cholesterol synthesis apparently due to a decrease in the activity of hydroxymethylglutaryl-CoA reductase. This is thought to result from a decrease in the activity of protein phosphatase I due to the increased phosphorylation and activation of a heat stable inhibitor by cAMP-dependent protein kinase. This mechanism could also contribute to the effects of glucagon on other hepatic enzymes.

Glucagon exerts a ketogenic action on the liver which is more pronounced in insulin-deficient states. This action is thought to be due mainly to the inhibition of acetyl-CoA carboxylase with resulting decrease in malonyl-CoA. Malonyl-CoA is an inhibitor of carnitine acyltransferase I which is the rate-limiting step for mitochondrial fatty acid oxidation. A decrease in malonyl-CoA is thus postulated to lead to overproduction of acetyl-CoA which is then condensed to form ketone bodies.

Glucagon increases cytosolic Ca^{2+} in liver cells at near-physiological and higher concentrations. The effect is due to the mobilization of intracellular Ca^{2+} stores and to the opening of plasma membrane Ca^{2+} channels, but the evidence that it involves a novel glucagon receptor is dubious. On the other hand, the effect is mimicked by cAMP analogues and other agents that elevate cAMP. The intracellular mobilization can be explained by an increase in inositol trisphosphate (IP_3) that appears to result from cAMP-dependent phosphorylation of the guanine nucleotide binding protein (G_p) or some other protein controlling the phospholipase C hydrolyzing phosphatidylinositol 4,5-bisphosphate (PIP_2). The mechanism by which glucagon controls the Ca^{2+} channel is unknown. High concentrations of glucagon inhibit Ca^{2+} transport and (Ca^{2+}-Mg^{2+})-ATPase in liver membrane vesicles by a mechanism that does not involve cAMP, but is inhibited by cholera toxin. Interestingly, this is because glucagon is proteolyzed by the plasma membranes to yield a peptide fragment which potently inhibits the (Ca^{2+}-Mg^{2+})-ATPase. The physiological significance of this finding, which might involve other tissues, awaits further developments.

Glucagon markedly potentiates the hepatic actions of other Ca^{2+}-mobilizing hormones, e.g., vasopressin and epinephrine. This can be attributed to increased IP_3 production and Ca^{2+} mobilization, and also to the opening of more plasma membrane Ca^{2+} channels. On the other hand, the actions of glucagon and other Ca^{2+}-mobilizing hormones on liver cell Ca^{2+} fluxes are inhibited or abolished by tumor promoting phorbol esters. There is evidence that this inhibition is exerted at the level of G_p or on another factor involved in the control of PIP_2 phospholipase C and may involve protein kinase C.

Glucagon activates hepatic pyruvate dehydrogenase and α-oxoglutarate dehydrogenase and its action is well correlated with increases in mitochondrial Ca^{2+} content. Ca^{2+} ions can activate α-oxoglutarate dehydrogenase directly and pyruvate dehydrogenase indirectly through activation of its specific phosphatase. The elevation in mitochondrial Ca^{2+} is marked in the presence of other Ca^{2+}-mobilizing hormones and is probably secondary to the rise in cytosolic Ca^{2+}, although other mechanisms may be involved.

Glucagon has many other rapid actions on hepatic processes, but the molecular mechanisms involved are obscure. These actions and the possible mechanisms involved are described in Section 10.

In addition to its rapid effects, glucagon has some slower actions due to changes in protein synthesis. It stimulates the synthesis of the gluconeogenic enzyme P-enolpyruvate carboxykinase due to an increase in functional mRNA. The cAMP regulatory element and the structural features of the gene have been defined and the regulatory element has a core sequence that is very similar to sequences in the 5' flanking regions of many other genes. Although there is evidence that cAMP-dependent protein kinase is involved in the regulation of the gene, the mechanism remains to be defined.

References

1. Wakelam, M.J.O., Murphy, G.J., Hruby, V.J. and Houslay, M.D. (1986) Nature 323, 68–71.
2. Williamson, J.R., Hansen, C.A., Verhoeven, A., Coll, K.A., Johanson, R., Williamson, M.T. and Filburn, C. (1987) In: Cell Calcium and the Control of Membrane Transport (Eaton, D.C. and Mandel, L.J., eds.) pp. 93–116. Rockefeller Press, New York.
3. Charest, R., Blackmore, P.F., Berthon, B. and Exton, J.H. (1983) J. Biol. Chem. 258, 8769–8773.
4. Sistare, F.D., Picking, R.A. and Haynes, R.C. (1985) J. Biol. Chem. 260, 12744–12747.
5. Blackmore, P.F. and Exton, J.H. (1986) J. Biol. Chem. 261, 11056–11063.
6. Friedmann, N. and Park, C.R. (1968) Proc. Natl. Acad. Sci. U.S.A. 61, 504–508.
7. Blackmore, P.F., Assimacopoulos-Jeannet, F., Chan, T.M. and Exton, J.H. (1979) J. Biol. Chem. 254, 2828–2834.
8. Whiting, J.A. and Barritt, G.J. (1982) Biochem. J. 210, 73–77.
9. Mauger, J.-P. and Claret, M. (1986) FEBS Lett. 195, 106–110.
10. Kraus-Friedmann, N. (1986) Proc. Natl. Acad. Sci. U.S.A. 83, 8943–8946.
11. Rodbell, M., Krans, H.M.J., Pohl, S.L. and Birnbaumer, L. (1971) J. Biol. Chem. 246, 1861–1871.
12. Bataille, D., Freychet, P., Kitabgi, P.E. and Rosselin, G. (1973) FEBS Lett, 30, 215–218.
13. Bataille, D., Freychet, P. and Rosselin, G. (1974) Endocrinology 95, 713–721.
14. Giorgio, N.A., Johnson, C.B. and Blecher, M. (1974) J. Biol. Chem. 249, 428–437.
15. Levey, G.S., Fletcher, M.A., Klein, I., Ruiz, E. and Schenk, A. (1974) J. Biol. Chem. 249, 2665–2673.
16. Goldfine, I.D., Roth, J. and Birnbaumer, L. (1972) J. Biol. Chem. 247, 1211–1218.
17. Welton, A.F., Lad, P.M., Newby, A.C., Yamamura, H., Nicosia, S. and Rodbell, M. (1977) J. Biol. Chem. 252, 5947–5950.
18. Herberg, J.T., Codina, J., Rich, K.A., Rojas, F.J. and Iyengar, R. (1984) J. Biol. Chem. 259, 9285–9294.
19. Iyengar, R. and Herberg, J.T. (1984) J. Biol. Chem. 259, 5222–5229.
20. Rodbell, M., Birnbaumer, L., Pohl, S.L. and Junby, F. (1971) Proc. Natl. Acad. Sci. U.S.A. 68, 909–913.
21. Lin, M.C., Wright, D.E., Hruby, V.J. and Rodbell, M. (1975) Biochemistry 14, 1559–1563.
22. Wright, D.E. and Rodbell, M. (1979) J. Biol. Chem. 254, 268–269.
23. Wright, D.E., Hruby, V.J. and Rodbell, M. (1978) J. Biol. Chem. 253, 6338–6340.
24. Epand, R.M., Rosselin, G., Hoa, D.H.B., Cote, T.E. and Laburthe, M. (1981) J. Biol. Chem. 256, 1128–1132.
25. Sonne, O., Berg, T. and Christoffersen, T. (1978) J. Biol. Chem. 253, 3203–3210.
26. Horwitz, E.M., Jenkins, W.T., Hoosein, N.M. and Gurd, R.S. (1985) J. Biol. Chem. 260, 9307–9315.
27. Horwitz, E.M., Wyborski, R.J. and Gurd, R.S. (1986) J. Biol. Chem. 261, 13670–13676.
28. Rouer, E., Desbuquois, B. and Postel-Vinay, M.-C. (1980) Mol. Cell. Endocrinol. 19, 143–164.
29. Barazzone, P., Gorden, P., Carpentier, J.-L., Orci, L., Freychet, P. and Canivet, B. (1980) J. Clin. Invest. 66, 1081–1093.
30. Canivet, B., Gorden, P., Carpentier, J.-L., Orci, L. and Freychet, P. (1981) Mol. Cell. Endocrinol. 23, 311–320.
31. Rodbell, M., Krans, H.M.J., Pohl, S.L. and Birnbaumer, L. (1971) J. Biol. Chem. 246, 1872–1876.
32. Rodbell, M., Birnbaumer, L., Pohl, S.L. and Krans, H.M.J. (1971) J. Biol. Chem. 246, 1877–1882.
33. Gilman, A.G. (1984) Cell 36, 577–579.
34. Birnbaumer, L. and Pohl, S.L. (1973) J. Biol. Chem. 248, 2056–2061.
35. Rodbell, M., Lin, M.C. and Salomon, Y. (1974) J. Biol. Chem. 249, 59–65.
36. Kimura, N. and Shimada, N. (1980) FEBS Lett. 117, 172–174.
37. Stadel, J.M., DeLean, A. and Lefkowitz, R.J. (1982) Adv. Enzymol. 53, 1–43.
38. Mumby, S.M., Kahn, R.A., Manning, D.R. and Gilman, A.G. (1986) Proc. Natl. Acad. Sci. U.S.A. 83, 265–269.

39. Roof, D.J., Applebury, M.L. and Sternweis, P.C. (1985) J. Biol. Chem. 260, 16242–16249.
40. Northup, J.K., Smigel, M.D., Sternweis, P.C. and Gilman, A.G. (1983) J. Biol. Chem. 11369–11376.
41. Bokoch, G.M., Katada, T., Northup, J.K., Hewlett, E.L. and Gilman, A.G. (1983) J. Biol. Chem. 258, 2072–2075.
42. Cassel, D. and Pfeuffer, T. (1978) Proc. Natl. Acad. Sci. U.S.A. 75, 2669–2673.
43. Northup, J.K., Sternweis, P.C., Smigel, M.D., Schleifer, L.S., Ross, E.M. and Gilman, A.G. (1980) Proc. Natl. Acad. Sci. U.S.A. 77, 6516–6520.
44. Katada, T. and Ui, M. (1982) Proc. Natl. Acad. Sci. U.S.A. 79, 3129–3133.
45. Cassel, D. and Selinger, Z. (1978) Proc. Natl. Acad. Sci. U.S.A. 75, 4155–4159.
46. Northup, J.K., Sternweis, P.C. and Gilman, A.G. (1983) J. Biol. Chem. 258, 11361–11368.
47. Lefkowitz, R.J., Cerione, R., Codina, J., Birnbaumer, L. and Caron, M.G. (1985) J. Membrane Biol. 87, 1–12.
48. Cerione, R., Codina, J., Benovic, J.L., Lefkowitz, R.J., Birnbaumer, L. and Caron, M.G. (1984) Biochemistry 23, 4519–4525.
49. Schramm, M. and Selinger, Z. (1984) Science 225, 1350–1356.
50. Pfeuffer, E., Drehev, R.-M., Metzger, H. and Pfeuffer, T. (1985) Proc. Natl. Acad. Sci. U.S.A. 82, 3086–3090.
51. Smigel, M.D. (1986) J. Biol. Chem. 261, 1976–1982.
52. Yeager, R.E., Heidemann, W., Rosenberg, G.B. and Storm, D.R. (1985) Biochemistry 24, 3776–2783.
53. Seamon, K. and Daly, J.W. (1981) J. Biol. Chem. 256, 9799–9801.
54. May, D.C., Ross, E.M., Gilman, A.G. and Smigel, M.D. (1985) J. Biol. Chem. 260, 15829–15833.
55. Cerione, R.A., Sibley, D.R., Codina, J., Benovic, J.L., Winslow, J., Neer, E.J., Birnbaumer, L., Caron, M.G. and Lefkowitz, R.J. (1984) J. Biol. Chem. 259, 9979–9982.
56. Pedersen, S.E. and Ross, E.M. (1982) Proc. Natl. Acad. Sci. U.S.A. 79, 7228–7232.
57. Brandt, D.R., Asano, T., Pedersen, S.E. and Ross, E.M. (1983) Biochemistry 22, 4357–4362.
58. Exton, J.H. (1982) In: Handbook of Experimental Pharmacology, Vol. 58/II (Kebabian, J.W. and Nathanson, J.A., eds.) pp. 3–87. Springer Verlag, Berlin.
59. Exton, J.H., Cherrington, A.D., Hutson, N.J. and Assimacopoulos, F.D. (1977) In: Glucagon: Its Role in Physiology and Clinical Medicine (Foa, P.P., Bajaj, J.S. and Foa, N.L., eds.) pp. 321–347, Springer Verlag, Berlin.
60. Flockhart, D.A. and Corbin, J.D. (1982) CRC Crit. Rev. Biochem. 12, 133–186.
61. Rannels, S.R. and Corbin, J.D. (1980) J. Biol. Chem. 255, 7085–7088.
62. Rannels, S.R. and Corbin, J.D. (1981) J. Biol. Chem. 256, 7871–7876.
63. Corbin, J.D., Rannels, S.R., Robinson-Steiner, A.M., Tigani, M.C., Doskeland, S.O., Suva, R. and Miller, J.P. (1982) Eur. J. Biochem. 125, 259–266.
64. Robinson-Steiner, A.M. and Corbin, J.D. (1983) J. Biol. Chem. 258, 1032–1040.
65. Ogreid, D., Doskeland, S.O. and Miller, J.P. (1983) J. Biol. Chem. 258, 1041–1049.
66. Ogreid, D., Ekanger, R., Suva, R.H., Miller, J.P., Sturm, P., Corbin, J.D. and Doskeland, S.O. (1985) Eur. J. Biochem. 150, 219–227.
67. Beebe, S.J., Holloway, R., Rannels, S.R. and Corbin, J.D. (1984) J. Biol. Chem. 259, 3539–3547.
68. Beebe, S.J., Blackmore, P.F., Chrisman, T.D. and Corbin, J.D. (1988) Methods Enzymol. 159, 118–139.
69. Sugden, P.H., Holladay, L.A., Reimann, E.M. and Corbin, J.D. (1976) Biochem. J. 159, 409–422.
70. Chan, K.-F.J. and Graves, D.J. (1984) In: Calcium and Cell Function Vol. 5 (W.Y. Cheung, ed.) pp. 1–31. Academic Press, New York.
71. Chrisman, T.D., Jordan, J.E. and Exton, J.H. (1982) J. Biol. Chem. 257, 10798–10804.
72. Grand, R.J.A., Shenolikar, S. and Cohen, P. (1981) Eur. J. Biochem. 113, 359–367.
73. Reimann, E.M., Titani, K., Ericsson, L.H., Wade, R.D., Fischer, E.H. and Walsh, K.A. (1984) Biochemistry 23, 4185–4192.

74. Skuster, J.R., Chan, K.-F.J. and Graves, D.J. (1980) J. Biol. Chem. 255, 2203–2210.
75. Chan, K.-F.J. and Graves, D.J. (1982) J. Biol. Chem. 257, 5948–5955.
76. Kee, S.M. and Graves, D.J. (1986) J. Biol. Chem. 261, 4732–4737.
77. Cohen, P. (1973) Eur. J. Biochem. 34, 1–14.
78. Hayakawa, T., Perkins, J.P. and Krebs, E.G. (1973) Biochemistry 12, 574–580.
79. Picton, C., Klee, C.B. and Cohen, P. (1980) Eur. J. Biochem. 11, 553–561.
80. King, M.M., Carlson, G.M. and Haley, B.M. (1982) J. Biol. Chem. 257, 14058–14065.
81. Cooper, R.H., Sul, H.S. and Walsh, D.A. (1981) J. Biol. Chem. 256, 8030–8038.
82. Fischer, E.H., Heilmeyer, L.M.G. Jr. and Haschke, R.H. (1971) Curr. Top. Cell. Regulat. 4, 211–251.
83. Roach, P.J., DePaoli-Roach, A.A. and Larner, J. (1978) J. Cyclic Nucleotide Res. 4, 245–257.
84. Soderling, T.R., Srivastava, A.K., Bass, M.A. and Khatra, B.S. (1979) Proc. Natl. Acad. Sci. U.S.A. 76, 2536–2540.
85. Soderling, T.R., Hickenbottom, J.P., Reimann, E.M., Hunkeler, F.L., Walsh, D.A. and Krebs, E.G. (1970) J. Biol. Chem. 245, 6317–6328.
86. Jett, M.F. and Soderling, T.R. (1979) J. Biol. Chem. 254, 6739–6745.
87. Embi, N., Parker, P.J. and Cohen, P. (1981) Eur. J. Biochem. 115, 405–413.
88. Juhl, H., Sheorain, V.S., Schworer, C.M., Jett, M.F. and Soderling, T.R. (1983) Arch. Biochem. Biophys. 222, 518–526.
89. Sheorain, V.S., Corbin, J.D. and Soderling, T.R. (1985) J. Biol. Chem. 260, 1567–1572.
90. Picton, C., Aitken, A., Bilham, T. and Cohen, P. (1982) Eur. J. Biochem. 124, 37–45.
91. Imazu, M., Strickland, W.G., Chrisman, T.D. and Exton, J.H. (1984) J. Biol. Chem. 259, 1813–1821.
92. Imazu, M., Strickland, W.G. and Exton, J.H. (1984) Biochim. Biophys. Acta 789, 285–293.
93. Parker, P.J., Aitken, A., Bilham, T., Embi, N. and Cohen, P. (1981) FEBS Lett. 123, 332–336.
94. Rylatt, D.B., Aitken, A., Bilham, T., Condon, G.D., Embi, N. and Cohen, P. (1980) Eur. J. Biochem. 107, 529–537.
95. Roach, P.J. (1981) Curr. Top. Cell. Regulat. 20, 45–105.
96. Brown, J.H., Thompson, B. and Mayer, S.E. (1977) Biochemistry 16, 5501–5508.
97. Brown, D.F., Reimann, E.M. and Schlender, K.K. (1980) Biochim. Biophys. Acta 612, 352–360.
98. Camici, M., DePaoli-Roach, A.A. and Roach, P.J. (1984) J. Biol. Chem. 259, 3429–3434.
99. Camici, M., DePaoli-Roach, A.A. and Roach, P.J. (1982) J. Biol. Chem. 257, 9898–9901.
100. Kaslow, H.R., Lesikar, D.D., Antwi, D. and Tan, A.W.H. (1985) J. Biol. Chem. 260, 9953–9956.
101. Akatsuka, A., Singh, T.J., Nakabayashi, H., Lin, M.C. and Huang, K.-P. (1985) J. Biol. Chem. 260, 3239–3242.
102. Bosch, F., Ciudid, C.J. and Guinovart, J.J. (1983) FEBS Lett. 151, 76–78.
103. Ciudad, C., Camici, M., Ahmad, Z., Wang, Y., DePaoli-Roach, A.A. and Roach, P.J. (1984) Eur. J. Biochem. 142, 511–520.
104. Lungstrom, O., Hjelmquist, G. and Engstrom, L. (1974) Biochim. Biophys. Acta 358, 289–298.
105. Ekman, P., Dahlqvist, U., Humble, E. and Engstrom, L. (1975) Biochim. Biophys. Acta 429, 374–382.
106. El-Maghrabi, M.R., Haston, W.S., Flockhart, D.A., Claus, T.H. and Pilkis, S.J. (1980) J. Biol. Chem. 255, 668–675.
107. Hjelmquist, G., Andersson, J., Edlund, B. and Engstrom, L. (1974) Biochem. Biophys. Res. Commun. 61, 559–563.
108. Feliu, J.E., Hue, L. and Hers, H.-G. (1976) Proc. Natl. Acad. Sci. U.S.A. 73, 2762–2766.
109. Pilkis, S.J., Riou, J.-P. and Claus, T.H. (1976) J. Biol. Chem. 261, 7841–7852.
110. Blair, J.B., Cimbala, M.A., Foster, J.L. and Morgan, R.H. (1976) J. Biol. Chem. 251, 3756–3762.
111. Van Berkel, T.J.C., Kruijt, J.K. and Koster, J.F. (1977) Eur. J. Biochem. 81, 423–432.
112. Hue, L., Blackmore, P.F. and Exton, J.H. (1981) J. Biol. Chem. 256, 8900–8903.
113. Van Schaftingen, E., Hue, L. and Hers, H.G. (1980) Biochem. J. 192, 887–895.

114. Van Schaftingen, E., Hue, L. and Hers, H.G. (1980) Biochem. J. 192, 897–901.
115. Pilkis, S.J., El-Maghrabi, M.R., Pilkis, J., Claus, T.H. and Cumming, D.A. (1981) J. Biol. Chem. 256, 3171–3174.
116. Van Schaftingen, E., Jett, M.F., Hue, L. and Hers, H.G. (1981) Proc. Natl. Acad. Sci. U.S.A. 86, 3483–3486.
117. Van Schaftingen, E., Davies, E.R. and Hers, H.G. (1981) Biochem. Biophys. Res. Commun. 103, 362–368.
118. El-Maghrabi, M.R., Claus, T.H., Pilkis, J. and Pilkis, S.J. (1982) Proc. Natl. Acad. Sci. U.S.A. 79, 315–319.
119. Furyua, E., Yokoyama, M. and Ueda, K. (1982) Proc. Natl. Acad. Sci. U.S.A. 79, 325–329.
120. El-Maghrabi, M.R., Claus, T.H., Pilkis, J., Fox, E. and Pilkis, S.J. (1982) J. Biol. Chem. 257, 7603–7606.
121. Van Schaftingen, E., Davies, D.R. and Hers, H.G. (1982) Eur. J. Biochem. 124, 143–149.
122. Pilkis, S.J., Regen, D.M., Stewart, H.B., Pilkis, J., Pate, T.M. and El-Maghrabi, M.R. (1984) J. Biol. Chem. 259, 949–958.
123. Cook, G.A., Nielsen, R.C., Haskins, R.A., Mehlman, M.A., Lakshamanan, M.R. and Veech, R.L. (1977) J. Biol. Chem. 252, 4421–4424.
124. Geelen, M.J.H., Benjamin, A.C., Christiansen, R.Z., Lepreau-Jose, M.J. and Gibson, D.M. (1978) FEBS Lett. 95, 326–330.
125. Witters, L.A., Kowoloff, E.M. and Avruch, J. (1979) J. Biol. Chem. 254, 245–248.
126. Holland, R., Witters, L.E. and Hardie, D.G. (1984) Eur. J. Biochem. 140, 325–333.
127. Janski, A.M., Srere, P.A., Cornell, N.W. and Veech, R.L. (1979) J. Biol. Chem. 254, 9365–9368.
128. Alexander, M.C., Palmer, J.L., Pointer, R.H., Koumijian, L. and Avruch, J. (1981) Biochim. Biophys. Acta 674, 37–47.
129. Ramakrishna, S., Pucci, D.L. and Genjamin, W.B. (1981) J. Biol. Chem. 256, 10213–10216.
130. Houston, B. and Nimmo, H.G. (1985) Biochim. Biophys. Acta 844, 233–239.
131. Beg, Z.H., Allman, D.W. and Gibson, D.M. (1973) Biochem. Biophys. Res. Commun. 54, 1362–1369.
132. Geelen, M.J.H., Harris, R.A., Beynen, A.C. and McCune, S.A. (1980) Diabetes 29, 1006–1022.
133. Ingebretsen, T.S., Geelen, M.J.H., Parker, R.A., Evenson, K.J. and Gibson, D.M. (1979) J. Biol. Chem. 254, 9986–9989.
134. Ingebretsen, T.S., Parker, R.A. and Gibson, D.M. (1981) J. Biol. Chem. 256, 1138–1144.
135. Ingebretsen, T.S., Blair, J., Guy, P., Witters, L. and Hardie, D.G. (1983) Eur. J. Biochem. 132, 275–281.
136. Huang, F.L. and Glinsmann, W.H. (1975) Proc. Natl. Acad. Sci. U.S.A. 72, 3004–3008.
137. Huang, F.L. and Glinsmann, W.H. (1976) Eur. J. Biochem. 70, 419–426.
138. Murphy, E., Coll, K., Rich, T.L. and Williamson, J.R. (1980) J. Biol. Chem. 255, 6600–6608.
139. Charest, R., Prpic, V., Exton, J.H. and Blackmore, P.F. (1985) Biochem. J. 227, 79–90.
140. Assimacopoulos-Jeannet, F.D., Blackmore, P.F. and Exton, J.H. (1977) J. Biol. Chem. 252, 2662–2669.
141. Blackmore, P.F., Brumley, F.T., Marks, J.L. and Exton, J.H. (1978) J. Biol. Chem. 253, 4851–4858.
142. Chan, T.M. and Exton, J.H. (1978) J. Biol. Chem. 253, 6393–6400.
143. Combettes, L., Berthon, B., Binet, A. and Claret, M. (1986) Biochem. J. 237, 675–683.
144. Mauger, J.-P., Poggioli, J. and Claret, M. (1985) J. Biol. Chem. 260, 11635–11642.
145. Poggioli, J., Mauger, J.-P. and Claret, M. (1986) Biochem. J. 235, 663–669.
146. Corvera, S., Huerta-Bahena, J., Pelton, J.T., Hruby, V.J., Trivedi, D. and Garcia-Sainz, J.A. (1984) Biochim. Biophys. Acta 804, 434–441.
147. Lynch, C.J., Blackmore, P.F., Charest, R. and Exton, J.H. (1985) Mol. Pharmacol. 28, 93–99.
148. Whipps, D.E., Arston, A.E., Pryor, H.J. and Halestrap, A.P. (1987) Biochem. J. 241, 835–845.
149. Bocckino, S.B., Blackmore, P.F. and Exton, J.H. (1985) J. Biol. Chem. 260, 14201–14207.

150. Lotersztajn, S., Epand, R.M., Mallat, A. and Pecker, F. (1984) J. Biol. Chem. 259, 8195–8201.
151. Lotersztajn, S., Mallat, A., Pavoine, C. and Pecker, F. (1985) J. Biol. Chem. 260, 9692–9698.
152. Lotersztajn, S., Pavoine, C., Mallat, A., Stengel, D., Insel, P.A. and Pecker, F. (1987) J. Biol. Chem. 262, 3114–3117.
153. Mallat, A., Pavoine, C., Dufour, M., Lotersztajn, S., Bataille, D. and Pecker, F. (1987) Nature 325, 620–622.
154. Cachelin, A.B., dePeyer, J.E., Kokubun, S. and Reuter, H. (1983) Nature 304, 462–464.
155. Curtis, B.M. and Catterall, W.A. (1985) Proc. Natl. Acad. Sci. U.S.A. 82, 2528–2532.
156. Hosey, M.M., Borsotto, M. and Lazdunski, M. (1986) Proc. Natl. Acad. Sci. U.S.A. 83, 3733–3737.
157. Pohl, S., Birnbaumer, L. and Rodbell, M. (1971) J. Biol. Chem. 246, 1849–1856.
158. Morgan, N.G., Charest, R., Blackmore, P.F. and Exton, J.H. (1984) Proc. Natl. Acad. Sci. U.S.A. 81, 4208–4212.
159. Assimacopoulos-Jeannet, F.D., Blackmore, P.F. and Exton, J.H. (1982) J. Biol. Chem. 257, 3759–3765.
160. Morgan, N.G., Blackmore, P.F. and Exton, J.H. (1983) J. Biol. Chem. 258, 5110–5116.
161. Staddon, J.M. and Hansford, R.G. (1986) Biochem. J. 238, 737–743.
162. Lynch, C.J., Charest, R., Bocckino, S.B. and Exton, J.H. (1986) J. Biol. Chem. 260, 2844–2851.
163. Cooper, R.H., Coll, K.E. and Williamson, J.R. (1985) J. Biol. Chem. 260, 3281–3288.
164. Cook, G.A., Nielsen, R.C., Hawkins, R.A., Mehlman, M.A., Lakshamanan, M.R. and Veech, R.L. (1977) J. Biol. Chem. 252, 4421–4424.
165. McGarry, J.D., Mannaerts, G.P. and Foster, D.W. (1977) J. Clin. Invest. 60, 265–270.
166. McGarry, J.D. and Foster, D.W. (1979) J. Biol. Chem. 254, 8163–8168.
167. Harano, Y., Kashiwagi, A., Kojima, H., Suzuki, M., Hashimoto, T. and Shigeta, Y. (1985) FEBS Lett. 188, 267–272.
168. Heimberg, M., Weinstein, I. and Kohout, M. (1969) J. Biol. Chem. 244, 5131–5139.
169. Tiengo, A. and Nosadini, R. (1983) In: Handbook of Experimental Pharmacology Vol. 66/I (Lefebvre, P.J., ed.) pp. 441–451. Springer Verlag, Berlin.
170. Exton, J.H. and Park, C.R. (1972) In: Handbook of Physiology, Sect. 7, Vol. 1, Endocrine Pancreas (Steiner, D.F. and Freinkel, N., eds.) pp. 437–455. American Physiological Society, Washington, DC.
171. Lefebvre, P.J. (1983) In: Handbook of Experimental Pharmacology Vol. 66/I (Lefebvre, P.J., ed.) pp. 419–440. Springer Verlag, Berlin.
172. Ashford, T.P. and Porter, K.R. (1962) J. Cell Biol. 12, 198–202.
173. Schworer, C.M. and Mortimore, G.E. (1979) Proc. Natl. Acad. Sci. U.S.A. 76, 3169–3173.
174. Ui, M., Exton, J.H. and Park, C.R. (1973) J. Biol. Chem. 248, 5350–5359.
175. Siess, E.A. and Wieland, O.H. (1978) FEBS Lett. 93, 301–306.
176. Mallette, L.E., Exton, J.H. and Park, C.R. (1969) J. Biol. Chem. 244, 5724–5728.
177. Moule, S.K., Bradford, N.M. and McGivan, J.D. (1987) Biochem. J. 241, 737–743.
178. LeCam, A. and Freychet, P. (1976) Biochem. Biophys. Res. Commun. 72, 893–901.
179. Fehlman, M., LeCam, A. and Freychet, P. (1979) J. Biol. Chem. 254, 10431–10437.
180. Kilberg, M.S., Barber, E.F. and Handlogten, M.E. (1985) Curr. Topics Cell. Regulat. 25, 133–163.
181. Williamson, J.R., Browning, E.T., Thurman, R.G. and Scholz, R. (1969) J. Biol. Chem. 244, 5055–5064.
182. Exton, J.H., Corbin, J.G. and Harper, S.C. (1972) J. Biol. Chem. 247, 4996–5003.
183. LaNoue, K.F., Strzelecki, T. and Finch, F. (1984) J. Biol. Chem. 259, 4116–4121.
184. Halestrap, A.P., Quinlan, P.T., Armston, A.E. and Whipps, D.E. (1985) In: Achievements and Perspectives in Mitochondrial Research, Vol. 1 (Quagliariello, E., ed.) pp. 469–480. Elsevier, Amsterdam.
185. Jensen, C.B., Sistare, F.D., Hamman, H.C. and Haynes, R.C. Jr. (1983) Biochem. J. 210, 819–827.
186. Haynes, R.C. and Pilking, R.A. (1984) J. Biol. Chem. 259, 13228–13234.

187. Oviasu, O.A. and Whitton, P.D. (1984) Biochem. J. 224, 181–186.
188. Assimacopoulos-Jeannet, F., McCormack, J.G. and Jeanrenaud, B. (1983) FEBS Lett. 159, 83–88.
189. Assimacopoulos-Jeannet, F., McCormack, J.G. and Jeanrenaud, B. (1986) J. Biol. Chem. 261, 8799–8804.
190. Adam, P.A.J. and Haynes, R.C. Jr. (1969) J. Biol. Chem. 244, 6444–6450.
191. Allan, E.H., Chisholm, A.B. and Titheradge, M.A. (1983) Biochem. J. 212, 417–426.
192. Denton, R.M. and McCormack, J.G. (1985) Am. J. Physiol. 249, E543–E554.
193. Blackmore, P.F., Dehaye, J.-P. and Exton, J.H. (1979) J. Biol. Chem. 254, 6945–6950.
194. Reinhart, P.H., Taylor, W.M. and Bygrave, F.L. (1982) Biochem. J. 208, 619–630.
195. Babcock, D.F., Chen, J.-L.J., Yip, B.P. and Lardy, H.A. (1979) J. Biol. Chem. 254, 8117–8120.
196. Joseph, S.K. and Williamson, J.R. (1983) J. Biol. Chem. 259, 10425–10432.
197. Baddams, H.M., Chang, L.B.F. and Barritt, G.J. (1983) Biochem. J. 210, 73–77.
198. Staddon, J.M. and Hansford, R.G. (1987) Biochem. J. 241, 729–735.
199. Williams, T.F., Exton, J.H., Friedmann, N. and Park, C.R. (1971) Am. J. Physiol. 221, 1645–1651.
200. Friedmann, N., Somlyo, A.V. and Somlyo, A.P. (1971) Science 171, 400–402.
201. Friedmann, N. (1972) Biochim. Biophys. Acta 274, 214–225.
202. Petersen, O.H. (1974) J. Physiol. 239, 647–656.
203. Kraus-Friedmann, N., Hummel, L., Radominska-Pyrek, A., Little, J.M. and Lester, R. (1982) Mol. Cell. Biochem. 44, 173–180.
204. Ihlenfeldt, M.J.A. (1981) J. Biol. Chem. 256, 2213–2218.
205. Haylett, D.G. (1976) Br. J. Pharmacol. 57, 158–160.
206. Jenkinson, D.H., Haylett, D.G. and Koller, K. (1978) In: Cell Membrane Receptors for Drugs and Hormones: A Multidisciplinary Approach (Bolis, L. and Straub, R.W., eds.) pp. 89–105. Raven Press, New York.
207. Weiss, S.J. and Putney, J.W. Jr. (1978) J. Pharmacol. Exp. Ther. 207, 669–676.
208. Burgess, G.M., Claret, M. and Jenkinson, D.H. (1981) J. Physiol. 317, 67–90.
209. Rosenfeld, M.G. and Barrieux, A. (1979) Adv. Cyclic Nucleotide Res. 11, 205–264.
210. Shrago, E., Lardy, H.A., Nordlie, R.C. and Foster, D.O. (1963) J. Biol. Chem. 238, 3188–3192.
211. Yeung, D. and Oliver, I.T. (1968) Biochemistry 7, 3231–3239.
212. Wicks, W.D. (1969) J. Biol. Chem. 252, 7202–7213.
213. Gunn, T.M., Tilghman, S.M., Hanson, R.W., Reshef, L. and Ballard, F.J. (1975) Biochemistry 14, 2350–2357.
214. Iynedjian, P.B. and Hanson, R.W. (1977) J. Biol. Chem. 252, 655–662.
215. Beale, E.G., Katzen, C.S. and Granner, D.K. (1981) Biochemistry 20, 4878–4883.
216. Cimbala, M.A., Lamers, W.H., Nelson, K., Monahan, J.E., Yoo-Warren, H. and Hanson, R.W. (1982) J. Biol. Chem. 257, 7629–7636.
217. Beale, E.G., Hartley, J.L. and Granner, D.K. (1982) J. Biol. Chem. 257, 2022–2028.
218. Salavert, A. and Iynedjian, P.B. (1982) J. Biol. Chem. 257, 13404–13412.
219. Chrapkiewitz, N.B., Beale, E.G. and Granner, D.K. (1982) J. Biol. Chem. 257, 14428–14432.
220. Lamers, W.H., Hanson, R.W. and Meisner, H.M. (1982) Proc. Natl. Acad. Sci. U.S.A. 79, 5137–5141.
221. Sasaki, K., Cripe, T.P., Koch, S.R., Andreone, T.L., Petersen, D.P., Beale, E.G. and Granner, D.K. (1984) J. Biol. Chem. 259, 15242–15251.
222. Wynshaw-Boris, A., Lugo, T.G., Short, J.M., Fourmier, R.E.K. and Hanson, R.W. (1984) J. Biol. Chem. 259, 12161–12169.
223. Short, J.M., Wynshaw-Boris, A., Short, H.P. and Hanson, R.W. (1986) J. Biol. Chem. 261, 9721–9726.
224. Wynshaw-Boris, A., Short, J.M. and Hanson, R.W. (1986) Bio Techniques 4, 104–119.
225. Beebe, S.J., Blackmore, P.F., Segaloff, D.L., Koch, S.R., Burks, D., Limbird, L.E., Granner, D.K. and Corbin, J.D. (1986) In: Hormones and Cell Regulation, Colloque INSERM Vol. 139, (Nunez, J., ed.) pp. 159–180. John Libbey Eurotext Ltd., London.

CHAPTER 13

Mechanism of action of growth hormone

MICHAEL WALLIS

Biochemistry Laboratory, School of Biological Sciences, University of Sussex, Falmer, Brighton BN1 9QG, Sussex, England

1. The growth hormone-prolactin family

Growth hormone (somatotropin; GH) is a protein hormone produced in specific cells (somatotrophs) of the pituitary gland. It comprises a single polypeptide chain of about 190 amino acids which folds, with formation of two disulphide bridges, to a compact tertiary structure (see also Addendum, p. 289). Amino acid sequences have been determined for GHs from several species [1,2] and these reveal a considerable amount of species variation; in particular, human GH shows extensive differences from the GHs of non-primate mammals, and this has been interpreted as indicating a rapid rate of evolution for the GH gene in the primates [3,4]. Differences in biological properties between human and non-primate GHs have also been observed, and will be considered later.

The amino acid sequences of GHs are homologous with those of prolactins [1,5]. Prolactin is also a protein hormone from the pituitary gland, produced in cells called lactotrophs. It comprises a single polypeptide chain of about 200 amino acid residues, which in some species, including man, may be partially glycosylated. As in the case of GH, prolactin has a compact tertiary structure, but contains three intrachain disulphide bridges, two of which are in homologous positions to those in GH. Primary structures have been determined for prolactins from various species; again there is considerable variation between species, and rates of evolution appear to have varied [3,6], but here it is the rodent hormones which stand out as being particularly different from those of other mammals. Comparison of the sequence of prolactin from any species with that of the corresponding GH reveals homology, with identical residues at about 25% of all positions [1]; such a degree of homology is highly significant and indicates that GH and prolactin are members of an evolutionary related family and probably arose as the consequence of duplication of a common ancestral gene, possibly early in chordate evolution.

Other members of the GH-prolactin family include the placental lactogens and a prolactin-like protein, proliferin. A placental lactogen is produced in large amounts by the human placenta. It is structurally closely related to human GH, but its growth-promoting activity, at least in the rat, is low. Its precise physiological role is not clear – possible functions include a role in the regulation of fetal growth. However, in at least one case, a woman who lacked placental lactogen due to a gene deletion gave rise to a perfectly normal child [7], suggesting that the protein may be of limited importance, at least under conditions of good nutrition. Placental lactogens have also been isolated and characterized from several non-primate mammals [8–11]. In rats and mice there are at least two placental lactogen-like proteins [8,11], whose amino acid sequences resemble that of prolactin more closely than that of GH, confirming earlier ideas [12] that the placental lactogens of lower mammals arose as the result of a gene duplication quite distinct from that which gave rise to the primate placental lactogen.

Another recently discovered member of the GH-prolactin family is the prolactin-like protein proliferin. This was first shown to to produced by actively proliferating (but not quiescent) mouse fibroblasts [13] but is now known to be present also in the mouse placenta, although it is distinct from the mouse placental lactogens.

2. Growth hormone and the control of somatic growth

The characteristic action of GH is the promotion of somatic growth. Animals and humans which lack GH show stunted growth, which can be restored by administering the hormone. Animals which have excessive GH display giantism. In man, after the epiphyses of the long bones have fused, excess GH causes acromegaly. Somatic growth is a complex phenomenon, however, which involves, directly or indirectly, many metabolic processes, and which is influenced by many factors, including nutrition and several different hormones. Because of this, and because somatic growth is inevitably a rather slow process, the mechanism of action of GH at the biochemical and physiological levels is not well understood.

GH has been shown to have many different metabolic effects. Which of these contribute directly to its growth-promoting activity is not clear. The hormone stimulates synthesis of proteins and nucleic acids in a wide range of tissues. It has actions on carbohydrate metabolism, including both diabetogenic and insulin-like actions, and also has effects on lipid metabolism, particularly in man. Overall, GH tends to switch metabolism from utilization of carbohydrate to utilization of fat [14], although there are marked differences between species in the relative importance of actions on different aspects of metabolism.

In addition to these various actions on rather general metabolic functions, GH also has some very specific effects. It has a general stimulatory effect on protein synthesis but also stimulates synthesis of some proteins selectively. Predominant among these is somatomedin C/insulin-like growth factor I (Sm-C/IGF-I), production of which is stimulated markedly in many tissues, especially the liver [15,16].

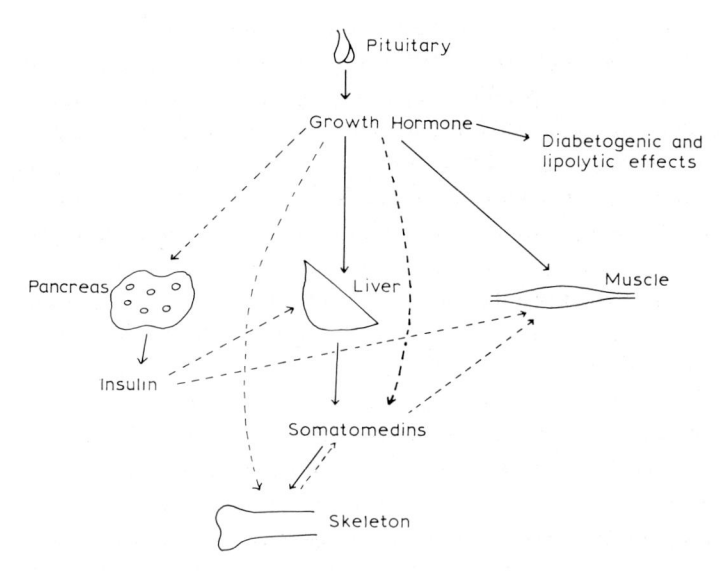

Fig. 1. An overview of the actions of growth hormone. Some of the effects of the hormone are direct, others are mediated by somatomedins.

Somatomedin C probably mediates many of the actions of GH. The hormone also has specific effects on the proliferation or differentiation of many cells, including fibroblasts and chondrocytes.

Consideration will be given to each of the many actions of GH in the following pages. Progress has been made in understanding the biochemistry of several of them, but we are still far from being able to explain precisely how they combine to give the overall effects of the hormone on somatic growth. Fig. 1 gives a simplified overview of current ideas.

3. Receptors for growth hormone

3.1. Distribution of growth hormone receptors

Sites which bind labelled GH are found in many different tissues, and in most cases a substantial proportion of the binding is specific (displaceable by unlabelled hormone) [17,18]. Specific binding sites for GH are found in particularly high concentrations in liver membranes, and this has been the most favoured source for detailed studies. The liver is a target organ for many actions of GH, including somatomedin production (see below), and this and various other factors support the idea that the binding sites are true receptors for the hormone, and they will be referred to as such here. However, it has not been formally demonstrated that these binding sites mediate the biological actions of GH, and some evidence suggests that at least some of the sites in liver do not mediate growth-promoting actions.

In many species the liver also contains receptors for lactogenic hormones [17,18]. Human GH, unlike GHs from non-primates, has both lactogenic and somatogenic properties and, since labelled human GH has been used widely for binding studies, this has led to some confusion. True GH receptors (somatogenic receptors) should bind somatogenic hormones (GHs) but not prolactins. Liver membranes prepared from the pregnant rabbit contain a high concentration of such receptors and have been studied intensively. Even here, however, the situation is complex, since GHs from some species bind rather poorly to the receptors while some prolactins can displace labelled GH. The explanation may lie partly in the presence of multiple receptor types (see below).

A large proportion of the GH receptors in rabbit liver is membrane-bound, though by no means all are associated with the plasma membrane. The role of receptors associated with intracellular membranes (including Golgi apparatus and endoplasmic reticulum) is not clear. They could represent newly-synthesized receptors en route for the plasma membrane or internalized receptors being recycled within the cell. There appears to be no significant difference between the binding properties of GH receptors associated with plasma membrane and those in intracellular membranes [19]. In addition to the membrane-bound receptors, rabbit liver also contains a soluble, cytosolic GH-binding protein which appears to be related to membrane-associated receptors [20]. Such a soluble form of a receptor is unusual for a polypeptide hormone, though a similar situation appears to exist in the case of prolactin [21]. The biological role of such soluble, cytosolic receptors is not clear, though it is interesting that they also occur in blood plasma in several species, including rabbit and man [22,23].

Although there has been emphasis on rabbit liver, GH receptors have been studied in liver from many other species and in many other tissues and cultured cell lines [17,18]. The human lymphocyte-derived cell line, IM9, has provided a useful source, though the physiological significance of binding of GH to these cells is not clear. The binding and structural properties of GH receptors vary considerably according to the tissue and species from which they are derived.

3.2. Heterogeneity of growth hormone receptors

Much evidence suggests that membrane-bound GH-binding proteins/receptors in rabbit liver are heterogeneous. The binding specificity is complex and is probably best explained in terms of multiple receptors, some of which are highly selective for rabbit GH, while others discriminate relatively poorly between GHs and prolactins from various species and bind rabbit GH poorly (Fig. 2) [18,19,24–26]. Studies using monoclonal antibodies to human GH also provide evidence for multiple receptors [27], as do cross-linking studies using ^{125}I-labelled hGH [28,29] and the binding properties of a shortened version of human GH, the 20K variant. The most defin-

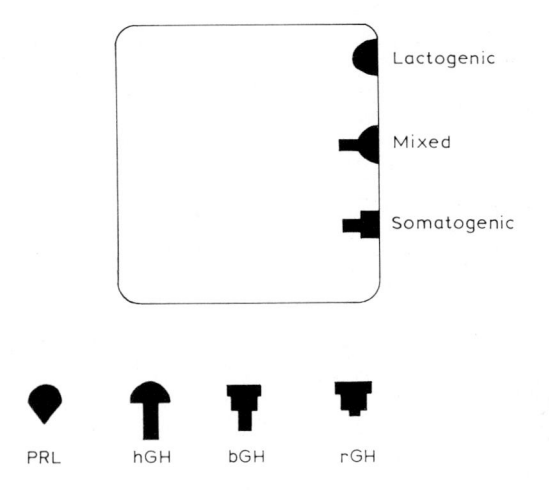

Fig. 2. Diagram illustrating the possible nature of the growth hormone and prolactin binding sites on the hepatocyte of the pregnant rabbit.

itive evidence for heterogeneity of receptors, however, is provided by the work of Barnard et al. [30,31], who have prepared a panel of monoclonal antibodies to rabbit liver GH receptors. The interaction of these with membrane-bound and soluble rabbit liver preparations provides evidence for three distinct sub-types of the GH receptor, which appear to be structurally related in that they share some common antigenic epitopes, but distinct in that some epitopes are not found in all three receptor types. The three receptor types are all found in microsomal membrane preparations from rabbit liver, but soluble receptors contain only two of the subtypes. How these multiple receptors detected using monoclonal antibodies relate to the heterogeneity revealed by other techniques is not clear. The biological significance of the multiple receptor types is unknown.

3.3. Structure and purification of growth hormone receptors

Information about the structure of the GH receptor has been provided by cross-linking studies using ^{125}I-labelled GH and various cross-linking agents, especially disuccinimidyl suberate. Cross-linking of ^{125}I-labelled human GH to rabbit liver receptors followed by solubilization and fractionation by SDS polyacrylamide gel electrophoresis revealed a major hormone-receptor complex of M_r about 80 000 [18,28,29,32,33] (Fig. 3). If a single hormone molecule per complex is assumed, this indicates a receptor protein component of about M_r 58 000 (Haeuptle et al. [34] identified an ^{125}I-labelled receptor component of rather higher M_r, Hughes et al. [35] a cross-linked component of rather lower M_r). Whether this corresponds to the entire receptor or a subunit of it is not clear. When the rabbit liver receptor was solubilized using detergent (Triton X-100) and then subjected to gel filtration, a

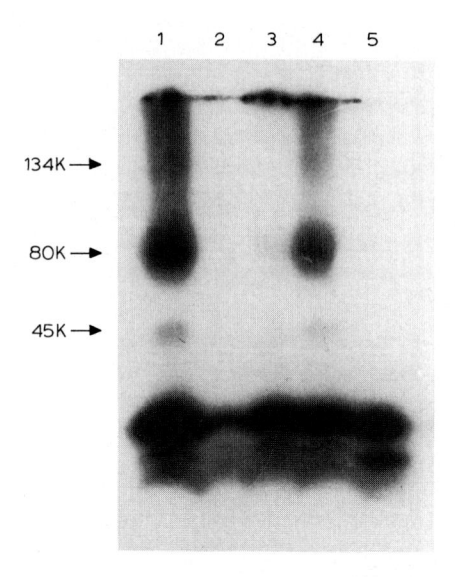

Fig. 3. Cross-linking of ^{125}I-labelled human GH to rabbit liver receptors. Labelled GH was incubated with receptors, cross-linked with disuccinimidyl suberate and then subjected to SDS polyacrylamide gel electrophoresis. 1. Total binding; 2–4. Displacement with 1 μg unlabelled human GH, bovine GH and ovine prolactin respectively; 5. ^{125}I-human GH without receptors. Note the major GH-receptor complex of $M_r \approx 80\,000$ (indicating a receptor protein or subunit of M_r 58 000) and various other components of larger and smaller M_r.

component of $M_r \approx 300\,000$ was detected [36] though this may represent an overestimate of the M_r due to associated detergent. In addition to the M_r 58 000 component, minor GH receptor components have been detected in rabbit liver preparations by cross-linking [28,29]. The precise pattern of cross-linked products varies somewhat when soluble or membrane-associated membrane preparations are compared [28]. Specificity studies did not reveal any marked differences in binding of GHs or prolactins from various species to the several receptor-components revealed by cross-linking [29].

Cross-linking has also been used to detect GH receptors/binding sites in other tissues and species. The main receptor component detected in rat liver and the human IM9 cell line (which binds human GH specifically) is rather larger ($M_r \approx 100\,000$) than that seen in rabbit liver [35,37]. Multiple cross-linked GH-receptor components have been detected in rat hepatocytes, but these are probably all related to aggregates of glycoprotein subunits of M_r 100 000 held together by disulphide bonds and non-covalent interactions [37].

Partial purification of membrane-bound GH receptors from rabbit liver has been achieved after solubilization with detergents [36,38]. Affinity chromatography on immobilized GH or lectins proved a particularly powerful way of isolating the receptors. However, antibodies raised against the partially-purified receptors failed to block the growth-promoting actions of the hormone.

3.4. Signal transduction following binding of growth hormone to its receptor

Little is known of the intracellular events that immediately follow binding of GH to its receptor. Some evidence suggests that inhibition of adenylate cyclase may be involved in mediating the actions of GH [26,39], but others have failed to obtain clear-cut effects [18]. GH has been reported to increase the activity of guanylate cyclase in a number of different tissues [40] though no effects were found on the content of this cyclic nucleotide in diaphragm [41]. Some evidence has been provided that the activated GH receptor may act as a protein kinase, but again others have obtained negative results. Binding of GH to rat hepatocytes leads to rapid (<10 min) phosphorylation of several cellular proteins, including one of M_r 46000, but the significance of this is not yet clear [42]. Overall it must be concluded that there is no consensus regarding second messengers or other mechanisms that may mediate the intracellular actions of GH.

3.5. Regulation of growth hormone receptor levels

The responsiveness of a target tissue to the actions of GH, or any other hormone, may be regulated by altering the number of receptors in that tissue. Levels of GH receptors in various tissues do vary; for example, hepatic GH binding increases markedly at puberty in many species, and also during pregnancy [18]. The factors that lead to such changes in receptor levels are poorly understood, though roles for insulin, oestrogens and GH itself in stimulating GH levels have been proposed [18]. Although GH may play an important role in stimulating levels of its own receptors in the long run, in short-term experiments it can lead to down-regulation of receptors, a phenomenon which has been studied particularly using the IM9 human lymphocyte-derived line. Down-regulation of receptors may also explain the refractory period which often follows stimulation of diaphragm or other tissues in vitro by GH [41].

The mechanisms whereby GH and other factors modulate receptor levels is not known. In medium- or long-term experiments use of protein synthesis inhibitors usually blocks such modulation, suggesting that de novo synthesis of receptors may be required, although involvement of other short-lived proteins cannot be ruled out. Rapid increase in receptor levels probably does not require synthesis of new receptors, and may be achieved by insertion of receptors that were previously stored intracellularly into the plasma membrane, and/or by activation of receptor proteins already in the plasma membrane in an inactive form. Down-regulation of receptors probably involves endocytosis, possibly followed by degradation [18].

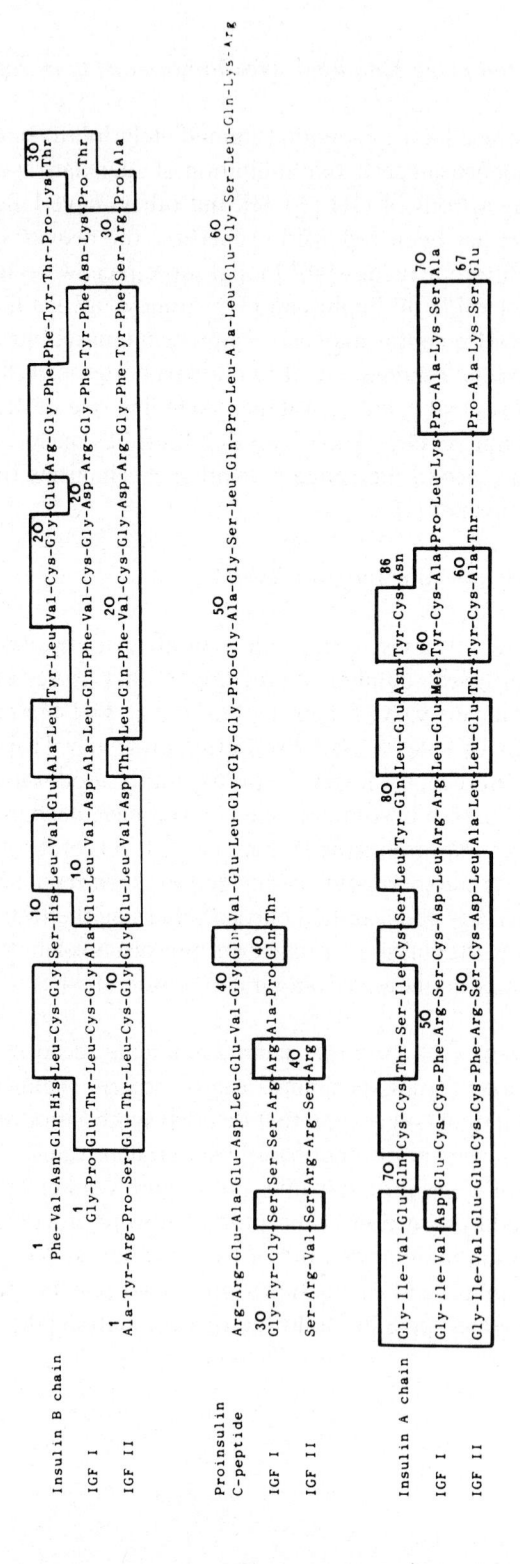

```
Insulin B chain              Phe-Val-Asn-Gln-His-Leu-Cys-Gly-Ser-His-Leu-Val-Glu-Ala-Leu-Tyr-Leu-Val-Cys-Gly-Glu-Arg-Gly-Phe-Phe-Tyr-Thr-Pro-Lys-Thr
                              1                                    10                            20                              30
IGF I                        Gly-Pro-Glu-Thr-Leu-Cys-Gly-Ala-Glu-Leu-Val-Asp-Ala-Leu-Gln-Phe-Val-Cys-Gly-Asp-Arg-Gly-Phe-Tyr-Phe-Asn-Lys-Pro-Thr
                              1                                    10                            20                              30
IGF II                       Ala-Tyr-Arg-Pro-Ser-Glu-Thr-Leu-Cys-Gly-Gly-Glu-Leu-Val-Asp-Thr-Leu-Gln-Phe-Val-Cys-Gly-Asp-Arg-Gly-Phe-Tyr-Phe-Ser-Arg-Pro-Ala

Proinsulin
C-peptide                    Arg-Arg-Glu-Ala-Glu-Asp-Leu-Glu-Val-Gly-Gln-Val-Leu-Gly-Gly-Gly-Pro-Gly-Ala-Gly-Ser-Leu-Gln-Pro-Leu-Ala-Leu-Glu-Gly-Ser-Leu-Gln-Lys-Arg
                                                            40                                    50                                    60
IGF I                        Gly-Tyr-Gly-Ser-Ser-Ser-Arg-Arg-Ala-Pro-Gln-Thr
                              30                            40
IGF II                       Ser-Arg-Val-Ser-Arg-Arg-Ser-Arg
                                                    40

Insulin A chain              Gly-Ile-Val-Glu-Gln-Cys-Cys-Thr-Ser-Ile-Cys-Ser-Leu-Tyr-Gln-Leu-Glu-Asn-Tyr-Cys-Asn
                                                  70                            80                            86
IGF I                        Gly-Ile-Val-Asp-Glu-Cys-Cys-Phe-Arg-Ser-Cys-Asp-Leu-Arg-Arg-Leu-Glu-Met-Tyr-Cys-Ala-Pro-Leu-Lys-Pro-Ala-Lys-Ser-Ala
                                                  50                            60                            70
IGF II                       Gly-Ile-Val-Glu-Glu-Cys-Cys-Phe-Arg-Ser-Cys-Asp-Leu-Ala-Leu-Leu-Glu-Thr------Tyr-Cys-Ala-Thr------Pro-Ala-Lys-Ser-Glu
                                                                              60                                                67
```

Fig. 4. The amino acid sequences of human proinsulin, somatomedin C/IGF-I and IGF-II. From Ref. 109.

4. Somatomedins/IGFs and the actions of growth hormone

As has been indicated, some of the most important actions of GH are mediated by somatomedins, particularly somatomedin C (IGF-I). The existence of a mediating factor (first called sulphation factor) was first postulated when it was recognized that GH had little direct effect on cartilage (and hence growth of long bones) in vitro, although the hormone did promote cartilage growth in vivo and a factor in the serum of GH-treated hypophysectomized rats actively stimulated cartilage growth in vitro [43]. The somatomedin hypothesis, developed to explain these observations, proposed that GH promoted formation of a factor, particularly in liver, which could bring about many of the actions of GH in peripheral tissues, especially cartilage (reviewed in Refs. 44–46).

4.1. The nature of somatomedins

The nature of the somatomedins is now quite well understood. Two human somatomedins have been well characterized, somatomedin C/IGF-I and IGF-II [45–47]. Both are single-chain polypeptides containing about 70 amino acid residues showing marked homology with insulin (Fig. 4). A 3-dimensional structure has been predicted for these proteins on the basis of their homology with insulin/proinsulin. They were isolated from human plasma, where they occur at concentrations that are rather low but nevertheless higher than in other tissues, by workers searching for (a) factors (somatomedins) that could stimulate cartilage growth and sulphation, and (b) factors with insulin-like activity that could not be suppressed by antibodies to insulin. It was only after many years that it was recognized that the two activities were due to the same molecules.

Somatomedins have now been isolated from many other species. The sequences of bovine and mouse somatomedins C have been reported. Rat somatomedin C/IGF-I differs from the human protein at a few residues, and the complete sequence of the gene that codes for it has been determined [47]. Rat IGF-II is probably identical to another polypeptide, multiplication stimulating activity (MSA), which is produced by a rat liver-derived cell line (BRL cells) and stimulates the growth of many cells in culture.

4.2. The actions of somatomedins

Somatomedin C stimulates cartilage growth in vitro, with specific effects on glycosaminoglycan synthesis (usually followed experimentally by incorporation of $^{35}SO_4$) and DNA synthesis/cell proliferation. In also acts as a growth factor for many cells in culture, including fibroblasts, chondrocytes and granulosa cells from the ovarian follicle. In the case of fibroblasts (and possibly other cell types) it appears to synergize with platelet-derived growth factor (PDGF) to promote growth, PDGF acting as a competence factor and somatomedin C as a progression factor [48]. IGF-II also acts as a growth factor for a variety of cell types, but probably plays a particularly important role in the fetus [49]. The physiological significance of the effects of somatomedins on proliferation (and possibly differentiation) of many cell types is not yet clear, though it is becoming clear that such actions may be important for various specific tissues in addition to cartilage, for example the ovarian granulosa cells. Receptors for somatomedin C/IGF-I and for IGF-II are found in a wide variety of cell types. Those for somatomedin C are structurally similar to the insulin receptor, with a 4-chain structure comprising two α chains and two β chains [50]. Indeed, there is some overlap between the actions of somatomedin C and insulin, each being able to bind to the receptors of the other hormone, in some tissues, with an affinity of about 1% that of the homologous interaction. (This probably explains the low insulin-like activity observed with somatomedin C and to some extent the ability of high concentrations of insulin to support the growth of some cell lines in vitro.) Like the insulin receptor, that for somatomedin C can act as a tyrosine kinase and this may underlie, at least in part, the mechanism of action of the polypeptide. The natural substrate(s) on which the tyrosine kinase acts, and which presumably mediate the actions of the hormone, are not known.

The receptor for IGF-II is quite different from that for somatomedin C or insulin, comprising one long polypeptide chain [50]. The mechanism by which IGF-II works is not clear.

4.3. Somatomedin-binding proteins

Somatomedins are unusual in that unlike most polypeptide hormones they circulate in association with specific binding proteins [51,52]. At least two such proteins are found in blood plasma, with molecular weights of approx. 40–60000 and 150000. The relationship between these has not been determined, though their partial purification has been reported. The function of these binding proteins is not fully understood, though they appear to greatly extend the half-life ($t_{1/2}$) of somatomedins – labelled somatomedin injected into animals has a $t_{1/2}$ of only about 10 min, whereas in association with binding protein $t_{1/2}$ appears to be about 3 h.

4.4. Synthesis and secretion of somatomedins

Somatomedins are synthesized as larger precursors, the structures of which in man have been determined by cloning corresponding cDNAs and nucleotide sequence determination [53,54]. The precursors are processed by enzymic cleavage to give the mature somatomedins, which are secreted from the cells of origin. There is little storage of somatomedins in the cells where they are produced; they are probably secreted from the cells in vesicles by the same route followed by other secreted hepatic proteins such as serum albumin.

4.5. Regulation of somatomedin production by growth hormone

GH stimulates the production of somatomedin C from liver and possibly other tissues. In the hypophysectomized rat circulating levels of somatomedin C are lowered to less than 5–10% of those in the intact animal [15,55]. Injection of GH restores the level in a dose-dependent fashion, with a maximal 20–30-fold stimulation. The bulk (>50%) of circulating somatomedin C in the rat is produced by the liver [15] and this is a major target organ for the effects of GH on production of the polypeptide. Somatomedins are also produced in a variety of other tissues, including kidney, lung, heart, muscle, testis, cartilage, brain and fat pads [15,56]. In the rat, hypophysectomy leads to a lowering of the somatomedin C levels in each of these tissues except for brain. Administration of GH restores the level towards that in the intact animal; the level in brain is unaffected [15] (Fig. 5).

GH also stimulates somatomedin C levels in a variety of other species, including mouse [57], sheep [58] and man [59]. In man and other species shortage of GH due to pituitary deficiency leads to lowered somatomedin C levels, and enhanced GH levels found in giantism or acromegaly are associated with elevated levels of somatomedin C.

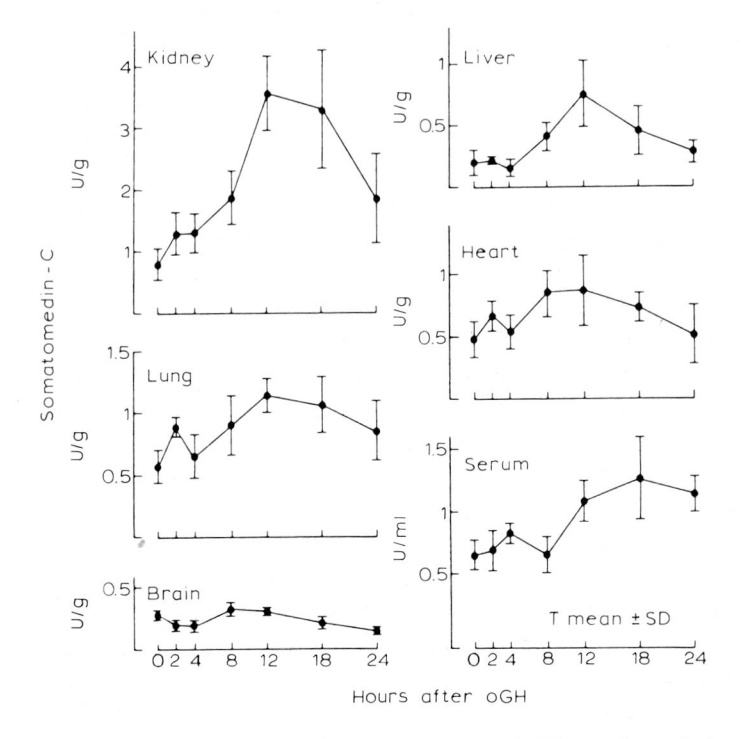

Fig. 5. Induction of somatomedin C/IGF-I in several different tissues in hypophysectomized rats injected with GH. From Ref. 15.

Actions of GH on somatomedin C production in vitro have also been demonstrated, using various different tissues. Thus the hormone has been shown to stimulate somatomedin C production by perfused rat liver [16], cultured rat hepatocytes [60,61], human fibroblasts [62] and human lymphoid cells [63] in vitro.

There is thus ample evidence supporting the idea that GH regulates somatomedin C levels. In many respects it appears to be the major factor affecting these levels, but it should be stressed that other hormones (including thyroxine and insulin) and nutritional factors are also of considerable importance. The effects of GH on IGF-II levels are less important than those on somatomedin C; placental lactogen may be more important here, in partial accordance with the idea that IGF-II is concerned particularly with the regulation of growth of fetal tissues.

GH may also stimulate production of at least some of the somatomedin-binding proteins. Production of binding proteins by hepatocytes in vitro has been demonstrated [60,61,64], as has its stimulation by GH. The 150K binding protein appears to be controlled by GH but not the 50K protein.

4.6. Biochemical mechanisms involved in the action of growth hormone on somatomedin C production

The mechanism by which GH stimulates somatomedin production is not well understood, but it is clear that an action on the expression of the gene is involved. GH causes a 5–10-fold elevation of the level of mRNA for somatomedin C in the liver of the GH-deficient (*lit/lit*) mouse [65] and hypophysectomized rat [66]. In *lit/lit* mice nuclear run-on assays showed that the regulation was occurring at least partly at the transcriptional level. GH had no effect on serum albumin mRNA levels, so induction of somatomedin C was specific. GH also had an effect on the size distribution of hepatic somatomedin C RNAs. Various extrahepatic tissues were found to contain mRNA for somatomedin C, but levels were mostly much lower than those in the liver, pancreas having the highest extrahepatic level at about 20% of the concentration in the liver. In none of the extrahepatic tissues in *lit/lit* mice were somatomedin C mRNA levels altered by GH treatment [65], a rather surprising result in view of the stimulation of somatomedin C levels by GH in extrahepatic tissues of the hypophysectomized rat (Ref. 15; see above). Differences between the animal models used may be important in this respect, though levels of somatomedin C mRNA have also been found to be very low in many non-hepatic tissues in the intact rat [67]. Induction of hepatic mRNA by GH in the *lit/lit* mouse was quite rapid, elevated levels of mRNA being observed after 2 h of administration of GH [65].

Although little progress has been made in determining the detailed mechanisms whereby GH induces somatomedin C production in the liver, recent developments are encouraging. It is now established that GH can stimulate expression of the gene for the growth factor in a rapid, specific and very substantial fashion, and the way is now open for studies on the nature of second messengers and other intracellular events involved in this effect.

4.7. Somatomedin C and somatic growth

The somatomedin hypothesis predicts that somatomedin C mediates the effects of GH on cartilage growth and possibly other aspects of general somatic growth. Somatomedin C is thought to act via the circulation, being produced in liver and possibly other tissues.

Although understanding of somatomedin C has increased markedly during the past few years, and the effects of GH on levels of this peptide have been amply demonstrated, several lines of evidence have led to questioning of the somatomedin hypothesis in its 'traditional' form outlined above. Firstly, difficulty has been encountered in demonstrating substantial effects of somatomedin C on growth. Growth-promoting effects have been demonstrated in dwarf mice [68] and hypophysectomized rats [69], but the effects were smaller than those obtained with GH. Furthermore, other authors have obtained only slight growth promotion with highly purified somatomedin C preparations in hypophysectomized rats [70]. A possible explanation for these poor growth-promoting effects is that the somatomedin was administered without binding protein; it normally circulates in association with binding proteins in intact animals, and this association may be necessary for full in vivo activity, possibly by enhancing half-life.

Another challenge to the 'classical' somatomedin hypothesis has come from the experiments of Isaksson et al. [71,72], which have been confirmed by others [73]. GH injected into the tibial growth plate of one leg of a hypophysectomized rat led to greater growth of the cartilage of that leg compared with the contralateral control. Had the GH been working via actions on liver (or other tissues) to produce elevated levels of circulating somatomedin C which then stimulated cartilage growth, it is difficult to explain why the injected leg should respond preferentially. The most direct interpretation of the experiment is that GH stimulated growth directly at the site of injection. In accordance with this idea Madsen et al. [74] have shown that GH can have a direct growth-promoting effect on chondrocytes from rabbit ear and rat rib growth cartilage.

A possible explanation of this direct action is that GH stimulated somatomedin C production by chondrocytes (or possibly other cells close to the growth plate), and that this then acted locally to stimulate growth. Isaksson and his colleagues have now extended their studies to show that chondrocytes can respond to GH by producing somatomedin C, in vivo, as demonstrated by immunofluorescent methods [75] and that the cells have GH receptors [76]. Others have shown that somatomedin C infused into the arterial supply of a rat hindlimb can induce growth of that hindlimb, like GH, and that antibodies to somatomedin C infused with GH can block the local growth-promoting effect of the GH [77]. It thus seems probable that growth of cartilage can be mediated at least partly by somatomedins produced locally in response to GH, and the same type of response may also occur in other tissues. The relative importance of locally produced somatomedins, acting as autocrines or paracrines, and of circulating somatomedins, acting as endocrines, remains to be established.

5. Actions of growth hormone on production of other specific proteins

The specific effects of GH on induction of somatomedin C (and its binding proteins) can be readily interpreted in terms of overall effects on somatic growth. The hormone also has effects on the production of other specific proteins, although here the connection with growth is less apparent.

GH has been shown to induce a number of enzymes concerned with amino acid metabolism in the liver of the hypophysectomized rat (in vivo or in perfused liver [78]). Induced enzymes include tyrosine aminotransferase, tryptophan oxygenase and ornithine decarboxylase. The effects are complex, particularly in relation to interaction with glucocorticoids, and in some experiments GH *lowered* enzyme levels induced by glucocorticoids, although given alone it led to induction of these enzymes.

Another protein, synthesis of which is stimulated specifically by GH, is the hepatic α_{2u} globulin [79,80], the biological function of which is not clear. This protein is synthesized in the liver of the male rat and appears in the urine of the animal. Hypophysectomy abolished the production of α_{2u} globulin, and normal production was restored in vivo by a combination of four hormones: a glucocorticoid, an androgen, thyroxine and GH. mRNA for α_{2u} globulin was absent from the liver of the hypophysectomized rat. Treatment with a glucocorticoid, androgen and thyroxine had only a small effect on mRNA levels, but treatment with GH in addition to these hormones greatly increased the levels of mRNA for α_{2u} globulin and production of the protein itself [80]. Thus it appears that here too the action of the hormone is at the level of induction of expression of a specific gene.

Induction of mRNAs for several other specific rat hepatic proteins by GH has also been demonstrated [81–83]. The effect could be demonstrated in vivo and in vitro and involved a relatively rapid induction with a 5-fold increase in mRNA levels within 4 h of the administration of GH, although synergism with cortisol (possibly and/or thyroxine) was necessary for a maximal response [83]. cDNAs corresponding to two of the induced proteins have been cloned [82,83] and found to have sequences homologous to those of a known family of serine protease inhibitors. One of these proteins was shown to be secreted as a heavily glycosylated serum protein, and to have potent anti-trypsin activity [83]. Regulation of the production of this protein by GH was shown to occur mainly at the transcriptional level [83].

6. Actions of growth hormone on protein metabolism

When GH is administered to an animal in vivo, it causes general nitrogen retention [84]. Levels of urinary nitrogen and blood urea fall. It is likely that these effects are due largely to a stimulation of protein synthesis which occurs in many tissues, especially skeletal muscle and liver. The effect is rapid, occurring within 30 min of injection of GH into hypophysectomized rats. It is not clear whether somatomedins play a part in mediating it (though the onset is rather too rapid to be explained entirely by somatomedin action), nor is it clear whether this general effect of GH on protein synthesis partly reflects the sum of specific effects on induction of a large number of different proteins, of which those discussed in the previous section would be examples. It seems most probable, however, that GH stimulates both the enhanced synthesis (mainly by increased transcription) of a moderate number of specific proteins and increased activity of the protein synthetic machinery in general.

6.1. Actions of growth hormone on protein synthesis in the liver

GH stimulates protein synthesis in the liver in vivo, and in perfused liver in vitro [85]. In perfused liver it also has effects on synthesis of most types of RNA and on amino acid uptake, and these may underlie much of the action on protein synthesis. For most effects on the liver there appears to be a delay of 15–60 min between the application of hormone and the first manifestation of the effect. Ribosomes from liver of hypophysectomized rats possess a lowered ability to carry out protein synthesis [86] and the defect can be reversed by administration of GH in vivo or in vitro. The effect may be at least partly on the rate of elongation of the growing polypeptide chain [85].

6.2. Actions on muscle

GH stimulates incorporation of labelled amino acids into protein in skeletal muscle (diaphragm) from hypophysectomized rats in vitro [41,84]. The effect is a consequence of stimulation of protein biosynthesis, although amino acid transport is also affected. Insulin and somatomedins also stimulate protein synthesis in diaphragm [44], but there are clear-cut differences between the actions of these hormones and of GH. Thus a 30 min lag period is characteristic of the action of GH, as is a refractory period after 2–3 h of stimulation.

Effects of GH on protein synthesis in diaphragm are relatively small (usually a 30–40% stimulation). Such a small in vitro effect may reflect a sufficiently large in vivo effect to bring about the actions of GH, however. Growth is a long-term process, and a 40% stimulation of protein synthesis, without an accompanying increase in protein breakdown, could be sufficient to give the sort of enhancement of growth of muscle seen in hypophysectomized animals treated with GH.

Whether somatomedins play a role in mediating the effects of GH on muscle is not clear. The direct in vitro effects described suggest that somatomedins are not involved, but it cannot be ruled out completely at this stage that the effects observed are in part caused by somatomedins released locally by the muscle, or associated cells, in response to GH.

GH also stimulates protein synthesis in heart muscle [87]. This effect seems to be associated with a fall in the concentration of cyclic AMP in the tissue [87], an effect which has also been shown by some workers for diaphragm [39] but not by others.

7. Actions of growth hormone on lipid and carbohydrate metabolism

In addition to its effects on the synthesis of proteins in general and on the induction of specific proteins, which can sometimes be related clearly to the growth-promoting actions of the hormone, GH has a variety of effects on lipid and carbohydrate metabolism (Table I). This topic has recently been reviewed very comprehensively [88].

TABLE I
Actions of growth hormone on carbohydrate and lipid metabolism

I Insulin-like actions (mostly rapid)
  (a) In vitro effects
    1. Increased glucose uptake, glucose oxidation, and glucose conversion to fatty acids in adipose tissue, adipocytes and muscle.
    2. Inhibition of adrenaline-stimulated lipolysis in adipose tissue.
  (b) In vivo effects
    1. Transient hypoglycaemia in hypophysectomized dog and rat and GH-deficient children.
    2. Transient fall in free fatty acids.

II Anti-insulin-like actions (mostly slow)
  (a) In vitro effects
    1. Decreased glucose oxidation and conversion to fatty acids in adipose tissue from dwarf mice (after 4 h) and in 3T3 adipocytes (after 24–48 h).
    2. Refractoriness to the insulin-like actions of GH.
    3. Increased lipolysis in adipose tissue (after 1 h).
  (b) In vivo effects
    1. Hyperglycaemia, especially in dogs.
    2. Impaired glucose tolerance.
    3. Lipolysis.
    4. Resistance to the actions of insulin.

In many cases GH shows directly opposite actions, usually insulin-like in the short-term, anti-insulin-like in the long-term. Many actions are most marked in GH-deficient animals or man.

7.1. Lipid metabolism

GH administered to hypophysectomized rats in vivo causes a drop in the level of plasma non-esterified fatty acids (NEFA), followed by a prolonged increase in this level [89]. This appears to be due to increased utilization of lipids – increased uptake of NEFA by muscle preceding increased output by adipose tissue. As a consequence GH diverts the energy metabolism of the organism from carbohydrate utilization to lipid utilization, and acts to oppose the effects of insulin. Actions of GH on lipid metabolism are particularly marked in man, where GH levels become elevated on fasting and presumably serve to help stimulate the increased lipid utilization seen in this condition. In contrast, in the rat, GH levels fall on fasting.

In vitro, GH has many effects on adipose tissue and cells, including stimulation of lipolysis and actions on glucose utilization [89,90]. Short-term effects in vitro are mainly insulin-like (increased utilization of glucose and amino acids, glycogen synthesis, antilipolytic actions, etc.) and may be mediated by mechanisms similar to those of insulin, including dephosphorylation of hormone-sensitive lipase [91]. In the longer term adipose tissue and cells in vitro become refractory to the insulin-like effects of GH, and counter-insulin effects predominate and reflect the main actions seen in vivo. Receptors for GH have been identified in adipose tissue [90,92].

7.2. Carbohydrate metabolism

In vivo GH has both insulin-like and diabetogenic effects. In hypophysectomized rats both effects of GH have been demonstrated, a transient hypoglycaemia being followed by prolonged hyperglycaemia [14]. The latter appears to result from reduced glucose uptake by peripheral tissues, and is often associated with elevated insulin levels due to either increased plasma glucose or possibly a direct effect of glucose on the pancreas [93]. However, diabetogenic effects of GH are often rather difficult to demonstrate in rats; indeed, hypophysectomy may lead to diabetogenic effects (elevated glucose levels and impaired glucose tolerance test) and administration of GH may reverse these effects [94]. In vivo diabetogenic actions of the hormone are more readily demonstrated in dogs or genetically obese (*ob/ob*) mice. There is some evidence that a fragment of GH has enhanced diabetogenic activity and may account for much of the activity in GH preparations [95], but this fragment has not been fully characterized, and other workers consider that the diabetogenic activity is associated primarily with intact GH [96].

The mechanism of the diabetogenic actions of growth hormone, beyond the observations that they involve lowered glucose uptake by peripheral tissues, is not well understood. It is unlikely that they are mediated by somatomedins, since these factors generally have insulin-like actions (though with much lower potency than insulin itself). Direct antagonism of glucose uptake may occur, or alternatively the effect may be caused by elevated levels of non-esterified fatty acids, resulting from the lipolytic actions of the hormone. These may act to inhibit glucose uptake by muscle, in accordance with the idea of a glucose/fatty acid cycle [97].

In vitro GH has been shown to stimulate glucose uptake and utilization in several tissues, including diaphragm and adipose tissue [41,90]. Diabetogenic effects of the hormone are more difficult to demonstrate in vitro, although some such effects have been obtained [90].

Some of the effects of GH may be a consequence of its stimulation of synthesis and secretion of insulin by the pancreas. Such effects have been demonstrated in various systems [93], as have actions on cell growth in islets of Langerhans. The pancreas is a site of production of somatomedin C, and the possibility that this factor may mediate the actions of GH on insulin production has not been excluded.

8. Actions of growth hormone on cellular differentiation and proliferation

As has been discussed, some of the actions of GH involve the induction of expression of specific proteins, but the effect is reversed on removal of the hormone. In a few cases, however, GH appears to induce a more permanent differentiation.

GH, together with other hormones, can induce differentiation of cells of the 3T3 mouse fibroblast line into adipocytes in vitro [98,99]. The change involves morphological alterations, enzymic differentiation, increased hormone responsiveness, triglyceride accumulation and a limited mitogenic response. Somatomedin C has no effect on this transformation.

GH also has actions on the differentiation of muscle. It stimulates the conversion of myoblasts to myotubes in vitro, with associated changes in the levels of creatine phosphokinase [99]. It is also possible that GH has direct effects on the differentiation of cartilage, acting synergistically with somatomedins to promote cartilage growth [72].

GH also has direct effects on the proliferation of a variety of cells and tissues in culture, including chondrocytes [74] (though others have been unable to confirm this effect [100]), haematopoietic progenitor cells [101], normal and leukaemic human T cells [102], smooth muscle [103] and rat pancreatic B-cells [104]. The physiological significance of these effects is not clear, and it has not been clarified to what extent somatomedin C/IGF-I released from the same cells may be mediating the mitogenic effects. It is clear, however, that enhanced cell division in many tissues follows administration of GH in vivo [72] and this may be a consequence of direct actions rather than (or as well as) somatomedin C/IGF-I mediated effects.

The mechanisms involved in mediating the direct actions of GH on cell proliferation are not well understood, but recent work on the induction of the c-*myc* protooncogene by GH has important implications [105]. Expression of this gene has been shown to follow mitogenic stimulation of various cell lines by a number of different factors, and the gene product has been implicated as necessary for the initiation of DNA synthesis [106]. Administration of GH to hypophysectomized rats led to a rapid induction of expression of the c-*myc* gene in the liver and kidney (maximal after only 1 h), an action which preceded induction of somatomedin C/IGF-I (first significant at 6 h), stimulation of thymidine incorporation (significant at 6 h) and elevation of circulating somatomedin C levels (significant at 3 h) [105]. Administration of somatomedin C had no effect on expression of c-*myc* in rat liver. The results strongly support the idea that GH can stimulate cell proliferation in at least some tissues in vivo independently of somatomedin C/IGF-I mediation, and that the induction of c-*myc* plays at least some part in the intracellular mechanism. It may be, of course, that a maximal proliferative response involves synergism between GH and somatomedin C.

9. Growth hormone and the control of lactation

Prolactin is usually considered to be the pituitary hormone of most importance in stimulating and maintaining milk production. In ruminants, however, GH also has a major role in promoting lactation [107] and it may also play some role in this respect in the rat [108]. The mechanisms involved are not clear, but altered partitioning of nutrients between the mammary gland and other organs may be involved rather than a direct effect on the synthesis of milk components [107]. As has been mentioned previously, human GH has lactogenic effects in many species, but this appears to be because it interacts (atypically for a GH) with lactogenic receptors. It is not clear that human GH has lactogenic effects in humans.

10. Applications of molecular biology to the study of the actions of growth hormone

The spectacular development of recombinant DNA techniques during the past decade has had a major impact on the study of GH. As yet this impact has been concerned mainly with the nature of the hormone itself, and of the genes that direct its production, and their regulation, though such techniques have also made a major contribution to studies of the actions of GH on transcription of specific genes, as discussed above. At least two additional lines of work are likely to contribute to our understanding of the action of GH and will be summarized here.

10.1. Protein engineering of growth hormone

There has been much work designed to study structure-function relationships in GH by producing variants of the hormone by chemical or enzymic modification [1]. Now that cDNAs for GHs from several species have been cloned and expressed in a variety of prokaryotic and eukaryotic species [109], it will be possible to produce a large number of GH variants altered in very precise ways by the techniques of site-directed mutagenesis and protein engineering. Among variants produced so far by this type of approach are a form of human GH lacking the N-terminal 13 amino acids of the normal hormone [110,111]. This analogue was able to inhibit the actions of GH in a number of test systems. It competed with human GH for binding to receptors but unlike the normal hormone was unable to induce receptor down-regulation. In another human GH analogue produced by recombinant DNA methods a single amino acid change was introduced, 1/2-Cys at residue 165 being replaced by Ala. This analogue showed immunological and biological activity that was similar to that of the pituitary-derived hormone [112], confirming earlier ideas derived from chemical modification studies [1] that the disulphide bridges of the hormone are not essential for retention of tertiary structure or biological activity.

10.2. Transgenic mice

Injection of rat or human GH genes, coupled to the metallothionein I promotor, has led to the production of transgenic mice with grossly elevated circulating GH levels [113,114]. Such animals provide models for studies on the effects of excess GH on the organism. Potentially the GH gene could be modified to allow expression only in selected tissues or could be introduced as variant forms. Exploration of the potential that such transgenic animals offer for the study of GH is at an early stage.

11. Potentiation of the actions of growth hormone by monoclonal antibodies

Several panels of monoclonal antibodies to human and bovine GH have been prepared and have been used to probe interactions between GH and its receptors [27,115–117]. When such antibodies were administered in vivo with GH, however, a surprising result was obtained [118–120]. Most of the antibodies enhanced the actions of GH to a greater or lesser degree. In some cases the effect was equivalent to an enhancement of the potency of the GH administered by more than 25-fold. The potentiation applied to growth-promoting effects, effects on in vivo sulphation of cartilage and on circulating somatomedin levels [121] (Fig. 6), and also some of the in vivo lactogenic actions of human GH. Effects on in vitro activity (Nb2 cell assay – a lactogenic assay – see Chapter 14) were not observed, though in a few cases a somewhat enhanced receptor-binding activity (in vitro) was observed [120].

The mechanism of this potentiation of the action of GH is not clear. It does not appear to be due to the bivalent nature of the antibodies, since monovalent (Fab)

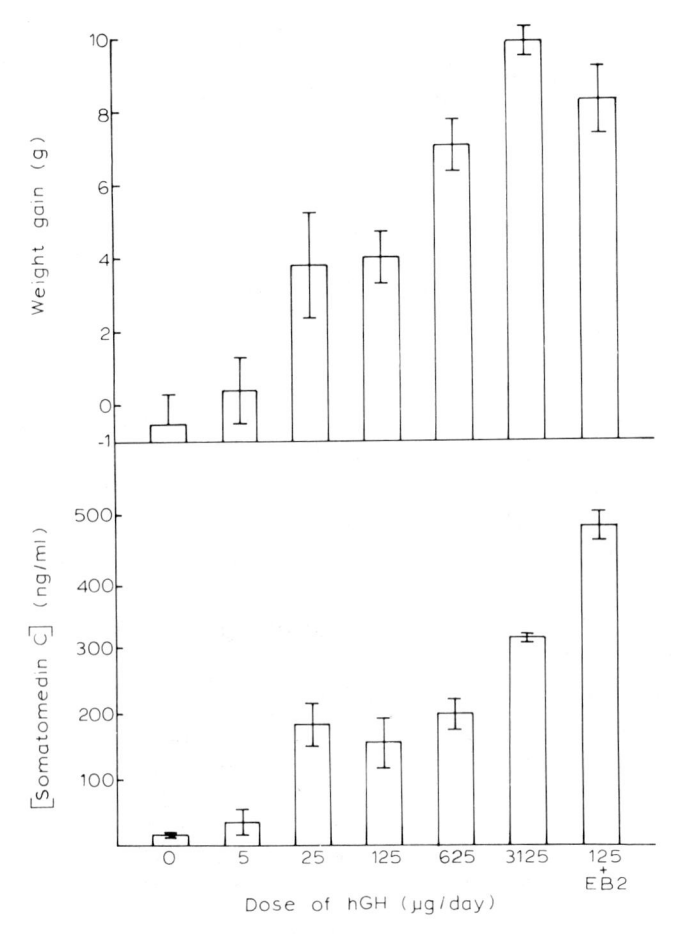

Fig. 6. Potentiation of the actions of GH by monoclonal antibodies. Increasing doses of human GH injected into hypophysectomized rats induced an increase in serum somatomedin concentration and growth in hypophysectomized rats after 24 h. Monoclonal antibody EB2 (right-hand column) markedly enhanced growth and somatomedin levels induced by 125 µg GH (centre column).

fragments were as active as intact antibodies. It is possible that the antibodies prolong the effective half-life of the hormone in the circulation, but some experiments suggested that although such an effect was observed, there was little correlation between effect on half-life and effect on biological activity. Enhanced receptor binding, resulting from conformational changes in the hormone, may contribute, as may restriction of binding to only a small population of biologically active receptors or prolonged life of the hormone after it has bound to target cells and possibly after internalization.

Whatever the mechanism underlying this potentiation effect it can produce a dramatic enhancement of the biological effects of the hormone. Such a potentiation is

12. Growth hormone variants

GH can occur in a variety of forms, and there has been much debate as to whether these forms vary significantly in their biological actions [1,95,123,124].

12.1. Naturally occurring variants

A particularly interesting variant is the 20K form of human GH, which has a single chain that is 15 residues shorter than that of the normal (22K) form [125]. The 20K variant lacks residues 32–46 of 22K GH and probably arises as a consequence of a variation in the processing of the mRNA precursor for the hormone [126,127] (Fig. 7). 20K GH retains full growth-promoting activity. It appears to show lowered insulin-like activity and at one time was thought to have little diabetogenic activity, though recent work suggests that it is not significantly different from normal (22K) GH in terms of diabetogenic effects [128]. The physiological significance of the 20K variant is not clear, but it does form a substantial fraction (approx. 10–15%) of the GH in the human pituitary and may well have a distinct function.

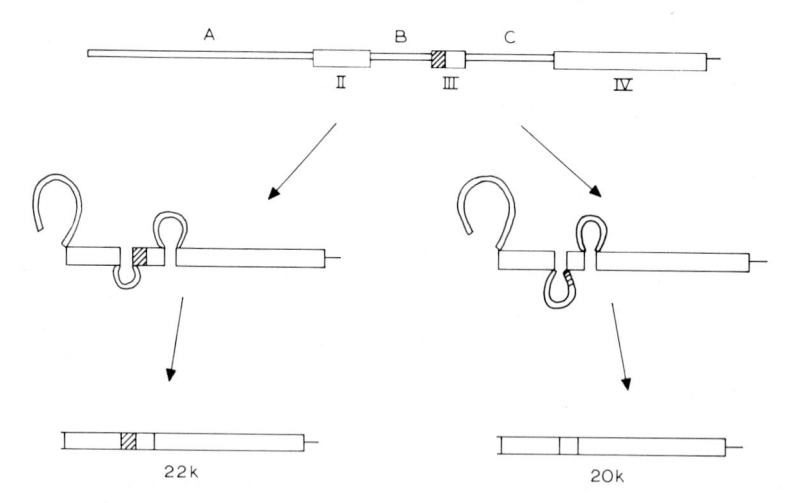

Fig. 7. Diagram showing the pathways by which a precursor of the mRNA for human GH (the product of a single gene) could give rise to mRNAs coding for either the 20K or the 22K variants of the molecule. A, B and C are the three introns. II, III, IV are three exons (coding region); the position of exon I is not shown. The shaded region represents the nucleotide sequence that codes for residues 32 to 46 in the 22K form of growth hormone, but which is included as part of intron B in processing to the mRNA for the 20K form.

A second type of naturally occurring variant of GH involves proteolysis of the molecule. Limited proteolysis of the human hormone gives rise to forms in which proteolytic 'nicks' are introduced into the large disulphide loop, usually in the region of residues 130–150 [1,123,129,130]. Some such cleaved forms have been reported to possess enhanced biological activity, and forms of the hormone with a potency 4–5-fold greater than that of the intact molecule have been described. Production of such superpotent forms has been difficult to reproduce, however. It has been suggested that enzymic processing of GH may occur in vivo, during or after secretion or even at the receptor site, with production of forms of the hormone of increased potency, but evidence for such processing remains tenuous.

Processing of GH could go further, with production of relatively small peptides which retain substantial or even enhanced activity, and it has been suggested that this may be important in the actions of the hormone in vivo [131]. In particular, fragments which retain the diabetogenic or insulin-like activities of GH have been reported [95,131]. What role these play in the normal physiological actions of the hormone is unclear.

Yet a further type of GH variant relates to the occurrence of dimers or higher oligomers of the hormone. These are found in the pituitary and circulation of man and animals. Their physiological significance, if any, is not known, but it is of potential significance that their relative proportions can vary according to the physiological and pathological state of the organism.

Several other variants of GH have been described, particularly for the human hormone, including potentially the product of a second GH gene [132]. It remains uncertain, however, as to whether any of these has real significance or whether the normal (22K) form of GH is sufficient to explain all significant effects of the hormone, the variant forms being simply 'side products' of little or no importance.

12.2. The multivalent nature of growth hormone

As has been discussed, GH has a wide range of different actions. There is considerable evidence that these various actions are not mediated by a common mechanism of action, but that they are associated with different regions or features of the GH molecule [1,133]. For example, modification of GH by chemical or enzymic means can lead to derivatives with altered relative properties in bioassays assessing the different biological actions. The idea has been proposed that GH is able to produce its various effects via interactions with different receptors, different regions of the hormone (or possibly different peptides produced from it) being able to interact with the appropriate receptors. The concept is by no means proven, nor is it universally accepted, but it does go some way towards explaining at least some of the varied actions of this complex hormone. It also offers promise that modification of GH, by chemical, enzymic or protein engineering techniques will be able to produce analogues with distinctively different patterns of activity compared to the native molecule.

13. Conclusions

It will be clear from what has been presented in this chapter that there is no clear consensus as to the mechanism of action of GH. The hormone has a large number of different actions, affecting most of the tissues of the body, and it is difficult to identify any one, or even a few, of these as of key importance. In the past 10–15 years emphasis has been placed on the role of somatomedin C as a mediator of the actions of GH, whereas previously direct actions of the hormone on protein synthesis and other metabolic processes received particular attention; however, the change of emphasis seems to have been determined by technical advances rather than any altered recognition of biological importance.

13.1. The multiple actions of growth hormone

It is common for hormones that regulate growth and development to elicit a number of different responses in their target tissues, including 'early' and 'late' effects. The set of responses is to some extent common to several different hormones, and has been referred to as the 'pleiotypic' response [134,135]. In this view different hormones differ mainly by acting on different tissues; the immediate responses with the target tissues are similar but the ultimate responses differ because of differences between tissues. One is tempted to view GH as a hormone which acts in this way but lacks a specific target tissue, and thus produces a pleiotypic response in most of the tissues of the organism. Such a view has several drawbacks, however. Thus, GH produces actions on target tissues that are different from those produced by other hormones which can be considered to produce a pleiotypic response; indeed, GH can have both insulin-like and diabetogenic actions on the same tissue, suggesting that it can, in the same cell, evoke both the intracellular responses characteristic of stimulation by insulin and a set of responses that is quite distinct from those produced by insulin. Another feature of the actions of GH that does not entirely accord with the notion of a pleiotypic response is the occurrence of multiple receptor types. It may be that to at least some extent the multiple actions of the hormone are mediated by distinct receptors.

GH clearly has effects on carbohydrate and lipid metabolism, the protein-synthetic machinery, expression of specific proteins, and cell differentiation and proliferation. For a hormone concerned with the overall control of somatic growth such a wide spread of activities is potentially physiologically meaningful. It is at least possible that GH mediates these various actions via several different intracellular mechanisms and that it will not be possible to identify any one action as of key importance.

13.2. The significance of somatomedin C/IGF-I

Although many actions of GH may be important, its actions on somatomedin C production have received particular attention in recent years. The role played by this growth factor remains uncertain, however. It is clear that somatomedins stimulate cell growth and a variety of other processes in many cell types in vitro and it seems likely that they do so also in vivo. It is also clear that GH stimulates somatomedin C production in the liver and probably many other tissues, although it is not the only regulatory factor. The relative importance of circulating and locally produced somatomedin C is not clear, however, nor is the role of the binding proteins with which it is normally associated in plasma. The binding proteins may play an important function in prolonging the half-life of somatomedins or potentiating their actions directly [136], but it cannot be ruled out that they sometimes serve to bind and inactivate somatomedins produced locally in the tissues, thus limiting the actions of these potent growth factors, in whose uncontrolled presence cell multiplication might get out of hand.

14. Addendum

In the few months since the main part of this chapter was written there have been several important developments in GH biochemistry.

The tertiary structure of recombinant DNA-derived pig GH has been reported [137]. The conformation includes a large proportion of α-helix (about 54%), 4 antiparallel α-helices being arranged in a left-twisted helical bundle. Several unrelated proteins also contain 4 α-helices arranged in this way, but the connections in GH are unusual and unlike those found elsewhere. In view of the marked homology between the amino acid sequences of members of the GH-prolactin protein family, it seems likely that a similar tertiary structure will be found in other GHs and in prolactins and placental lactogen.

cDNAs corresponding to the GH receptor in rabbit and human liver have been cloned [138]. The nucleotide sequences of these allow derivation of the amino acid sequences of the corresponding receptors. These contain 620 amino acid residues, plus a putative signal peptide. A single, short hydrophobic region near the centre of the sequence probably represents a trans-membrane domain, which would leave an N-terminal extracellular domain of 246 residues, which is probably heavily glycosylated, and a C-terminal intracellular domain of 350 residues, presumably involved in signal transduction. The sequences of the rabbit and human receptors are very similar (differing at only about 16% of all residues). The N-terminal sequence of the rabbit receptor is very similar to that of the soluble GH-binding protein found in rabbit serum; the latter may represent mainly the extracellular domain of the receptor, released by proteolysis or synthesized on a truncated mRNA.

Expression of the cDNA for the rabbit GH receptor and of the extracellular domain of the human receptor was obtained in transfected monkey COS-7 cells. The proteins so obtained bound ^{125}I-labelled human GH, and showed a specificity similar to that of the receptors obtained from normal tissues. Thus, the rabbit receptor could bind human GH, bovine GH and to a lesser extent ovine prolactin, while the human receptor bound human GH but not bovine GH or ovine prolactin. This provides strong evidence that these receptors mediate the growth-promoting actions of GH, since biological responses show a similar specificity.

The cloning of the GH receptor will have a major impact on approaches to the mechanism of action of the hormone. It provides sequences for the 'receptor' in rabbit and human liver, shows that differences between species are less than expected and demonstrates that a single polypeptide (after glycosylation) is all that is needed for GH binding (though additional subunits may be necessary for full biological functioning). Expression of the receptor in cultured mammalian cells provides a basis for structure-function studies using site-directed mutagenesis. The mechanism of signal transduction mediated by the receptor remains unclear, however. The receptor sequence is not related to known tyrosine-kinase growth-factor receptors. How a single trans-membrane sequence transmits the signal from extracellular to intracellular domains is particularly difficult to understand; possibly aggregation of receptors moving in the fluid mosaic structure of the membrane is required [139].

Acknowledgements

I thank Mrs Eileen Willis for skilful preparation of the manuscript and the Medical Research Council and Cancer Research Campaign for research support.

References

1. Wallis, M. (1978) In: Chemistry and Biochemistry of Amino Acids, Peptides and Proteins, Vol. 5 (Weinstein, B., ed.) pp.213–320. Dekker, New York.
2. Nicoll, C.S., Mayer, G.L. and Russell, S.M. (1986) Endocr. Rev. 7, 169–203.
3. Wallis, M. (1981) J. Mol. Evol. 17, 10–17.
4. Miller, W.L. and Eberhardt, N.L. (1983) Endocr. Rev. 4, 97–130.
5. Nicoll, C.S. (1982) Perspect. Biol. Med. 25, 369–381.
6. Wallis, M. (1984) In: Prolactin Secretion: A Multidisciplinary Approach (Mena, F. and Valverde-R., C., eds.) pp. 1–16. Academic Press, Orlando.
7. Wurzel, J.M., Parks, J.S., Herd, J.E. and Nielsen, P.V. (1982) DNA 1, 251–257.
8. Colosi, P., Ogren, L., Thordarson, G. and Talamantes, F. (1987) Endocrinology 120, 2500–2511.
9. Chan, J.S.D., Robertson, H.A. and Friesen, H.G. (1976) Endocrinology 98, 65–76.
10. Murthy, G.S., Schellenberg, C. and Friesen, H.G. (1982) Endocrinology 111, 2117–2124.
11. Duckworth, M.L., Kirk, K.L. and Friesen, H.G. (1986) J. Biol. Chem. 261, 10871–10878.

12. Wallis, M. (1975) Biol. Rev. 50, 35–98.
13. Linzer, D.I.H. and Nathans, D. (1984) Proc. Natl. Acad. Sci. USA 81, 4255–4259.
14. Altszuler, N. (1974) In: Handbook of Physiology, Section 7: Endocrinology, Vol. 4, Part 2 (Knobil, E. and Sawyer, W.H., eds.), pp. 233–252, American Physiological Society, Washington DC.
15. D'Ercole, A.J., Stiles, A.D. and Underwood, L.E. (1984) Proc. Natl. Acad. Sci. USA 81, 935–939.
16. McConaghey, P. and Sledge, C.B. (1970) Nature 225, 1249–1250.
17. Wallis, M. (1980) In: Cellular Receptors for Hormones and Neurotransmitters (Schulster, D. and Levitzki, A., eds.) pp. 163–183. Wiley, Chichester.
18. Hughes, J.P., Elsholtz, H.P. and Friesen, H.G. (1985) In: Polypeptide Hormone Receptors (Posner, B.I., ed.) pp. 157–199. Dekker, New York.
19. Webb, C.F., Cadman, H.F. and Wallis, M. (1986) Biochem. J. 236, 657–663.
20. Ymer, S.I., Stevenson, J.L. and Herington, A.C. (1984) Biochem. J. 221, 617–622.
21. Ymer, S.I. and Herington, A.C. (1986) Biochem. J. 237, 813–820.
22. Ymer, S.I. and Herington, A.C. (1985) Mol. Cell. Endocrinol. 41, 153–161.
23. Baumann, G., Stolar, M.W., Amburn, K., Barsano, C.P. and DeVries, B.C. (1986) J. Clin. Endocrinol. Metab. 62, 134–141.
24. Cadman, H.F. and Wallis, M. (1981) Biochem. J. 198, 605–614.
25. Hughes, J.P. (1979) Endocrinology 105, 414–420.
26. Moore, W.V., Kover, K. and Hung, C.H. (1986) In: Human Growth Hormone (Raiti, S. and Tolman, R.A., eds.) pp. 475–498. Plenum Medical, New York.
27. Thomas, H., Green, I.C., Wallis, M. and Aston, R. (1987) Biochem. J. 243, 365–372.
28. Ymer, S.I. and Herington, A.C. (1987) Biochem. J. 242, 713–720.
29. Wallis, M., Daniels, M. and Webb, C.F. (1987) J. Endocr. 112, Suppl., Abstr. 110.
30. Barnard R., Bundesen, P.G., Rylatt, D.B. and Waters, M.J. (1985) Biochem. J. 231, 459–468.
31. Barnard, R. and Waters, M.J. (1986) Biochem. J. 237, 885–892.
32. Tsushima, T., Murakami, H., Wakai, K., Isozaki, O., Sato, Y. and Shizume, K. (1982) FEBS Lett. 147, 49–53.
33. Ymer, S.I. and Herington, A.C. (1984) Endocrinology 114, 1732–1739.
34. Haeuptle, M.T., Aubert, M.L., Djiane, J. and Kraehenbuhl, J.P. (1983) J. Biol. Chem. 258, 305–314.
35. Hughes, J.P., Simpson, J.S.A. and Friesen, H.G. (1983) Endocrinology 112, 1980–1985.
36. Waters, M.J. and Friesen, H.G. (1979) J. Biol. Chem. 254, 6815–6825.
37. Yamada, K., Lipson, K.E. and Donner, D.B. (1987) Biochemistry 26, 4438–4443.
38. Tsushima, T., Sasaki, N., Imai, Y., Matsuzaki, F. and Friesen, H.G. (1980) Biochem. J. 187, 479–492.
39. Albertsson-Wikland, K. and Rosberg, S. (1982) Endocrinology 111, 1855–1861.
40. Vesely, D.L. (1981) Am. J. Physiol. 240, E79–E82.
41. Ahrén, K., Albertsson-Wikland, K., Isaksson, O. and Kostyo, J.L. (1976) In: Growth Hormone and Related Peptides (Pecile, A. and Muller, E.E., eds.) pp. 94–103. Excerpta Medica, Amsterdam.
42. Yamada, K., Lipson, K.E., Marino, M.W. and Donner, D.B. (1987) Biochemistry 26, 715–721.
43. Salmon, W.D. and Daughaday, W.H. (1957) J. Lab. Clin. Med. 49, 825–836.
44. Froesch, E.R., Schmid, C., Schwander, J. and Zapf, J. (1985) Annu. Rev. Physiol. 47, 443–467.
45. Preece, M.A. and Holder, A.T. (1982) In: Recent Advances in Endocrinology and Metabolism, Vol. 2 (O'Riordan, J.L.H., ed.), pp. 47–73, Churchill-Livingstone, New York.
46. Humbel, R.E. (1984) In: Hormonal Proteins and Peptides, Vol. XII (Li, C.H., ed.) pp. 57–79. Academic Press. New York.
47. Shimatsu, A. and Rotwein, P. (1987) J. Biol. Chem. 262, 7894–7900.
48. Stiles, C.D., Capone, G.T., Scher, C.D., Antoniades, H.N., van Wyk, J.J. and Pledger, W.J. (1979) Proc. Natl. Acad. Sci. USA 76, 1279–1283.
49. Brown, A.L., Graham, D.E., Nissley, S.P., Hill, D.J., Strain, A.J. and Rechler, M.M. (1986) J. Biol. Chem. 261, 13144–13150.

50. Rechler, M.M. and Nissley, S.P. (1985) Annu. Rev. Physiol. 47, 425–442.
51. Smith, G.L. (1984) Mol. Cell. Endocrinol. 34, 83–89.
52. Mottola, C., MacDonald, R.G., Brackett, J.L., Mole, J.E., Anderson, J.K. and Czech, M.P. (1986) J. Biol. Chem. 261, 11180–11188.
53. Jansen, M., van Schaik, F.M.A., Ricker, A.T., Bullock, B., Woods, D.E., Gabbay, K.H., Nussbaum, A.L., Sussenbach, J.S. and Van den Brande, J.L. (1983) Nature 306, 609–611.
54. Bell, G.I., Merryweather, J.P., Sanchez-Pescador, R., Stempien, M.M., Priestley, L., Scott, J. and Rall, L.B. (1984) Nature 310, 775–777.
55. Maes, M., Underwood, L.E. and Ketelslegers, J.-M. (1986) Endocrinology 118, 377–382.
56. Han, V.K.M., D'Ercole, A.J. and Lund, P.K. (1987) Science 236, 193–197.
57. Holder, A.T. and Wallis, M. (1977) J. Endocr. 74, 223–229.
58. Pell, J.M., Blake, L.A., Elcock, C., Hathorn, D.J., Jones, A.R., Morrell, D.J. and Simmonds, A.D. (1987) J. Endocr. 112, Suppl., Abstr. 63.
59. Copeland, K.C., Underwood, L.E. and Van Wyk, J.J. (1980) J. Clin. Endocrinol. Metab. 50, 690–697.
60. Spencer, E.M. (1979) FEBS Lett. 99, 157–161.
61. Scott, C.D., Martin, J.L. and Baxter, R.C. (1985) Endocrinology 116, 1102–1107.
62. Clemmons, D.R., Underwood, L.E. and Van Wyk, J.J. (1981) J. Clin. Invest. 67, 10–19.
63. Palmer, J.M. and Wallis, M. (1987) J. Endocr. 112, Suppl., Abstr. 61.
64. Morris, D.H., Schalch, D.S. and Monty-Miles, B. (1981) FEBS Lett. 127, 211–224.
65. Mathews, L.S., Norstedt, G. and Palmiter, R.D. (1986) Proc. Natl. Acad. Sci. USA 83, 9343–9347.
66. Roberts, C.T., Brown, A.L., Graham, D.E., Seelig, S., Berry, S., Gabbay, K.H. and Rechler, M.M. (1986) J. Biol. Chem. 261, 10025–10028.
67. Murphy, L.J., Bell, G.I. and Friesen, H.G. (1987) Endocrinology 120, 1279–1282.
68. Holder, A.T., Spencer, E.M. and Preece, M.A. (1981) J. Endocr. 89, 275–282.
69. Schoenle, E., Zapf, J., Humbel, R.E. and Froesch, E.R. (1982) Nature 296, 252–253.
70. Skottner, A., Clarke, R.G., Robinson, I.C.A.F. and Fryklund, L. (1987) J. Endocr. 112, 123–132.
71. Isaksson, O.G.P., Jansson, J.-O. and Gause, I.A.M. (1982) Science 216, 1237–1239.
72. Isaksson, O.G.P., Eden, S. and Jansson, J.-O. (1985) Annu. Rev. Physiol. 47, 483–499.
73. Russell, S.M. and Spencer, E.M. (1985) Endocrinology 116, 2563–2567.
74. Madsen, K., Friberg, U., Roos, P., Edén, S. and Isaksson, O. (1983) Nature 304, 545–547.
75. Nilsson, A., Isgaard, J., Lindahl, A., Dahlström, A., Skottner, A. and Isaksson, O.G.P. (1986) Science 233, 571–574.
76. Eden, S., Isaksson, O.G.P., Madsen, K. and Friberg, U. (1983) Endocrinology 112, 1127–1129.
77. Schlechter, N.L., Russell, S.M., Spencer, E.M. and Nicoll, C.S. (1986) Proc. Natl. Acad. Sci. USA 83, 7932–7934.
78. Korner, A. and Hogan, B.L.M. (1972) In: Growth and Growth Hormone (Pecile, A. and Muller, E.E., eds.) pp. 98–105. Excerpta Medica, Amsterdam.
79. Roy, A.K., Chatterjee, B., Demyan, W.F., Milin, B.S., Motwani, N.M., Nath, T.S. and Schiop, M.J. (1983) Rec. Prog. Horm. Res. 39, 425–461.
80. Kulkarni, A.B., Gubits, R.M. and Feigelson, P. (1985) Proc. Natl. Acad. Sci. USA 82, 2579–2582.
81. Liaw, C., Seelig, S., Mariash, C.N., Oppenheimer, J.H. and Towle, H.C. (1983) Biochemistry 22, 213–221.
82. Yoon, J.-B., Towle, H.C. and Seelig, S. (1987) J. Biol. Chem. 262, 4284–4289.
83. Le Cam, A., Pages, G., Auberger, P., Le Cam, G., Leopold, P., Benarous, R. and Glaichenhaus, N. (1987) EMBO J. 6, 1225–1232.
84. Kostyo, J.L. and Nutting, D.F. (1974) In: Handbook of Physiology, Section 7: Endocrinology, Vol. 4, Part 2 (Knobil, E. and Sawyer, W.H., eds.) pp. 187–210. American Physiological Society, Washington, DC.
85. Jefferson, L.S., Robertson, J.W. and Tolman, E.L. (1972) In: Growth and Growth Hormone (Pecile, A. and Muller, E.E., eds.). pp. 106–123. Excerpta Medica, Amsterdam.

86. Korner, A. (1961) Biochem. J. 81, 292–297.
87. Mowbray, J., Davies, J.A., Bates, D.J. and Jones, C.J. (1975) Biochem. J. 152, 583–592.
88. Davidson, M.B. (1987) Endocr. Rev. 8, 115–131.
89. Goodman, H.M. and Schwartz, J. (1974) In: Handbook of Physiology, Section 7: Endocrinology, Vol. 4, Part 2 (Knobil, E. and Sawyer, W.H., eds.) pp. 211–231. American Physiological Society, Washington DC.
90. Goodman, H.M., Grichting, G. and Coiro, V. (1986) In: Human Growth Hormone (Raiti, S. and Tolman, R.A., eds.) pp. 499–512. Plenum Medical, New York.
91. Bjorgell, P., Rosberg, S., Isaksson, O. and Belfrage, P. (1984) Endocrinology 115, 1151–1156.
92. Gorin, E., Grichting, G. and Goodman, H.M. (1984) Endocrinology 115, 467–475.
93. Bennett, L.L. and Curry, D.L. (1976) In: Growth Hormone and Related Peptides (Pecile, A. and Muller, E.E., eds.) pp. 116–126. Excerpta Medica, Amsterdam.
94. Penhos, J.C., Castillo, L, Voyles, N., Gutman, R., Lazarus, N. and Recant, L. (1971) Endocrinology 88, 1141–1149.
95. Lewis, U.J., Frigeri, L.G., Sigel, M.B., Tutwiler, G.F. and Vanderlaan, W.P. (1986) In: Human Growth Hormone (Raiti, S. and Tolman, R.A., eds.) pp. 439–447. Plenum Medical, New York.
96. Kostyo, J.L. (1986) In: Human Growth Hormone (Raiti, S. and Tolman, R.A., eds.) pp. 449–453. Plenum Medical, New York.
97. Randle, P.J., Garland, P.B., Hales, C.N. and Newsholme, E.A. (1963) Lancet 1, 785–789.
98. Morikawa, M., Nixon, T. and Green, H. (1982) Cell 29, 783–789.
99. Nixon, B.T. and Green, H. (1984) Proc. Natl. Acad. Sci. USA 81, 3429–3432.
100. Jones, K.L., Villela, J.F. and Lewis, U.J. (1986) Endocrinology 118, 2588–2593.
101. Golde, D.W. (1980) In: Growth Hormone and Other Biologically Active Peptides (Pecile, A. and Muller, E.E., eds.) pp. 52–62. Excerpta Medica, Amsterdam.
102. Mercola, K.E., Cline, M.J. and Golde, D.W. (1981) Blood 58, 337–340.
103. Ledet, T. (1976) Diabetes 25, 1011–1017.
104. Rabinovitch, A., Quigley, C. and Rechler, M.M. (1983) Diabetes 32, 307–312.
105. Murphy, L.J., Bell, G.I. and Friesen, H.G. (1987) Endocrinology 120, 1806–1812.
106. Iguchi-Ariga, S.M.M., Itani, T., Kiji, Y. and Ariga, H. (1987) EMBO J. 6, 2365–2371.
107. Bines, J.A., Hart, I.C. and Morant, S.V. (1980) Br. J. Nutr. 43, 179–188.
108. Madon, R.J., Ensor, D.M., Knight, C.H. and Flint, D.J. (1986) J. Endocrinol. 111, 117–123.
109. Wallis, M., Howell, S.L. and Taylor, K.W. (1985) The Biochemistry of the Polypeptide Hormones, Chapter 18, Wiley, Chichester.
110. Gertler, A., Shamay, A., Cohen, N., Ashkenazi, A., Friesen, H.G., Levanon, A., Gorecki, M., Aviv, H., Hadary, D. and Vogel, T. (1986) Endocrinology 118, 720–726.
111. Ashkenazi, A., Vogel, T., Barash, I., Hadary, D., Levanon, A., Gorecki, M. and Gertler, A. (1987) Endocrinology 121, 414–419.
112. Tokunaga, T., Tanaka, T., Ikehara, M. and Ohtsuka, E. (1985) Eur. J. Biochem. 153, 445–449.
113. Palmiter, R.D., Brinster, R.L., Hammer, R.E., Trumbauer, M.E., Rosenfeld, M.G., Birnberg, N.C. and Evans, R.M. (1982) Nature 300, 611–615.
114. Evans, R.M., Swanson, L. and Rosenfeld, M.G. (1985) Rec. Progr. Hormone Res. 41, 317–337.
115. Cadman, H.F., Wallis, M. and Ivanyi, J. (1982) FEBS Lett. 137, 149–152.
116. Aston, R. and Ivanyi, J. (1985) Pharmac. Ther. 27, 403–424.
117. Retegui, L.A., De Meyts, P. and Masson, P.L. (1982) In: Protides of the Biological Fluids, Vol. 29 (Peeters, H., ed.) pp. 827–832. Pergamon Press, Oxford.
118. Holder, A.T., Aston, R., Preece, M.A. and Ivanyi, J. (1985) J. Endocrinol. 107, R9–R12.
119. Aston, R., Holder, A.T., Preece, M.A. and Ivanyi, J. (1986) J. Endocrinol. 110, 381–388.
120. Aston, R., Holder, A.T., Ivanyi, J. and Bomford, R. (1987) Mol. Immunol. 24, 143–150.
121. Wallis, M., Daniels, M., Ray, K.P., Cottingham, J.D. and Aston, R. (1987) Biochem. Biophys. Res. Commun. 149, 187–193.

122. Holder, A.T., Aston, R., Rest, J.R., Hill, D.J., Patel, N. and Ivanyi, J. (1987) Endocrinology 120, 567–573.
123. Lewis, U.J. (1984) Annu. Rev. Physiol. 46, 33–42.
124. Chawla, R.K., Parks, J.S. and Rudman, D. (1983) Annu. Rev. Med. 34, 519–547.
125. Lewis, U.J., Bonewald, L.F. and Lewis, L.J. (1980) Biochem. Biophys, Res. Commun. 92, 511–516.
126. Wallis, M. (1980) Nature 284, 512.
127. De Noto, F.M., Moore, D.D. and Goodman, H.M. (1981) Nucleic Acids Res. 9, 3719–3730.
128. Ader, M., Agajanian, T., Finegood, D.T. and Bergman, R.N. (1987) Endocrinology 120, 725–731.
129. Li, C.H. and Bewley, T.A. (1976) In: Growth Hormone and Related Peptides (Pecile, A. and Muller, E.E., eds.) pp. 14–32. Excerpta Medica, Amsterdam.
130. Kostyo, J.L., Mills, J.B., Reagan, C.R., Rudman, D. and Wilhelmi, A.E. (1976) In: Growth Hormone and Related Peptides (Pecile, A. and Muller, E.E., eds.) pp. 33–40. Excerpta Medica, Amsterdam.
131. Bornstein, J. (1978) Trends Biochem. Sci. 3, 83–86.
132. Seeburg, P.H. (1982) DNA 1, 239–249.
133. Paladini, A.C., Pena, C. and Retegui, L.A. (1979) Trends Biochem. Sci. 4, 256–260.
134. Hershko, A., Mamont, P., Shields, R. and Tomkins, G.M. (1971) Nature New Biol. 232, 206–211.
135. Tata, J.R. (1980) Biol. Rev. 55, 285–319.
136. Elgin, R.G., Busby, W.H. and Clemmons, D.R. (1987) Proc. Natl. Acad. Sci. USA 84, 3254–3258.
137. Abdel-Meguid, S.S., Shieh, H.-S., Smith, W.W., Dayringer, H.E., Violand, B.N. and Bentle, L.A. (1987) Proc. Natl. Acad. Sci. USA 84, 6434–6437.
138. Leung, D.W., Spencer, S.A., Cachianes, G., Hammonds, R.G., Collins, C., Henzel, W.J., Barnard, R., Waters, M.J. and Wood, W.I. (1987) Nature 330, 537–543.
139. Wallis, M. (1987) Nature 330, 521–522.

CHAPTER 14

Mechanism of action of prolactin

MICHAEL WALLIS

Biochemistry Laboratory, School of Biological Sciences, University of Sussex, Falmer, Brighton, Sussex BN1 9QG, England

1. Lactogenic hormones

Pituitary prolactin is a protein hormone comprising a single polypeptide chain of about 200 amino acid residues and 3 intrachain disulphide bridges [1,2]. It is structurally homologous with growth hormone (GH) as is discussed in Chapter 13 (Fig. 1). Amino acid sequences of prolactins from various different species have been described [1–3]; they show a considerable amount of species variation, which is most notable in the case of the prolactins of the rat and mouse, which differ from prolactins of other mammals at about 40% of all amino acid residues [3,4].

The main actions of prolactin in mammals are concerned with the mammary gland (see below). Lactogenic and mammogenic actions are also shown by several other hormones, including the placental lactogens found in man and many other mammalian species [5–7]. The physiological role of these placental hormones has not been fully established, but it does seem likely that their biochemical mode of action will prove to be similar in some respects to that of prolactin. Their structural relationships to GH and prolactin are discussed in Chapter 13.

The similarity in structure between GH and prolactin is associated with some overlap in biological activity. Thus prolactin has growth-promoting activity in some species (though this is usually much lower than that of GH), and can stimulate somatomedin production [8,9]. A particularly striking example of this overlap is shown in the case of human GH. In addition to its growth-promoting effects, the human hormone has lactogenic activity which, in some mammalian systems, is approximately equal to that of prolactin [10]. In other assay systems, for example the pigeon crop sac, human GH is less active than prolactin [10]. At one time it was suggested that human GH fulfilled the role of a lactogenic hormone in man, and that a distinct human prolactin did not exist. It is now clear, however, that a separate prolactin occurs in man, as in other mammals [11]. Whether human GH is lactogenic in humans, and plays any physiological role as a lactogenic hormone is not clear. In non-primate mammals GH does not possess lactogenic activity in the large

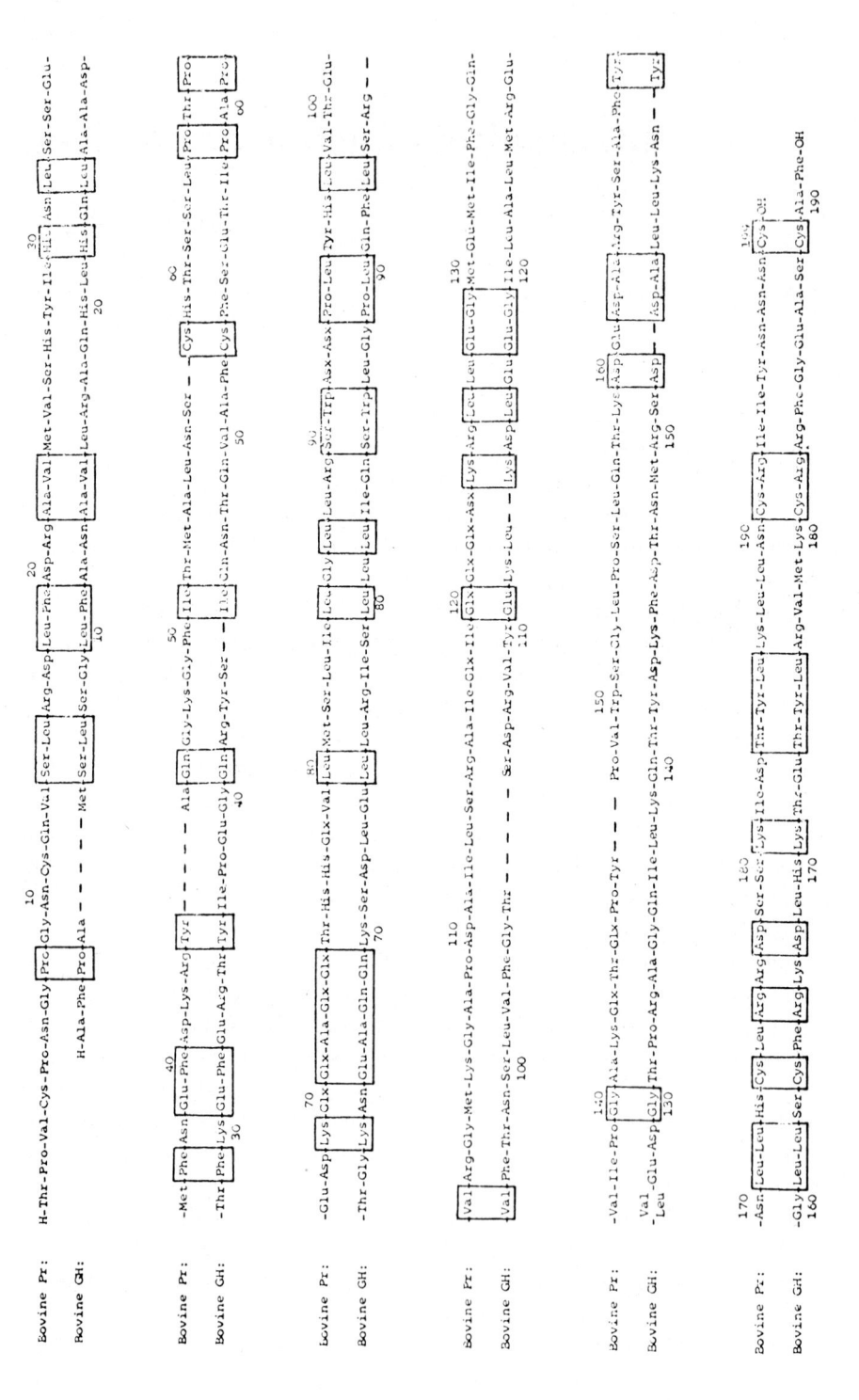

Bovine Pr: H-Thr-Pro-Val-Cys-Pro-Asn-Gly-Pro-Asn-Cys-Gln-Val-Ser-Leu-Arg-Asp-Leu-Phe-Asp-Arg-Ala-Val-Met-Val-Ser-His-Tyr-Ile-His-Asn-Leu-Ser-Ser-Glu-
Bovine GH: H-Ala-Phe-Pro-Ala- - - - - - - - - -Met-Ser-Leu-Ser-Gly-Leu-Phe-Ala-Asn-Ala-Val-Leu-Arg-Ala-Gln-His-Leu-His-Gln-Leu-Ala-Ala-Asp-

Bovine Pr: -Met-Phe-Asn-Glu-Phe-Asp-Lys-Arg-Tyr- - - - - Ala-Gln-Gly-Lys-Gly-Phe-Ile-Thr-Met-Ala-Leu-Asn-Ser-   -Cys-His-Thr-Ser-Ser-Leu-Pro-Thr-Pro-
Bovine GH: -Thr-Phe-Lys-Glu-Phe-Glu-Arg-Thr-Tyr-Ile-Pro-Glu-Gly-Gln-Arg-Tyr-Ser- - - Ile-Gln-Asn-Thr-Gln-Val-Ala-Phe-Cys-Phe-Ser-Glu-Thr-Ile-Pro-Ala-Pro-

Bovine Pr: -Glu-Asp-Lys-Glx-Ala-Glx-Thr-His-His-Glx-Val-Leu-Met-Ser-Leu-Ile-Leu-Gly-Leu-Leu-Arg-Ser-Trp-Asx-Asx-Pro-Leu-Tyr-His-Leu-Val-Thr-Glu-
Bovine GH: -Thr-Gly-Lys-Asn-Glu-Ala-Gln-Gln-Lys-Ser-Asp-Leu-Glu-Leu-Leu-Arg-Ile-Ser-Leu-Leu-Leu-Ile-Gln-Ser-Trp-Leu-Gly-Pro-Leu-Gln-Phe-Leu-Ser-Arg- -

Bovine Pr: Val-Arg-Gly-Met-Lys-Gly-Ala-Pro-Asp-Ala-Ile-Leu-Ser-Arg-Ala-Ile-Glx-Ile-Glx-Glx-Asx-Lys-Leu-Leu-Gly-Met-Glu-Met-Ile-Phe-Gly-Gln-
Bovine GH: Val-Phe-Thr-Asn-Ser-Leu-Val-Phe-Gly-Thr- - - Ser-Asp-Arg-Val-Tyr-Glu-Lys-Leu- - -Lys-Asp-Leu-Glu-Glu-Gly-Ile-Leu-Ala-Leu-Met-Arg-Glu-

Bovine Pr: -Val-Ile-Pro-Gly-Ala-Lys-Glx-Thr-Glx-Pro-Tyr- - - Pro-Val-Trp-Ser-Gly-Leu-Pro-Ser-Leu-Gln-Thr-Lys-Asp-Glu-Asp-Ala-Arg-Tyr-Ser-Ala-Phe-Tyr-
Bovine GH: Val-Leu-Gly-Gly-Thr-Pro-Arg-Ala-Gly-Gln-Ile-Leu-Lys-Gln-Thr-Tyr-Asp-Lys-Phe-Asp-Thr-Asn-Met-Arg-Ser-Asp-Ala-Ala-Leu-Leu-Lys-Asn- -Tyr-

Bovine Pr: -Asn-Leu-Leu-His-Cys-Leu-Arg-Arg-Asp-Ser-Ser-Lys-Ile-Asp-Thr-Tyr-Leu-Lys-Leu-Leu-Asn-Cys-Arg-Ile-Ile-Tyr-Asn-Asn-Asn-Cys-OH
Bovine GH: -Gly-Leu-Leu-Ser-Cys-Phe-Arg-Lys-Asp-Leu-His-Lys-Thr-Glu-Thr-Tyr-Leu-Arg-Val-Met-Lys-Cys-Arg-Arg-Phe-Gly-Glu-Ala-Ser-Cys-Ala-Phe-OH

Fig. 1. Comparison of the amino acid sequences of bovine prolactin and growth hormone. Identical sequences are enclosed in 'boxes'.

majority of in vitro lactogenic assays. It is established that GH may promote lactogenesis in ruminants [12], and possibly in the rat [13], but it appears to do this via a mechanism distinct from that involved in the action of prolactin.

2. The biological actions of prolactin

2.1. Actions on the mammary gland

In mammals the main actions of prolactin appear to be promotion of milk production and growth of the mammary gland [14–16]. The hormone has been shown to initiate or maintain milk production, in vivo, in a variety of mammals, including rat, mouse and rabbit, provided the animal is in a suitable 'primed' state. Its actions on the mammary gland involve interaction with a variety of other hormones, which depend on the species and may include insulin (and/or somatomedin C/IGF-I), corticosteroids, thyroid hormones, growth hormone and oestrogens [14–16] (Fig. 2). Progesterone blocks the action of prolactin, at least in some circumstances. The actions of prolactin involve both growth of the gland, particularly the secretory tissue, and induction and maintenance of production of milk components, including proteins such as casein and lactalbumin, carbohydrates (especially lactose) and fats. Production of carbohydrates and fats is enhanced via induction of the appropriate enzymes. Prolactin also has actions on the salt and water relationships associated with milk production [17].

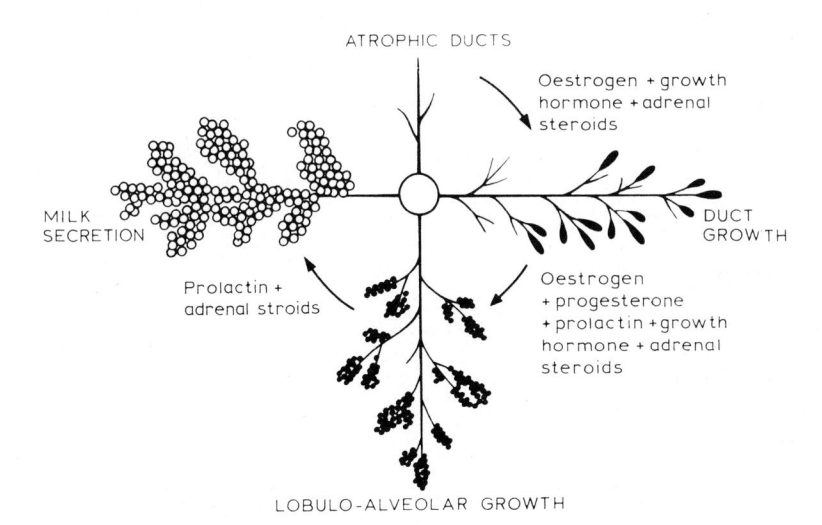

Fig. 2. The action of hormones on growth and milk secretion in the mammary glands of hypophysectomized-ovariectomized-adrenalectomized rats. From Ref. 14.

2.2. Other actions in mammals

In addition to its effects on the mammary gland, prolactin has a wide range of other effects in mammals. The hormone has luteotrophic actions in rodents and some other mammals [18], playing a role in maintaining the corpus luteum during early pregnancy. In different circumstances, however, prolactin has an anti-gonadal action in some female mammals; it may be at least partly responsible for the lowered fertility seen during lactation in several mammalian groups, including women [19]. Prolactin may also be involved in the control of testicular function in some male mammals [20].

Prolactin appears to play a role in regulating the immune system in mammals, though the significance of this is not yet clear. Receptors for the hormone have been

TABLE I
Some actions of prolactin

Elasmobranchs
Control of salt and water permeability.
Teleosts
Osmoregulatory actions.
Parental behaviour (nest-building, buccal incubation of eggs, maintenance of brood pouch) in some species.
Dispersion of pigment xanthophores in some species.
Metabolic effects.
Stimulation of epidermal mucus production.
Amphibians
Eft water drive in newts.
Osmoregulatory actions.
Stimulation of moulting.
Larval growth; antagonizes effects of thyroxine on metamorphosis.
Limb regeneration.
Reptiles
Osmoregulatory actions.
Effects on growth, metabolism and tail regeneration.
Epidermal sloughing.
Anti-gonadotropic actions.
Birds
Stimulation of crop-sac milk production in pigeons and doves.
Stimulation of salt gland secretion.
Maintenance of broodiness.
Antigonadal effects.
Metabolic effects.
Mammals
Stimulation of mammary growth and development of milk secretion.
Actions on the immune system.
Luteotropic actions.
Behavioural effects.
Control of testicular function.

detected in cells of the immune system [21]. The hormone may enhance the stimulation of lymphocyte mitogenesis by lectins such as concanavalin A; indeed it has been suggested that a prolactin-like protein is produced by splenocytes in response to stimulation by mitogens, and may promote proliferation [22]. Prolactin also has marked mitogenic effects on a rat lymphoma-derived cell line (Nb2); this effect is discussed further below. Prolactin has a wide range of other effects in mammals [8], but the physiological significance of these is not clear. Various behavioural effects have been attributed to the hormone, in the rat, man and other animals, including effects on maternal behaviour, sexual activity and drug withdrawal symptoms [23].

2.3. Actions in lower vertebrates

A very wide range of effects of prolactin has been described in non-mammalian vertebrates, some of which are summarized in Table I, and the comparative endocrinology of prolactin has been studied in considerable detail [8,24]. Such effects of prolactin are usually associated with secondary sexual activities, growth, or salt and water relationships. They cannot be considered in detail here but a few actions of particular biochemical interest will be mentioned briefly. In some fish the properties of the homologue of prolactin appear to be somewhat different from those of the mammalian hormone, and the alternative term 'paralactin' has sometimes been used. Among the many actions described for prolactin in fish those on osmoregulation are of particular interest; the hormone plays an important role in maintaining osmoregulation in fresh-water fish. Osmoregulatory actions are also important in most other vertebrate groups; for example, prolactin stimulates the secretion of NaCl from the salt gland in some seabirds. Another important action in birds is stimulation of the production of crop-sac 'milk' in the pigeon. Cells from the wall of the crop sac are sloughed off to form a milk-like substance that is regurgitated to feed the young. Proliferation of the cells required for this 'milk' is initiated and maintained by prolactin. This action forms the basis of an important bioassay for prolactin, and has been the subject of some biochemical studies. The overall effect may be enhanced by synergism between prolactin and a growth factor (synlactin) related to insulin-like growth factor, production of which may also be under the control of prolactin [25] (see below).

3. Receptors for prolactin

3.1. Characterization of receptors

Binding sites for prolactin and other lactogenic hormones are widely distributed in mammalian tissues [26,27]. Discrimination by such binding sites between GH and prolactin is not always complete, and the frequent use of ^{125}I-labelled human GH

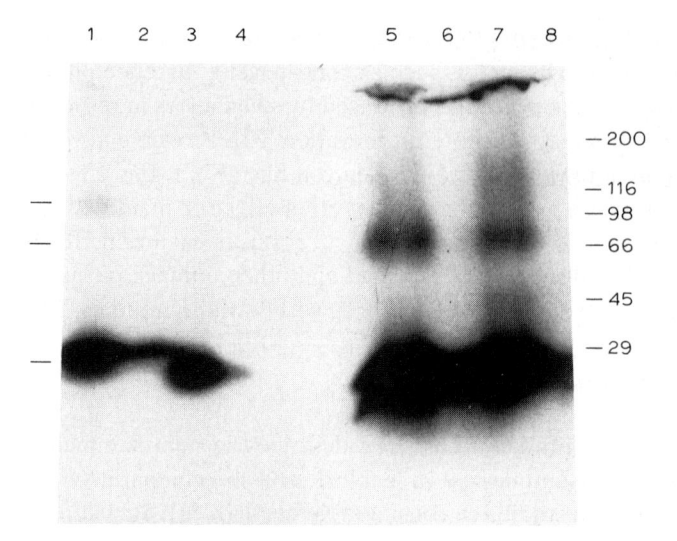

Fig. 3. Crosslinking of ^{125}I-labelled human growth hormone to lactogenic receptors from rat liver and the Nb2 cell line. Hormone-receptor complexes were cross-linked with disuccinimidyl suberate, fractionated by SDS gel electrophoresis and detected by autoradiography. Lanes 1–4, Nb2 cells; lanes 5–8, rat liver. Lines on the left indicate the positions of the hormone-receptor complex for Nb2 cells (102 kDa), rat liver (71 kDa) and uncomplexed hormone (hGH). Numbers on the right are M_r values for standards (kDa). Samples were incubated with (lanes 2,4,6,8) or without (lanes 1,3,5,7) unlabelled hGH and electrophoresed under reducing (lanes 1,2,5,6) or non-reducing (lanes 3,4,7,8) conditions. From Ref. 36.

(which has both lactogenic and somatogenic activity) has tended to complicate the interpretation of binding studies (see also Chapter 13). Most work has concentrated on binding sites (putative receptors) in mammary gland and in liver, though prolactin receptors in a variety of other tissues have also been characterized.

Prolactin receptors from rabbit mammary gland bind labelled prolactin and other lactogenic hormones (including human GH) but not non-primate GHs [26–28]. The receptors from this source can be solubilized from membrane preparations by the use of non-ionic detergents such as Triton X-100, with retention of binding properties similar to those of the membrane-bound receptors. The M_r of the solubilized receptor was approximately 220 000, as assessed by gel filtration [29], though it is difficult to assess fully the contribution of the associated detergent. Purification of prolactin receptor solubilized from rabbit mammary gland has been carried out, using affinity chromatography on hormone attached to a solid phase [29]. Overall purification of about 1500-fold was achieved, but the material obtained in this way still showed considerable heterogeneity.

Prolactin receptors from a number of other tissues, including rat and mouse liver [30,31], have been solubilized using various detergents and then partially characterized using gel filtration. The apparent M_r found appears to depend on the type of detergent used for solubilization, and in one study a prolactin-binding 'subunit' of M_r about 37 000 was obtained for CHAPS-solubilized prolactin receptor from mouse liver [31,32]. M_r values in the range 36 000 – 45 000 have also been found for the recognition subunits of rabbit and rat receptors using cross-linking of labelled hormone to receptor followed by SDS polyacrylamide gel electrophoresis [26,33–36] (Fig. 3). Others have found a more complex pattern of prolactin-binding subunits using cross-linking techniques in rat liver and Leydig cells [37,38]. The prolactin receptor isolated from rat ovary appeared to contain subunits of M_r 41 000 and 88 000 [39].

Prolactin and other lactogenic hormones can stimulate the proliferation of Nb2 cells, as has been mentioned previously. Binding sites/receptors for prolactin have been detected on this cell line. Interestingly, these receptors appear to have a considerably higher affinity for prolactin than do those on rat liver [40]. They also appear to have a larger recognition subunit as detected by cross-linking (Fig. 3), and a different organization of subunits, since the cross-linked receptors behave quite differently on SDS gel electrophoresis if reducing agent is omitted, whereas the behaviour of rat liver receptors is unchanged [36,41].

Polyclonal antibodies have been prepared against partially purified rabbit mammary gland receptors [42–44], and have been used to demonstrate that the receptors really are involved in mediating the biological actions of prolactin. Anti-receptor antibodies that blocked binding of labelled prolactin were able to inhibit prolactin-stimulated casein synthesis and amino acid transport in rabbit mammary gland explants in vitro [43,45] (Fig. 4). They had no effect on the actions of insulin

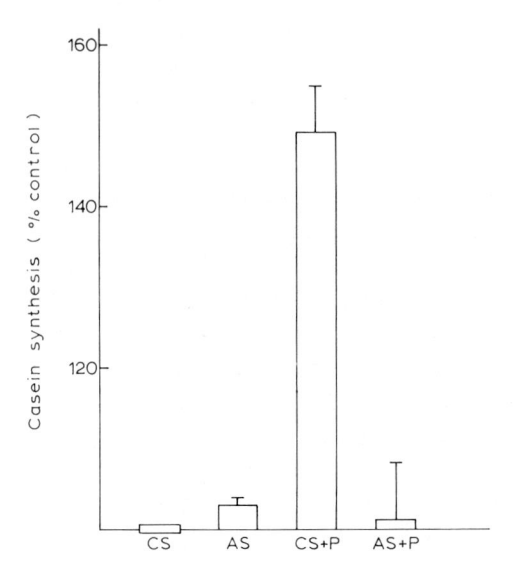

Fig. 4. Antibodies to the prolactin receptor block the actions of the hormone on casein synthesis in rabbit mammary tissue explants. CS, control serum; AS, antiserum to receptor; P, prolactin. Data from Ref. 45.

in the same explants. The antibodies also inhibited milk production in lactating rats [46], markedly inhibited a luteolytic action of prolactin on the ovary and caused elevation of circulating prolactin levels. Interestingly they were also able to block the actions of prolactin on proliferation of Nb2 cells [40], suggesting that despite the differences between receptors from Nb2 cells and those in liver, the Nb2 cell receptors share antigenic determinants with rabbit mammary gland prolactin receptors.

In the absence of prolactin some anti-receptor antibodies were able to mimic partially the actions of prolactin in both Nb2 cells and mammary gland explants [26,40,43]. Some evidence suggested that the biochemical mechanisms involved in the action of such antibodies are similar to those mediating the actions of prolactin itself [43]. For the response of Nb2 cells to stimulation by anti-prolactin receptor antibodies it has been demonstrated that univalent (Fab) fragments of the antibodies were without effect [40], suggesting that the cross-linking and aggregation of receptors induced by divalent antibodies is necessary for manifestation of prolactin-like actions. The polyclonal antibodies to the prolactin receptor could also mimic the biological actions of the hormone on rat mammary tumour explants [47] and rat hepatocytes [48].

More recently, monoclonal antibodies to the rabbit prolactin receptor have been described [49,50]. These showed strong species specificity, competed with ^{125}I-labelled prolactin for binding to receptors, and provided evidence for receptor heterogeneity in some tissues [49]. Two of the monoclonal antibodies could block the actions of prolactin on casein synthesis in rabbit mammary gland explants, and one of them could inhibit milk production in vivo [50]. A third antibody could mimic the actions of prolactin on both casein and DNA synthesis in mammary explants. For this effect bivalency was essential, and down-regulation of receptors was much less than that produced by prolactin itself [50].

These monoclonal antibodies to the rabbit prolactin receptor showed little or no cross-reaction with the rat receptor, and monoclonal antibodies have therefore been prepared also to the receptor from rat liver [51]. Two such antibodies were produced, which recognized prolactin receptors from several different rat tissues but did not cross-react with receptors from other species. Studies with these antibodies confirmed the M_r of the prolactin receptor (or its subunit) as 42 000 – 46 000, close to the value obtained by cross-linking techniques (see above). Biological effects of these monoclonal antibodies were not reported, although stimulatory actions of polyclonal antibodies to the rat prolactin receptor have been described [47,48,52].

Another approach to the preparation of antibodies against the prolactin receptor has involved the production of anti-idiotypic antibodies against anti-prolactin antibodies [53]. Such antibodies were able to compete with prolactin for binding to the receptor.

Work with antibodies to receptors has provided important information about the mechanism of action of prolactin, and should provide further insights in future. It provides confirmation that at least some (though not necessarily all) of the binding sites involved really are mediators of biological activity. It demonstrates that at least some of the actions of prolactin cannot be mediated by internalized fragments of the hormone. Such antibodies can also be used to study the physiological importance of various actions of prolactin and the relatedness of the various receptors that have been described, and are potentially valuable for purification of receptors.

In addition to the membrane-associated receptors for prolactin that have been described here, soluble binding proteins for the hormone have also been reported [54,55]. These have properties similar (but not identical) in several respects to the membrane-bound receptors, but their biological significance is not clear.

3.2. Regulation of prolactin receptors

The levels of prolactin receptors in target tissues are subject to regulation by a variety of hormonal and other factors (reviewed in Refs. 26 and 56). The mechanisms involved remain to be elucidated fully. Regulation at the level of receptor synthesis is undoubtedly important (particularly in view of the fairly rapid turnover rate observed for the receptor: $t_{1/2}$ of 35–50 min in rat liver [57,58]), but regulation by exposure of previously synthesized but 'hidden' receptors may also play a role, as may regulation of internalization and degradation/recycling of receptors. It has been demonstrated that a substantial proportion of the prolactin receptors in rat liver is associated with intracellular membranes (rather than the plasma membrane). These may represent receptors that have been internalized and are en route to recycling or degradation. Alternatively, they could be newly synthesized receptors destined for insertion into the plasma membrane, or a store of receptors available for such insertion when circumstances require [59,60].

Prolactin receptors are known to be internalized following binding of the hormone [59,61], leading to 'down-regulation'. The mechanisms involved in such internalization are similar to those applying in the case of other polypeptide hormones, and have been discussed in Volume 18A (Chapter 9). The significance of the internalization of prolactin is not fully understood. It seems likely that following internalization there is a separation of hormone and receptor, the hormone being degraded and the receptor being at least partly recycled for further use. In this view, internalization represents primarily a mechanism for switching off the activity of the hormone-receptor complex and removing the hormone after it has carried out its main function. However, it is also possible that internalization of the hormone-receptor complex plays an important part in mediating the actions of the hormone.

The interaction of factors regulating levels of prolactin receptors [26,56] in a variety of tissues is complex and not fully understood. The topic will not be considered in detail here. The actions of prolactin itself on the process will be discussed briefly, however. In the short term, prolactin often leads to down-regulation of its own receptors, presumably as a consequence of internalization following binding of hormone to receptor. In the longer term, however, the hormone usually leads to up-regulation of its receptors, as has been demonstrated in a range of tissues [26,56,62]. GH, too, may lead to an increase in the level of prolactin receptors in some tissues [63].

4. Biochemical mode of action of prolactin on the mammary gland

The main actions of prolactin in female mammals involve stimulation of mammary growth and function, and it is in this area that the bulk of work on the biochemical mode of action of the hormone has been concentrated.

4.1. Actions on mammary gland differentiation and development

The development of the mammary gland both during puberty and pregnancy is under the control of a number of hormones [15,16,64–66], but these differ according to the species. During puberty oestrogens, and possibly progesterone, play a central part in the development of the gland, and insulin, GH and/or prolactin (possibly via the mediation of somatomedins or related polypeptides) may also be necessary at this stage. Development during puberty mainly involves growth of the ductal system of the gland, relatively few secretory alveoli being present prior to the first pregnancy. During pregnancy some additional branching of the ductal system occurs, together with formation of the secretory alveoli which actually produce the milk components. Hormones involved in development of the gland during pregnancy include oestrogens, progesterone, insulin, glucocorticoids and prolactin and in some species thyroid hormones, GH and possibly epidermal growth factor (EGF). In some animals placental lactogen may partly or completely take over the role of prolactin in the later stages of pregnancy.

Secretion of milk itself is inhibited during late pregnancy by progesterone and possibly oestrogens. Progesterone may also inhibit prolactin secretion. At parturition, levels of these steroid hormones, especially progesterone, fall and this plus elevated prolactin levels leads to initiation of lactation. In some species, such as the rabbit, administration of prolactin alone directly into the mammary gland can lead to initiation of milk production in a suitably primed (pregnant or pseudo-pregnant) animal, emphasizing the specific and central role played by prolactin in initiation of lactation (though the presence of a variety of other, endogenous, hormones is necessary for such initiation to be demonstrated). Once lactation has been initiated, prolactin plays an important part in maintaining it, although it is notable that in lactating ruminants GH appears to be more important in this respect than prolactin.

A considerable amount of work has been carried out on the sequence of events occurring at the cellular level during development and differentiation of the mammary gland, using explants of mammary gland from mouse and other species, cul-

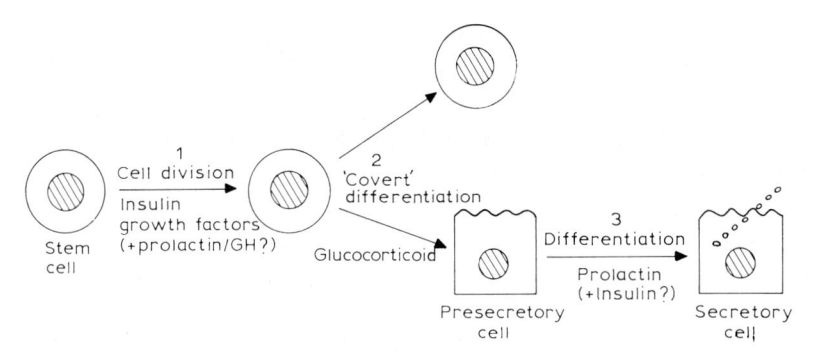

Fig. 5. The possible roles of hormones in promoting proliferation and differentiation of mouse mammary epithelial cells in vitro. From Ref. 132.

tured in vitro [64–69]. If explants from the mammary gland of a mid-pregnant mouse (prior to initiation of the production of milk proteins) are cultured in vitro they fail to survive unless hormones are added to the culture medium. Insulin, added to the medium at quite high concentrations, allows survival of the explants, with DNA synthesis and cell division but no production of milk proteins. It may be that the physiologically important hormone at this stage is in fact somatomedin C (insulin-like growth factor I) or a related polypeptide rather than (or as well as) insulin [70–72]. Further development in vitro requires the presence of a glucocorticoid – induction of specific milk proteins appears to require a combination of prolactin and glucocorticoid. Explants that have been cultured in the presence of insulin and a corticosteroid for 2–3 days will respond to the further addition of prolactin by producing substantial quantities of milk proteins such as casein and lactalbumin.

On the basis of these in vitro studies it has been suggested that a distinction can be made between a series of hormonally controlled stages in the development of the mammary gland, which is summarized diagrammatically in Fig. 5. First there is a phase of proliferation and growth induced by insulin and/or somatomedin-like peptides. A cycle of cell division induced by these hormones seems to be necessary before the mammary epithelial cells can differentiate and start milk production. In vitro insulin stimulates a variety of functions associated with cell division and growth, including DNA and RNA synthesis, synthesis and post-translational modification of histones and other nuclear proteins and production of RNA and DNA polymerases [67]. After the phase of proliferation mammary cells appear to undergo a 'covert differentiation' induced by corticosteroids. This may include stimulation of production of the apparatus (rough endoplasmic reticulum, Golgi apparatus, etc.) needed by any secretory cell. Finally terminal differentiation of the mammary epithelial cell occurs, under the influence of prolactin, in which the cell begins to synthesize and secrete milk proteins and other components of milk.

Culture of whole mammary gland from immature female mice primed with oestrogen and progesterone has enabled hormonal influences on mammary growth and development of the lobular-alveolar structure to be dissociated further from those on induction of milk proteins [66]. For the former process, culture for several days in medium containing insulin, prolactin, estrogen, progesterone and GH is required; subsequently milk protein (casein) synthesis can be induced by insulin, prolactin and cortisol. This work demonstrates that the actions of prolactin can be separated into those on development of the mammary gland and those on induction of milk protein production. It also suggests that corticosteroids are required in the process to synergize with prolactin in the induction of specific milk proteins rather than as inducers of a general differentiation of the secretory cells as had been suggested from work with mammary explants from pregnant mice [67]. It is clear that the details of interaction between prolactin and glucocorticoids in this system are complex and still to be elucidated fully. A further complicating factor is the observation [73] that prolactin may be required for maintenance of glucocorticoid receptors in mammary tissue.

4.2. Effects on synthesis of milk proteins

Among its effects on the epithelial cells of the mammary alveoli prolactin stimulates the production of the major milk proteins casein and lactalbumin [64,66–69,74–76]. Mouse mammary explants incubated in vitro with insulin and a corticosteroid produce very little of these proteins, but addition of prolactin to the culture medium induces their synthesis 10–20-fold within about 24 h. As has already been discussed, there is some uncertainty about the importance of glucocor-

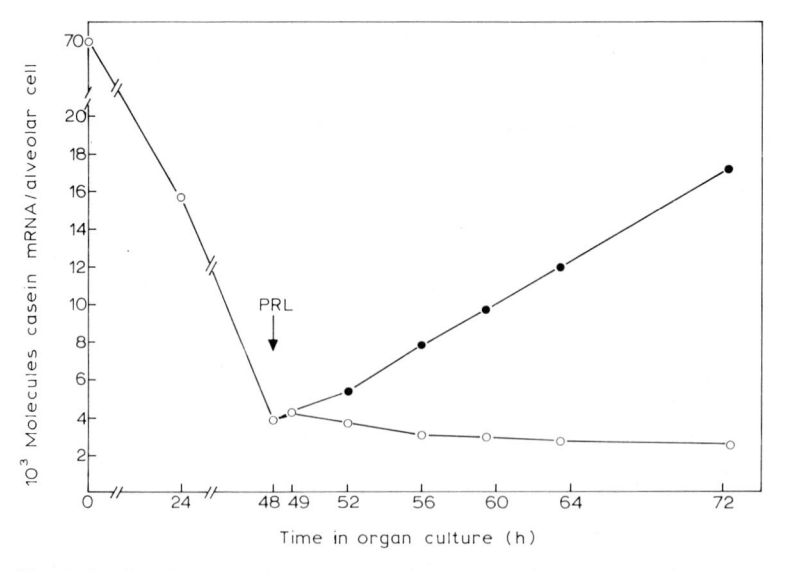

Fig. 6. Casein mRNA accumulation in mammary gland tissue from pregnant rats incubated in vitro in the presence of insulin and cortisol with or without prolactin (PRL). From Ref. 74.

ticoids in the induction, but it seems most probable that a glucocorticoid is required. Progesterone can block the induction of milk protein synthesis, probably by blocking binding of glucocorticoid to its receptor.

Induction of milk proteins appears to involve increased transcription of the appropriate genes. Levels of mRNA for casein increase substantially [66,69,74–77] (Fig. 6), though to a lower extent than the increase in casein production, suggesting that an increased rate of translation of mRNA (or possibly decreased protein degradation) may also be important. The increased levels of mRNA may be partly due to decreased degradation as well as increased transcription [74].

4.3. Effects on other milk components

In addition to its effects on synthesis of milk proteins themselves, prolactin also induces the production in the mammary gland of other proteins/enzymes involved in production of milk components. Lactose synthetase and enzymes of lipid metabolism, including acetyl-CoA carboxylase and lipoprotein lipase, are among these inducible enzymes [75].

Prolactin also stimulates the sodium/potassium ATPase of the mammary gland, leading to uptake of potassium ions and extrusion of sodium ions from the epithelial cells [78]. Via this and other effects the hormone probably plays an important part in regulating salt and water relationships in the mammary gland. Milk secretion involves a considerable loss of both water and salts from the mother, and control of these is of crucial importance.

4.4. Second messengers in the action of prolactin

The nature of the mechanisms coupling binding of prolactin to its receptor at the plasma membrane to intracellular responses is poorly understood. The effects of the hormone are seen within the cell, including the nucleus, and a second messenger mechanism or its equivalent is therefore expected. Considerable effort has been devoted to investigating possible second messengers for prolactin [26,68,79,80], but there is little clear-cut evidence in favour of any one.

A role for cyclic AMP as a mediator for the actions of prolactin has been proposed, but the bulk of the evidence available suggests that it is not directly involved. It is possible that levels of cyclic AMP-dependent protein kinase are limiting, rather than cyclic AMP itself, and that these can be increased by prolactin, presumably by induction of the appropriate genes [67]. Cyclic AMP derivatives do not mimic the actions of prolactin in in vitro systems. Indeed, they may inhibit some actions of the hormone, including stimulation of synthesis of milk proteins, fatty acids, DNA and RNA in mammary gland explants [79–81].

Cyclic GMP can have some prolactin-like actions in rat mammary tissue, including stimulation of RNA synthesis and elevation of levels of casein mRNA but not stimulation of casein synthesis [80,82]. However, the effects of this cyclic nucleotide are too small to make it a likely candidate as the sole mediator of the actions of prolactin.

Prostaglandins B_2, E_2 and $F_{2\alpha}$ also have some prolactin-like actions, particularly in the stimulation of RNA synthesis in mouse mammary gland, but they cannot stimulate casein synthesis [80,83]. They can elevate synthesis of milk proteins, however, if combined with polyamines (especially spermidine), though again the magnitude of the response is too small to indicate that this combination could provide the main second messenger for prolactin action. A role for polyamines in the action of prolactin is also supported by observations that the hormone is able to stimulate ornithine decarboxylase (an enzyme involved in polyamine production) in various tissues including mammary explants [84,85]. Phospholipases A_2 and C can also stimulate ornithine decarboxylase in mammary explants, suggesting a possible role for these enzymes in the actions of prolactin [85].

Cytoskeletal elements may also be involved in mediating the effects of prolactin on the mammary gland. Colchicine, which disrupts intracellular microtubules, completely blocks the effect of prolactin on casein mRNA production [86]. However, griseofulvin, another microtubule-disrupting drug, has no such effect.

A series of studies has been reported suggesting that a messenger for the actions of prolactin on mammary gland could be produced by incubating prolactin with mammary membrane preparations in vitro [87,88]. Partial purification of this putative messenger was reported, but the work proved difficult to repeat [89] and it is unclear whether the factor involved is a true prolactin messenger.

Involvement of products of cleavage of phosphatidylinositol bisphosphate and calcium ions as mediators of the actions of prolactin has also been suggested [90], as have changes in the ion flux into the cell brought about by changes in activity of sodium-potassium ATPase [79,91], but the importance of these factors is not clear. The hormone has marked effects on synthesis of phospholipids [92] and on sodium-potassium ATPase, but these may reflect the altering needs of the mammary gland as it begins to secrete milk, rather than stimulation of second messengers.

Most consideration of possible second messengers for the action of prolactin assumes that the hormone will act at the plasma membrane. There is considerable evidence, however, that following binding to its receptor prolactin may be internalized (see above). Such internalization is usually considered to be concerned mainly with degradation of the hormone, but it is possible that it is involved also in the actions of the hormone. Some direct actions of prolactin on intracellular components have been described (e.g. actions on isolated nuclei [93]) and it is possible that receptor-mediated endocytosis provides a mechanism whereby prolactin is moved to sites within the cell at which it (or fragments of it) act directly. Alternatively, internalization could lead to the introduction into the cell of prolactin-containing vesicles in which the action of hormone-receptor complexes is prolonged and brought into close contact with intracellular structures or components which are modified by action of the hormone. A number of studies have indicated that prolactin may be processed to smaller, well-defined fragments at its target tissues.

Various studies argue against a critical role for internalization and processing in the actions of prolactin, however. Thus, prolactin bound to large particles, such as Sephadex beads, can bring about the normal actions of the hormone [94], although in this form it could not be internalized (unless removed from the beads). Furthermore, various agents, such as chloroquine, which block the normal processing pathway, do not prevent the actions of the hormone [86,95]. Finally, as has been discussed, antibodies to prolactin receptors can in some circumstances show prolactin-like effects [26,40,43,50], making it unlikely that the actions could be due to fragments of the hormone acting at intracellular sites.

5. *Actions of prolactin on the pigeon crop sac*

As has been mentioned, in pigeons and doves the young are fed by a milk-like fluid regurgitated from the crop. This 'milk' is formed by a sloughing of cells from the wall of the crop sac, a process which is stimulated by prolactin [96,97]. The hormone stimulates both an increase in proliferation of the cells which contribute to the 'milk' and a change in the proteins and lipids contained within the cells [97,98], and these two types of response appear to be controlled independently [99]. The effect is a direct one, and prolactin can produce local effects when injected in the immediate vicinity of one side of the crop.

There has been some work on the nature of the second messengers that mediate the mitogenic actions of prolactin on the crop sac. A role for cyclic GMP has been suggested [100]. Injection of this nucleotide with a small quantity of prolactin into the skin above the crop potentiated the actions of the hormone, although when injected alone the cyclic nucleotide had no effect.

5.1. Synlactin and the actions of prolactin

Although prolactin has direct effects on the pigeon crop sac, its actions when given locally are qualitatively different, and in some cases less marked, than when it is given systemically. On the basis of this and various other observations Nicoll and his colleagues [25,101] have proposed that the mitogenic actions of prolactin on the tissue involve a synergism between prolactin itself, acting directly, and another factor, 'synlactin', produced elsewhere in the body. GH and prolactin, given systemically, could markedly potentiate the action of a small amount of prolactin injected locally over the crop sac, although various other pituitary hormones were ineffective. Serum from pigeons treated systemically with prolactin potentiated the actions of prolactin when given locally with the hormone, but had no effect when given alone. Such experiments suggested that prolactin (and/or GH) could induce the production of synlactin at a site distant from the crop sac, and thus promote mitogenesis in the crop sac both directly and indirectly.

The source and nature of synlactin have been studied further. Medium produced by incubating liver slices from pigeon or pregnant rat or mouse in vitro was able to synergize with prolactin, suggesting the liver as a major source of synlactin [25,102,103]. Somatomedin C and proinsulin were also able to synergize with prolactin in induction of pigeon crop sac growth, suggesting that synlactin is a somatomedin-like polypeptide. However, it is unlikely that synlactin is in fact somatomedin C, since livers of pregnant rats produce far more synlactin than those of non-pregnant animals, although they produce approximately equal quantities of somatomedin C.

These studies have led to the formulation of the 'synlactin hypothesis' [25,101,104] (Fig. 7), equivalent in some ways to the somatomedin hypothesis proposed for the actions of GH. This postulates that in promoting mitogenesis prolactin can have both direct and indirect effects, the latter being mediated by synlactin. The hypothesis is proposed to operate not only in relation to the actions of prolactin on the pigeon crop sac, but also to mitogenic actions on mammary gland. Such mitogenic actions are difficult to demonstrate in vitro, although they are clear from various in vivo studies. Mitogenic effects of somatomedin C have been demonstrated using mam-

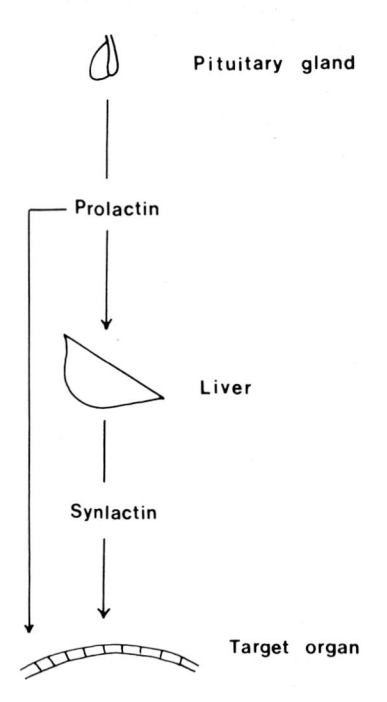

Fig. 7. The synlactin hypothesis proposes that prolactin acts on target tissues both directly and via a mediating peptide, synlactin, produced in liver and possibly other tissues.

mary explants in vitro, and production of a mitogenic factor has been shown to be associated specifically with liver from pregnant or lactating mammals, but not non-pregnant ones.

6. Actions of prolactin on the immune system

There were several relatively early reports that prolactin can have actions on cells and tissues of the immune system. More recently a variety of effects of the hormone on lymphocyte-related cells have been described and investigated in some detail.

6.1. Nb2 cell proliferation

The Nb2 cell line was derived from a rat lymphoma and can be grown in vitro indefinitely. Its growth is dependent on the presence of serum; in particular, in the presence of 10% horse serum and 10% fetal calf serum, proliferation of Nb2 cells is rapid, but if the fetal calf serum is omitted proliferation almost ceases. A series of studies showed that the requirement for fetal calf serum could be replaced by prolactin or other lactogenic hormones (human GH has frequently been used) [105]. It is not clear that the component in fetal calf serum that provides the stimulant for growth of the cell line actually is prolactin [106], but it is established that under appropriate conditions (presence of horse serum) lactogenic hormones act as specific mitogens for the line (Fig. 8). The response is very sensitive, 1 ng/ml giving a max-

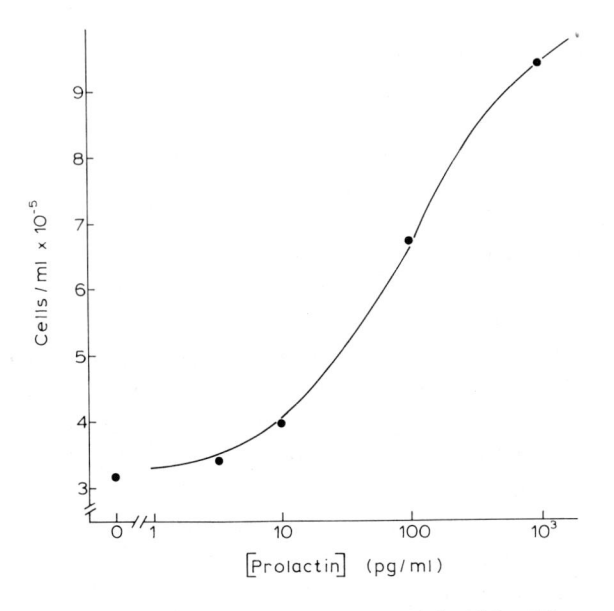

Fig. 8. Prolactin stimulates proliferation of the Nb2 cell line.

imal response in most systems, and has been established as the basis of a very sensitive biological assay for lactogenic hormones [107].

The mechanism of the mitogenic actions of prolactin on Nb2 cells has been studied. Receptors for the hormone on the cell line have been investigated using binding studies [40] and cross-linking techniques. As indicated above, binding studies showed that the affinity of the receptor found in Nb2 cells was much higher than that found in rat liver or mammary gland. Cross-linking of receptors to ^{125}I-labelled prolactin or human GH indicated that the hormone-receptor complex differed considerably from that obtained with rat liver receptors [36,41]. An M_r of 110 000 was determined for the Nb2 cell hormone-receptor complex, compared with $\approx 65 000$ for that from liver. Furthermore, the crosslinked hormone-receptor complex obtained from Nb2 cells disappeared when subjected to SDS polyacrylamide electrophoresis in the absence of reducing agent, whereas the complex with rat liver membranes was unchanged under these conditions, indicating a clear difference in the overall structure of the two receptors. To what extent the presence of such an 'unusual' lactogenic receptor in Nb2 cells underlies the ability of these cells to grow as tumours in rats is not known. Interestingly, in rat ovary, prolactin-receptor complexes of both $M_r \approx 65 000$ and 110 000 have been detected [39].

The nature of the second messenger(s) involved in the action of prolactin on Nb2 cells has been studied [108]. Levels of intracellular cyclic AMP do not change when the cells are treated with prolactin, and cyclic AMP analogues and phosphodiesterase inhibitors do not induce proliferation. Similarly, no clear evidence could be obtained for a clear-cut involvement of prostaglandins in Nb2 cell proliferation.

A possible involvement of polyamines in the response has been suggested. Prolactin stimulated ornithine decarboxylase activity in Nb2 cells [109]. A specific inhibitor of this enzyme partially blocked the actions of lactogenic hormones on Nb2 cell proliferation [110], and addition of polyamines (putrescine, spermidine or spermine) restored normal growth. However, polyamines alone had no effect on Nb2 cell proliferation, suggesting that they are not the sole or major factors involved in mediating the actions of prolactin on cell growth.

Phorbol esters can give a small stimulation of Nb2 cell proliferation, suggesting that activation of protein kinase C may play a part in mediating the effects of lactogenic hormones [108,111]. Elevation of Ca^{2+} levels may also be involved, and it is possible that stimulation of proliferation may involve activation of both arms of the phosphatidylinositol bisphosphate pathway, with activation of protein kinase C and elevation of intracellular Ca^{2+} levels. A mechanism of this sort has been invoked for the stimulation of proliferation of normal lymphocytes by lectin mitogens. However, recent studies failed to find any stimulation of turnover of phosphatidylinositol phosphates [112], which makes such a mechanism less likely. Prolactin did lead to an increase in phosphatidylcholine levels in Nb2 cells [113], by lowering turnover, but it is not clear whether this relates to second messenger activity.

The effects of lactogenic hormones on Nb2 cell proliferation are associated with expression of the proto-oncogene c-*myc* [114]. Expression of this gene occurred within 15 min of exposure of the cells to the lactogenic hormone (human GH), and peaked after about 3 h. If the lactogenic hormone is removed (by washing) within 4 h of first application, the proliferative response of the cells (seen after 40 h) is prevented although the maximal c-*myc* expression has been achieved at 3 h. However, expression of c-*myc* falls off much more rapidly if the lactogenic hormone is removed than if it is maintained in the medium.

Expression of c-*myc* (the product of which is thought to be involved in induction of DNA synthesis [115]) is seen in response to many growth factors including GH acting on the liver [116]. The response usually differs significantly from that seen in Nb2 cells, however, in that for many growth factors a brief exposure is all that is necessary to induce sustained expression of c-*myc* and growth whereas in the case of lactogenic hormones acting on Nb2 cells, prolonged stimulation is necessary. It has been suggested that for stimulation of mitogenesis in a mouse cell line (BALB/c 3T3 cells) a brief exposure to a 'competence factor' such as PDGF renders the cells 'competent' to proceed through the cell cycle in the presence of a 'progression factor' such as IGF [117]. In the case of Nb2 cells a lactogenic hormone may act as both a competence factor and a progression factor. Alternatively the lactogenic hormone may have a permissive role, allowing the cells to respond to other mitogenic factors present in the horse serum.

Although the Nb2 cell system is proving a valuable one for investigating the mechanism whereby prolactin acts as a mitogen, it is possible that it may not provide a very good model for many major actions of the hormone on the mammary gland. The marked difference between receptors on Nb2 cells and on liver and mammary gland cells suggests that the response of Nb2 cells may be quite different from that of 'normal' tissues. Nevertheless, the cells provide a very sensitive bioassay for lactogenic hormones, and have been used to characterize variants of human GH with inhibitory properties (see Chapter 13).

6.2. Other tissues and cells of the immune system

A variety of effects of prolactin has been described on normal tissues and cells of the immune system, deriving from rats, mice and humans [21,22]. The nature of the effects varies; they can be either stimulatory or inhibitory, depending on the system studied and the concentration of prolactin. Thus, in the rat, hypophysectomy leads to impairment of the immune response which can be restored by injection of prolactin [118]. On the other hand, excessively high levels of prolactin can also give rise to an impaired immune response.

Prolactin has been shown to enhance the stimulation of T lymphocyte proliferation induced by mitogens in vitro [119]. A possible role for cyclic GMP in the process has been suggested. The effect of the hormone depends on the amount of mitogen used. At maximal doses of mitogen prolactin does not enhance proliferation further. At submaximal doses it does so, but at very low levels of mitogen the hormone may in fact be inhibitory. Prolactin can also enhance the production of antibodies by spleen cells from immunized mice [119], showing effects at very low hormone concentrations.

A variety of studies on the binding of labelled prolactin to cells or membrane preparations from immune tissues have been described, with rather variable results. In some cases, at least, substantial binding has been shown to membrane fractions from spleen and thymus and to human lymphocytes [21,120,121]. It has been reported that this binding can be enhanced or blocked by the immunosuppressive drug cyclosporin A [21,120–122], and a mechanism for this drug involving such inhibition of prolactin binding has been proposed.

Cyclosporin A may also block some of the actions of prolactin, including effects on ornithine decarboxylase activity in vivo in spleen and thymus [21], providing further support for the idea that the actions of this immunosuppressant may be at least partly explained by interference with the normal function of prolactin.

7. Variants of prolactin

7.1. Fragments

A variety of studies suggest that prolactin may be cleaved specifically either before or during secretion from the pituitary gland, or at target organs. Thus ^{125}I-labelled prolactin incubated with preparations of mammary tissue or spleen can be converted to specific fragments. Prolactin secreted from the pituitary gland has been shown to occur in several different forms, showing different bioassayable:immunoassayable ratios. Mittra [123,124] has reported that prolactin may be secreted from the rat pituitary in a specifically modified (cleaved) form, and that the relative amounts of cleaved and intact hormone may vary according to the physiological state of the animal. It was suggested that the cleaved prolactin may have biological properties that differed from those of the uncleaved hormone, with markedly enhanced activity. Others have also suggested specific cleavage of prolactin at the target organ [125]. How these modified prolactins relate to each other is not clear, and their biological significance remains unknown.

7.2. Glycosylated prolactins

Prolactin from various species, including sheep and man, can occur in a partially glycosylated form [126,127]. The glycosylated form occurs in the circulation [128], but its biological role is not clear.

8. Prolactin and mammary cancer

Induction of mammary tumours in rats with a variety of carcinogens is dependent on prolactin [129]. There has been much debate about a possible link between prolactin and human breast cancer, though the evidence remains equivocal [129,130]. Prolactin has been shown to have effects on cells of mammary tumours in vivo and in vitro and a considerable amount of biochemical work has been carried out [131]. Space precludes a full discussion of the topic here.

9. Conclusions

It will be clear from what has been presented in this chapter that the mechanism of action of prolactin remains far from clear. Much progress has been made in the past few years, particularly in the characterization of receptors for prolactin and in understanding the induction of expression of the genes for milk proteins that are controlled by prolactin. The signal transduction events which link these two processes remain unclear, however. A few of the main aspects of prolactin activity that need explanation at the biochemical level will be summarized here.

(a) Prolactin has a bewilderingly wide range of actions, on a very large number of tissues in many different vertebrate species. The receptors by which it acts may vary considerably from tissue to tissue, and the biological and biochemical events regulated by the hormone are extremely variable. Biochemical studies have been largely confined to actions on mammary tissue and to a lesser extent pigeon crop sac and Nb2 cells, but it is clear that for a full understanding of the hormone they will have to be extended to many other tissues.

(b) It seems clear that prolactin has several different actions on each of its target tissues. In particular, its actions on mammary gland and pigeon crop sac have been shown to fall into two groups – actions on cell proliferation and development and actions on terminal differentiation, including induction of expression of specific proteins. Such a division probably applies to many trophic hormones. It is noticeable that in the case of prolactin acting on mammary gland, at least, each type of action requires synergism with several other hormones, but the set of synergistic factors differs.

(c) Many of the actions of prolactin on lactogenesis (as opposed to lobuloalveolar development) can be explained in terms of induction of expression of specific genes (caseins, α-lactalbumin, etc.). Whether all of the actions of prolactin can be explained in terms of regulation of gene expression is unclear, however.

Acknowledgements

I thank Mrs Eileen Willis for skilful preparation of the manuscript and the Medical Research Council and Cancer Research Campaign for research support.

References

1. Wallis, M. (1978) In: Chemistry and Biochemistry of Amino Acids, Peptides and Proteins, Vol. 5 (Weinstein, B., ed.) pp. 213–320. Dekker, New York.
2. Nicoll, C.S., Mayer, G.L. and Russell, S.M. (1986) Endocr. Rev. 7, 169–203.
3. Wallis, M. (1981) J. Mol. Evol. 17, 10–18.
4. Wallis, M. (1984) In: Prolactin Secretion: A Multidisciplinary Approach (Mena, F. and Valverde-R., R.C., eds.) pp. 1–16. Academic Press, Orlando.
5. Forsyth, I.A. (1986) J. Dairy Sci. 69, 886–903.
6. Colosi, P., Ogren, L., Thordarson, G. and Talamantes, F. (1987) Endocrinology 120, 2500–2511.
7. Duckworth, M.L., Kirk, K.L. and Friesen, H.G. (1986) J. Biol. Chem. 261, 10871–10878.
8. Nicoll, C.S. (1982) Perspect. Biol. Med. 25, 369–381.
9. Holder, A.T. and Wallis, M. (1977) J. Endocr. 74, 223–229.
10. Forsyth, I.A. (1968) In: Growth Hormone (Pecile A. and Müller, E.E., eds.) pp. 364–370. Excerpta Medica, Amsterdam.
11. Frantz, A.G. (1978) New Engl. J. Med. 298, 201–207.
12. Bines, J.A., Hart, I.C. and Morant, S.V. (1980) Br. J. Nutr. 43, 179–188.
13. Madon, R.J., Ensor, D.M., Knight, C.H. and Flint, D.J. (1986) J. Endocrinol. 111, 117–123.
14. Cowie, A.T. (1984) In: Hormonal Control of Reproduction (Austin, C.R. and Short, R.V., eds.) 2nd Edn., pp. 195–231. Cambridge University Press, Cambridge.
15. Elias, J.J. (1980) In: Hormonal Proteins and Peptides, Vol. 8 (Li, C.H., ed.) pp. 37–74. Academic Press, New York.
16. Forsyth, I.A. (1983) In: Biochemistry of Lactation (Mepham, T.B., ed.) pp. 309–349. Elsevier, Amsterdam.
17. Horrobin, D.F. (1980) Fed. Proc. 39, 2567–2570.
18. Smith, M.S., McLean, B.K. and Neill, J.D. (1976) Endocrinology 98, 1370–1377.
19. Short, R.V. (1984) In: Hormonal Control of Reproduction (Austin, C.R. and Short, R.V., eds.) 2nd Edn., pp. 115–152. Cambridge University Press, Cambridge.
20. Bartke, A. (1980) Fed. Proc. 39, 2577–2581.
21. Russell, D.H., Kibler, R., Matrisian, L., Larson, D.F., Poulos, B. and Magun, B.E. (1985) In: Prolactin. Basic and Clinical Correlates (MacLeod, R.M., Thorner, M.O. and Scapagnini, U., eds.) pp. 375–384. Liviana Press, Padova.
22. Montgomery, D.W., Zukoski, C.F., Shah, G.N., Buckley, A.R., Pacholczyk, T. and Russell, D.H. (1987) Biochem. Biophys. Res. Commun. 145, 692–698.
23. Scapagnini, U., Drago, F., Continella, G., Spadaro, F., Pennisi, G. and Gerendai, I. (1985) In: Prolactin. Basic and Clinical Correlates (Macleod, R.M., Thorner, M.O. and Scapagnini, U., eds.) pp. 583–590, Liviana Press, Padova.
24. Clarke, W.C. and Bern, H.A. (1980) In: Hormonal Proteins and Peptides, Vol. 8 (Li, C.H., ed.) pp. 105–197. Academic Press, New York.
25. Nicoll, C.S., Anderson, T.R., Hebert, N.J. and Russell, S.M. (1985) In: Prolactin. Basic and Clinical Correlates (Macleod, R.M., Thorner, M.O. and Scapagnini, U., eds.) pp. 393–410. Liviana Press, Padova.

26. Hughes, J.P., Elsholtz, H.P. and Friesen, H.G. (1985) In: Polypeptide Hormone Receptors (Posner, B.I., ed.) pp. 157–199. Dekker, New York.
27. Wallis, M. (1980) In: Cellular Receptors for Hormones and Neurotransmitters (Schulster, D. and Levitzki, A., eds.) pp. 163–183. Wiley, Chichester.
28. Shiu, R.P.C., Kelly, P.A. and Friesen, H.G. (1973) Science 180, 968–971.
29. Shiu, R.P.C. and Friesen, H.G. (1974) J. Biol. Chem. 249, 7902–7911.
30. Bonifacino, J.S., Sanchez, S.H. and Paladini, A.C. (1981) Biochem. J. 194, 385–394.
31. Liscia, D.S., Alhadi, T. and Vonderhaar, B.K. (1982) J. Biol. Chem. 257, 9401–9405.
32. Liscia, D.S. and Vonderhaar, B.K. (1982) Proc. Natl. Acad. Sci. USA 79, 5930–5934.
33. Katoh, M., Djiane, J. and Kelly, P.A. (1985) Endocrinology 116, 2612–2620.
34. Borst, D.W. and Sayare, M. (1982) Biochem. Biophys. Res. Commun. 105, 194–201.
35. Hughes, J.P., Simpson, J.S.A. and Friesen, H.G. (1983) Endocrinology 112, 1980–1985.
36. Webb, C.F. and Wallis, M. (1988) Biochem. J. 250, 215–219.
37. Haldosen, L.A. and Gustafsson, J.A. (1987) J. Biol. Chem. 262, 7404–7411.
38. Bonifacino, J.S. and Dufau, M.L. (1985) Endocrinology 116, 1610–1614.
39. Mitani, M. and Dufau, M.L. (1986) J. Biol. Chem. 261, 1309–1315.
40. Shiu, R.P.C., Elsholtz, H.P., Tanaka, T., Friesen, H.G., Gout, P.W., Beer, C.T. and Noble, R.L. (1983) Endocrinology 113, 159–165.
41. Ashkenazi, A., Cohen, R. and Gertler, A. (1987) FEBS Lett. 210, 51–55.
42. Shiu, R.P.C. and Friesen, H.G. (1976) Biochem. J. 157, 619–626.
43. Djiane, J., Houdebine, L.-M. and Kelly, P.A. (1981) Proc. Natl. Acad. Sci. USA 78, 7445–7448.
44. Katoh. M., Djiane, J., Leblanc, G. and Kelly, P.A. (1984) Mol. Cell. Endocrinol. 34, 191–200.
45. Shiu, R.P.C. and Friesen, H.G. (1976) Science 192, 259–261.
46. Bohnet, H.G., Shiu, R.P.C., Grinwich, D. and Friesen, H.G. (1978) Endocrinology 102, 1657–1661.
47. Edery, M., Djiane, J., Houdebine, L.M. and Kelly, P.A. (1983) Cancer Res. 43, 3170–3174.
48. Rosa, A.A.M., Djiane, J., Houdebine, L.M. and Kelly, P.A. (1982) Biochem. Biophys. Res. Commun. 106, 243–249.
49. Katoh, M., Djiane, J. and Kelly, P.A. (1985) J. Biol. Chem. 260, 11422–11429.
50. Djiane, J. Dusanter-Fourt, I., Katoh, M. and Kelly, P.A. (1985) J. Biol. Chem. 260, 11430–11435.
51. Katoh, M., Raguet, S., Zachwieja, J. Djiane, J. and Kelly, P.A. (1987) Endocrinology 120, 739–749.
52. Ferland, L.H., Rosa, A.A.M. and Kelly, P.A. (1984) Can. J. Physiol. Pharmacol. 62, 1429–1433.
53. Amit, T., Barkey, R.J., Gavish, M. and Youdim, M.B.H. (1986) Endocrinology 118, 835–843.
54. Ymer, S.I. and Herington, A.C. (1986) Biochem. J. 237, 813–820.
55. Ymer, S.I., Kelly, P.A., Herington, A.C. and Djiane, J. (1987) Mol. Cell. Endocrinol. 53, 67–73.
56. Shiu, R.P.C. and Friesen, H.G. (1981) In: Receptor Regulation (Lefkowitz, R.J., ed.) pp. 67–81. Chapman and Hall, London.
57. Baxter, R.C. (1985) Endocrinology 117, 650–655.
58. Savoie, S., Rindress, D., Posner, B.I. and Bergeron, J.J.M. (1986) Mol. Cell. Endocrinol. 45, 241–246.
59. Posner, B.I., Josefsberg, Z. and Bergeron, J.J.M. (1979) J. Biol. Chem. 254, 12494–12499.
60. Bergeron, J.J.M., Cruz, J., Khan, M.N. and Posner, B.I. (1985) Annu. Rev. Physiol. 47, 383–403.
61. Posner, B.I., Bergeron, J.J.M., Josefsberg, Z., Khan, M.N., Khan, R.J., Patel, B.A., Sikstrom, R.A. and Verma, A.K. (1981) Rec. Prog. Horm. Res. 37, 539–582.
62. Posner, B.I., Kelly, P.A. and Friesen, H.G. (1975) Science 188, 57–59.
63. Baxter, R.C., Zaltsman, Z. and Turtle, J.R. (1984) Endocrinology 114, 1893–1901.
64. Topper, Y.J. and Freeman, C.S. (1980) Physiol. Rev. 60, 1049–1106.
65. Cowie, A.T., Forsyth, I.A. and Hart, I.C. (1980) The Hormonal Control of Lactation. Springer-Verlag, Berlin.
66. Banerjee, M.R. and Antoniou, M. (1985) In: Biochemical Actions of Hormones, Vol. 12 (Litwack, G., ed.) pp. 237–288. Academic Press, Orlando.

67. Turkington, R.W. (1972) In: Biochemical Actions of Hormones, Vol. 2 (Litwack, G., ed.) pp. 55–80. Academic Press, New York.
68. Shiu, R.P.C. and Friesen, H.G. (1980) Annu. Rev. Physiol. 42, 83–96.
69. Ray, D.B., Jansen, R.W., Horst, I.A., Mills, N.C. and Kowal, J. (1986) Endocrinology 118, 393–407.
70. Imagawa, W., Spencer, E.M., Larson, L. and Nandi, S. (1986) Endocrinology 119, 2695–2699.
71. Prosser, C.G., Sankaran, L., Hennighausen, L. and Topper, Y.J. (1987) Endocrinology 120, 1411–1416.
72. Winder, S.J. and Forsyth, I.A. (1986) J. Endocrinol. 108, Suppl. Abstr. 141.
73. Schneider, W. and Shyamala, G. (1985) Endocrinology 116, 2656–2662.
74. Rosen, J.M. (1981) In: Prolactin (Jaffe, R.B., ed.) pp. 85–126. Elsevier, New York.
75. Neville, M.C. and Berga, S.E. (1983) In: Lactation (Neville, M.C. and Neifert, M.R. eds.) pp. 141–177. Plenum Press, New York.
76. Houdebine, L.-M., Djiane, J., Dusanter-Fourt, I., Martel, P., Kelly, P.A., Devinoy, E. and Serveley, J.-L. (1985) J. Dairy Sci. 68, 489–500.
77. Hobbs, A.A., Richards, D.A., Kessler, D.J. and Rosen, J.M. (1982) J. Biol. Chem. 257, 3598–3605.
78. Falconer, I.R. and Rowe, J.M. (1977) Endocrinology 101, 181–186.
79. Oka, T. (1983) In: Biochemistry of Lactation (Mepham, T.B., ed.) pp. 381–396. Elsevier, Amsterdam.
80. Rillema, J.A. (1980) Fed. Proc. 39, 2593–2598.
81. Sepag-Hagar, M., Greenbaum, A.L., Lewis, D.J. and Hallowes, R.C. (1974) Biochem. Biophys. Res. Commun. 59, 261–268.
82. Matusik, R.J. and Rosen, J.M. (1980) Endocrinology 106, 252–259.
83. Rillema, J.A. (1976) Endocrinology 99, 490–495.
84. Frazier, R.P. and Costlow, M.E. (1982) Exp. Cell. Res. 138, 39–45.
85. Rillema, J.A., Wing, L.-Y.C. and Foley, K.A. (1983) Endocrinology 113, 2024–2028.
86. Houdebine, L.M. and Djiane, J. (1980) Mol. Cell. Endocrinol. 17, 1–15.
87. Teyssot, B., Houdebine, L.-M. and Djiane, J. (1981) Proc. Natl. Acad. Sci. USA 78, 6729–6733.
88. Kelly, P.A., Djiane, J., Katoh, M., Ferland, L.H. Houdebine, L.-M., Teyssot, B. and Dusanter-Fourt, I. (1984) Rec. Prog. Horm. Res. 40. 379–439.
89. Houdebine, L.-M., Djiane, J., Kelly, P.A., Katoh, M., Dusanter-Fourt, I. and Martel, P. (1984) In: Endocrinology (Labrie, F. and Proulx, L., eds.) pp. 203–206. Excerpta Medica, Amsterdam.
90. Cameron, C.M. and Rillema, J.A. (1983) Endocrinology 113, 1596–1600.
91. Falconer, I.R., Forsyth, I.A., Wilson, B.M. and Dils, R. (1978) Biochem. J. 172, 509–516.
92. Rillema, J.A., Foley, K.A. and Etindi, R.N. (1985) Endocrinology 116, 511–515.
93. Chomczynski, P. and Topper, Y.J. (1974) Biochem. Biophys. Res. Commun. 60, 56–63.
94. Turkington, R.W. (1970) Biochem. Biophys. Res. Commun. 41, 1362–1367.
95. Djiane, J., Kelly, P.A., and Houdebine, L.M. (1980) Mol. Cell. Endocrinol. 18, 87–98.
96. Riddle, O. (1963) J. Natl. Cancer Inst. 31, 1039–1110.
97. Pukac, L.A. and Horseman, N.D. (1984) Endocrinology 114, 1718–1724.
98. Anderson, T.R., Mayer, G.L., Herbert, N. and Nicoll, C.S. (1987) Endocrinology 120, 1258–1264.
99. Horseman, N.D. and Nollin, L.J. (1985) Endocrinology 116, 2085–2089.
100. Anderson, T.R., Mayer, G.L. and Nicoll, C.S. (1981) J. Cyclic Nucleotide Res. 7, 225–233.
101. Anderson, T.R., Pitts, D.S. and Nicoll, C.S. (1984) Gen. Comp. Endocrinol. 54, 236–246.
102. Mick, C.C.W. and Nicoll, C.S. (1985) Endocrinology 116, 2049–2053.
103. Nicoll, C.S., Hebert, N.J. and Russell, S.M. (1985) Endocrinology 116, 1449–1453.
104. Anderson, T.R., Rodriguez, J., Nicoll, C.S. and Spencer, E.M. (1983) In: Insulin-Like Growth Factors/Somatomedins (Spencer, E.M., ed.) pp. 71–78. De Gruyter, Berlin.
105. Gout, P.W., Beer, C.T. and Noble, R.L. (1980) Cancer Res. 40, 2433–2436.
106. Webb, C.F., Ray, K.P., Watkins, D.C. and Wallis, M. (1986) J. Endocrinol. 108, Suppl. Abstr. 224.

107. Tanaka, T., Shiu, R.P.C., Gout, P.W., Beer, C.T., Noble, R.L. and Friesen, H.G. (1980) J. Clin. Endocrinol. Metab. 51, 1058–1063.
108. Friesen, H.G., Gertler, A., Walker, A. and Elsholtz, H. (1985) In: Prolactin. Basic and Clinical Correlates (MacLeod, R.M., Thorner, M.O. and Scapagnini, U., eds.) pp. 315–326. Liviana Press, Padova.
109. Richards, J.F., Beer, C.T., Bourgeault, C., Chen, K. and Gout, P.W. (1982) Mol. Cell. Endocrinol. 26, 41–49.
110. Elsholtz, H.P. and Shiu, R.P.C. (1982) 64th Annu. Meet. Endocrin. Soc. Abstract 866.
111. Gertler, A., Walker, A. and Friesen, H.G. (1985) Endocrinology 116, 1636–1644.
112. Gertler, A. and Friesen, H.G. (1986) Mol. Cell. Endocrinol. 48, 221–228.
113. Ko, K.W.S., Cook, H.W. and Vance, D.E. (1986) J. Biol. Chem. 261, 7846–7852.
114. Fleming, W.H., Murphy, P.R., Murphy, L.J., Hatton, T.W., Matusik, R.J. and Friesen, H.G. (1985) Endocrinology 117, 2547–2549.
115. Iguchi-Ariga, S.M.M., Itani, T., Kiji, Y. and Ariga, H (1987) EMBO J. 6, 2365–2371.
116. Murphy, L.J., Bell, G.I. and Friesen, H.G. (1987) Endocrinology 120, 1806–1812.
117. Stiles, C.D., Capone, G.T., Scher, C.D., Antoniades, H.N., Van Wyk, J.J. and Pledger, W.J. (1979) Proc. Natl. Acad. Sci. USA 76, 1279–1283.
118. Nagy, E., Berczi, I. and Friesen, H.G. (1983) Acta Endocrinol. 102, 351–357.
119. Spangelo, B.L., Hall, N.E. and Goldstein, A.L. (1985) In: Prolactin, Basic and Clinical Correlates (MacLeod, R.M., Thorner, M.O. and Scapagnini, U., eds.) pp. 343–349. Liviana Press, Padova.
120. Russell, D.H., Matrisian, L., Kibler, R., Larson, D.F., Poulos, B. and Magun, B.E. (1984) Biochem. Biophys. Res. Commun. 121, 899–906.
121. Russell, D.H., Kibler, R., Matrisian, L., Larson, D.F., Poulos, B. and Magun, B.E. (1985) J. Immunol. 134, 3027–3031.
122. Hiestand, P.C., Mekler, P., Nordmann, R., Grieder, A. and Permmongkol, C. (1986) Proc. Natl. Acad. Sci. USA 83, 2599–2603.
123. Mittra, I. (1980) Biochem. Biophys. Res. Commun. 95, 1750–1759.
124. Mittra, I. (1980) Biochem. Biophys. Res. Commun. 95, 1760–1767.
125. Wong, V.L.Y., Compton, M.M. and Witorsch, R.J. (1986) Biochim. Biophys. Acta 881, 167–174.
126. Lewis, U.J., Singh, R.N.P., Lewis, L.J., Seavey, B.K. and Sinha, Y.N. (1984) Proc. Natl. Acad. Sci USA 81, 385–389.
127. Lewis, U.J., Singh, R.N.P., Sinha, Y.N. and VanderLaan, W.P. (1985) Endocrinology 116, 359–363.
128. Champier, J., Claustrat, B., Sassolas, G. and Berger, M. (1987) FEBS Lett. 212, 220–224.
129. Clifton, K.H. and Furth, J. (1980) In: Hormonal Proteins and Peptides, Vol. VIII (Li, C.H., ed.) pp. 75–103, Academic Press, New York.
130. Moore, D.H., Moore, D.H. (II) and Moore, C.T. (1983) Adv. Cancer Res. 40, 189–253.
131. Shiu, R.P.C., Murphy, L.C., Tsuyuki, D., Myal, Y., Lee-Wing, M. and Iwasiow, B. (1987) Rec. Prog. Horm. Res. 43, 277–303.
132. Wallis, M., Howell, S.L. and Taylor, K.W. (1985) The Biochemistry of the Polypeptide Hormones. Wiley, Chichester.

CHAPTER 15

Structure and function of the receptor for insulin

MILES D. HOUSLAY and MICHAEL J.O. WAKELAM

Molecular Pharmacology Group, Department of Biochemistry, University of Glasgow, Glasgow G12 8QQ, Scotland

1. Introduction

Insulin is commonly referred to as an 'anabolic' hormone as it stimulates a variety of synthetic pathways, e.g. for glycogen and protein synthesis, whilst attenuating the functioning of degradative ones, e.g. for lipolysis, glycogenolysis and protein degradation. These various processes occur in distinct cellular compartments and in each case insulin exerts its effect by controlling the activity of a key regulatory protein, often by affecting its phosphorylation state. Insulin thus exerts widespread and disparate effects.

Target cells for the action of insulin express on their cell surface insulin receptors. Like all receptors, this protein has two basic functions. These allow it to recognize the occurrence of insulin at the cell surface and also to generate a signal within the cell. The latter triggers all of the appropriate events characteristic of the action of this hormone.

Whilst much is known about the physiological actions of insulin and the defects in its synthesis that occur in Type-I diabetes, it is only recently that we have begun to appreciate the nature and properties of the insulin receptor itself.

2. Insulin receptor structure

Insulin receptors comprise only a very small number of the total protein molecules found embedded in the plasma membrane of target cells. In the key target tissues, liver and fat, there are, for example, about 50 000–100 000 copies per cell. Thus, sensitive assays utilizing ^{125}I-monoiodinated radio-iodinated insulin are required for the detection and characterization of insulin receptors. In many cells, however, the specific binding of insulin occurs over a rather extended range of ligand concentrations. This is thought to result from either negative co-operativity, displayed by a homogeneous population of receptors or multiple populations of receptors expressing somewhat different affinities for insulin [1,2]. The recent cloning of the gene for the insulin receptor [3,4], however, implies that there is but a single gene for the receptor. Nevertheless, the possibility cannot be completely excluded that more than a single population of receptors exists, each with a unique binding activity, for the receptor is extensively glycosylated and, as glycosylation can exert potent alterations in insulin binding [2], it is quite possible that a single receptor protein population, exhibiting various degrees of glycosylation, might result in species able to bind insulin with very different affinities.

The insulin receptor can be solubilized using detergents, such as Triton X-100, and then purified using immobilized lectin columns, which take advantage of the glycoprotein nature of the receptor. Final resolution can then be achieved employing affinity columns of immobilized insulin (see e.g. Refs. 5 and 6). More recently, however, purification has been achieved using a 'one-step' process utilizing an immobilized monoclonal antibody directed against the insulin receptor itself [7,8]. The purified, holomeric protein has a molecular mass of around 360 kDa and is now known to be comprised of two α-subunits (circa 95 kDa) and two β-subunits (circa 130 kDa). These subunits are held together by covalent interactions. Partial reduction of the molecule can lead to 'half-molecules' ($\alpha\beta$) being produced [9–12].

The constituent subunits of the insulin receptor can also be identified by either linking ^{125}I-insulin, covalently, to the insulin receptor with a cross-linking reagent such as disuccinimidyl suberate or using a photo-reactive (azido) I^{125}-labelled insulin molecule to interact covalently with the insulin receptor [9–12]. Using these methods it has been demonstrated that the insulin binding site lies on the 130 kDa α subunit of the insulin receptor which is found associated with the 95 kDa β subunit of the receptor. The receptor complex (Fig. 1), consisting of two α and two β subunits held together by disulphide bridges [$(\alpha\text{-}\beta)_2$], is thus disposed vectorially in the place of the membrane. The α subunits, which provide the major binding domain for insulin, are exposed solely at the external surface, whereas the β subunits are exposed at both the extra-cellular and the cytosol surface of the membrane [13]. The stoichiometry of binding of insulin to its receptor is still unclear, with values of between 1 and 4 molecules of insulin/receptor being noted.

The receptor, particularly the β subunit, is extremely sensitive to proteolysis [14] and can give rise to the apparent association of peptides of lower molecular mass being associated with receptor preparations. Nevertheless, there are indications, from both immunoprecipitation [15,16] and cross-linking [17] studies, that other distinct protein subunits may be associated with the insulin receptor. It is possible that these proteins represent species that are functionally or structurally capable of interacting with the insulin receptor, yet are not covalently attached to the receptor itself. Hence their association with the receptor would be expected to be easily disrupted by the manipulative processes used in purifying the solubilized receptor. The notion that receptor-associated proteins exist, however, deserves further consideration in view of the possibility that they may provide part of the signal generation mechanism. Furthermore, there is evidence from radiation inactivation studies [18] that a protein species capable of attenuating the binding of insulin is found associating with the receptor.

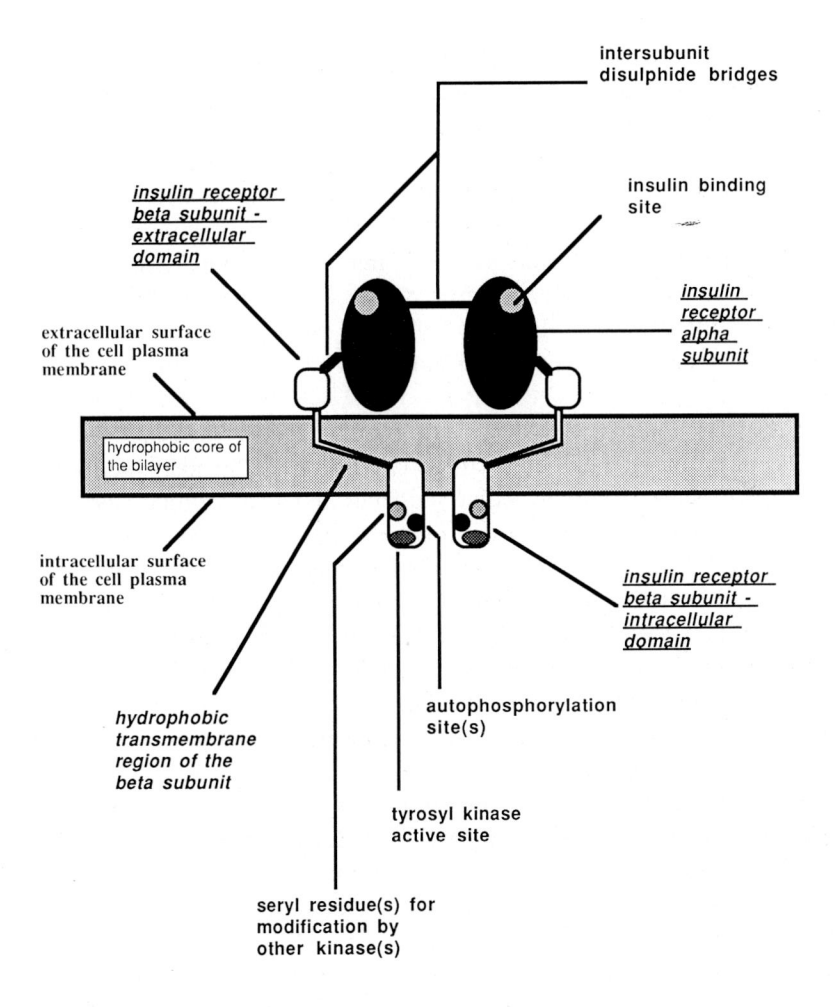

Fig. 1. Structure of the insulin receptor. Demonstrates the disposition of the α and β subunits in the plasma membrane. It identifies the putative insulin-binding sites on the α subunit. The tyrosyl kinase site is located on the β subunit as are three autophosphorylation sites and one site for seryl phosphorylation.

3. Cloning of the gene for the insulin receptor

Cloning of the insulin receptor gene [3,4] has shown that a single gene codes for the entire α and β subunits of the receptor in the haploid human genome. The gene actually codes for a precursor molecule which starts with a nucleotide sequence which is responsible for directing the mRNA to membrane-bound ribosomes for biosynthesis [see 19]. This coding region specifies a 27-amino-acid N-terminal signal sequence, which is typical of many integral proteins. This 'signal sequence' is cleaved very early on during the biosynthesis of the receptor. Such a sequence is followed by that for the α subunit and then a sequence specifying the precursor-processing enzyme cleavage site which allows for the cleavage and separation of the β subunit, whose sequence follows on behind it, from those of the α subunit. As implied from biosynthetic studies [20,21] a single mRNA codes for the entire receptor, with the receptor precursor-protein being processed in the endoplasmic reticulum to yield the separate α and β subunits. Based upon the cDNA sequence the molecular masses of the unmodified receptor subunits are actually 82 kDa for the α subunit and 70 kDa for the β subunits. It would appear, however, that these subunits are heavily glycosylated at most, if not at all, the potential N-linked sites that are identified. This accounts for the differences in observed molecular masses, determined using SDS polyacrylamide gel electrophoresis, from those derived from the predicted sequences.

It is of interest to note that there is a slight difference between the sequence of the human insulin receptor gene cloned by these two groups. This takes the form of a 22-nucleotide difference within the coding sequence and an additional block of nucleotides found in the data of one of the groups [4] accounting for an extra 12 amino acids. The basis for these discrepancies remains to be ascertained. However, one group has inserted their cloned gene into both a hamster cell line and *Xenopus* oocytes, allowing for its successful expression [22]. Indeed, in the hamster cell line, expression was accompanied by the occurrence of high-affinity insulin binding, insulin-stimulated autophosphorylation of its β-subunit (vide infra) and insulin-stimulated glucose transport. Thus a gene has been cloned which allows for the normal synthesis, processing and functional expression of the receptor.

It is of interest that although a single gene appears to code for the insulin receptor inhuman placenta, multiple species of mRNA were detected [3,4] and, indeed, in 3T3-L1 fibroblasts at least two forms of mRNA, coding for the insulin receptor, have been identified. Thus, it may be that heterogeneous mRNA, coupled with heterogeneous glycosylation leads to the production of slightly different receptor molecules with perhaps differing abilities to bind insulin and produce intracellular signals.

4. Insulin receptor internalization

A phospholipid bilayer containing cholesterol forms the basis of the cell plasma membrane into which protein molecules are inserted. The bilayer has a viscosity akin to a light machine oil which gives it an inherent flexibility and also allows for the movement of proteins within the two dimensions of the bilayer. Thus proteins inserted in the bilayer can rotate and migrate laterally within the two dimensions of the bilayer unless specifically restrained by cytoskeletal or other proteins [19]. The insulin receptor is no different and has been shown to be able to migrate over the cell surface. However, when insulin receptors bind insulin they cluster together rapidly within a minute or so and migrate to coated pits whereupon they are internalized (Fig. 2). The endocytosed receptors are then found in endosomes and other intracellular vesicle fractions. However, once the internalized receptor has entered the endosome it has two choices, either to be degraded by transfer to the lysosomes or to be recycled intact back to the plasma membrane [23].

Endocytosis thus offers a means of allowing insulin itself to enter the cell, allowing an 'activated' insulin-bound receptor to enter the cell and a means of reducing the number of receptors on the cell surface through degradation in the lysosomes in response to challenge of cells with insulin. There appears thus to be the potential of desensitizing a cell to the action of insulin by down-regulating receptor number. However, it should be noted that many of the rapid actions that insulin exerts on metabolic processes are mediated by extremely high-affinity receptors and low receptor occupancy, perhaps with less than 10% of the total cell surface receptors being occupied by insulin. Thus, unless down-regulation is particularly extensive or selective in the 'population' of receptors affected, it is unlikely to lead to a rapid desensitizing response occurring in minutes. Nevertheless, chronic treatment of hepatoma cells with insulin has been shown to lead to a loss of insulin receptors, an increase in affinity of the residual population for insulin and desensitization of insulin's action on these cells [24].

The molecular mechanism underlying the clustering and endocytosis of insulin receptors remains to be ascertained. It is possible that insulin itself might act as a bivalent cross-linking ligand, although this is perhaps rather unlikely. The implication is that endocytosis results from a change in the receptor itself which occurs upon occupancy. One suggestion [25] has been that receptor autophosphorylation (vide infra), which leads to negatively charged phosphate groups being attached to the receptor, might allow it to interact with a polyvalent cationic membrane protein and hence cluster. Site-specific mutagenesis experiments to delete the autophosphorylation sites have shown their fundamental importance in this process.

Endocytosis thus also allows the transfer of an occupied an presumably active receptor from the plasma membrane compartment of the intracellular, cytosolic milieu. As it goes through the route of receptor-mediated endocytosis it has the opportunity of interacting with a collection of distinct membrane-vesicle proteins. Thus,

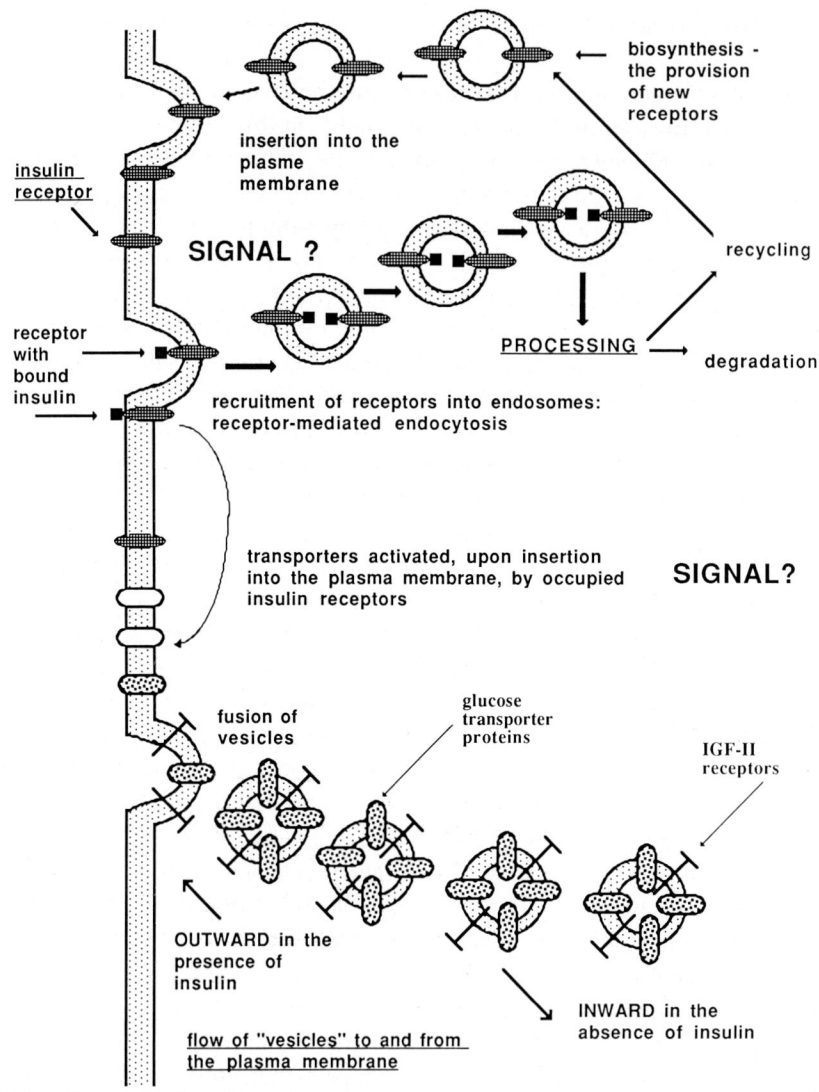

Fig. 2. Internalization of insulin receptors and the recruitment of glucose carriers and IGF-II receptors. Occupied insulin receptors are constantly being recycled to and from the plasma membrane with a half-time of 20 min. Presumably, a very slow basal rate also occurs. Receptors can be down-regulated by degradation and up-regulated by increased synthesis. Internalization of the insulin receptor triggers glucose transporters and IGF-II receptors to be 'recruited' to the plasma membrane in adipocytes. Whether these two proteins are in the same or different vesicles remains to be seen.

such a route offers a mechanism for signal transduction. Support for this comes from recent evidence [26–29] suggesting that when insulin stimulates glucose transport in adipocytes it triggers the translocation of glucose carriers from an internal vesicle pool to the plasma membrane. This suggests that the internalization of the insulin-receptor complex could provide the signal or 'second message' for this process (vide infra).

An indication that receptor endocytosis might serve as a signal for certain events has come from using lysosomotropic agents. These agents perturb the processing and recycling of insulin receptors but do not prevent their internalization per se. Thus insulin receptors accumulate in endosome vesicles of cells treated with lysosomotropic agents [30–32]. Such compounds have been shown to prevent insulin stimulating a specific high-affinity form of intracellular cyclic AMP phosphodiesterase in hepatocytes [33]. As the ability of insulin to stimulate this 'dense-vesicle' enzyme, in both adipocytes and hepatocytes, is both energy- and temperature-dependent it is possible that receptor-processing may provide a signal for activation of this enzyme.

Certain insulin receptor antibody preparations have also been demonstrated to be able to mimic particular actions of insulin in adipocytes, e.g. on lipolysis and glucose oxidation. These antibodies are bivalent, expressing two identical sites capable of binding to insulin receptors and causing them to cross-link together, whereupon they are internalized. In one instance it has been demonstrated that the univalent Fab' fragments of a particular polyclonal antibody preparation were ineffective unless they were cross-linked together to form a bivalent reagent, allowing for the restoration of the insulin-mimicking action [34]. This might imply that receptor aggregation and internalization was a necessary signal to generate an intracellular response. However, more recent studies [35] using panels of monoclonal antibodies directed against specific sites on the insulin receptor indicate that biological effectiveness is primarily related to some particular conformation of the antibody. This is in fact the most probable interpretation of experiments using the polyclonal antiserum as (i) such antisera appear to function well in adipocytes but not in all tissues and (ii) various anti-receptor antisera have now been described which elicit the receptor-mediated endocytosis of the insulin receptor but do not mimic other actions of insulin on adipocytes [36]. Indeed, a monoclonal antibody has been shown [37] to bind as a 'competitive inhibitor' of insulin's binding to the receptor. This caused endocytosis but did not mimic other actions of insulin. One can conclude that antibodies which mimic the action of insulin must cause a conformational change in the receptor itself such that an 'activated-receptor state' is elicited and stabilized by the antibody.

5. Insulin's stimulation of glucose transport

It has been well-documented that insulin stimulates glucose transport in adipocytes by increasing the V_{max} for the reaction rather than by exerting any dramatic effect on the K_m for this process. This action is rapid and reversible and is not due to the biosynthesis of new carrier molecules [28,38].

Proposals have been made [26–29] that glucose carriers are to be found not only in the plasma membrane but also in an internal vesicle pool, whereupon, as a consequence of insulin receptor occupancy, they are translocated to the plasma membrane (Fig. 2). This stimulatory action of insulin on glucose transport in adipocytes follows an almost identical time-course to that seen for the internalization of the insulin receptor itself [39] and, furthermore, both processes are temperature- and energy-dependent. Identification of glucose carriers in two distinct membrane fractions was done either by reconstitution into liposomes [26] or by utilizing the fact that glucose can displace [^{3}H]cytochalasin B from membranes [27,28]. The binding of cytochalasin is presumed to reflect glucose binding sites on the receptor; however, such an interpretation should be treated with a degree of caution. The identity of these vesicle fractions has yet to be unequivocally resolved as has their association with insulin transporters using a specific antibody for identification purposes.

Internalization of insulin receptors per se does not, however, appear to be a sufficient signal to stimulate glucose transport [28,39]. This has been demonstrated in adipocytes where receptor internalization, but not glucose transport, can be triggered by Tris buffers in the absence of insulin. Intriguingly, although glucose transport was not increased by exposure of adipocytes to Tris buffers the translocation of, presumably inactive, glucose carriers to the plasma membrane was seen to be induced by such treatment [28]. Various independent pieces of evidence have led to the suggestion that insulin might activate glucose transport by a two-step process [28,40]. The first step requires the recruitment of inactive glucose carriers to the plasma membrane and the second step is a mechanism which elicits their activation. Indeed, there is now evidence which suggests the presence of two distinct forms of glucose carrier molecules [41]. Furthermore, an increasing body of evidence now suggests that there appear to be as yet undefined mechanism(s), for the activation of glucose carriers in the plasma membrane, which do not involve translocation [42]. The evidence [43] that the glucose transporter from red blood cells can be phosphorylated by protein kinase C implies that one route exploited may be that of transporter phosphorylation. However, recent evidence suggests that the glucose transporter in 3T3-L1 adipocytes is not phosphorylated in response to insulin [44]. The purported insulin-stimulated translocation of glucose carriers appears to be a highly specific event as other 'marker' enzymes in the plasma membrane and various membrane fractions were not redistributed in response to insulin. In fact the only other species which seemed to respond as did glucose carriers was that of the receptors for insulin-like growth factor-II (IGF-II receptors), whose shuttling between an internal vesicle pool and the plasma membrane was stimulated by the application of either IGF-II or insulin to target cells [45,46]. This process appeared to be triggered by an indirect mechanism as insulin does not bind to IGF-II receptors. It is thus tempting to suggest that all of these specific effects occur as a consequence of not only the binding of insulin to its receptor but also the internalization of the occupied receptor itself.

6. Insulin-like growth factors (IGFs)

Insulin-like growth factors (IGFs) or somatomedins are in a family of polypeptides with growth-promoting activities and which can also modulate some of the short-term metabolism effects of insulin [47,48].

IGF-I and II have a proinsulin-like primary structure. Their major site of production is the liver, from which they are secreted continuously rather than being stored. Thus serum serves as their reservoir where they are bound non-covalently to specific carrier proteins [48].

There are distinct receptors for both IGF-I and IGF-II. The IGF-I receptor is similar in structure to that for the insulin receptor, having a disulphide bridge-linked subunit $(\alpha\text{-}\beta)_2$ structure [49–52]. The α subunit has a molecular mass of 130 kDa which is capable of binding IGF-I. The 95 kDa β subunit of the IGF-I receptor, like that for the insulin receptor, exhibits a tyrosyl kinase activity. In marked contrast, however, the IGF-II receptor is a monomeric protein of molecular mass 220 kDa [53,54] with no known intrinsic activity.

Insulin can bind to the IGF-I receptor, albeit with low affinity, and activate its tyrosine kinase activity. However, insulin does not bind to IGF-II receptors. Nevertheless, the binding of insulin either to its own receptors or to those for IGF-I actually stimulates the upregulation of IGF-II receptors. This is achieved by the recruitment of IGF-II receptors to the plasma membrane in the same way that glucose transporters are recruited by insulin action. The binding of IGF-I to its receptors also triggers the rapid internalization of the occupied receptor, with its subsequent down-regulation [45,46] in a dose-dependent manner just as is seen for the receptors for insulin and other species which express a tyrosyl kinase activity. This action is, however, homologous as IGF-I only causes the down-regulation of its own receptors and not those for insulin. The same is true for the action of insulin, where internalization is specific for the insulin receptor.

Thus IGF-I receptors, and not those for IGF-II, can mimic the short-term, acute actions of insulin, certainly as far as stimulation of glucose and amino acid transport goes [55]. The quest for a role for IGF-II receptors is often complicated by the fact that cells that can be studied also express IGF-I receptors which can interact with the test ligands. It has been suggested that IGF-II receptors may promote DNA synthesis. However, studies done on H-35 cells, which do not express IGF-I receptors, imply that IGF-II receptors do not stimulate DNA synthesis, at least in these cells [56]. The biological function of IGF-II receptors, and whether they interact with G-proteins, thus remains to be ascertained.

7. Insulin receptor tyrosyl kinase activity

In contrast to those protein kinases which are stimulated by cAMP, Ca^{2+}/calmodulin and diacylglycerol [57] and which modify serine/threonine residues on target proteins, the kinase activity associated with the insulin receptor modifies tyrosyl residues on its target proteins [58]. Tyrosyl phosphate is relatively uncommon and accounts for less than 1% of the total protein-bound phosphate in cells [59]. However, this rare activity is in fact exhibited by various retrovirus gene products and receptors for a number of growth factors, e.g. epidermal growth factor (EGF) and platelet-derived growth factor (PDGF). It is believed that such a modification is related to the ability of these ligands to affect cellular growth control although, to date, no functional change has been attributed to tyrosine phosphorylation of a cellular protein. However, site-specific mutagenesis experiments have indicated that the functioning of a number of these proteins is abolished if the autophosphorylation sites are deleted [60].

Insulin receptor preparations solubilized, using detergents, from a number of sources have been shown to express tyrosyl kinase activity [58,61–66]. However, the majority of studies on this kinase activity have been performed using placental membranes as this tissue provides an unusually rich source of insulin receptors. In all systems examined to date, insulin has been shown to elicit the autophosphorylation of its 95 kDa β subunit. This can readily be identified by separation on SDS-PAGE with subsequent autoradiography or by immunoprecipitation using anti-insulin receptor antibodies. Thus, the binding of insulin to its α subunit leads to a conformational change in the receptor, effecting a stimulation of the kinase activity associated with its β subunit.

As well as causing autophosphorylation, this receptor-associated kinase activity can act on a variety of exogenous substrates (Fig. 3), including tyrosine-containing peptides that bear analogy to the peptide sequence surrounding the site of tyrosine phosphorylation on rous sarcoma virus-derived protein $pp60^{src}$ [61–66]. Unoccupied receptors catalysed a basal rate of phosphorylation of the exogenous species and the presence of insulin by eliciting, predominantly, an increase in the V_{max} of the reaction. Of the exogenous proteins that have been tested, histone H2b, casein, angiotensin-II, progesterone receptor and actin were effective substrates. In comparison with the tyrosyl kinase activity of the EGF receptor it is clear that they have similar, but by no means identical, substrate specificities [54].

The insulin-receptor kinase was also shown to be capable of phosphorylating the

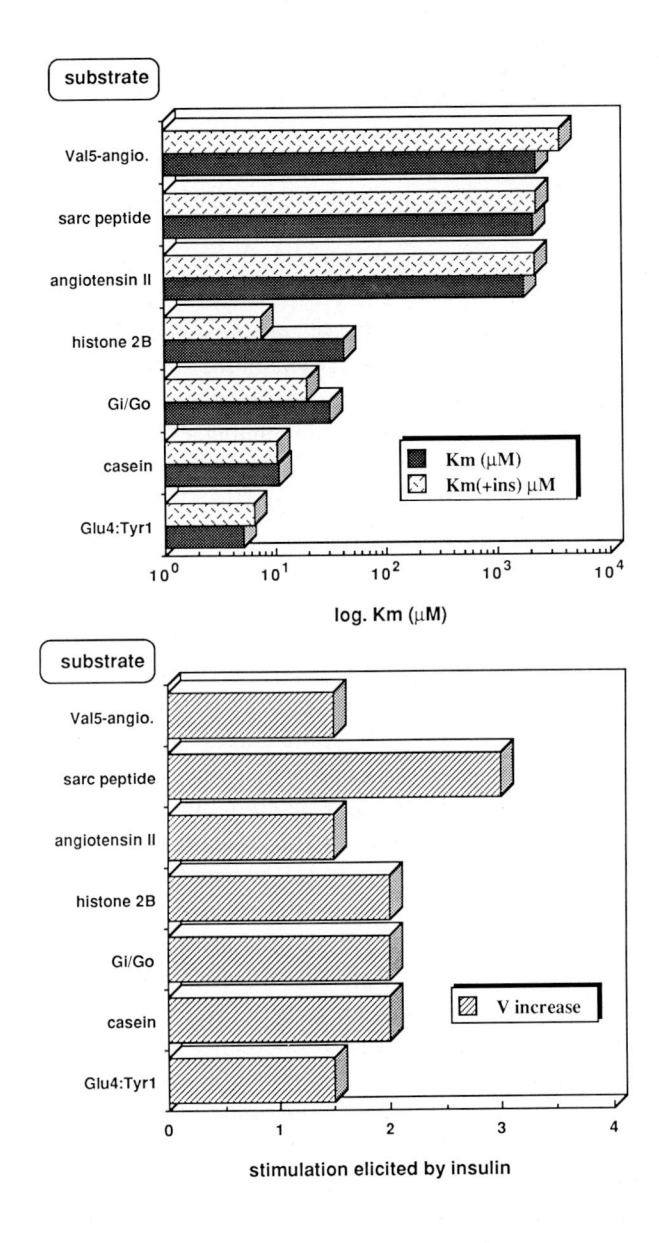

Fig. 3. Artificial substrates for the insulin receptor tyrosyl kinase. Shows the K_m values exhibited by basal and insulin-stimulated kinase activities together with the insulin-stimulated increase in V_{max} for a variety of substrates. These include angiotensin and its modified derivative (VAL-5), also synthetic peptides of Glu:Tyr and the so-called 'sarc'-peptide, which bears the sequence around the tyrosyl autophosphorylation site of the sarc protein. Data are also given for a G-protein mixture of G_i/G_o. These studies (referred to in the text) all employed soluble, purified insulin receptor preparations. No evidence has yet been presented for tyrosyl phosphorylation of substrates using isolated membrane preparations containing insulin receptors.

heavy chain of the IgG fraction of anti-pp60^{src} antibody. In this respect the activity of the insulin receptor-kinase resembles that of pp60^{src} and that of the EGF receptor [60]. There is, however, no evidence to suggest that the insulin receptor can phosphorylate antibodies directed against itself. This is perhaps because the various auto-antibodies, and the majority of monoclonal antibodies that have been studied, are directed against extracellular sites on the receptor, rather than at any intracellular sites close to the kinase domain on the β subunit.

Half-maximal effects for insulin stimulation of the receptor tyrosyl kinase occurred at around 5–10 nM insulin and, from analysis of insulin specific-binding studies, it appears that there is a non-linear relationship between the stimulation of receptor autophosphorylation by insulin and receptor occupancy. In contrast, a linear relationship has been noted between steady-state insulin binding and the ability of solubilized receptor kinase preparations to phosphorylate various exogenous substrates. The kinetics of coupling between insulin binding and the ability to elicit kinase activation as regards both exogenous substrates and its autophosphorylative capability requires further study.

The similarity between the functioning of the receptors for both insulin and EGF has been demonstrated by the construction [67] of chimeric receptors using cloned human EGF and human insulin receptor cDNAs. The generated chimeric receptors, expressed in COS-7 cells, were made up of the extracellular portion of the α subunit of the insulin receptor 'fused' to the transmembrane and cytoplasmic portions of the α subunit of the EGF receptor, together with the β subunit of the EGF receptor. Exposure of this chimeric receptor to insulin resulted in binding and the subsequent autophosphorylation of the EGF receptor-derived cytoplasmic domain. Not surprisingly this chimeric receptor exhibited a K_m for ATP which reflected that of the EGF receptor rather than that expected of the insulin receptor tyrosyl kinase. Such experiments demonstrate that insulin and EGF (and presumably IGF-I) share a common mechanism of signal transfer between their α and β subunits and across the transmembrane spanning regions of the β subunits of their receptors.

Analysis of tyrosyl kinase activity in vitro is often performed under rather unusual assay conditions which are far from those experienced physiologically. For example, in all instances detergent-solubilized, rather than membrane-bound, insulin receptor preparations are used. There are normally incubated with insulin at room temperature for a number of hours, as a prelude to performing the actual phosphorylation experiment, which itself is often carried out on ice, for anything from a few minutes to an hour at very low ATP concentrations. Indeed, until very recently, it was believed that this receptor-kinase exhibited an absolute requirement for Mn^{2+} in the assay, rather than Mg^{2+}. However, as ATP concentrations are elevated in order to approach physiological levels, tyrosine phosphorylation was seen to proceed at comparable rates using either Mg^{2+} or Mn^{2+} [55]. As the K_m for ATP is reported to be around 50 μM then it may be that high concentrations of ATP are either triggering some modification of the receptor kinase or that ATP itself is binding to a regulatory site.

Tyrosyl phosphorylation of the solubilized insulin receptor also appears to affect its functioning per se, in that autophosphorylation leads to an increase in the receptor tyrosyl kinase activity expressed towards exogenous substrates [68]. As autophosphorylation occurs at multiple sites on the β subunit it will be necessary to determine which of these sites are of regulatory significance. Such studies require that results obtained using solubilized receptor preparations be compared with those using intact cells where tyrosyl phosphorylation of the receptor appears to be very small in comparison with the phosphorylation of the receptor on serine residues [69].

Of particular interest are recent observations that the tyrosyl kinase activity of the insulin receptor can be attenuated by phosphorylation of the receptor on serine residues with cAMP-dependent kinase and protein kinase C [61,69,70]. The sites of phosphorylation for the action of these two kinases remain to be determined, however.

In intact cells insulin has been shown to stimulate receptor autophosphorylation. However, much smaller amounts of phosphotyrosine were found compared to that seen using purified, solubilized receptor preparations and, indeed, the predominant phosphorylation actually occurred on serine residues [69,71]. It has been suggested that there may be an insulin-stimulated, phosphoseryl-specific kinase which is loosely associated with the receptor and is activated by the receptor tyrosyl kinase [61]. This might account for the insulin-stimulated phosphoseryl kinase activity observed in both intact cells and when using crude, solubilized receptor preparations. Such an activity might also provide a mechanism for insulin's ability to enhance serine phosphorylation on target proteins in plasma membranes and elsewhere in the cell [25,78]. Nevertheless, (auto)-phosphorylation of the insulin receptor on tyrosine residues has been shown to occur immediately upon receptor occupancy by insulin and to precede any phosphorylation on serine residues [69]. Indeed, it remains to be seen as to whether the seryl phosphorylation occurs as a direct insulin-stimulation event or mediated by other kinases (cAMP/C-kinase) as a consequence of insulin-stimulated autophosphorylation on tyrosyl residues.

The reduced level of insulin-stimulated receptor autophosphorylation upon tyrosine residues in intact cells compared to that seen using solubilized receptors may also be due in part to the fact that additional sites become available on the solubilized receptors. Alternative explanations are that other membrane components might alter the specificity of the receptor-kinase in situ in the membrane [40] or that the action of other kinases on the receptor may attenuate the autophosphorylative activity. In any event a detailed analysis of the phosphorylation sites on the receptor in both solubilized and intact cell systems is required.

The phosphorylated receptor appears to be capable of being dephosphorylated by the action of endogenous cytosolic phosphatases, as incubation of cells with vanadate, which inhibits the action of tyrosine-phosphate-specific phosphatases, increased the phosphorylation state of the receptor [60,61]. The significance of this observation has yet to be ascertained.

It has also been suggested that the receptor/kinase activity of the receptor might be involved with some other process in the membrane such as inositol lipid phosphorylation leading to an increased rate of PtdIns 4,5-P_2 phosphorylation. Although this has been demonstrated using a solubilized receptor preparation, the rates of inositol lipid phosphorylation are extremely slow [73].

The search for physiological targets of action of tyrosyl kinases has been very unrewarding. No clear target for the action of insulin, other than the receptor itself, has been identified [61]. However, evidence has been presented which suggests that a membrane glycoprotein of molecular mass 110–120 kDa may provide a substrate for the receptor kinase. The identity and function of this species, however, remain to be elucidated [61,74].

It has thus to be clearly demonstrated whether the phosphotyrosyl kinase activity of the insulin receptor is directly involved in mediating the central metabolic effects of insulin; also, whether receptor auto-phosphorylation per se is a prerequisite in the chain of steps through which insulin exerts its actions on cellular metabolism. Certainly, by analogies with the tyrosyl kinase activities expressed by growth factor receptors and oncogene kinases [60] it is tempting to suggest that the receptor tyrosyl kinase activity is related to the growth-promoting activity of insulin. A number of investigators have attempted to ascertain whether the insulin receptor tyrosyl kinase activity plays a prime role in signal transduction. The results are, however, equivocal. Thus a particular anti-insulin receptor antibody preparation was found not to alter receptor phosphorylation despite its clear ability to stimulate glucose transport in adipocytes [75]. This could be interpreted as implying that there is no *direct* relationship between the phosphorylation of the insulin receptor and the stimulation of glucose transport. However, small changes in receptor phosphorylation, perhaps at a specific regulatory site, could not be ruled out. However, more recently a monoclonal antibody directed against the insulin receptor was found capable of mimicking the action of insulin in stimulating glucose transport but did not enhance the tyrosyl kinase activity of the receptor [36]. Receptor autophosphorylation has also been shown to be inhibited by Tb^{3+}, binding to a Ca^{2+} regulatory site on the receptor, without exerting any significant effect on the ability of insulin to stimulate glucose transport in adipocytes [75]. Taken at their face value these experiments appear to dissociate receptor autophosphorylation from insulin's stimulation of glucose transport. However, such observations might merely be due to a change in the conformation of the insulin receptor, caused either by the antibody or by Tb^{3+}, such that amino acids at specific sites were not available for phosphorylation. Indeed, in these experiments, a precise analysis of phosphorylation sites, and types of phosphoamino acids occurring, was not performed. Such experiments are essential for the critical evaluation of these studies. Nevertheless, the most convincing evidence for tyrosyl kinase involvement in the action of insulin is the recent demonstration that injection of a monoclonal antibody which inhibited the tyrosyl kinase into a variety of different cell types actually inhibited a spectrum of insulin-stimulated processes [77]. This is powerful evidence for a functional role of tyrosyl kinase activity in the molecular mechanism of insulin's action.

The receptor-kinase activity of the insulin receptor appears to be rather insensitive to insulin, with half-maximal stimulatory effects occurring at 2–8 nM insulin. Whilst this is comparable to the dose dependence of long-term effects of insulin it contrasts with insulin's rapid effects on, for example, glucose transport and cyclic AMP metabolism, which occur at some two orders of magnitude lower concentration. Of course, it is possible that the coupling between receptor kinase activation and the post-receptor mechanisms for other processes is amplified so that there need not be a linear correlation between kinase activation and effector stimulation. Indeed as the kinase is an enzyme which, once activated, can exert continuing effects then the activation of but a few receptors would be expected to lead to amplification of the response and perhaps, through its action on (phosphorylation of) particular components, could give rise to different apparent dose-effect curves for various 'downstream' processes which are regulated by insulin.

The expression of tyrosine kinase activity by the insulin receptor has, however, allowed the first biochemical defect of the receptor, distal to ligand binding, to be characterized [78]. This has been demonstrated in a particular patient exhibiting a form of insulin resistance. In this regard it should be noted that a large proportion of so-called type II diabetics actually exhibit normal or even elevated insulin levels. However, their tissues do not respond adequately to this hormone and thus such diabetics are classed as being 'insulin-resistant'. In most instances receptor number and affinity are normal in such patients and the lesion thus appears to be at a post-receptor/signal-generation level. The mononuclear blood cells of this patient [78] actually exhibited normal insulin binding yet were defective in the functioning of the insulin receptor tyrosyl kinase as regards both insulin-stimulated autophosphorylation and phosphorylation of exogenous substrates. A decreased extent of autophosphorylation of insulin receptors has also been noted in streptozotocin-diabetic rats [79]. In this instance, animals were treated with the drug streptozotocin, which destroys pancreatic β cells, in order to reduce the ability of the animals to produce insulin. Such animals become diabetic, with reduced insulin levels and elevated blood glucose levels. This is considered to be an 'animal model' of type I, insulin-dependent diabetes. However, there is evidence for insulin-resistance occurring in such animals [70–82] and also in insulin-dependent (type I) diabetic human beings [83,84]. It will be of interest then to determine whether the reduced autophosphorylative activity associated with the insulin receptor either results from a defect at the level of the kinase itself or is related to a defect in the 'coupling' between the α subunit and the β subunit of the receptor, whereby the activity of the kinase is decreased. Alternatively, an alteration in 'availability' of the autophosphorylation site on the receptor may have occurred. Nevertheless, that such changes in autophosphorylative activity do occur might be taken to imply a role for the kinase activity of the receptor in the mechanism by which insulin exerts at least certain of its actions on intracellular events. They do not necessarily implicate the kinase activity per se in regulating the rapid metabolic actions of insulin, as in these insulin-resistant states a defect in the functioning of the kinase may be merely one reflection of a receptor whose functioning has been 'crippled' by some modification in its structure. Nevertheless, that such changes have been noted in insulin-resistant diabetic states shows very clearly that the functioning of the insulin-resistant diabetic states shows very clearly that the functioning of the insulin receptor is fundamentally altered in such states.

8. Insulin and its action on guanine nucleotide regulatory proteins

Many hormones, which bind to cell surface receptors, stimulate their effector systems to generate an intracellular second message by interacting with specific guanine nucleotide regulatory proteins (G-proteins) [85–89].

Three distinct G-proteins have been fully characterized. These are transducin (T), which allows rhodopsin to stimulate a high-affinity cyclic GMP phosphodiesterase upon photoactivation in retinal rods; G_s, which allows receptors to stimulate the activity of adenylate cyclase; and G_i, which allows receptors to inhibit adenylate cyclase activity. These three G-proteins contain three non-identical subunits (α, β and γ). It is their α subunits which provide the binding site for GTP as well as distinct sites for interaction with specific receptor and effector systems and for Mg^{2+}. The β subunits associated with all three regulatory proteins are apparently identical although T has a different γ subunit.

The *ras* oncogene and proto-oncogenes have also been shown to code for a protein of molecular mass 21 kDa, called p21. This too exhibits GTP-binding and GTPase activity, although it does not appear to be capable of interacting with either β or γ subunits isolated from G_s, G_i or T. Evidence presented recently strongly suggests that this G-protein can regulate the stimulation of the phospholipase controlling the production of diacylglycerol and inositol 1,4,5-trisphosphate upon hormonal stimulation [90].

Four more guanine nucleotide regulatory proteins, G_o, G_x, G_h and $G_{25\alpha}$, have also been described recently, although a function has yet to be ascribed to them. Indeed, molecular biological and immunological studies indicate that there is a wide family of G-proteins involved in signal transduction mechanisms.

Activation of G-proteins by occupied receptors results usually in the dissociation of the G-protein complex, releasing an activated α subunit. In the case of G_s, the activated α subunit interacts with and activates the catalytic unit of adenylate cyclase by increasing the affinity of the enzyme for Mg^{2+}, whereas the activated α subunit of transducin stimulates cyclic GMP phosphodiesterase (PDE) activity by causing the dissociation of a γ subunit from the PDE, hence relieving an inhibitory effect. G_i appears to exert its inhibitory effect on adenylate cyclase in two distinct ways. Firstly, by releasing an activated α subunit. This inhibits adenylate cyclase by binding to the catalytic unit of this enzyme on a different site from that where the activated α subunit of G_s binds. Secondly, activation of G_i releases its β-γ complex, which has a structure identical to that found associated with G_s. This complex then inhibits G_s dissociation through mass action and hence inhibits the action of stimulatory receptors on adenylate cyclase.

The α subunits of all of these various G-proteins exhibit a GTPase activity which is important for their functioning. Indeed, it is believed that the hydrolysis of GTP is related to the 'turn-off' or de-activation of the dissociated and activated α subunits.

G-protein identification has been aided considerably by the ability of their α subunits to be specifically ADP-ribosylated by either cholera or pertussis toxins. Cholera toxin ribosylates both transducin and G_s, causing dissociation of the holomeric form of these G-proteins to release a permanently activated and ADP-ribosylated α subunit. In contrast, pertussis toxin (islet activating protein, IAP) ribosylates the α subunit of the holomeric forms of G_i, G_o, G_h and transducin preventing their dissociation and, in the case of G_i, blocking the action of inhibitory hormones on adenylate cyclase activity. In the case of G_i, G_o and transducin, it has been clearly demonstrated that pertussis toxin can only ribosylate the holomeric form of these G-proteins. Cholera toxin, however, can ADP-ribosylate the free α subunits of G_s and transducin equally well, if not better, than when they form part of the holomeric complex.

Human platelets have provided a useful model system for studying G-protein functioning, as receptor-stimulated GTPase activities are easily detected and quantitated. Thus platelets express a functional G_i, G_o and 'G_p', the G-protein believed to be involved in controlling inositol phospholipid metabolism and which may be equivalent to the *ras* proto-oncogene p21 [90–92]. However, it has recently been demonstrated that insulin can stimulate a high-affinity GTPase activity in human platelets [93]. This activity is controlled by high-affinity insulin receptors and is distinct from the G-proteins G_s, G_i and G_p. However, like G_s and transducin, this novel GTPase appears to be affected by cholera toxin, indicating that it may reflect the activity of the putative G_{ins}, a novel G-protein suggested to mediate certain of insulin's actions [94–97]. Indeed, evidence has been presented to show that G_{ins} is activated by treatment with cholera toxin and, in liver plasma membranes, the α subunit of G_{ins} was identified as a 25 kDa species whose ribosylation by cholera toxin was inhibited by the presence of insulin [95]. No such effect of insulin was apparent on the ribosylation of G_s by cholera toxin. Such an action is reminiscent of that of ligands which activate G_i, to release its α subunit and hence prevent its ribosylation by pertussis toxin. The action of insulin in preventing the cholera toxin-mediated ribosylation of G_{ins} was attenuated in certain 'insulin-resistant states' and after treatment of hepatocytes with phorbol esters. Such factors are reminiscent of the attenuation of receptor tyrosyl kinase functioning (vide supra). It has been suggested that only certain of insulin's actions may be controlled through the functioning of G_{ins} [96,97] based on studies of the ability of insulin to inhibit adenylate cyclase activity and to activate a peripheral plasma membrane cyclic AMP phosphodiesterase.

Activation of the peripheral plasma membrane phosphodiesterase in rat liver can be observed by treating either intact hepatocytes [98] or isolated liver plasma membranes [99] with insulin. The activation process appears to be dependent upon micromolar concentrations of both cyclic AMP and MgATP as well as insulin. Indeed, activation ensues as a result of the phosphorylation of the plasma membrane phosphodiesterase on serine residues. The mechanism of activation and the precise nature of the kinase involved is, however, unclear. Nevertheless, employing isolated membranes the activation and phosphorylation of this enzyme could be triggered by non-hydrolysable guanine nucleotide analogues [99]. This is consistent with the suggestion that a G-protein controls the activation of this enzyme species. Clearly, G_s cannot be involved, as glucagon, which functions through G_s, does not activate this enzyme, and G_i does not appear to be involved, as pertussis toxin-treatment of cells does not prevent activation of this enzyme by insulin [100]. However, treatment of hepatocytes with cholera toxin led to the activation of this enzyme. Such an effect cannot be due to G_s activation both for the reasons described above and because it can be demonstrated, in broken membranes and in whole cells, that activation was not caused by any rise in cyclic AMP concentrations that occurred through cholera toxin also activating adenylate cyclase, through G_s. This suggests that the G-protein involved is, like G_s, a substrate for ribosylation and activation by cholera toxin: properties attributed to G_{ins}.

Brief exposure of hepatocytes to glucagon prior to challenge with insulin was found to block the stimulatory action of insulin on this peripheral phosphodiesterase [98]. This action bears a remarkable similarity to the action of glucagon in eliciting the desensitization or uncoupling of adenylate cyclase in hepatocytes [101]. Similar time- and dose-dependencies were evident as well as it being demonstrated that agents such as adenosine, N^6-phenylisopropyladenosine (PIA) and islet activating protein all prevent glucagon exerting both of these effects [100,102]. These actions of glucagon are mediated by a fraction of high-affinity glucagon receptors (called GR1 receptors) that are distinct from those (called GR2 receptors) which act to stimulate adenylate cyclase through G_s [103]. Indeed, we have now shown that GR1 receptors actually stimulate inositol phospholipid metabolism to produce the second messengers, inositol trisphosphate and diacylglycerol [103]. Thus both the desensitization and 'blockade' of insulin's action by glucagon are cyclic AMP-independent effects. Desensitization appears to be caused by a perturbation at the level of regulation of guanine nucleotides, as the functioning of G_s can be seen to be altered. It is possible, therefore, that such a process, triggered by glucagon through GR1 receptors, causes the inactivation of G_{ins}.

If insulin does exert certain functions through a distinct G-protein, called G_{ins}, it might explain the various reports which have indicated that insulin may inhibit adenylate cyclase activity. Indeed, in liver membranes it can be clearly demonstrated that insulin does inhibit adenylate cyclase in a GTP-requiring fashion through very-high-affinity receptors [94]. However, at high concentrations of the stimulatory hormone glucagon this effect of insulin is abolished. This is consistent with insulin inhibiting adenylate cyclase through release of β-γ subunits from G_{ins}, where gross stimulation of G_s overwhelms this effect as per a competitive inhibitor where $[G_s]$ is much greater than $[G_{ins}]$. This action of insulin is also lost in cells made 'insulin-resistant' by brief exposure to insulin. As such it is reminiscent of G_{ins} functioning determined by ribosylation and by phosphodiesterase activation. The failure to observe insulin's inhibition of adenylate cyclase in assays based upon α-$[^{32}P]$ATP utilization may well be due to the high concentrations of unlabelled cyclic AMP added. These will activate cAMP-kinase in the assays and hence (vide supra) can be expected to phosphorylate the insulin receptor and attenuate its functioning in isolated systems.

Other evidence for the involvement of a G-protein in the action of insulin has come from studies by Walaas and co-workers [104]. They have demonstrated that insulin stimulated the activity of a cyclic AMP-dependent protein kinase activity in sarcolemma membranes. As this effect of insulin was enhanced if micromolar concentrations of GTP-binding protein were present, they suggested that a guanine nucleotide regulatory protein was involved in the hormonal control of this kinase. Indeed, cholera toxin also appeared to obliterate this action of insulin, as it did the effect of insulin on liver adenylate cyclase and the peripheral plasma membrane cyclic AMP phosphodiesterase in liver.

It is possible that G_{ins} might alter the functioning or even the specificity of the receptor kinase. Indeed, the loss of a 'regulatory protein' may offer an explanation for the changes both in residues phosphorylated and in the kinase activity observed comparing solubilized receptor preparations with those seen in whole cells (vide supra). In this regard the catalytic unit of adenylate cyclase when freed from interaction with G_s, by detergent solubilization, exhibits an altered Me^{2+} dependency with a preference for MnATP over MgATP [85]. This is because G_s conveys Mg^{2+} sensitivity to the enzyme. One could draw an analogy to this with the insulin receptor/kinase which shows a preference for MgATP when solubilized. Perhaps a regulatory protein could alter the functioning and Me^{2+} dependency of the insulin receptor kinase.

G_{ins} may also contribute to the mechanism whereby insulin activates glucose transport in adipocytes and heart. This has been suggested to be a two-step process where the first step involves the recruitment of inactive glucose carriers from an internal vesicle pool to the plasma membrane and the second step involves the activation of the newly inserted carriers in the plasma membrane (vide supra). It is possible that this second step may be controlled by G_{ins}. The reasons for thinking that this may be so are related to observations that, under certain conditions, glucagon and isoprenaline can each inhibit insulin-stimulated glucose transport, i.e. create a state of apparent 'insulin-resistance' [102, 105–108] through a mechanism which does not involve any block in the translocation of glucose carriers to the plasma membrane. Furthermore, this inhibitory effect of glucagon and β-adrenoceptor agonists was abolished if either adenosine or R-type adenosine analogues were present in the external medium. These actions are of course similar to those seen for the effect of glucagon in causing desensitization and blocking the effects of insulin on both the plasma membrane phosphodiesterase and on adenylate cyclase in hepatocytes by causing an 'insulin-resistant' state. Given the similarities between the effects of adenosine and glucagon in modulating the action of insulin on glucose transport it is possible that G_{ins} may be involved in the control of the activation/activity of glucose carriers in the plasma membrane.

Other evidence that such a regulatory protein might be involved in mediating the stimulatory action of insulin on glucose transport comes from studies on barnacle muscle [109]. There it was shown that glucose transport could be stimulated by the introduction of non-hydrolysable GTP analogues into the giant muscle cells through a cyclic AMP-independent process. This result is compatible with the direct activation of a regulatory species such as G_{ins}, by-passing the insulin receptor.

Recent studies have shown that in rats made diabetic with either streptozotocin or alloxan, insulin failed to inhibit adenylate cyclase activity in liver [110]. This is very similar to the attenuation of functioning of insulin receptor tyrosyl kinase activity in streptozotocin-diabetic rats. Clearly this is a post-receptor effect as there is little, if any, change in insulin receptor number or affinity in these diabetic animals. Intriguingly, treatment of streptozotocin-diabetic animals with the hypoglycaemic drug metformin, which is used to treat insulin-resistant diabetics [111,112], completely restored the ability of insulin to inhibit adenylate cyclase. This suggests that the biguanide drug metformin can correct post-receptor defects in the insulin receptor signalling system.

It remains to be seen whether the insulin receptor can activate G_{ins} either directly or indirectly and whether the tyrosyl kinase activity of the receptor is involved. However, it is of interest that recently purified G_o and G_i have been shown to be capable of being phosphorylated by the insulin receptor tyrosyl kinase [113]. Such phosphorylation occurred on both the α and the β subunits. However, phosphorylation only occurred using the intact holomeric form of these G-proteins and not the activated, dissociated subunits. Other G-proteins may also be phosphorylated. However, whether any functional changes ensue and whether such reactions can occur in vivo will have to be determined.

Other evidence of interaction of insulin with the G-protein system comes from evidence that in rats made diabetic with either streptozotocin or alloxan, functional G_i activity, in liver, appeared to disappear. The use of a specific antibody to quantitate G_i showed that the level of G_i in liver plasma membranes had fallen to below 10% of that of normal animals [114]. The loss of this key G-protein may explain some of the hormonal and other abnormalities seen in diabetes.

9. An intracellular 'mediator' of insulin's action

It has been suggested that insulin might exert actions on target cells by stimulating the production of an intracellular chemical factor or 'mediator' in an analogous fashion to the production of second messengers by other hormones (see Refs. 115–117). This concept arose from the initial observations that treatment of skeletal muscle with insulin appeared to generate a substance which inhibited cyclic AMP-dependent protein kinase. Other reports followed this providing preliminary chromatographic separation of a 1–2 kDa 'mediator' fraction. This 'mediator' appeared to (i) stimulate mitochondrial pyruvate dehydrogenase activity, (ii) stimulate glycogen synthase phosphoprotein phosphatase and (iii) inhibit cyclic AMP-dependent protein kinase.

Initial analysis of the partially purified material indicated that it interacts with ninhydrin, and the field was dominated, during the seventies, by the possibility that the 'mediator' could be a peptide or 'peptide-like' substance. Indeed, 'mediator' fractions were shown to contain various amino acids although there remained controversy over the purported protease-sensitivity of this material. However, 'mediator' production from adipocyte membranes was reported to be blocked by specific protease inhibitors, which indicated that a membrane-bound protease, acting at an arginine residue, might lead to the production of the 'mediator'. However, such inhibitors were found not to block the actions of insulin in intact cells. Indeed, proponents of such a peptide hypothesis even suggested that the mediator might be produced by the action of a protease on the receptor itself. If this were to occur then one might expect insulin treatment to lead to a rapid 'desensitization', due to the inactivation of receptors by proteolysis where resensitization would be dependent upon the synthesis of new insulin receptors. There is no evidence to support such an action.

'Mediator' preparations have also been claimed to contain various carbohydrate and phospholipid components as well. Many of these will exert non-specific effects on the various enzyme systems studied, undoubtedly adding to the confusion.

Recently, however, there has been an upsurge in interest in the 'insulin second messenger' field with the partial purification and characterization of two novel 'mediator' substances (Fig. 4). Rather than being peptides, these substances, called GIPs, are claimed to be complex carbohydrate-phosphate substances containing glucosamine and, interestingly, inositol. Such substances appear to activate membrane-bound cyclic AMP phosphodiesterase activity in adipocytes and mitochondrial pyruvate dehydrogenase activity, as per insulin's action on intact cells [118,119].

It was shown, however, that the ability of insulin to release GIPs could be mimicked by an inositol phospholipid-specific phospholipase C. This enzyme is distinct from the phospholipidase C which acts to break down polyphosphatidylinositols as it acts specifically on phosphatidylinositol and has been used to release proteins which are anchored to membranes by the covalent attachment to phosphatidylinositol

[120]. The specificity is such that it is entirely feasible that the action of such an enzyme is to release a glycan-inositol phosphate compound. However, the nature and mechanism of action of such a compound have yet to be resolved.

If insulin does act to stimulate such a specific phospholipase then one might well expect diacylglycerol (DAG) to be produced. Indeed, this has been shown to be the

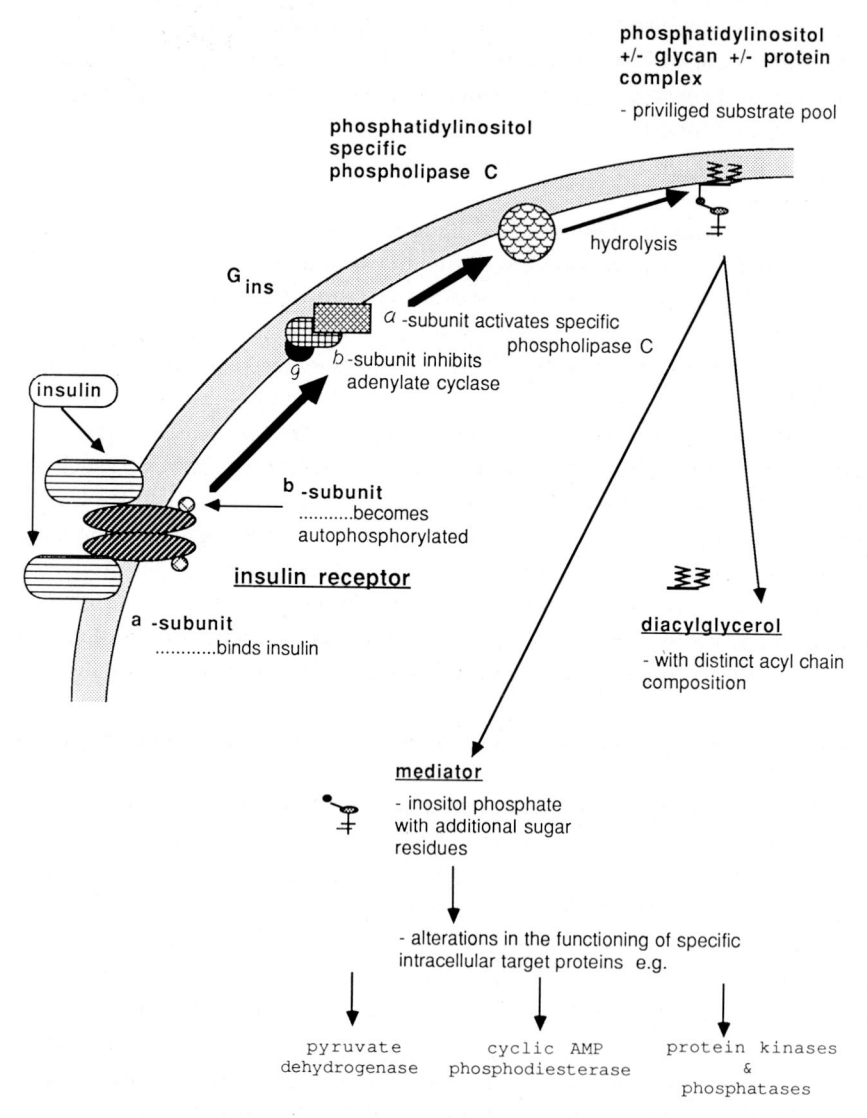

Fig. 4. Proposed production of a 'glycan-inositol phosphate' 'mediator' by insulin. This summarizes Saltiel and Cuatrecasas' concept for the production of novel mediators through insulin's action. It may, by analogy with the receptor-mediated stimulation of inositol phospholipid metabolism, involve a G-protein. Here it is postulated that the putative G_{ins} may play such a role.

case and, furthermore, the fact that the DAG produced has an unusual fatty acyl (myristoyl) chain associated with it clearly shows that this enzyme acts on a distinct pool of inositol-containing phospholipid. Thus the DAG produced by insulin's action is distinct from that produced by the action of hormones which stimulate polyphosphatidylinositol metabolism. As it has recently been demonstrated that multiple forms of protein kinase C exist, it is entirely possible that the DAG produced through the action of insulin might activate a specific isoenzyme.

There is now considerable evidence to suggest that hormones which stimulate inositol phospholipid metabolism do so through a distinct G-protein [90–92]. It is thus possible that insulin might activate a G-protein in order to stimulate the proposed phosphatidylinositol-specific phospholipase C claimed to produce the GIPs insulin 'mediator' (Fig. 4).

10. Concluding remarks

The control of cellular kinases and phosphatases appears to be of undoubted importance to the action of insulin. Whether such effects are determined either directly or indirectly by the receptor tyrosyl kinase, 'G_{ins}', the second messengers GIP and DAG and by modification of other G-proteins, perhaps even all of these, remains to be determined (Fig. 5).

Some time ago, one of us suggested that insulin might exert its actions through a 'multipathway mechanism', whereby the membrane-bound receptor generated a number of distinct signals [87]. There appears to be an increasing body of evidence consistent with this proposal (Fig. 5). Certainly it would offer an explanation for the plethora of actions of insulin and diversity of (selective) insulin-resistant conditions, where malfunction of one or more of these pathways occurs.

It remains to be seen whether all receptor molecules can 'couple' to different signalling systems or whether there may be distinct populations of insulin receptors exerting different actions. Indeed the anomalous binding properties of the insulin receptor, where saturation is achieved only over a number of orders of magnitude of insulin, coupled with observations that particular insulin-mediated processes exhibit different concentration dependencies upon insulin, might support a view for distinct signalling systems coupled to the receptor. The basis for this might lie in the existence of multiple related mRNA molecules coding for the receptor and differences in receptor glycosylation.

Molecular biological and immunological approaches together with the availability of purified receptor and other components should therefore lead to the full elucidation of the functioning of this receptor in normal and disease states.

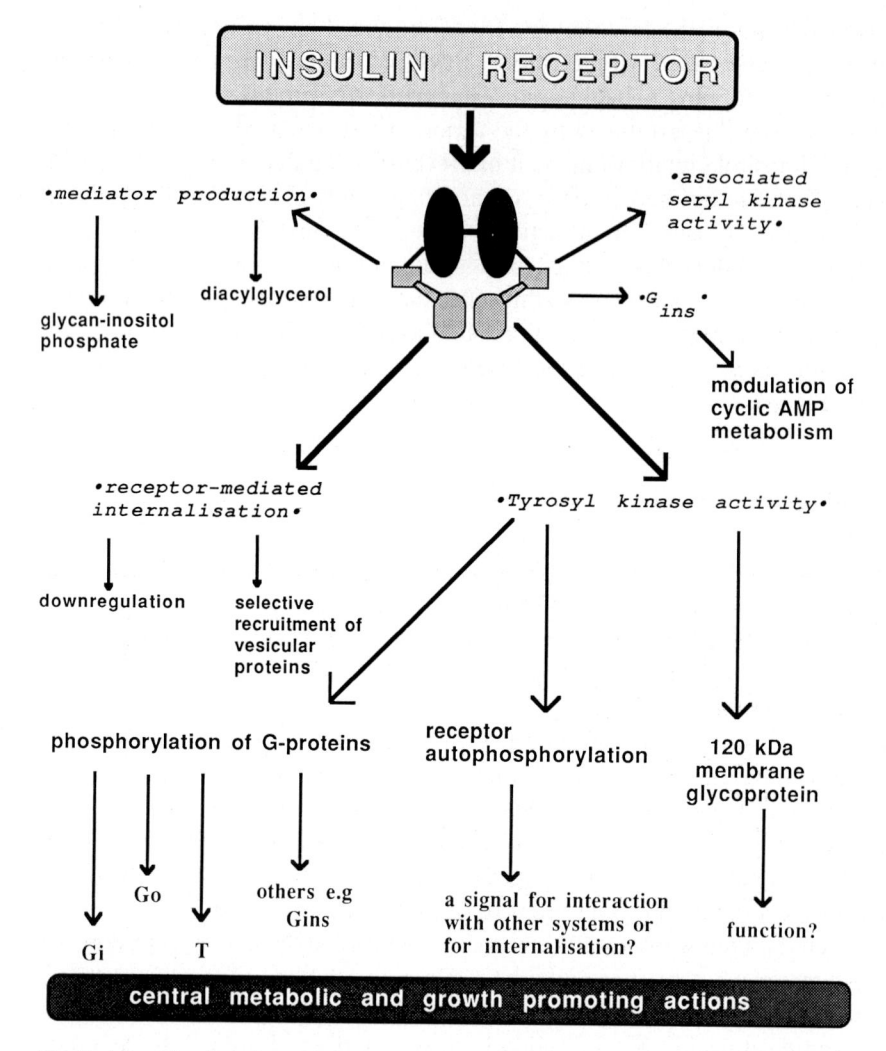

Fig. 5. A 'multipathway' mechanism of action for insulin. Insulin exerts a plethora of actions on target tissues and it is extremely difficult to visualize a single 'second message' acting to achieve all such effects. Here it is suggested that a number of signals emanate from the plasma membrane as a consequence of insulin binding to its receptor. This may also offer a molecular explanation for 'selective' insulin-resistant states noted in type-I and type-II human diabetes as well as experimentally in animals. In such circumstances lesions could occur at particular points on this 'multipathway'.

Acknowledgements

Work in the authors' laboratory was supported by grants from the Medical Research Council (UK), British Diabetic Association, Scottish Home and Health Department, SERPA and California Metabolic Research Foundation (USA).

References

1. De Meyts, P., Bianco, A.R. and Roth, J. (1976) J. Biol. Chem. 251, 1877–1888.
2. Kahn, C.R. (1976) J. Cell Biol. 70, 261–286.
3. Ullrich, A., Bell, J.R., Chen, E.Y., Herrera, P., Petruzzelli, L.M., Dull, T.J., Gray, A., Coussens, L., Liao, Y.C., Tsubokawa, M., Mason, A., Seeburg, P.H., Grunfeld, C., Rosen, P.M. and Ramachandran, J. (1985) Nature (Lond.) 313, 756–761.
4. Ebina, Y., Ellis, L., Jarnagan, K., Edery, M., Graf, L., Clauser, E., Ou, J.-H., Masiarz, F., Kan, Y.W., Goldfine, I.D., Roth, R.A. and Rutter, W.J. (1985) Cell 40, 747–758.
5. Fujita-Yamaguchi, Y. (1984) J. Biol. Chem. 259, 1206–1211.
6. Finn, F.M., Titus, G., Horstman, D. and Hofman, K. (1984) Proc. Natl. Acad. Sci. USA 81, 7328–7332.
7. Soos, M.A., Siddle, K., Baron, M.D., Heward, J.M., Luzio, J.P., Bellatin, J. and Lennox, E.S. (1986) Biochem. J. 235, 199–208.
8. O'Brien, R.M., Soos, M.A. and Siddle, K. (1986) Biochem. Soc. Trans. 14, 316–317.
9. Pilch, P.F. and Czech, M.P. (1980) J. Biol. Chem. 255, 1722–1731.
10. Czech, M.P. (1982) Cell 31, 8–10.
11. Yip, C.C., Moule, M.L. and Yeung, C.W.T. (1982) Biochemistry 21, 2940–2945.
12. Boyle, T.R., Campana, J., Sweet, L.J. and Pessin, J.E. (1985) J. Biol. Chem. 260, 8593–8600.
13. Hedo, J.A. and Simpson, I.A. (1984) J. Biol. Chem. 259, 11083–11089.
14. Massague, J., Pilch, P.F. and Czech, M.P. (1981) J. Biol. Chem. 256, 3182–3190.
15. Kasuga, M., Karlsson, F.A. and Khan, C.R. (1982) Science 215, 185–186.
16. Kasuga, M., Zick, Y., Blith, D.L., Karlsson, F.A., Haring, H.V. and Khan, C.R. (1982) J. Biol. Chem. 257, 9891–9894.
17. Baron, M.D. and Sonksen, P.H. (1983) Biochem. J. 212, 79–84.
18. Harmon, J.T., Kahn, C.R., Kempner, E.S. and Schlegel, W. (1980) J. Biol. Chem. 255, 2412–3419.
19. Houslay, M.D. and Stanley, K.K. (1982) Dynamics of Biological Membranes: influence on synthesis, structure and function. John Wiley, London.
20. Ronnett, G.V., Knutson, V.P., Kohawki, R.A., Simpson, T.L. and Lane, D.M. (1984) J. Biol. Chem. 259, 4566–4573.
21. Hedo, J.A. and Simpson, I.A. (1985) Biochem. J. 232, 71–78.
22. Ebina, Y., Edery, M., Ellis, L., Standring, D., Beaudoin, J., Roth, R.A. and Rutter, W.J. (1985) Proc. Natl. Acad. Sci. USA 82, 8014–8018.
23. Fehlmann, M., Carpentier, J.-L., Van Obberghen, E., Freychet, P., Thamm, P., Saunders, D., Brandenburg, D. and Orci, L. (1982) Proc. Natl. Acad. Sci. USA 79, 5921–5925.
24. Crettaz, M. and Khan, C.R. (1984) Diabetes 33, 477–485.
25. Houslay, M.D. (1981) Biosci. Rep. 1, 19–34.
26. Suzuki, K. and Kono, T. (1980) Proc. Natl. Acad. Sci. USA 77, 2542–2545.
27. Cushman, S.W. and Wardsala, L.J. (1980) J. Biol. Chem. 255, 2542–2545.
28. Cushman, S.W., Wardsala, L.J., Simpson, I.A., Karnieli, E., Hissin, H.J., Wheeler, T.J., Hinkle, P. and Salans, L.B., (1983) in Hormones and Cell Regulation (Dumont, J.E., Nunez, J. and Denton, R.M., eds.) Vol. 7 pp. 73–84. Elsevier, Amsterdam.
29. Lienhard, G.E. (1983) Trends Biochem. Sci. 8, 125–127.
30. Heidenreich, K.A., Brandenburg, D., Berhanu, P. and Olefsky, J.M. (1984) J. Biol. Chem. 259, 6511–6515.
31. Posner, B.I., Patel, B.A., Khan, M.N. and Bergeron, J.J.M. (1982) J. Biol. Chem. 257, 5789–5799.
32. Khan, M.N., Savoie, S., Khan, R.J., Bergeron, J.J.M. and Posner, B.I. (1985) Diabetes 34, 1025–1030.
33. Wilson, S.R., Wallace, A.V. and Houslay, M.D. (1983) Biochem. J. 216, 245–248.
34. Khan, C.R., Baird, K.L., Jarrett, D.B. and Flier, J.S. (1978) Proc. Natl. Acad. Sci. USA 75, 4209–4213.

35. Taylor, R., Soos, M.A., Wells, A., Argyraki, M. and Siddle, K. (1987) Biochem. J. 242, 123–129.
36. Zick, Y., Rees-Jones, R.W., Taylor, S.I., Gorden, P. and Roth, J. (1984) J. Biol. Chem. 259, 4396–4400.
37. Roth, R.A., Cassell, D.J., Maddox, B.A. and Goldfine, I.D. (1983) Biochem. Biophys. Res. Commun. 115, 245–252.
38. Siegel, J. and Olefsky, J.M. (1980) Biochemistry 19, 2183–2190.
39. Sonne, O. and Simpson, I.A. (1984) Biochim. Biophys. Acta 804, 404–413.
40. Houslay, M.D. (1985) in Molecular Aspects of Cellular Regulation, Vol. 4 (Cohen, P. and Houslay, M.D., eds.) pp. 279–333. Elsevier, Amsterdam.
41. Horuk, R., Matthaei, S., Olefsky, J.M., Baly, D.L., Cushman, S.W. and Simpson, I.A. (1986) J. Biol. Chem. 261, 1823–1828.
42. Hyslop, P.A., Kuhn, C.E. and Sauerheber, R.D. (1985) Biochem. J. 232, 245–254.
43. Witters, L.A., Vater, C.A. and Lienhard, G.E. (1985) Nature (Lond.) 315, 777–778.
44. Gibbs, E.M., Allard, W.J. and Lienhard, G.E. (1986) J. Biol. Chem. 261, 16597–16603.
45. Wardzala, L.J., Simpson, I.A., Rechler, M.M. and Cushman, S.W. (1984) J. Biol. Chem. 259, 8378–8383.
46. Oka, Y., Mottola, C., Oppenheimer, C.L. and Czech, M.P. (1984) Proc. Natl. Acad. Sci. USA 81, 4028–4032.
47. Zapf, J., Rinderkrecht, E., Humbel, R.E. and Froesch, E.R. (1978) Metabolism 27, 1803–1828.
48. Zapf, J., Schoenle, E., Waldvogel, M., Sand, I. and Froesch, E.R. (1981) Eur. J. Biochem. 113, 605–609.
49. Massague, J. and Czech, M.P. (1982) J. Biol. Chem. 257, 5038–5045.
50. Bhaumick, B., Bala, R.M. and Hollenberg, M.D. (1981) Proc. Natl. Acad. Sci, USA 78, 4279–4283.
51. Jacobs, S., Hazum, E. and Cuatrecasas, P. (1980) J. Biol. Chem. 255, 6937–6940.
52. Kasuga, M., Van Obberghen, E., Nissley, P.P. and Rechler, M. (1982) Proc. Natl. Acad. Sci. USA 79, 1864–1868.
53. Oppenheimer, C.L. and Czech, M.P. (1983) J. Biol. Chem. 258, 8539–8541.
54. August, G.P., Nissley, S.P., Kasuga, M., Lee, L., Greenstein, L. and Rechler, M.M. (1983) J. Biol. Chem. 258, 9033–9036.
55. Yu, K.T. and Czech, M.P. (1984) J. Biol. Chem. 259, 3090–3095.
56. Mottola, C. and Czech, M.P. (1984) J. Biol. Chem. 259, 12705–12715.
57. Cohen, P. (1982) Nature (Lond.) 296, 613–620.
58. Kasuga, M., Karlsson, F.A. and Kahn, C.R. (1981) Science 215, 185–187.
59. Hunter, T. (1980) Cell 22, 647–648.
60. Foulkes, J.G. and Rosner, M.R. (1985) In: Molecular Mechanisms of Transmembrane Signalling (Cohen, P. and Houslay, M.D., eds.) pp. 217–244. Elsevier, Amsterdam.
61. Gammeltoft, S. and Van Obberghen, E. (1986) Biochem. J. 235, 1–11.
62. Casnellie, J.E., Harrison, M.L., Pike, L.J., Hellstrom, K.E. and Krebi, E.G. (1982) Proc. Natl. Acad. Sci, USA 79, 282–286.
63. Stadtmauer, L.A. and Rosen, P.M. (1983) J. Biol. Chem. 258, 6682–6685.
64. Pike, L.J., Huenzel, E.A., Basnellie, J.E. and Krebs, E.G. (1984) J. Biol. Chem. 259, 9913–9921.
65. Nemenoff, R.A., Kwok, Y.C., Shulman, G.I., Blackshear, P.J., Osathanondh, R. and Avruch, J. (1984) J. Biol. Chem. 259, 5058–5065.
66. Klein, H.H., Freidenberg, G.R., Cordera, R. and Olefsky, J.M. (1985) Biochem. Biophys. Res. Commun. 127, 254–263.
67. Riedel, H., Dull, T.J., Schlessinger, J. and Ullrich, A. (1986) Nature 324, 68–70.
68. Yu, K.-T. and Czech, M.P. (1984) J. Biol. Chem. 259, 5277–5286.
69. White, M.F., Takayama, S. and Kahn, C.R. (1985) J. Biol. Chem. 260, 9470–9478.
70. Pessin, J.E., Gitomer, W., Oka, Y., Oppenheimer, C.L. and Czech, M.P. (1983) J. Biol. Chem. 258, 7386–7394.

71. Kasuga, M., Fujita-Yamaguchi, Y., Blithe, D.L., White, M.F. and Kahn, C.R. (1983) J. Biol. Chem. 258, 10973–10980.
72. Denton, R.M., Brownsey, R.W. and Belsham, G.J. (1981) Diabetologia 21, 347–362.
73. Machicao, E. and Wieland, O.H. (1984) FEBS Lett. 175, 113–116.
74. Sadoul, J.L., Peyron, J.F., Ballotti, R., Debont, A., Fehlmann, M. and Van Obberghen, E. (1985) Biochem. J. 227, 887–892.
75. Simpson, I.A. and Hedo, J.A. (1984) Science 223, 1301–1304.
76. Plewhe, W.E., Williams, P.F., Caterson, I.D. Harrison, L.C. and Turtle, J.R. (1983) Biochem. J. 214, 361–366.
77. Morgan, D.O. and Roth, R.A. (1987) Proc. Natl. Acad. Sci. USA 84, 41–45.
78. Grunberger, G., Zick, Y. and Gordon, P. (1984) Science 223, 932–934.
79. Kadowaki, T., Kasuga, M., Akanuma, Y., Fzaki, O. and Takaka, F. (1984) J. Biol. Chem. 259, 14208–14216.
80. Kobayashi, M. and Olefsky, J.M. (1979) Diabetes 28, 87–95.
81. Karnieli, E., Zarnowski, M.J., Hissin, P.J., Simpson, I.A., Salans, L.B. and Cushman, S.W. (1981) J. Biol. Chem. 256, 4772–4777.
82. Reavan, G.M., Sageman, W.S. and Swenson, R.S. (1977) Diabetologia 13, 459–462.
83. Caro, J.F., Ittoop, O., Pories, W.J., Meelheim, D., Flickinger, E.G., Thomas, F., Jenquin, M., Silverman, J.F., Khaznie, P.G. and Sinha, M.K. (1986) J. Clin. Invest. 78, 249–258.
84. Tessari, P., Nosadini, R., Trevison, R., De Kreutzenberg, S.V., Inchiostro, S., Duner, E., Biolo, G., Marescotti, M.C., Tiengo, A. and Crepaldi, G. (1986) J. Clin. Invest. 77, 1797–1804.
85. Birnbaumer, L., Codina, J., Mattera, R., Verione, R.A., Hildebrandt, J.D., Sunya, T., Rojas, F.J., Caron, M.J., Lefkowitz, R.J. and Iyengar, R. (1985) Mol. Aspects Cell. Regul. 4, 131–182.
86. Gilman, A.G. (1984) Cell 36, 577–579.
87. Houslay, M.D. (1983) Nature (Lond.) 303, 133.
88. Houslay, M.D. (1984) Trends Biochem. Sci. 9, 39–40.
89. Northup, J.K. (1985) Mol. Aspects Cell. Regul. 4, 91–116.
90. Wakelam, M.J.O., Davies, S.A., Houslay, M.D., McKay, I., Marshall, C.J. and Hall, A. (1986) Nature (Lond.) 323, 173–176.
91. Houslay, M.D., Bojanic, D. and Wilson, A. (1986) Biochem. J. 234, 737–740.
92. Houslay, M.D. Bojanic, D., Gawler, D., O'Hagan, S. and Wilson, A. (1986) Biochem. J. 238, 109–113.
93. Gawler, D. and Houslay, M.D. (1987) FEBS Lett. 216, 94–98.
94. Heyworth, C.M. and Houslay, M.D. (1983) Biochem. J. 214, 547–552.
95. Heyworth, C.M. Whetton, A.D., Wong, S., Martin, B.R. and Houslay, M.D. (1985) Biochem. J. 228, 593–603.
96. Houslay, M.D. and Heyworth, C.M. (1983) Trends Biochem. Sci, 8, 449–452.
97. Houslay, M.D. (1986) Biochem. Soc. Trans. 14, 183–193.
98. Heyworth, C.M., Wallace, A.V. and Houslay, M.D. (1983) Biochem. J. 214, 99–110.
99. Heyworth, C.M., Rawal, S. and Houslay, M.D. (1983) FEBS Lett. 154, 87–91.
100. Heyworth, C.M., Grey, A.-M., Wilson, S.R., Hanski, E. and Houslay, M.D. (1986) Biochem. J. 235, 154–149.
101. Heyworth, C.M. and Houslay, M.D. (1983) Biochem. J. 214, 93–98.
102. Wallace, A.V., Heyworth, C.M. and Houslay, M.D. (1984) Biochem. J. 222, 177–182.
103. Wakelam, M.J.O., Murphy, G.J., Hruby, V.J. and Houslay, M.D. (1986) Nature (Lond.) 323, 68–71.
104. Walaas, O., Horn, R.S., Lystad, E. and Adler, L. (1981) FEBS Lett. 128, 133–136.
105. Londos, C. and Honnor, R.C. (1985) In: Adenosine: receptors and modulation of cell function (Stefanovich, V., Rudolphi, K. and Schubert, P., eds.) IRC Press, Oxford.
106. Kashiwagi, A., Huecksteadt, T.P. and Foley, J.E. (1983) J. Biol. Chem. 258, 13685–13692.
107. Londos, C., Honnor, R.C. and Dhitton, G.S. (1985) J. Biol. Chem. 260, 15139–15145.

108. Smith, U., Kuroda, M. and Simpson, I.A. (1984) J. Biol. Chem. 259, 8758–8763.
109. Baker, P.F. and Carruthers, A. (1983) J. Physiol. 336, 397–431.
110. Gawler, D., Milligan, G. and Houslay, M.D. (1987) Biochem. J. 249, 537–542.
111. Lord, J.M., Puah, J.A., Atkins, T.W. and Bailey, C.J. (1985) J. Pharm. Pharmacol. 37, 821–823.
112. Lord, J.M., Atkins, T.W. and Bailey, C.J. (1983) Diabetologia 23, 108–113.
113. O'Brien, R.M., Houslay, M.D., Milligan, G. and Siddle, K. (1987) FEBS Lett., in press.
114. Gawler, D., Milligan, G., Spiegel, A.M., Unson, C.G. and Houslay, M.D. (1987) Nature (Lond.) 327, 229–232.
115. Larner, J. (1982) J. Cyclic Nucl. Res. 8, 289–296.
116. Larner, J. (1983) Am. J. Med. 74, 38–51.
117. Larner, J. (1984) Trends. Pharmacol. Sci. 5, 67–70.
118. Saltiel, A.R., Fox, J.A., Sherline, P. and Cuatrecasas, P. (1986) Science 233, 967–972.
119. Saltiel, A.R. and Cuatrecasas, P. (1986) Proc. Natl. Acad. Sci. USA 83, 5793–5797.
120. Low, M.G., Ferguson, M.A.J., Futerman, A.H. and Silman, I. (1986) Trends Biochem. Sci. 11, 212–214.

CHAPTER 16

A comparison of the structures of single polypeptide chain growth factor receptors that possess protein tyrosine kinase activity

W.J. GULLICK

Institute of Cancer Research, Chester Beatty Laboratories, Cell and Molecular Biology Section, Protein Chemistry Laboratory, Fulham Road, London SW3 6JB, England

1. Introduction

This chapter compares and contrasts the structural features and functional activities of some of the known growth factor receptors (GFRs) that are apparently composed of a single polypeptide chain. Over the last two years the complete primary structure of about ten glycoproteins have been described which are either known or thought to be involved in the transduction of growth-regulatory signals across cell membranes. In some cases this primary structural information is complemented by some understanding of knowledge concerning the function of the molecules, but in other cases almost nothing is known about their physiological roles. However, analysis of this structural information has revealed some apparent relationships between GFRs so that small families can be defined.

The epidermal growth factor (EGF) receptor and the c-erbB-2 protein are clearly structurally related and constitute one family of GFRs. The platelet-derived growth factor (PDGF) receptor and the colony-stimulating factor (CSF-1) receptor, although also composed of a single polypeptide chain, are considered as a separate family since they possess several features that distinguish them from the EGF receptor and c-erbB-2 proteins.

2. The EGF receptor and the c-erbB-2 protein

The EGF receptor was first isolated by Downward et al. [1] in sufficient quantity to determine directly the primary amino acid sequence of several short polypeptide fragments. Since the molecule has an apparent molecular weight of 175 000 it was clearly too large to determine its complete sequence by this approach. Rather, the amino acid information obtained was used to predict a single long oligonucleotide sequence which was synthesized and used to identify clones from both a human placental and an A431 human vulval tumor cell cDNA library [2]. This approach has been subsequently employed in cloning the majority of the GFRs. Several overlapping clones were isolated and the complete primary amino acid sequence of the molecule was deduced from the DNA sequence data. All of the directly determined peptide sequences were contained within the sequence predicted by the cDNA clones. The open reading frame encodes a molecule of 1210 amino acids. Since the N-terminus had been determined by direct protein sequencing the signal sequence could be defined as residues 1–24, which have the typical structure of basic and polar residues flanking a hydrophobic central region. After removal of the signal sequence, the polypeptide of 1186 amino acids has a predicted molecular weight of 131360 but this is increased by post-translational N-linked glycosylation at about eleven sites, giving the mature receptor an apparent molecular weight of 175000 [3]. Roughly one-third of the sugar side-chains are high mannose and two-thirds are of the complex type [4]. Analyses of the hydrophobicity of the sequence suggested that the molecule has a single transmembrane spanning region of 23 amino acids near its centre. The N-terminal 621 amino acids of the receptor are thus extracellular and the remaining C-terminal 542 residues are cytoplasmic (Fig. 1), as determined by using antibodies raised to synthetic peptide sequences [5–7]. The extracellular region of the molecule is very rich in cysteine residues, which are concentrated in two blocks of about 170 amino acids each (Fig. 1). The spacing of the cysteines in the two clusters is rather similar, suggesting that they may have been generated by a duplication of one or more exons of the EGF receptor gene.

The intracellular sequences of the receptor can be subdivided into several dis-

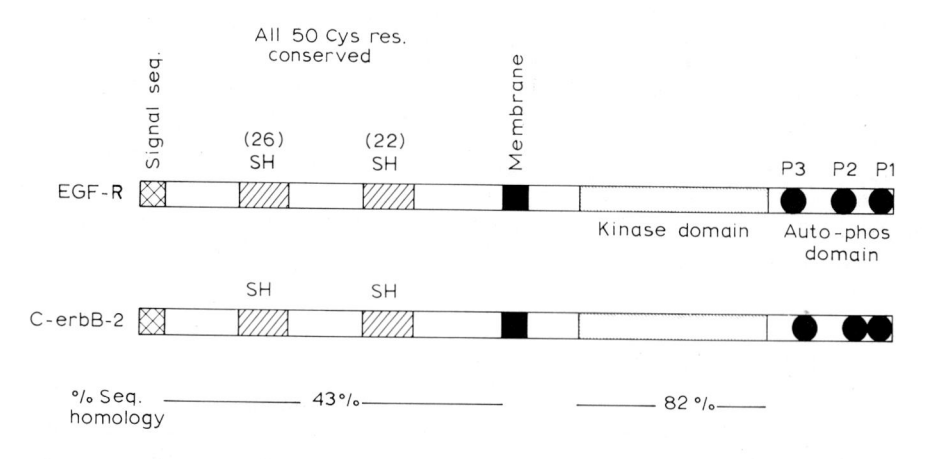

Fig. 1. A diagrammatic comparison of the structures of the human EGF receptor and the human c-erbB-2 protein. SH, cysteine-rich sequence; P1, P2, P3, the three mapped sites of autophosphorylation of the EGF receptor.

crete structures. Just inside the cytoplasmic face of the cell membrane, in the centre of a region very rich in basic amino acids, at position 654 is a threonine residue. This residue, which is ten amino acids away from the membrane, is known to be phosphorylated by protein kinase C (PKC) [8–10], which has the effect of reducing the receptor's affinity for EGF and reducing its rate of ligand-stimulated tyrosine kinase activity [11]. A similar analysis of the related pp60$^{c\text{-}src}$ molecule has shown that it too can be phosphorylated by PKC, in this case on a serine residue at position 12 which is also surrounded by basic amino acids. pp60$^{c\text{-}src}$ is thought to be anchored to the cytoplasmic face of the cell membrane by myristic acid which is covalently coupled to the α-amine group of the N-terminal glycine amino acid [12]. Gould et al. [13] have proposed that the serine residue in pp60$^{c\text{-}src}$ is in the same position relative to the cell membrane as is threonine 654 in the EGF receptor (see Fig. 2). Mutant forms of pp60$^{v\text{-}src}$ in which the N-terminal residue was altered to alanine or glutamic acid [14], which are neither myristilated nor membrane-associated, are also not phosphorylated by PKC, emphasizing the requirement for the close apposition of the two molecules on the inner surface of the plasma cell membrane for phosphorylation at this site to occur.

The EGF receptor possesses intrinsic kinase activity and will transfer the terminal phosphate of ATP specifically to the parahydroxyl group of tyrosine residues in certain substrate proteins. The region that encodes this enzyme activity lies between residues 690 and 940 of the cytoplasmic domain (Fig. 1). These sequences are the central, common structural feature shared by the GFRs which possess ligand-stimulated tyrosine kinase activity. Towards the N-terminal end of this region are residues involved in binding the substrate ATP, but it is not known which residues interact with substrate proteins. This is the only known activity of the EGF

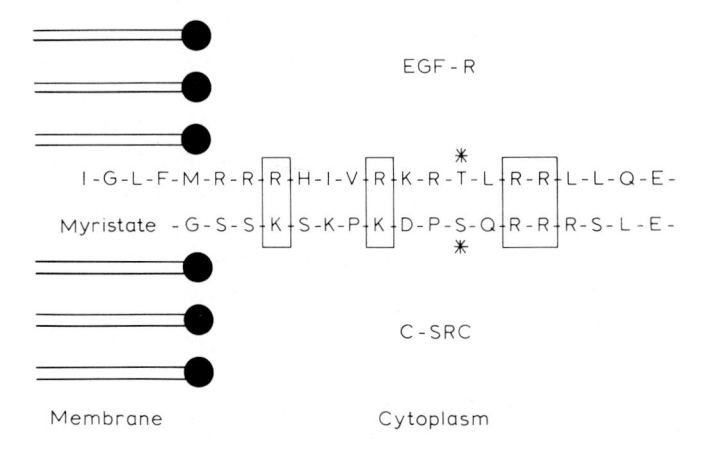

Fig. 2. A comparison of the sequences surrounding the sites of phosphorylation by protein C of the human EGF receptor and the human c-*src* protein. Taken from Gould et al. [13]. The single-letter amino acid code is used. *, serine (S) or threonine (T) at equivalent positions.

receptor and, since enzyme activity is stimulated by ligand binding, it seems likely that it is responsible for transmembrane signalling. In agreement with this, mutations in the receptor which inactivate its kinase activity deprive cells of their ability to respond mitogenically to EGF binding [15].

C-terminal to the kinase domain the receptor appears to be proteolytically sensitive, suggesting that these sequences are exposed [6]. The most C-terminal domain of the molecule is, however, less sensitive to proteolysis and may be more highly folded. The receptor not only catalyses the phosphorylation of exogenous substrate proteins but can also modify itself by phosphorylating three tyrosine residues within the cytoplasmic domain at positions 1068, 1148 and 1173 [1]. The functional significance of this autophosphorylation is not clear. Some reports suggest that autophosphorylation leads to a three-fold stimulation of catalytic rate [16], while others have found no effect [6,10].

Although in total four sites of phosphorylation have been defined, phosphopeptide mapping of the EGF receptor isolated from cultured cells suggests that several other sites are modified, including serine residues. The enzymes responsible for this and the effect on the receptor are not known.

Recently a second molecule has been described that has many similarities in structure to the EGF receptor. Originally isolated as an oncogene from a chemically-induced rat neuroblastoma [17,18], the rat gene has been called either onc-*neu* for the transforming version or c-*neu* for its normal cellular cognate. Subsequently the equivalent human gene has been isolated by cDNA cloning and has been called either HER2/neu [19] or c-erbB-2 [20]. Overall the protein which it encodes has a remarkable structural similarity to the EGF receptor (Fig. 1). The c-erbB-2 precursor protein is composed of 1255 residues with a predicted N-terminal signal sequence of 20 amino acids. The mature protein is, like the EGF receptor, heavily glycosylated, running on SDS polyacrylamide gels with a molecular weight of 185–190 000. The sequence of the c-erbB-2 protein can be aligned with that of the human EGF receptor and is essentially co-linear, with the exception of a small additional sequence of 40 amino acids near its C-terminus. The putative extracellular domain of the c-erbB-2 molecule also has two clusters of cysteine residues, all of which have exactly the same relative spacing as those in the EGF receptor. Overall, however, the two extracellular domains are only 43% identical in sequence (Fig. 1). Thus the proteins are clearly distinct and are in fact encoded by genes on different chromosomes; the EGF receptor gene is on chromosome seven and c-erbB-2 is on chromosome seventeen. Interestingly, the transmembrane regions of the two molecules have no significant sequence homology.

It is not known whether c-erbB-2 is a GFR since no ligand for it has yet been identified, but it may possess protein tyrosine kinase activity [21,22]. The sequence of the cytoplasmic domain encodes a tyrosine kinase-like domain very homologous to that of the EGF receptor. This homology (82%) is much higher than that seen between most of the members of the *src* gene family (25–40%) and emphasizes the close structural relationship of the two molecules. Both kinase domains are of the 'contiguous' type, contrasting with those in the PDGF receptor and c-*fms* molecules (see below).

C-terminal to the c-erbB-2 kinase domain is a region of about 40 amino acids not shared with the EGF receptor. This sequence is equivalent in position to the proteolytically sensitive region of the EGF receptor and it may be that it is also rather exposed in the c-erbB-2 protein, perhaps 'looped out' from between the kinase domain and the autophosphorylation site domain. If the c-erbB-2 protein does indeed possess a ligand which can stimulate its kinase activity, this region is a candidate for being involved in influencing the molecule's substrate specificity.

Downstream of this region the two proteins are once again fairly homologous. Of particular interest is that the three tyrosine residues, known to be sites of ligand-stimulated autophosphorylation in the EGF receptor, are conserved in the c-erbB-2 protein. The c-erbB-2 molecule does become phosphorylated on tyrosine residues in cells, but the identity of the tyrosines is not known.

Recently the mutation that activates the c-*neu* gene to generate onc-*neu* has been defined [23]. A single base transversion mutation converts a valine residue to a glutamic acid residue in the transmembrane region about five amino acids inside the extracellular face of the plasma cell membrane (Fig. 3). Clearly this newly introduced hydrophilic residue is an unlikely one to be found under normal circumstances in this hydrophobic environment. The nature of the activating mutation should help in understanding how GFRs normally transduce proliferative signals across the cell membrane. Two theories as to how this process works are currently favoured, both assuming that ligand binding alters the conformation of a receptor's extracellular domain. The first proposes that this conformational change is propagated across the cell membrane by the transmembrane sequence, altering the conformation of the intracellular domains and thereby increasing the rate of catalysis of the kinase. This model would therefore represent an intramolecular activation. The second hypothesis suggests that ligand binding alters the aggregation state of the receptor by altering the affinity of interaction between the extracellular domains of two receptors. Ligand binding influences association of the cytoplasmic domains and the model

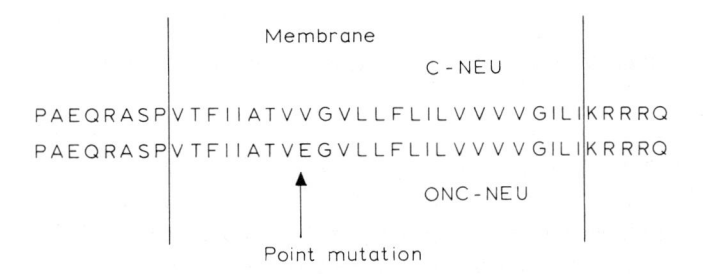

```
                    Membrane
                      C-NEU

PAEQRASP|VTFIIATVVGVLLFLILVVVVGILI|KRRRQ
PAEQRASP|VTFIIATVEGVLLFLILVVVVGILI|KRRRQ
                 ↑
                      ONC-NEU

                Point mutation
```

Fig. 3. The transmembrane sequences of the rat c-*neu* and onc-*neu* proteins showing the position of the activating mutation.

proposes that this alters their conformation and therefore their kinase activity. This ligand-induced alteration of the equilibrium state of receptor clustering or polymerization might therefore be called intermolecular activation. Recently reports have appeared showing that receptor aggregation activates the kinase [24,25]; however, one report has suggested that disaggregation may lead to activation [26] and thus the issue is yet to be unambiguously decided. The activating mutation in onc-*neu* can be assimilated into either model. The intramolecular hypothesis might argue that the hydrophilic glutamic acid residue in onc-*neu* is energetically unfavoured within the cell membrane and would be more stable if positioned at the more hydrophilic membrane surface and would thus 'pull' on the transmembrane sequence and mimic the activation normally achieved by ligand building. The intermolecular model might propose that the glutamic acid residue in the non-aqueous membrane interior is uncharged even at neutral pH and can therefore form two hydrogen bonds with an equivalently positioned glutamic acid residue in another onc-*neu* protein. This interaction would encourage and stabilize dimerization and thereby activate the two molecules. Clearly both these hypotheses must be examined experimentally. Finally, it is not even clear whether onc-*neu* does indeed possess a more active tyrosine kinase than c-*neu* or whether changes in kinase activity cause cell transformation.

In conclusion, it is clear that the structures of the EGF receptor and c-erbB-2 proteins possess considerable similarities in their organization. Both molecules have extracellular domains arranged into two regions rich in cysteine residues which are probably important in determining their three-dimensional structure, and in conferring an ability to interact with specific ligands. Both molecules have highly homologous kinase domains which are formed by a contiguous sequence of amino acids. Towards their c-termini both molecules have three tyrosine residues in equivalent positions which are known to be sites of autophosphorylation in the EGF receptor. It will be interesting to see whether the ligand (if any) for the c-erbB-2 protein in any way resembles EGF. Experiments to date, however, indicate that the ligand is not any of the related family of EGF-like molecules which includes transforming growth factor, type α or vaccinia virus growth factor.

3. *Platelet-derived growth factor receptor and colony-stimulating factor 1 receptor*

Recently the complete primary structures of the receptors for PDGF [27] and for the haemopoietic growth factor, CSF-1 [28], have been determined. The PDGF receptor consists of a single polypeptide chain with an apparent molecular weight of 185 000 on SDS polyacrylamide gels. The cell surface receptor has an affinity for its ligand of approximately 10^{-9} M. Binding of PDGF to its receptor stimulates tyrosine kinase activity which is intrinsic to the receptor. However, although similar in these respects to the EGF receptor, the PDGF receptor differs in being able to indirectly stimulate the breakdown of the phosphoglycolipid phosphatidylinositol. The products of this reaction are diacylglycerol and inositol trisphosphate, which activate protein kinase C and release calcium ions from sequestered intracellular stores respectively. Although phosphatidylinositol breakdown is not normally associated with the EGF receptor it has been reported in cells overexpressing EGF receptors [29,30]. The PDGF receptor is commonly expressed in cells of mesodermal origin, whereas the EGF receptor is expressed in both mesodermal and epidermal cell types.

The PDGF receptor was isolated by a two-stage procedure which involved two forms of affinity chromatography, neither of which was entirely specific for the receptor. BALB/c3T3 cells expressing about 10^5 receptors per cell were first incubated with PDGF to stimulate autophosphorylation of the receptor. The cells were then lysed with detergent and the extract was applied to a column of the lectin, wheat germ agglutinin, which binds to N-acetylglucosamine residues. After elution with the free sugar the partially purified material was run through a column containing an immobilized anti-phosphotyrosine antibody and the specifically bound material was eluted with phenylphosphate [27]. The resultant material was essentially pure and was therefore fragmented with trypsin and the separated peptides were used for direct protein sequence analysis. These peptide sequences were then employed to predict oligonucleotide probes which were used to select cDNA clones from libraries prepared from mouse placenta and NR6 mouse fibroblasts. In this way the complete coding sequence of the 1098 amino acids of the molecule was determined [27]. The PDGF receptor has an N-terminal signal sequence of 31 amino acids. Hydrophobicity plots predict that the protein has a single transmembrane spanning sequence of 25 amino acids starting at residue 500. Thus, by analogy with the transmembrane distribution of the EGF receptor, 500 amino acids of the mature PDGF receptor molecule are extracellular and 542 are intracellular (Fig. 4).

The extracellular sequence of the receptor has only ten cysteine residues, six of which are roughly equally spaced and in the N-terminal half of the domain while the remaining four are relatively closer together and more adjacent to the transmembrane region (Fig. 4). Thus there are no blocks of multiple cysteine residues reminiscent of those in the EGF receptor, and c-erbB-2 proteins (and the insulin receptor and the IGF-I receptor). There are 11 extracellular potential sites of N-linked glycosylation, and since the mature molecule has an apparent molecular weight about 65 000 higher than that predicted for the protein alone, probably most if not all of these sites are modified.

The intracellular sequence of the PDGF receptor has several interesting features. Immediately adjacent to the cell membrane are a cluster of five basic amino acids which form the 'stop-transfer' sequence. There are serine residues at positions 12, 13, 15, 17 and 18 in from the cell membrane, but it is not known if these are ever phosphorylated. The particularly striking feature of the cytoplasmic domain of the PDGF receptor molecule is that it encodes a tyrosine kinase domain that is 'split'

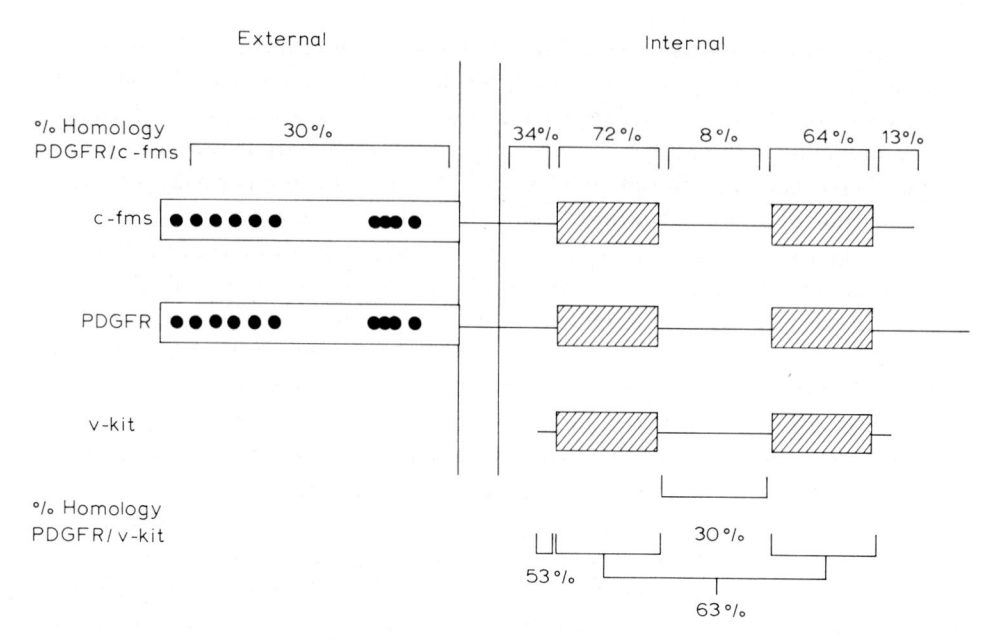

Fig. 4. A diagrammatic comparison of the structures of the human c-*fms* protein, the human PDGF receptor and the v-*kit* oncogene of the avian virus HZ4. Solid dots represent the position of cysteine residues in the extracellular domains of the proteins.

into two halves. The more N-terminal half (residues 572–662) contains residues implicated in nucleotide binding, referred to as the ATP binding site. The sequence that connects the two kinase domains is 104 amino acids long and contains eight tyrosine residues and one cysteine residue but has no discernible special features. The more C-terminal kinase sequence extends from residue 766 to 919. The 'intrakinase' sequence is relatively hydrophilic and may be looped out of the kinase domain and play a role in substrate binding [27]. One possibility is that it may confer the PDGF receptor's ability to interact with systems involved in stimulating phosphatidylinositol breakdown. The sequences C-terminal to the end of the kinase domain extend for a further 148 amino acids and contain four tyrosine residues. None of these is within sequences particularly homologous to known autophosphorylation sites although tyrosine 989 is preceded by an acidic residue which is usually found at an autophosphorylation site. Of the 29 tyrosine residues within the putative cytoplasmic domain, a total of three are preceded by acidic residues and these are at positions 660, 719 and 989. It is not presently known if any of these residues is ever phosphorylated.

The feline CSF-1 receptor is a protein of apparent molecular weight of 165 000 which can be phosphorylated on tyrosine residues in an immune complex assay. Recently, the murine CSF-1 receptor has been purified and shown to possess intrinsic tyrosine kinase activity [31]. The receptor is expressed at relatively high levels in mature differentiated mononuclear phagocyte cells [28].

It is now known that the McDonough strain of feline sarcoma virus which encodes the oncogene v-*fms*, a 140000 molecular weight transmembrane glycoprotein, has acquired a fragment of the feline CSF-1 receptor gene. Thus c-*fms*, the cellular homologue of v-*fms*, is the feline CSF-1 receptor [28,32]. A comparison of the structures of v-*fms* and c-*fms*, although outside the scope of this chapter, has proved very interesting [32,33] and has revealed some information as to the normal properties of the c-*fms* proto-oncogene/ CSF-1 receptor.

A comparison of the structures of human c-*fms* [34] and the PDGF receptor reveals considerable, but regional, sequence homology. The percentage sequence homologies are shown in Fig. 4 and are taken from Yarden et al. [27]. The extracellular domains of the two receptors have an overall sequence homology of 30%; however, all ten cysteine residues in the sequence of the PDGF receptor are found in equivalent positions in the c-*fms* molecule. This concordance emphasizes both the relationship of the two molecules and the importance of these residues in determining the structure of the two proteins. A comparison of the ligands for the two receptors shows that they have some gross structural similarities. Human PDGF is composed of two chains called A and B, which are disulphide-bonded together. It is not known whether the dimeric structure of biologically active PDGF is a hetero or homo dimer. The molecular weight of mature, processed PDGF is about 32000 but at least one and possibly both chains are glycosylated [35]. CSF-1 is also a dimer of two disulphide-bonded chains of about molecular weight 14000 which are also variably glycosylated [36]. Human CSF-1, however, has no structural homology to human PDGF [37].

The transmembrane regions of the two receptors show little sequence homology, with only 4 out of 25 identities. The c-*fms* sequence has a stretch of ten contiguous leucine residues at its more C-terminal end. Generally there appears to be little homology between the transmembrane domains of growth factor receptors but it remains to be shown, as some have suggested, that this hydrophobic sequence is of little consequence to the receptor's structure and function. Obviously, that the transmembrane region sequence can be very important is exemplified by the location of the activating mutation of onc-*neu* within this region.

The cytoplasmic domains of c-*fms* and the PDGF receptor have regions of considerable homology. The short sequence of 47 amino acids between the membrane and the kinase domain of PDGF receptor is 34% homologous with the c-*fms* sequence (Fig. 4). c-*fms* has no threonine or serine residue 10 or 12 amino acids in from the cell membrane; the nearest is a serine at position 18. Within the first 14 residues inside the cell membrane, 6 are either lysine or arginine, representing the 'stop-transfer' region common to all protein tyrosine kinase growth factor receptors. The c-*fms* kinase domain is also split into two halves in essentially identical positions to that of the PDGF receptor. The two moieties of the c-*fms* kinase domain are 72% (N-terminal) and 64% (C-terminal) homologous to the PDGF receptor (Fig. 4). Interestingly, however, the 'intra-kinase' sequence in c-*fms* is rather smaller, being composed of 70 amino acids, and has no significant homology to that of the PDGF receptor (8%). This sequence in c-*fms* may, therefore, confer some substrate specificity to its kinase. The simple presence of a split kinase domain is not sufficient to couple receptor activation to stimulation of phosphatidylinositol lipid hydrolysis, however, since CSF-1 does not induce PI breakdown in murine macrophages [38]. It would be particularly interesting to observe the properties of molecules whose intra-kinase domains were either deleted or exchanged. These experiments are not easy, however, particularly with regard to the analysis of the biological functions of the two receptors.

The c-terminal sequence of c-*fms*, downstream of the kinase domain, has little homology to the PDGF receptor (13%). There are two tyrosine residues in this region at positions 924 and 970, the former being preceded by an aspartic acid residue. Feline c-*fms* is probably phosphorylated at four sites on tyrosine [28] but the mapping of their positions in the primary structure has not yet been published.

Another molecule shows considerable similarity to both c-*fms* and in particular the PDGF receptor. This is the v-*kit* oncogene encoded by the HZ4 feline retrovirus [39]. An outline of the relationship of the v-*kit* gene product with that of the PDGF receptor is shown in the bottom half of Fig. 4. The v-*kit* protein is apparently a truncated version of a cellular proto-oncogene molecule whose N-terminus can be aligned to position 541 of the human PDGF receptor. The v-*kit* gene product also has a split kinase domain highly homologous (63%) to the PDGF receptor. The 'intra-kinase' sequence of 79 amino acids in v-*kit* shows significant homology with that of the PDGF receptor (30%). Nothing is presently known of the structure or function of the c-*kit* proto-oncogene, but possibly it encodes a growth factor receptor which forms a third member of the PDGF receptor/c-*fms* family.

4. Summary

It is clear from inspection of their respective sequences that the EGF receptor and the c-erbB-2 proteins and the PDGF receptor and CSF-1 receptors form respective pairs of highly related molecules. The EGF receptor/c-erbB-2 molecules also have features in common with the PDGF receptor/CSF-1 receptor molecules as well as some striking differences. Unfortunately, comparison of their function is less easy, since the c-erbB-2 protein has no known ligand or biological activity and the c-*fms*/CSF-1 receptor is not functionally well characterized. It will be interesting as these more physiological properties of the molecules are revealed to observe whether the structural relationships described here correlate with their functions.

References

1. Downward, J., Parker, P. and Waterfield, M.D. (1984) Nature (Lond.) 311, 483–485.
2. Ullrich, A., Coussens, L., Hayflick, J.S., Dull, T.J., Gray, A., Tam, A.W., Lee, J., Yarden, Y., Libermann, T.A., Schlessinger, J., Downward, J., Mayes, E.L.V., Whittle, N., Waterfield, M.D. and Seeburg, P.H. (1984) Nature 309, 418–425.
3. Mayes, E.L.V. and Waterfield, M.D. (1984) EMBO J. 3, 531–537.
4. Cummings, R.D., Manglesdorf Soderquist, A. and Carpenter, G. (1985) J. Biol. Chem. 260, 11944–11952.
5. Kris, R., Lax, I., Gullick, W., Waterfield, M.D., Ullrich, A., Fridkin, M. and Schlessinger, J. (1985) Cell 40, 619–625.
6. Gullick, W.J., Downward, J. and Waterfield, M.D. (1985) EMBO J. 4, 2869–2877.
7. Gullick, W.J., Downward, J., Foulkes, J.G. and Waterfield, M.D. (1986) Eur. J. Biochem. 25, 4268–4275.
8. Davis, R.J. and Czech, M.P. (1985) Proc. Natl. Acad. Sci. USA 82, 1974–1978.
9. Hunter, T., Ling, N. and Cooper, J.A. (1985) Nature 311, 480–483.
10. Downward, J., Waterfield, M.D. and Parker, P.J. (1985) J. Biol. Chem. 260, 14538–14546.
11. Cochet, C., Gill, G.N., Meisenhelder, J., Cooper, J.A. and Hunter, T. (1984) J. Biol. Chem. 259, 2553–2558.
12. Schultz, A.M., Henderson, L.E., Oroszlan, S., Garber, E.A. and Hanafusa, H. (1985) Science 227, 427–429.
13. Gould, K.L., Woodgett, J.R., Cooper, J.A., Buss, J.E., Shalloway, D. and Hunter, T. (1985) Cell 42, 849–857.
14. Kamps, M.P., Buss, J.E. and Sefton, B.M. (1985) Proc. Natl. Acad. Sci. USA 82, 4625–4628.
15. Prywes, R., Livneh, E., Ullrich, A. and Schlessinger, J. (1986) EMBO J. 5, 2179–2190.
16. Bertics, P.J. and Gill, G.N. (1985) J. Biol. Chem. 260, 14642–14644.
17. Shih, C., Padhy, L.C., Murray, M. and Weinberg, R.A. (1981) Nature 290, 261–264.
18. Bargmann, C.I., Hung, M.-C. and Weinberg, R.A. (1986) Nature 319, 226–230.
19. Schechter, A.L., Hung, M.-C., Vaidyanathan, L., Weinberg, R.A., Yang-Feng, T.L., Francke, U., Ullrich, A. and Coussens, L. (1985) Science 229, 976–978.
20. Yamamoto, T., Ikawa, S., Akiyama, T., Semba, K., Nomura, N., Miyajima, N., Saito, T. and Toyoshima, K. (1986) Nature 319, 230–234.
21. Stern, D.F., Heffernan, P.A. and Weinberg, R.A. (1986) Mol. Cell. Biol. 6, 1729–1740.
22. Akiyama, T., Sudo, C., Ogawara, H., Toyoshima, K. and Yamamoto, T. (1986) Science 232, 1644–1646.
23. Bargmann, C.I., Hung, M.-C. and Weinberg, R.A. (1986) Cell 45, 649–657.
24. Yarden, Y. and Schlessinger, J. (1987) Biochemistry 26, 1434–1442.
25. Yarden, Y. and Schlessinger, J. (1987) Biochemistry 26, 1443–1451.
26. Biswas, R., Basu, M., Sen-Majumdar, A. and Das, M. (1985) Biochemistry 24, 3795–3802.
27. Yarden, Y., Escobedo, J.A., Kuang, W.-J., Yang-Feng, T.L., Daniel, T.O., Tremble, P.M., Chen, E.Y., Ando, M.E., Harkins, R.N., Francke, U., Fried, V.A., Ullrich, A. and Williams, L.T. (1986) Nature 323, 226–232.
28. Scherr, C.J., Rettenmier, C.W., Sacca, R., Roussel, M.F., Look, A.T. and Stanley, E.R. (1985) Cell 41, 665–676.
29. Sawer, S.T. and Cohen, S. (1981) Biochemistry 20, 6280–6286.
30. Pike, L.J. and Eakes, A.T. (1987) J. Biol. Chem. 262, 1644–1651.
31. Yeung, Y.G., Jubinsky, P.T., Sengupta, A., Yeung, D.C.Y. and Stanley, E.R. (1987) Proc. Natl. Acad. Sci. USA 84, 1268–1271.
32. Roussel, M.F., Dull, T.J., Rettenmier, C.W., Ralph, P., Ullrich, A. and Scherr, C.J. (1987) Nature 325, 549–552.

33. Browning, P.J., Bunn, H.F., Cline, A., Shuman, M. and Nienhuis, A.W. (1986) Proc. Natl. Acad. Sci. USA 83, 7800–7804.
34. Coussens, L., Van Beveren, C.V., Smith, D., Chen, E., Mitchell, R.L., Isacke, C.M., Verma, I.M. and Ullrich, A. (1986) Nature 320, 277–280.
35. Heldin, C.-H., Wasteson, A. and Westermark, R. (1985) Mol. Cell. Endocrinol. 39, 169–187.
36. Das, S.K. and Stanley, E.R. (1982) J. Biol. Chem. 257, 13679–13684.
37. Kawasaki, E.S., Ladner, M.B., Wang, A.M., Van Arsdell, J., Warren, M.K., Coyne, M.Y., Schweickart, V.L., Lee, M.-T., Wilson, K.J., Boosman, A., Stanley, E.R., Ralph, P. and Mark, D.F. (1985) Science 230, 291–296.
38. Whetton, A.D., Monk, P.N., Consalvey, S.D. and Downes, C.P. (1986) EMBO J. 5, 3281–3286.
39. Besmer, P., Murphy, J.E., George, P.C., Qiu, F., Bergold, P.J., Lederman, L., Snyder, H.W., Brodeur, D., Zuckerman, E.E. and Hardy, W.D. (1986) Nature 320, 415–421.

Subject index

Acetyl-CoA carboxylase, glucagon, 245
ACTH,
  action, 193
  adrenal zones, 196, 203, 207
  calcium, 109, 206
  cholesterol, 197
  cyclic AMP, 194
  cyclic GMP, 205
  cytochrome $P450$, 196
  G-proteins, 195, 204
  labile proteins, 199
  protein kinase C, 207
  steroidogenesis, 195
  trophic effects, 201, 202
Action of,
  ACTH, 193
  angiotensin II, 211
  CRF, 113
  dopamine, 113
  glucagon, 231
  GnRH, 135
  GRF, 113
  insulin, 321
  LH, 155
  LHRH, 113
  prolactin, 298
  SRIF, 113
  TRH, 113
  vasopressin, 113
  VIP, 113
Adenylate cyclase, G-proteins, 5
ADP ribosylation, G-proteins, 17
Adrenal zones, ACTH, 196, 203, 207
Aldosterone secretion, calcium, 103
Angiotensin II,
  action, 211
  calcium, 109, 219
  cyclic AMP, 215
  G-proteins, 214, 216
  inositol trisphosphate, 216, 217
  maintenance of response, 224
  phospholipase C, 216
  receptor regulation, 213
  receptors, 212
  second messengers, 222
Annexin-fold family, calcium binding proteins, 77
Arachidonic release, inositol phospholipids, 59
Arachidonic acid derivatives and,
  CRF, 128
  GRF, 129
  LH, 165
  LHRH, 128
  SRIF, 129
  TRH, 126
  vasopressin, 128
  VIP, 127
  dopamine, 127
Aromatase, FSH, 187
ATP citrate lyase, glucagon, 245

Brain, G-proteins, 10

C-erbB-2 protein, EGF, 349
Calcium,
  ACTH, 109, 206
  aldosterone secretion, 103
  angiotensin II, 109, 219
  cellular metabolism, 94
  CRF, 121
  cyclic AMP, 103
  dopamine, 120
  endoplasmic reticulum, 97
  glucagon, 105, 245
  GnRH, 143
  GRF, 122
  inositol trisphosphate, 52
  insulin secretion, 106
  LH, 166
  LHRH, 120
  messenger generation, 99
  mitochondria, 98
  muscle contraction, 102
  phosphoinositides, 100
  plasma membrane channels, 95
  SRIF, 122

transient, 65
TRH, 118
vasopressin, 122
VIP, 119
Calcium binding proteins, 63
　Annexin-fold family, 77
　cell growth, 87
　EF domain family, 74
　egg fertilization, 86
　G-proteins, 35
　intermediary metabolism, 83
　muscle contraction, 81
　secretion, exocytosis, 84
　signal transduction, 67
　structure and function, 69
Calcium mobilizing agonists,
　inositol phospholipids, 51
　glucagon, 250
Carbohydrate metabolism, GH, 281
Cell growth, calcium binding proteins, 87
Cellular differentiation, GH, 282
Cholera toxin, G-proteins, 18
Cholesterol,
　ACTH, 197
　FSH, 188
　LH, 169
Cloning, G-proteins, 20
CRF,
　action, 113
　arachidonic acid derivatives, 128
　calcium, 121
　cyclic AMP, 117
　inositol trisphosphate, 125
Cyclic AMP,
　ACTH, 194
　angiotensin II, 215
　calcium, 103
　CRF, 117
　CSF-1, PDGF, 354
　CSF-1, receptor, 354
　dopamine, 115
　FSH, 184
　glucagon, 235
　GnRH, 142
　GRF, 117
　LH, 164, 167
　LHRH, 116
　SRIF, 117
　TRH, 114
　vasopressin, 117

　VIP, 114
Cyclic GMP, ACTH, 205
Cytochrome P450, ACTH, 196

Deglycosylation,
　FSH, 183
　LH, 156
Desensitization, LH, 171
Diacylglycerol, 52
　GnRH, 146
Dopamine,
　action, 113
　arachidonic acid derivatives, 127
　calcium, 120
　cyclic AMP, 115
　inositol trisphosphate, 124
Down regulation, LH, 171

EF domain family, calcium binding proteins, 74
EGF,
　c-erbB-2 protein, 349
　receptor tyrosine kinase, 351
　receptor, 349
Egg fertilization, calcium binding protein, 86
Endocytosis, GnRH, 138
Endoplasmic reticulum, calcium, 97

Fertilization, inositol trisphosphate, 57
FSH,
　aromatase, 187
　cholesterol, 188
　cyclic AMP, 184
　deglycosylated, 183
　gap junctions and microvilli, 186
　granulosa cell differentiation, 185
　inhibin, 188
　LH receptor, 185
　lipoprotein receptors, 186
　plasminogen activator, 189
　prolactin receptors, 185
　receptors, 182
　steroidogenesis, 186
　structure, 181
　two cell theory, 187

G-proteins,
　α subunits, 23
　β subunits, 31
　ACTH, 195, 204
　adenylate cyclase, 5

ADP ribosylation, 17
angiotensin II, 214, 216
brain (G_o), 10
calcium channels, 35
cholera toxin, 18
cloning, 20
glucagon, 233
GnRH, 147
inositol phospholipids, 50
insulin, 336
ion channels, 32
K^+ channels, 13
neurotransmitters, 2
peptide hormones, 3
pertussis toxin, 18, 19
phospholipase, 11
prostanoids, 4
purification, 6
subunit structure, 21
Gap junctions and microvilli, FSH, 186
GH,
  carbohydrate metabolism, 281
  cellular differentiation, 282
  lactation, 283
  lipid metabolism, 281
  monoclonal antibodies, 284
  muscle, 279
  prolactin, 265
  protein engineering, 283
  protein metabolism, 279
  receptor regulation, 271
  receptor cloning, 289
  receptor purification, 269
  receptors 267
  signal transduction, 271
  somatic growth, 266
  somatomedins, 273
  specific proteins, 278
  transgenic mice, 284
  variants, 286
Glucagon,
  acetyl-CoA carboxylase, 245
  action, 231
  ATP citrate lyase, 245
  calcium, 105, 245
  calcium mobilizing agonists, 250
  cyclic AMP, 235
  G-proteins, 233
Gluconeogenesis, 244
  glucagon, 244

Glucose transport, insulin, 328
Glycogen synthase, 241
Glycogen synthase, glucagon, 241
GnRH, see also LHRH
  action, 135
  calcium, 143
  cyclic AMP, 142
  diacylglycerol, 146
  endocytosis, 138
  G proteins, 147
  inositol trisphosphate, 145
  protein kinase C, 147
  receptor, 137
  receptor regulation, 141
  structure, 135
Granulosa cell differentiation, FSH, 185
GRF,
  action, 113
  arachidonic acid derivatives, 129
  calcium, 122
  cyclic AMP, 117
  inositol trisphosphate, 126
Growth factor, receptors, 349
GTP analogues, 5

Immune system, prolactin, 298, 311
Inhibin, FSH, 188
Inositol phospholipids,
  arachidonic acid release, 59
  calcium mobilizing agonists, 51
  G-proteins, 50
  phospholipase C, 49
  structure, 48
Inositol trisphosphate,
  angiotensin II, 216, 217
  calcium mobilization, 52
  CRF, 125
  dopamine, 124
  fertilization, 57
  GnRH, 145
  GRF, 126
  ionophores, 56
  LH, 164
  LHRH, 124
  metabolism, 54
  oncogenes, 59
  proliferation, 56
  SRIF, 126
  TRH, 123
  vasopressin, 125

VIP, 124
Insulin,
  action, 321
  G-proteins, 336
  glucose transport, 328
  intracellular mediator, 341
  -like growth factors, 329
  phospholipase C, 341
  phospholipids, 341
  receptor structure, 321
  receptor gene cloning, 324
  receptor internalization, 325
  receptor tyrosyl kinase activity, 330
Insulin secretion, calcium, 106
Intermediary metabolism, calcium binding proteins, 83
Ion channels, G-proteins, 32
Ionophores, inositol trisphosphate, 56

K^+ channels, G-proteins, 13

Labile proteins,
  ACTH, 199
  LH, 169
Lactation, GH, 283
LH,
  action, 155
  arachidonic acid metabolites, 165
  calcium, 166
  cholesterol, 169
  cyclic AMP, 164, 167
  deglycosylated, 156
  desensitization, 171
  down regulation, 171
  inositol trisphosphate, 164
  labile proteins, 169
  phosphoproteins, 168
  receptor, 157
  receptor, FSH, 185
  receptor, recycling, 159
  receptor regulation, 160
  steroidogenesis, 166, 172
  structure, 156
  transducing systems, 163
  trophic effects, 173
LHRH, see also GnRH
  action, 113
  arachidonic acid derivatives, 128
  calcium, 120
  cyclic AMP, 116

  inositol trisphosphate, 124
Lipid metabolism, GH, 281
Lipoprotein receptors, FSH, 186

Mammary cancer, prolactin, 314
Mammary gland, prolactin, 296, 304
Milk proteins, prolactin, 306
Mitochondria, calcium, 98
Monoclonal antibodies, GH, 284
Muscle, GH, 279
Muscle contraction,
  calcium, 102
  calcium binding proteins, 81

Neurotransmitters, G-proteins, 2

Oncogenes, inositol trisphosphate, 59
Ovary, prolactin, 297

PDGF, receptor, 354
PDGF, CSF-1, 354
Peptide hormones, G-proteins, 3
Pertussis toxin, G-proteins, 18, 19
Phorbol esters, glucagon, 252
Phosphoinositides, calcium, 100
Phospholipase, C,
  angiotensin II, 216
  G-proteins, 11
  inositol phospholipids, 49
  insulin, 341
Phospholipids, insulin, 341
Phosphoproteins, LH, 168
Phosphorylase kinase, glucagon, 239
Pigeon crop sac, prolactin, 309
Plasma membrane channels, calcium, 95
Plasminogen activator, FSH, 189
Prolactin,
  actions, 298
  GH, 265
  immune system, 298, 311
  lower vertebrates, 299
  mammary gland, 296, 304
  mammary cancer, 314
  milk proteins, 306
  ovary, 297
  pigeon crop sac, 309
  receptor regulation, 303
  receptor internalization, 303
  receptors, 299
  second messengers, 307

structure, 296
variants, 314
Prolactin receptors, FSH, 185
Proliferation, inositol trisphosphate, 56
Prostanoids, G-proteins, 4
Protein kinase A, glucagon, 236
Protein kinase C, 52
  ACTH, 207
  GnRH, 147
  LH, 165
Protein engineering, GH, 283
Protein metabolism, GH, 279
Pyruvate kinase, glucagon, 242

Receptor
  cloning, GH, 289
  gene cloning, insulin, 324
  internalization, insulin, 325
  internalization, prolactin, 303
  purification, GH, 269
  recycling, LH, 159
  structure, insulin, 321
  tyrosine kinase activity, insulin, 330
  tyrosine kinase activity, EGF, 351
Receptors for,
  angiotensin II, 212
  CSF-1, 354
  EGF, 349
  FSH, 182
  GH, 267
  glucagon, 232
  GnRH, 137
  growth factors, 349
  LH, 157
  PDGF, 354
  prolactin, 299
Receptor regulation,
  angiotensin II, 213
  GH, 271
  GnRH, 141
  LH, 160
  prolactin, 303

Second messengers,
  angiotensin II, 222
  prolactin, 307
Secretion, exocytosis, calcium binding proteins, 84
Signal transduction,
  calcium binding proteins, 67

GH, 271
Somatic growth, GH, 266
Somatomedins, GH, 273
Specific proteins, GH, 278
SRIF,
  action, 113
  arachidonic derivatives, 129
  calcium, 122
  cyclic AMP, 117
  inositol trisphosphate, 126
Steroidogenesis and,
  ACTH, 195
  FSH, 186
  LH, 166, 172
Structure of,
  FSH, 181
  GnRH, 135
  inositol phospholipids, 48
  LH, 156
  prolactin, 296
Structure and function, calcium binding proteins, 69
Subunit structure, G-proteins, 21

Transducin, 8
Transducing systems, LH, 163
Transgenic mice, GH, 284
Transmembrane signalling, 1, 47, 63
TRH,
  action, 113
  arachidonic acid derivatives, 126
  calcium, 118
  cyclic AMP, 114
  inositol trisphosphate, 123
Trophic effects,
  ACTH, 201, 202
  LH, 173
Two cell theory, FSH, 187

Variants of,
  GH, 286
  prolactin, 314
Vasopressin,
  action, 113
  arachidonic acid derivatives, 128
  calcium, 122
  cyclic AMP, 117
  inositol trisphosphate, 125
VIP,
  action, 113

arachidonic acid derivatives, 127
calcium, 119

cyclic AMP, 114
inositol trisphosphate, 124